Edward Teller

Edward Teller

The Real Dr Strangelove

Peter Goodchild

Harvard University Press
Cambridge, Massachussetts

In memory of Douglas and May Goodchild, my parents

Library of Congress Cataloging-in-Publication Data

Goodchild, Peter
Edward Teller, the real Dr. Strangelove / Peter Goodchild.
p. cm.
"First published in Great Britain in 2004 by Weidenfeld & Nicoloson"—T.p. verso.
Includes bibliographical references and index.
ISBN 0-674-01669-6 (alk. paper)
1. Teller, Edward, 1908– 2. Physicists—United States—Biography. 3. Atomic
bomb—United States—History. I. Title: Edward Teller. II. Title
QC16.T37G66 2004
623.4'5119'092—dc22
[B]   2004054257

Contents

Illustrations

Photographs reproduced courtesy of:

University of California, Lawrence Livermore National Laboratory Archives and Research Center, 1–3, 10, 11, 20, 22–24, 27, 33–36, 38, 40– 43, 47, 48; Niels Bohr Archive, Copenhagen, 4–7; photograph by Francis Simon, courtesy AIP Emilio Segrè Visual Archives, Francis Simon Collection, 8; AIP Emilio Segrè Visual Archives, Stein Collection, 9; AIP Emilio Segrè Visual Archives, 12; Ernest Orlando Lawrence Berkeley National Laboratory, courtesy AIP Emilio Segrè Visual Archives, 14; photograph by Alan W. Richards, Princeton NJ, courtesy AIP Emilio Segrè Visual Archives, 25; Getty Images–Time Life, 13, 30, 37; Courtesy of Los Alamos National Laboratory Archives, 15–19, 21, 26; University Archives Bancroft Library, University of California, Berkeley, 28 (UARC PIC 13:990); the Bancroft Library University of California, Berkeley, 32 (BANC PIC 2002.171); © Bettmann/Corbis, 29, 45; © Corbis, 46; Jon Brenneis, 31; Ronald Reagan Library, 44. Every effort has been made to contact copyright holders of material reproduced and quoted in this book, but any omissions will be restituted at the earliest opportunity.

Glossary of Characters

ABRAHAMSON, JAMES A.: US Army Lieutenant General in charge of SDI Organisation.

AGNEW, HAROLD: American experimental physicist who flew on Hiroshima mission and worked on Mike shot. Third director of Los Alamos who invited Teller back to the laboratory.

ALVAREZ, LUIS: American experimental physicist, close to Lawrence, who gave evidence against Oppenheimer at the latter's security hearing.

ANDERSON, MARTIN: senior fellow at Hoover Institution, later Reagan's campaign aide and economic and domestic policy assistant.

ARNESON, GORDON: State Department's specialist on atomic matters.

BATZEL, ROGER: American chemist who worked on the Plowshare Gnome project. Director of Livermore, 1971–88.

BENDETSEN, KARL: industrialist and former Under-Secretary of the Army. Member of Committee on the Present Danger and of Reagan's close circle. Founder member of High Frontier.

BETHE, HANS: German émigré theoretical physicist and Nobel laureate for work on the mechanism of stars. First met and befriended Teller when both were students in Munich. They fell out when he was appointed head of theoretical division at wartime Los Alamos, a post Teller thought should be his.

BORDEN, WILLIAM LISCUM: attorney who became executive director of JCAE. His letter accusing Oppenheimer of espionage precipitated 1954 hearing.

BRADBURY, NORRIS: American experimental physicist, second director of Los Alamos, 1945–71. Experienced growing antagonism from Teller over his lack of support for the Super.

BROWN, HAROLD: American theoretical physicist and, at age twenty-four, Livermore's first director of fission weapons development. Became the laboratory's second director and subsequently the Pentagon's director of research and engineering and Defense Secretary.

BUSH, VANNEVAR: American science administrator who directed wartime Office of Scientific Research and Development.

BYRNES, JAMES: Truman's Secretary of State, 1945–47, representing him in discussions over A-bomb's use on Japan.

CANAVAN, GREGORY: US Air Force officer who became a Hertz Fellow, researching into lasers. Worked with Teller on SDI projects, initiating Brilliant Pebbles.

CHADWICK, SIR JAMES: British physicist and Nobel laureate for the discovery of the neutron. Led the British Mission at wartime Los Alamos.

CHAPLINE, GEORGE: American theoretical physicist who developed the initial concept for the nuclear-pumped X-ray laser.

CHEVALIER, HAAKON: Berkeley professor of Romance languages who was a friend of Oppenheimer's and major catalyst in his security problems.

CHRISTY, ROBERT: American physicist who proposed the solid 'Christy' core for the implosion bomb. Family shared a house in post-war Chicago with the Tellers, and he precipitated Teller's scientific 'exile'.

COLGATE, STIRLING: American astrophysicist and friend of Teller's, whose diagnostics on Livermore's early test failures allowed the laboratory to take speedy corrective action.

COMPTON, ARTHUR HOLLY: American experimental physicist, Nobel laureate and wartime director of the Metallurgical Laboratory at Chicago. He had to contend with the unrest over the A-bomb's use against Japan.

CONANT, JAMES: American chemist, President of Harvard University and wartime head of OSRD. As a member of the General Advisory Committee, fought hard to prevent development of the Super.

COWAN, GEORGE: American radiochemist who analysed fall-out from Soviet and US tests, and worked on diagnostics of the Mike shot. Became a member of PSAC, and later established the Santa Fe Institute.

DEAN, GORDON: attorney, at one time Brien McMahon's law partner, and second chairman of AEC who, on resignation, renewed Oppenheimer's consultancy for a further year.

DOBÓ, STEPHEN AND GEORGE: Edward's cousins. Stephen committed suicide aged fourteen.

DONNAN, GEORGE FREDERICK: British biochemist who organised placement of scientists escaping the Nazis through the Academic Assistance Council. Brought Teller to University College, London.

DYSON, FREEMAN: British theoretical physicist, designer of the safe Triga reactor and member of Institute of Advanced Study at Princeton.

EINSTEIN, ALBERT: theoretical physicist, formulated the special theory of relativity and the equation $E=mc^2$. Signed letter to Roosevelt, drafted by Szilard, Wigner and Teller, warning of the implications of the discovery of fission.

ELTENTON, GEORGE: British petroleum engineer, fervent Communist and espionage agent who engineered approaches to Oppenheimer.

EUCKEN, ARNOLD THOMAS: German physical chemist who, in 1930, invited Teller to Göttingen as his assistant.

EVANS, WARD V.: American chemist and member of the Board investigating Oppenheimer whose minority report argued for the reinstatement of his clearance.

EVERETT, CORNELIUS: American mathematician who, with Ulam, assessed the viability of Teller's classic Super.

FERMI, ENRICO: Italian theoretical and experimental physicist, Nobel laureate, co-inventor, with Szilard, of the nuclear reactor, and friend of Teller. Suggested the possibility of a fission-triggered fusion bomb.

FINLETTER, THOMAS J.: Secretary of the Air Force under Truman who was alienated by Oppenheimer and excited by Teller's second laboratory proposal.

FLETCHER, JAMES C.: Former head of NASA, who chaired the Defensive Technologies Study Team of sixty-seven scientists which investigated SDI for Reagan.

FOSTER, JOHN S.: American physicist, first head of Livermore's fission devices group and fourth director of the laboratory, 1961–65. Supported Teller's resistance to Kennedy's plans for a test-ban treaty.

FRANCK, JAMES.: German émigré theoretical physicist, Nobel laureate for experimental work supporting Bohr's atomic model. Employed Teller at Göttingen and helped in his escape from Germany.

FRIEMAN, EDWARD A.: Defence consultant and chairman of the team from the White House Science Council charged with assessing the potential of the X-ray laser.

FROMAN, DAROL: American physicist who managed the 1948 Sandstone tests. Post-war associate director at Los Alamos who oversaw Teller and the Family Committee.

FUCHS, KLAUS: German émigré theoretical physicist and fervent anti-Nazi who came to Los Alamos with the British Mission. As Soviet espionage agent he passed on information about fission and fusion bombs.

GAMOW, GEORGE: Russian émigré theoretical physicist and Nobel laureate. Invited Teller to join him in US at George Washington University. Later worked on the Super design at Los Alamos.

GARRISON, LLOYD: attorney and Oppenheimer's chief counsel during his security hearing.

GARWIN, RICHARD: American theoretical physicist and Fermi's protégé who worked on the Super and on consolidating the design of the radiation–implosion configuration. Later prominent in the Science Wars opposing SDI.

GOLD, HARRY: industrial chemist and Soviet espionage agent working as Fuchs's courier.

GOLDBERGER, MARVIN L.: American physicist, graduate student of Teller's at Chicago. Befriended Oppenheimer after the hearing and later fell out with Teller. Became President of California Institute of Technology.

GRAHAM, DANIEL O.: US Army lieutenant general, campaign adviser to Reagan who promoted the idea of anti-missile defence. Founder of High Frontier but squeezed out by Teller.

GRAY, GORDON: attorney, former Secretary of the Army and chairman of the security board that investigated Oppenheimer.

GREEN, HAROLD: AEC attorney who drafted the list of charges against Oppenheimer which reflected Teller's complaints about him.

GRIGGS, DAVID: American geophysicist, Chief Scientist to the Air Force, and friend of Teller who supported his bid for a second laboratory.

GROVES, LESLIE R.: US Army general, responsible for the wartime Manhattan Project and subsequently the Armed Forces' Special Weapons Project. Chose Oppenheimer as scientific director of Los Alamos.

HAGELSTEIN, PETER: American physicist recruited to Livermore's 'O' Group in 1975, aged twenty. Resisted work on defence projects but was drawn in as he formulated a more promising concept than Chapline's for the nuclear-pumped X-ray laser.

HARKÁNYI-SCHÜTZ, EDE (SUKI): Mici Teller's brother and one of Teller's close friends during his last years at the Minta. Died in Mauthausen concentration camp.

HEISENBERG, WERNER: German theoretical physicist and Nobel laureate, Teller's professor at Leipzig. Became a leading figure in the German A-bomb project.

HESLEP, CHARTER: AEC public information officer to whom Teller unburdened himself shortly before his testimony at the Oppenheimer hearing.

HESZ, MAGDA: born in the US, she was the Teller family's nanny in Budapest, whose childhood had been spent in the US. She remained in touch, acting as go-between for Teller and his family in Hungary after the war.

HOFFMANN, FREDERIC DE: Austrian theoretical physicist and Teller's protégé and general factotum. His company, General Atomic, developed the safe Triga reactor and he chaired a committee highly critical of SDI.

HOLLOWAY, MARSHALL: American physicist appointed by Bradbury as director of Mike, the first megaton thermonuclear blast, rather than Teller.

JACKSON, HENRY 'SCOOP': Democrat senator and member of the Joint Committee on Atomic Energy who backed Teller's 'clean' bomb programme.

JOHNSON, GERALD: American physicist who, in 1955, took charge of all Livermore's testing programmes, including the Chariot project.

KAPITZA, PETER: Soviet theoretical physicist and Nobel laureate who studied under Rutherford at Cambridge. Pleaded successfully with Stalin for Lev Landau's release from prison.

KENNAN, GEORGE: US diplomat whose lengthy telegrams from Moscow became the basis of Truman's 'containment' policy towards Russia. Provided an important overview of the Soviet situation for the GAC meeting in October 1949.

KERR, DONALD M.: Director of Los Alamos who led heavy criticism of Livermore's X-ray laser programme.

KEYWORTH, GEORGE A., II: American physicist and protégé of Teller's who was appointed Reagan's scientific adviser.

KHARITON, YULI BORISOVICH: Soviet physicist who served as director to Sakharov and Zeldovich on the Soviet H-bomb. Teller unsuccessfully proposed him for a Fermi Award.

KIDDER, RAY E.: American physicist and Livermore veteran whose assessments of the X-ray laser displeased Teller.

KIRZ, EMMI: Edward Teller's older sister who remained in Hungary throughout the war and after under communist rule. Came to US in 1959 and died in 2001.

LANDAU, LEV: Russian physicist, Nobel laureate and friend of Teller's during their studies in Europe. His persecution during the Stalinist purges contributed to Teller's deep mistrust of the Soviets.

LANSDALE, JOHN, JR: Attorney and wartime intelligence officer with the Manhattan Project. Among officers who interviewed Oppenheimer over the Chevalier incident.

LÁSZLÓ, TIBOR: Teller's school friend who perished in a Nazi concentration camp.

LATTER, ALBERT: American theoretical physicist with Rand Corporation who discovered 'Latter' holes for concealing underground tests. Joint author with Teller of 'Our Nuclear Future'.

LAWRENCE, ERNEST ORLANDO: American experimental physicist and 'impresario' of 'big science'. Nobel laureate for inventing the cyclotron. Provided crucial backing to the US A-bomb programme, then to the Super and Teller's second laboratory.

LIBBY, WILLARD F.: American chemist, Nobel laureate for development of radiocarbon dating, political conservative and friend of Teller's. Member of the GAC who supported Teller's second laboratory and fought protests over fall-out dangers.

LILIENTHAL, DAVID: former director of the Tennessee Valley Authority and first AEC chairman. Close to Oppenheimer, he resisted development of the Super.

McCARTHY, JOSEPH: US Republican senator whose anti-Communist witch-hunts gave his name to an era.

McFARLANE, ROBERT C.: deputy national security adviser under Reagan who had a strong political interest in promoting SDI.

McMAHON, BRIEN: US Democrat senator and sponsor of the McMahon Act bringing nuclear power under civilian control. Member, then chairman, of the JCAE who encouraged development of the Super.

MARKS, HERBERT: attorney, whose wife was Oppenheimer's assistant, who defended Oppenheimer at his hearing. Supporter of radical causes.

MARSHALL, GEORGE C.: wartime US Army Chief of Staff, later Truman's Secretary of State and Secretary of Defense who initiated the Marshall Plan.

MAYER, MARIA GOPPERT: German émigré theoretical physicist who worked on the Manhattan Project. Nobel laureate and Teller's confidante.

MEESE, EDWIN: Long-time friend of Reagan's who became his powerful White House counsellor. A major point of contact in the administration for Bendetsen and High Frontier.

METROPOLIS, NICHOLAS: Greek-American mathematician who pioneered the use of computers at post-war Los Alamos and provided early optimistic assessments of the Super.

MILLS, MARK: American theoretical physicist who was head of Livermore's theoretical group and director-designate when killed in a helicopter accident at Eniwetok.

MULLER, HERMAN J.: American geneticist and Nobel laureate. In the 1950s his classic thirty-year-old studies of mutations in fruit flies caused by radiation took on a new significance in the debate over fall-out.

MURRAY, THOMAS: engineer and inventor who ran the subway in wartime New York. AEC Commissioner from 1950 when Teller persuaded him of the need for a second laboratory.

NEUMANN, JOHN VON: Hungarian émigré mathematician and consultant at wartime Los Alamos, helping to develop implosion fission bomb. Pioneered digital computers, using early models to confirm difficulties with Teller's Super design.

NICHOLS, KENNETH D.: engineer, US Army officer and aide-de-camp to Groves during the war. Appointed general manager of AEC by Strauss in 1953, and organised Oppenheimer's hearing.

NITZE, PAUL H.: Director of State Department Policy Planning Staff. Major Washington figure for five decades and became Reagan's leading arms control adviser.

OLIPHANT, MARCUS: Australian physicist who discovered thermonuclear fusion reaction with Rutherford. In 1941 successfully provoked moribund US A-bomb programme into action.

OPPENHEIMER, FRANK: American experimental physicist and Robert's younger brother. Card-carrying Communist whose left-wing activities created problems for both himself and Robert.

OPPENHEIMER, JULIUS ROBERT: American theoretical physicist and wartime director of Los Alamos. As chairman of the GAC, was suspected by Teller and others of deliberately obstructing development of thermonuclear weapons.

PANOFSKY, WOLFGANG: Polish-born physicist who advised President Kennedy against taking technical objections to a test ban too seriously.

PASH, BORIS: US Army security officer who interviewed Oppenheimer about the Chevalier incident. Later ran Alsos Mission into Germany to discover state of their A-bomb programme.

PAULING, LINUS: American chemist, Nobel laureate for Chemistry (1954) and Peace (1963). Organised petition among scientists supporting a test ban, proselytised on future effects of fall-out and debated with Teller.

PEARSON, DREW: Popular political columnist and heavy critic of Teller.

PEIERLS, RUDOLPH: German émigré theoretical physicist and member of wartime British Mission at Los Alamos. Brought Fuchs to the laboratory.

PHILLIPS, GENEVIEVE: Edward Teller's secretary at Livermore for more than forty years.

PLACZEK, GEORGE: Czech émigré physicist who collaborated with Teller at Göttingen. Later advised Bethe against H-bomb work.

RABI, ISIDOR I.: American experimental physicist, Nobel laureate, member and eventually chairman of GAC. Worked for international control of nuclear weapons and resisted development of H-bomb. One of Teller's most virulent critics.

RESTON, JAMES: Influential *New York Times* journalist and Oppenheimer supporter.

RICHTMYER, ROBERT: American theoretical physicist and post-war director of Los Alamos's theoretical division. Worked with Teller to develop 'Alarm Clock' thermonuclear device.

ROBB, ROGER: attorney and well-known trial lawyer who presented government case at Oppenheimer hearing.

ROSEN, LOUIS: American experimental physicist who identified the presence of fusion in the George test.

ROSENBLUTH, MARSHALL: American theoretical physicist who worked on the Mike design. Internationally acclaimed expert on plasma physics.

ROTBLAT, SIR JOSEPH: Polish-born physicist and one of the few scientists to leave Los Alamos during the war. Researched into fall-out from Bravo test. Helped found Pugwash Conferences on Science and World Affairs and awarded Nobel Peace Prize in 1995.

SACHS, ALEXANDER: economist who supported early efforts to start a US A-bomb programme.

SAKHAROV, ANDREI: Soviet theoretical physicist, inventor of the 'layer cake' H-bomb and co-inventor of the Soviet radiation–implosion H-bomb. Following internal exile, played influential role in late 1980s in shaping Gorbachev's disarmament initiatives.

SEABORG, GLENN T.: American radiochemist and Nobel laureate as co-discoverer of plutonium. Member of the GAC, then chairman of the AEC during Kennedy's presidency.

SERBER, ROBERT: American theoretical chemist who directed design of Little Boy. Close associate of Oppenheimer's.

SHEPLEY, JAMES: *Time* journalist who, with a colleague, Clay Blair, wrote *The Hydrogen Bomb*, a controversial book lionising Teller.

STRAUSS, LEWIS: American financier, wartime admiral and Eisenhower's unlikely appointee as adviser on nuclear affairs and chairman of the AEC. Resisted attempts at disarmament, and prominent in the efforts to displace Oppenheimer. Close colleague of Teller's.

SZILARD, LEO: Hungarian émigré physicist, co-inventor, with Fermi, of the nuclear reactor. Organised the Einstein letter to Roosevelt pleading case for an A-bomb programme and resisted the bomb's use on Japan without warning. Friendship with Teller withstood many disagreements.

TATLOCK, JEAN: Oppenheimer's lover who introduced him to pre-war left-wing politics. Suffered depression and eventually committed suicide.

TISZA, LÁZLÓ (LACI): joint winner of the Eotvos maths prize with Teller. Friendship lasted to the end of Teller's life.

ULAM, FRANÇOISE: French wife of Stanislaw and 'computer' at Los Alamos working on her husband's Super calculations.

ULAM, STAN: Polish-born mathematician who shared credit with Teller for the radiation–implosion hydrogen bomb.

UREY, HAROLD: American chemist and Nobel laureate for the discovery of deuterium. Worked on gaseous diffusion method for separating out U-235 during the war. Friend of Teller's.

WATKINS, JAMES D.: US admiral, Chief of Navy Operations and devout Catholic who questioned morality of mutually assured destruction and supported SDI's defensive premise with the Joint Chiefs of Staff.

WECHSLER, JACOB J.: American engineer at Los Alamos who developed the cryogenic system for Mike.

WEINBERGER, CASPAR W.: Secretary of Defense under Reagan, one of the few to earn Teller's respect. Though never convinced of SDI's viability, made pragmatic political use of it.

WEIZSÄCKER, CARL FRIEDRICH: German physicist and friend of Teller's who shared lodgings with him in Copenhagen. Leading figure in German A-bomb programme. The friendship nevertheless survived until his death.

WHEELER, JOHN ARCHIBALD: American theoretical physicist who developed fission theory with Bohr. Friend of Teller's who collaborated with him on thermonuclear research, 1950–51, and establishment of Livermore.

WIESNER, JEROME: American electrical engineer and advocate of arms control. Became Kennedy's science adviser in 1961 and successfully organised opposition to Teller's attempts to block the Limited Test Ban Treaty.

WIGNER, EUGENE: Hungarian theoretical physicist and Nobel laureate. Friend of Teller's and early pioneer of US A-bomb who helped prepare the Einstein letter to Roosevelt. Shared Teller's deep mistrust of Soviets.

WOOD, LOWELL L., JR: American physicist who headed Livermore's experimental 'O' Group. Campaigned vigorously for the X-ray laser and, subsequently, Brilliant Pebbles. Described as 'another son' to Teller.

YORK, HERBERT F.: American theoretical physicist and first director of Livermore. Moved on to senior positions in Washington. Increasingly critical of what he saw as Teller's messianic approach to the arms race with the Russians.

Acknowledgements

This book is not an authorised account, and work was begun before any approach had been made to Dr Teller. When the approach was made, I was uncertain whether he would either agree to meet or feel able to collaborate in any way. This uncertainty was based to some extent on my dealings with him in 1980 when researching a BBC drama series about Robert Oppenheimer, reinforced by the experience of colleagues who had approached him over the years and by that of other authors, some of whom had received a decisive 'No!' to their requests for an interview. However, Dr Teller took the trouble to read my book on Oppenheimer and then agreed, first to meet me and then to spend four afternoons in conversation. Those conversations took place without any request from him to vet the resulting material in any way. I also received a great deal of help – and kindness – from Judith Shoolery, who was Dr Teller's editor and friend for the last quarter century. I am grateful to both.

During my research, I interviewed some forty people among Dr Teller's friends, colleagues and family, who gave freely of their time and their advice, and I offer them my grateful thanks. I am particularly indebted to those whose interviews are quoted in the book and who, in many cases, advised on sections of the manuscript. These include: Harold Agnew, Harold and Jean Argo, Hans and Rose Bethe, Greg Canavan, George Chapline, Stirling and Rosie Colgate, Hugh de Witt, Richard Garwin, Marvin L. Goldberger, Peter Hagelstein, Chuck Hansen, Norris Keeler, Ray Kidder, George Maenchen, Louis Rosen, Marshall Rosenbluth, Sir Joseph Rotblat, Paul Teller, Françoise Ulam, Jay Wechsler, and Herb York. Hugh de Witt, Richard Garwin and Chuck Hansen also assembled portfolios of reference material for me.

There are others, too, to whom I owe a particular debt of thanks. Roger Meade, the archivist at Los Alamos, and his assistant Linda Sandoval helped by making available material in his laboratory's possession, and he also spent hours debating the whole subject area with me and then commenting on the manuscript. So, too, did my colleagues Peter Jones and Martin Cook, and also Herb York, whose broad historical knowledge is augmented by invaluable personal experience. It was Martin who bore the additional burden of searching for illustrations. Sadly, we were unable to use any of the Teller family's photographs, so I am particularly grateful to him for assembling such a comprehensive collection. I am obliged to Professor Peter Winlove of the physics department at Exeter University for his help on the technical aspects of the manuscript; to Maxine Trost, Pat Rhiner and Beverly Bull of the Lawrence Livermore Laboratory; to Linda Sandoval of the Los Alamos National Laboratory; and to the Los Alamos Historical Society, for their assistance in accessing material in their respective archives.

In covering such a full and important life as Edward Teller's, I have used a wide range of primary and secondary sources. However, there are two books in particular that provided an inspiration for sections of this biography. The first was *Teller's War*, William Broad's fascinating account of the Star Wars saga, and the second was Dan O'Neill's *Firecracker Boys*, an excellent and full account of the Chariot project.

At various times I have needed additional secretarial assistance, and Nokomis Suffield, Wendy Kirk, Judith Travell and Gill Meredith deserve my thanks for helping out, as does Jim Mothersole for his technical back-up. My grateful thanks also go to my editors at Weidenfeld and Nicolson – to Peter Tallack whose enthusiasm saw the project to commission, and to Richard Milner who took over and saw it through the first draft stage. But special thanks are due to Tom Wharton who has worked through the final text with such commitment and clarity of overview.

This book and its subject have dominated my existence for nearly three years, resulting in all the boorishness and other symptoms of a true obsession. No one has been more the victim of this than my wife, Penny, and she deserves my thanks for the way she has provided both practical support and a ready and sympathetic ear throughout.

Finally, I would like to acknowledge an intellectual debt. I owe much to my late colleague, Robert Reid, who, when I worked for him at the BBC, first kindled what has become a lifelong interest in the nuclear arms race. I am pleased to have this opportunity to put my gratitude to him on record.

Asterisks in the text refer to additional information material contained in the Notes and References section at the end of the book.

Introduction

On 5 February 1965 Columbia Pictures took space on the front page of the entertainment industry's trade journal, *Variety*, to proclaim the extraordinary popular success of one of their latest releases, *Dr Strangelove – Or How I Learned To Stop Worrying And Love The Bomb*.

FLASH: STANLEY KUBRICK'S DR STRANGELOVE BREAKS EVERY OPENING-WEEK RECORD IN HISTORY OF VICTORIA THEATRE (NEW YORK), BARONET THEATRE (NEW YORK), COLUMBIA THEATRE (LONDON).

The film was to go on to break all attendance records across the US and, by the end of the year, had earned a place in *Variety*'s list of All-Time Top Grossers, an extraordinary achievement for a film whose subject was the Cold War and the nuclear arms race.

It was a black comedy, richly ironic, at the same time both cruel and funny. It tells the story of a US Air Force General, Jack Ripper, who is obsessed by fears of a communist conspiracy to pollute his 'vital bodily fluids' by fluoridation, and orders the bombers under his command to launch a nuclear attack on the Soviet Union. Once the bombers are on their way, he seals off his base and refuses all communication, even from the US President. It emerges that, if the bombers reach their target, they will trigger a Soviet Doomsday device programmed automatically, and irreversibly, to destroy all living things on Earth.

In an acting *tour de force* Peter Sellers plays three of the principal roles, each of whom is struggling to avert the catastrophe. He plays the US President, Merkin Muffley, whose best attempts to avert disaster depend on placating a drunken Soviet Premier; the British exchange officer, Group Captain Mandrake, isolated at the bomber base with the psychotic Ripper and attempting to obtain the crucial code to stop the bombers; and the title role, Dr Strangelove, the ex-Nazi scientist advising the President.

The film had its genesis during one of the most critical moments in the Cold War, the Cuban Missile Crisis. During those thirteen days in October

1962, not only world leaders but the public as well had looked into the abyss, having to confront the stark reality of a nuclear war. They had seen movie footage of what the blast from a nuclear bomb did to the kind of houses many of them occupied and there had been several well-publicised scare stories about the effect of radioactive fall-out. At the time, *Life* magazine carried a cover story under the headline: 'How You Can Survive Fall-Out. 97 out of 100 Can be Saved', which advised that the best remedy for radiation sickness was hot tea or a solution of baking powder; but everyone was painfully aware of how inadequate the civil defence measures available would be, and there was a deep and widespread fear and anger at this vulnerability and helplessness.

Everything depended on deterrence – the fear the Russians had of what would happen to them if they did attack – and everyone had watched as this act of faith was tested to the limit before the Russians backed down. It was against this background that Stanley Kubrick had begun to research and to script his film. Initially his intention had been a straight adaptation of the book *Red Alert*, by a former RAF navigator, Peter George; but as Kubrick had researched more, he had found that he could best conceive of the film as a nightmare comedy. 'Following this approach,' he later said, 'I found that it never interfered with presenting well-reasoned arguments . . . In the context of impending world destruction, hypocrisy, misunderstanding, lechery, paranoia, euphemism, patriotism, heroism and even reasonableness can evoke a grisly laugh.'

So the characters had become caricatures, graced with names loaded with irreverent and often sexual innuendo. President Merkin Muffley's name is derived from terms for female pubic hair, while Mandrake is a plant root used as an aphrodisiac. Then there are General Buck Turgidson, Colonel Guano, Russian Ambassador Desadesky and Major 'King' Kong. The dialogue is equally pregnant-to-bursting with irony.

'You can't fight in here, this is the War Room!' says the President when Turgidson tries to prevent the Russian ambassador from sneaking photos of the War Room's giant strategic map.

'Do you realise that in addition to fluoridated water,' Ripper tells Mandrake in an attempt to explain the nature of the Russian threat provoking his action, 'why there are studies under way to fluoridate salt, flour, fruit juices, soup, sugar, milk, ice cream. Ice cream, Mandrake? Children's ice cream! . . . You know when fluoridation began? . . . 1946. 1946, Mandrake. How does that coincide with your post-war Commie conspiracy, uh?'

And later in the film, Strangelove produces a justification for the Russians' fully automated unstoppable Doomsday machine that sounds all too probable: 'Deterrence is the art of producing in the mind of the enemy the fear to attack. And so, because of the co-ordinated and irrevocable decision-making process which rules out human meddling, the Doomsday Machine is terrifying. It's simple to understand. And completely credible and convincing' – or it would have been, it emerges, if only the Russians had actually told anybody about it.

It is a satire, bordering at times on farce, and yet so sensitive was the mood at the time – its release was delayed for several months because of the assassination of President Kennedy in November 1963 – that Columbia Films had to include this disclaimer at the beginning of the film: 'It is the stated position of the United States Air Force that their Safeguards would prevent the occurrence of such events as are depicted in this film. Furthermore, it should be noted that none of the characters portrayed in this film are meant to represent any real persons living or dead.'

At first sight the second half of this disclaimer seems redundant, so extreme do the caricatures of, in particular, the military seem to be; but there was one character over whom there was speculation as to whether he was based on a real character, and that was Dr Strangelove himself. As played by Peter Sellers, he is an extraordinary character – cold, rational, and confined to a wheelchair with a mechanical arm likely to spring into a Nazi salute or attempt to strangle its owner. He is a variant on the 'mad scientist' stereotype, a direct descendant of Dr Frankenstein; but there is just enough significant detail provided about him to raise the possibility of his being drawn from life. He is German, an ex-Nazi, and he is a cold and calculating nuclear strategist, having connections with the 'Bland Corporation' – a very thinly veiled reference to the Rand Corporation, a think tank set up by Douglas Aircraft that specialised in nuclear strategy.

No one individual well known at the time precisely matched these characteristics, and Kubrick himself never offered an identification; but working with these clues, there has been a sport continuing ever since in trying to provide the most likely inspiration for the character. Four names have stood the test of time and frequently appear as the likely progenitors of Strangelove:

Werner von Braun, the German rocket pioneer, a Nazi brought out of Germany after the war to lead US rocket development.

Herman Kahn, the American nuclear strategist, who worked at Rand and, in 1960, shot to fame with his book *On Thermonuclear War*.

Henry Kissinger, the strategist and future Secretary of State, a German by birth who fled the Nazis. He is probably a later addition to the list, as he was a relatively unknown academic at the time the film was made.

Edward Teller was another who fled the Nazis. More a middle-European contender, he was a Hungarian by birth who spent nearly a decade studying at German universities before emigrating to the US. Unlike the others, he did have an obvious disability and he was 'father' of the H-bomb as well as a nuclear strategist who had both advised, and opposed, presidents.

This literal matching of characteristics is, however, ultimately a self-defeating exercise. Strangelove is clearly a composite, at first sight an updating of the Dr Frankenstein model. But Kubrick went much further than this. He used Strangelove to personalise the theme that imbued the whole of his film: the dangers inherent in the obsessions and mania incubated within the secure and closed world of nuclear politics. As such, Strangelove accrued such relevance that he became not just a variation of a stereotype but an archetype in his own right. At the time the film was made he provided a focus for the fear and the dissent which, by the mid-sixties, had come to replace the patriotic apathy of the 1950s. The motives of the scientists at the heart of the arms race were now open to question and doubt. The biologist and essayist Lewis Mumford praised Kubrick for having placed a scientist at the centre of his 'scientifically organised nightmare of mass extermination' and for then having identified his 'ultimate strategy of nuclear gamesmanship for precisely what it would be: an act of treason against the human race'.

However, the link with Dr Strangelove was a stigma, which, in varying degrees, the shortlist of four each had to bear, enhancing their links with the nuclear nightmare, stereotyping and demonising them, devaluing their opinions and their actions.

Of this group, Edward Teller has arguably suffered most, if for no other reason than his longevity, but he also continued his association right up to his death in September 2003 with new nuclear strategies, including President George W. Bush's 'Son of Star Wars' programme. The Strangelove link was always an irritant for Teller, and journalists who raised it in interviews often received short shrift. As recently as 1999 a journalist interviewing Teller for the *Scientific American* had the temerity to mention it and the ninety-one-year-old scientist threatened to throw him out of the office.

The fact is, however, that Teller's life can be seen as a rich variation on the theme of manic obsession, which lies at the heart of Kubrick's film and was personalised in the character of Dr Strangelove. For more than half a century now he has been recognised for 'fathering' the H-bomb. His struggle, against the odds, in making both the initial crucial discovery and then pushing through the development of the weapon and establishing a laboratory of his own, the Livermore, to specialise in its development, has over the years been a source of bitter controversy. He was lionised in magazines such as *Time, Newsweek,* and *Life* for what they saw as a struggle of Churchillian proportions but others saw as dangerous and self-aggrandising extravagance.

He was seen as riding on the wave of growing anti-Soviet paranoia to play a leading role in opposing any attempts to cool the arms race. One of President Eisenhower's prime personal objectives for his administration was the achievement of a test ban as a first step towards disarmament. Using a continuing string of technical objections, generated by the Livermore, as to why a test ban would be impossible to maintain with an enemy as devious as the Soviets, Teller managed to frustrate Eisenhower's efforts. That frustration found expression in the President's farewell speech, in which he referred to the danger of a scientific–technological elite hijacking public policy. Eisenhower let it be known that those he had in mind when he made this comment were von Braun and Edward Teller.

Eisenhower's successor, John F. Kennedy, was to face similar opposition. He also set out working towards a complete test ban, but at much the same time as Kubrick was shooting *Dr Strangelove* in England Kennedy was having to reconcile himself to the enormous influence Teller had, both in the Pentagon and in the Senate, which he was using to foment opposition to the President's aims. As he and his frustrated aides were forced to reduce their ambitions to a partial test ban, they were to describe the campaigning scientist as a cross between Billy Sunday, the baseball star turned blazing-fisted evangelist, and John L. Lewis, the frighteningly aggressive leader of the Union of Mineworkers.

Teller was certainly no *éminence grise* in the Strangelove mould. In the fifties he waged a high-profile battle against those who, like the Nobel scientist Linus Pauling, were warning about the persistent dangers of even low-level radioactive fall-out. Teller argued that their claims were both exaggerated and dishonest. In the sixties it was the controversies surrounding the 'peaceful' uses of atomic energy. Teller was a main proponent of what were seen as cavalier and dangerous schemes using multi-megaton nuclear explosions for enormously ambitious civil-engineering projects such as a sea-level Panama Canal and a vast harbour in northern Alaska; and in the seventies, when he seemed to be losing his footing in Washington, he was nevertheless highly vocal in defence of nuclear power following the disaster at Three Mile Island.

During these decades there was a slow and tortuous rapprochement between the two superpowers, which took concrete form in the various Strategic Arms Limitation Treaties and in the Anti-Ballistic Missile Treaty, and yet Teller was seen to maintain his obsessions in an unreconstructed form. He resisted all the treaties and was devastated, particularly by the ABM Treaty of 1972, which threatened his last great vision: the impregnable defensive shield against nuclear-missile attack.

It was a vision in which the fragile balance of deterrence through fear of mutually assured destruction (MAD) was replaced by the objective of assured survival. 'Better to save lives than to avenge them,' was how President Reagan summarised the objective when it began to take shape as the Strategic Defence Initiative or 'Star Wars', as it became popularly known.

Such freedom from the tyranny of the arms race, which had been implicit in MAD, was at first sight both enormously seductive and estimable. Whole populations would no longer be the hostages of nuclear war. However, just as in hand-to-hand combat, a shield on its own may offer only protection, but when used with a weapon it becomes an integral part of attack, so it was with the Star Wars missile shield. To many it looked as if it was not only a vastly expensive technical gamble but also the most cunning ploy yet by which the US military could expand its power base worldwide under the guise of ending for ever the dread of a mutual holocaust. This was certainly how it was seen in the Soviet Union – one of the reasons why, in 1984, they came close to launching a pre-emptive attack on the West.

Given this broad outline of Edward Teller's life, it is not difficult to see how the mantle of Dr Strangelove would seem to fit so comfortably on his shoulders. It was the Nobel physicist Isidor Rabi who declared that Teller had been 'a danger to all that is important' and that 'it would have been a better world without Teller'. This is a truly damning statement from a man who was deeply involved in arms-race politics for almost as long as Teller himself and knew its inherent contradictions and ambivalences. But then so did another Nobel physicist, Eugene Wigner, and he declared Teller to be 'a great man of vast imagination' and one of the 'most thoughtful statesmen of science'.

I should say now that the outline above is not only broad but simplified, and it is partial. In it I have highlighted those elements that create a match with Kubrick's fictional obsessive; but there are many other elements in Edward Teller's life that complicate, confuse and even confound this picture.

There is Edward Teller's own personal background, so important in understanding the motives that drove him for more than sixty years and earned him the reputation of being unstoppable, a force of nature.

There are the personalities with whom he worked, those he befriended, those he alienated. The list includes men like Robert Oppenheimer, Lewis Strauss, Hans Bethe, Norris Bradbury, Herb York. They played a crucial role in shaping a career that shifted and changed direction dramatically throughout his long life.

There is the world of US and international politics, which make such definite demands on those who, like Edward Teller, hope to make a serious impact and to achieve action. The interaction between these factors makes up the reality of the remarkable man who remained a potent force in arms-race politics for such a long time. I interviewed Teller's colleagues and friends, some fifty of them. I very seldom heard anything other than strong feelings. There were admiration and contempt, true affection and deep hatred. I revisited incidents, some of them occurring more than half a century ago and found emotions both powerful and fresh.

Dr Teller himself, ninety-four years old when I talked with him and physically frail though mentally frighteningly alert, was as emotional in his replies as anyone. The hurt of the struggle to develop a hydrogen bomb and the subsequent Oppenheimer affair and its aftermath was a constant companion. Over the years he wrote and spoke a great deal about those experiences, but his own version of events, written very much in justification, became varnished truths in the retelling, increasingly at odds with accounts from other sources. This has created problems in writing about him, and others have found him much easier to describe than to understand.

There are, however, two sources, which have only recently become available, that do help in achieving that understanding. The first is his own recently published *Memoirs*, his last comprehensive assessment of a controversial and emotionally riven life. His special versions of certain controversies still make their appearance, some with yet another coat of varnish. However, he does write with painful honesty about the difficulties of his early years and provides a level of insight that has not previously been available. Then, some years ago, there emerged a collection of correspondence, which, until very recently, Teller himself had not realised still existed. For two decades, from the late thirties, Teller corresponded with the physicist Maria Gopperts Mayer. He destroyed her letters but, unknown to him, she kept his. They are intimate and full of insight and they add a level of understanding to events in the forties and fifties, those two turbulent decades in Teller's life.

Teller will certainly remain a controversial figure, and will no doubt continue to be caricatured and demonised. His association with Dr Strangelove is but one example. I hope that his portrayal in this book provides a better understanding of the reality of being one of the most powerful scientists of the twentieth century – a *real* Dr Strangelove.

1
War, Revolution, Peace and Maths

During the summer of 1916, when he was eight years old, Edward Teller's family had holidayed with his mother's family in the Hungarian town of Lugos. It was here, at a small lido beside the River Temes, that he and his sister Emmi had learnt to swim. However, while other children were jumping, diving into the water and playing, his mother, Teller recalled, 'was too worried by the dangers of the river and the in-experience of her children to be satisfied with watching: throughout our swimming excursions she sat rigidly holding the end of the cords she had tied around the waist of each of us ... The people in her home town were well acquainted with my mother's tremendous capacity for worry.'

This memory, fresh after more than eighty years, says much about the pathological protectiveness of his childhood, which was to have such a profound effect on Edward Teller throughout his long life. His was the privileged childhood of the Budapest professional classes in the closing years of the Austro-Hungarian Empire, one spent in large apartments near the Courts of Justice, apartments where his father also had his law office. It was a childhood of games in the nearby city park in the company of a governess, of Sunday walks in the Buda hills with his father and older sister Emmi, of games in the ruined cloisters of St Margaret's nunnery on the island in the middle of the Danube.

Then, each summer, the family left the city for the countryside, along with the governess and domestic staff, often travelling to Lugos, his mother's home town on the Romanian border. His earliest memory was of one such holiday in 1910: 'I hear myself repeating two words: *igazán, igazán* (really, really) and *igen, igen* (yes, yes). Although two words are a tiny vocabulary for a two-and-a-half-year-old, I remember my feeling of pride and the approval of my parents.'

His initial progress in learning to speak was slow, so slow, in fact, that his maternal grandfather, Ignaz Deutsch, warned his parents that they should face up to the possibility that they had a retarded child. However, at the age of four he began, overnight, speaking in complete and structured sentences – his sister recalled that it was as if a dam had been breached. His own later explanation was the bilingual confusion created by a father speaking Hungarian and a mother who, as her maiden name implied, spoke German as her first language.

Edward Teller was born at home on 15 January 1908. His Jewish father, Max, had read law at Budapest University. The word most commonly used in describing him was 'gentle'. He seemed to be self-effacing as well, writing in his diary, 'A lawyer needs better weapons than I possess to get along in life.' Yet when his parents both died while he was still a student, he not only assumed responsibility for three younger sisters – as was the expectation at the time – but also went on to establish a successful law practice. He became the associate editor of the Hungarian Law Journal, performing the more routine editorial tasks, and mixed with a circle of friends and acquaintances that included senior academic and government ministers. He remained a bachelor until all his sisters were married and only then began seriously to look for a wife.

He met Ilona Deutsch at a friend's house in Budapest on 15 January 1904. She was twenty, he was then thirty-two. Ilona was petite, only five feet tall, blonde and already a pianist of near-professional standard who spoke Hungarian, French, Spanish and Italian in addition to her native tongue. Max was immediately smitten: 'Now, perhaps, I have found the right one. If the eyes are the mirror of the soul, she is the embodiment of kindness and gentleness . . . In the first five minutes I felt as though we had been friends for a long time . . . Since I took leave of her I have not done any work. I cannot think while she is away.' Within sixteen days, they were engaged and four months after that they were married by the Chief Rabbi of Lugos.

At that time there were a million Jews amongst Hungary's population of 20 million and they were among the most assimilated in Europe. They had supported and gained the trust of the Magyars during the 1848 revolt and had now come to occupy a pre-eminent position among the professional class. Although Jews represented only 6 per cent of the population, half the journalists and lawyers were Jewish and more than half the doctors. It was not, therefore, their Jewishness that marked out Ilona's family but the fact that they were 'new rich' – merchants. Ignaz Deutsch was a banker, the owner of a brewery, a textile mill and a factory, one of the wealthiest men in Lugos. Their house was one of the most imposing; yet in the stratified Hungarian society of the time their social status was practically non-existent. Teller wrote:

> Perhaps as a consequence, my grandmother had strict rules in her home about what constituted respectful conduct towardsher. That rigidity distanced her from her daughters (as well as her grandchildren). My mother and aunt were always dutiful but I never saw an exchange between them and their mother that suggested warmth . . .
>
> My grandfather was different . . . of those in my family, I liked him best . . . Within the family, everyone (except his wife) liked him a lot. My grandfather was hard of hearing and used an old-fashioned ear trumpet; and my grandmother claimed that his deafness had ruined her social life. During the last five or six years of his life, she hardly spoke to him. Although he was always pleasant and kind, I cannot remember hearing him laugh; I suspect he was deeply sad. My mother was so fond of her father that to say she loved him profoundly is an understatement. I suspect the extreme devotion she showed toward me was partly a transfer of her love for him.

These powerful tensions and the accompanying atmosphere within her own family must have affected Ilona, and, whether or not there was a direct transfer of her profound feelings for her father across to her son, there can be no doubt that he was his mother's favourite. 'My mother doted on me and made no effort to hide her feelings,' Teller recalled. 'By the time I was six or seven years old, I was uncomfortable with my mother's preferential treatment.' He had become a serious-minded child without a sense of humour, worried about doing anything intentionally wrong, afraid of the dark, and the cumulative effect of his mother's oppressive care weighed very heavily. During their summer holidays at Lugos the Teller family actually stayed not with Ilona's parents but with her sister, Margaret Dobó. It was there that Edward witnessed at first hand the profound damage that can result from obsessive favouritism:

> My mother and her sister Margaret were opposites: my mother was thin, my aunt, plump, my mother was sad, my aunt, laughing, my mother absurd in her devotion to her children, my aunt, perfect in her selfishness. Two of my Dobó cousins were as close to me as brothers. George, a year younger than I, and Stephen, two years younger . . . One of them, Stephen, was Aunt Margaret's favourite child. Perhaps one of the reasons that I felt so close to my cousins was that I had a similar discomfort. But the dynamics of the Dobó family were more difficult. My mother may have treated Emmi as if she were less

important than I, but Emmi clearly had the love and approval of both
our parents. Aunt Margaret idolised sweet-tempered Stephen, and in her
eyes he could do no wrong. George, on the other hand, could do no
right. As a result, George became more and more prickly and sour and
failed to get along comfortably with anyone . . .

I could not help feeling sorry for George, even though he spread his
misery widely. However, I felt equally sorry for Stephen, who was as
unhappy with his lion's share of praise and attention as George was with
the absence of approval. Five years later, those tensions led to tragedy.
An itinerant peddler came through Lugos, selling handguns. Stephen
surreptitiously purchased one. He committed suicide with it at the age
of fourteen. Although I was saddened by his death, and neither anticipated nor understood it, I was not surprised by it.

Edward's interpretation of and empathy with the roots of his cousin's
fate do underline what a burden his mother's concern for him was. His
comparison between her and her sister not only described her as 'absurd
in her devotion to her children' but also as someone who was essentially
'sad'. He believed also that he himself had inherited her capacity for worry.

One of the earliest incidents that convinced him of this occurred at the
age of about five, when his father gave him a small mirror to play with. He
showed him how to catch the reflections of sunlight with it and Edward
spent a contented morning directing the sun's beams on to the walls and
into the windows of the building opposite – the Courts of Justice. Later
that day, at dinner time, there was a knock at the door and his father left
the room. When he returned it was to say that the caller had been a
policeman reporting that on several occasions beams of reflected sunlight
from their apartment had scorched the pate of a Supreme Court judge.
'Both my parents were kind,' he recalled. 'But even though they laughed
about my "crime", I was almost overwhelmed by my guilt. As a small
child, I had an almost chronic bad conscience. I do not believe it was
justified, but I worried most of the time that some absurdity or another
was an offence. My mother was a worrier, and I may have worried in
imitation.'

His early memories of his mother were also intertwined with his first
experiences with music. She was a fine pianist but, in Edward's view, had
lacked the self-assurance needed to play in public. Certainly, her early
hopes of a professional career came to an end with her marriage and, until
the early onset of arthritis prevented her playing, she contented herself
with performing for her immediate family and fostering the real talent that
Edward had demonstrated. For a while she transferred her own dreams of
a professional career to him but, while making good progress, his heart

seemed not to be in it. 'My love of music grew from listening to my mother play,' he wrote. 'My love of the mountains may have been inspired by my father's enjoyment of them. My interest in mathematics was self-generated ... Soon after our summer vacation in 1912, when I was about four and a half years old, I began consistently spending time thinking about numbers.'

When he was put to bed he indulged in a secret game. Working from the knowledge that a minute consists of sixty seconds, he set out to discover how many seconds there were in an hour, a day, a year. The fact that he came up with different answers each time only added to the excitement. 'A child cannot know how his days and nights will determine his years – how they may make him fit or not fit into his world ... Whether Intellectual Independence leads to success, misfortune, or both is not clear. But finding the consistency of numbers is the first memory I have of feeling secure.'

However safe such games made him feel, though, they could not have disguised the political and social ferment in the world outside, certainly not from a child as intelligent as Edward. The Austro-Hungarian Empire was in its death throes. Less than fifty years before, in 1867, the Habsburg rulers of the conglomerate Austrian Empire had at last given in to the nationalism of their Hungarian subjects. Under a dual monarchy, where Franz Josef ruled as Emperor of Austria and King of Hungary, the Hungarians achieved a level of independence; but it was a continually strife-ridden period with Hungary's ethnic minorities – the Slovenes, Czechs, Slovaks and Romanians – all nursing their own nationalist ambitions. Only the economic pre-eminence of the capital, Budapest, and the last vestiges of a feudal society held the disparate groups together. On the sidelines of these disputes loomed the ever-present threat of the old enemy, Russia. Much as the French and the Germans were the 'bogeymen' to British children, so were the Russians to those of Hungary.

The final disastrous polarisation in these disputes came with the assassination in 1914 of Archduke Ferdinand, the heir to the throne of Austro-Hungary – a killing engineered by the Serbs. In the ensuing conflict, Austro-Hungary was backed by the Germans, while the Serbs were backed by France and the 'bogeyman', Russia. On a walk in the mountains close to Budapest, Edward and his father came upon some trenches, which, his father explained, had been dug as a defence against Hungary's old enemy, the Russians. These trenches were to the west of the Danube. It was not lost on Edward that, by the time they were in use, Budapest would have been overrun, and with it the Teller home and those of their friends.

During the war Max Teller had a map on the wall of his office that he continually updated as the various fronts moved back and forth. Edward was able to watch as the Russian front was first beaten back by the Germans, then, after regrouping, moved back against the much weaker Austro-

Hungarian army. He recalled the despair when they pushed more than a hundred miles into Hungary itself, before yet again being repelled by the Germans. The Russians were a continuing rapacious threat to the increasingly vulnerable Hungarians and they were to remain so even after the war was over.

The end of the war was a disaster for the Austro-Hungarian Empire and for Hungary itself. Certainly the country gained its independence from Austria, but under the Treaty of Trianon it was reduced from a nation of some 20 million people to one of only 8 million. Half of those who were ethnic Hungarians were to live under foreign rule. In the redistribution of territory, Romania gained more than the new Hungarian nation retained.

It was a situation ripe for revolution, a revolution that brought to power a Communist government of 'people's commissars' with Bela Kun at its head.

Originally an NCO in the Hungarian army, Kun had been captured in 1915 and taken back to Russia as a prisoner of war. While there, he had been converted to Bolshevism and then, at the end of the war, sent back to Hungary to propagate Party views and radicalise the government there. He had very soon been jailed for incitement to riot, but his popularity had sky-rocketed when a journalist reported that he had been beaten by the police. The outcry had been so great that the existing government was forced to resign and Kun had emerged from prison triumphant. In a short time, he had proclaimed a draconian 'dictatorship of the proletariat'. Industry was nationalised, private houses were declared the property of the state, and the courts of law were suspended and replaced by revolutionary tribunals.

'The Communists overturned every aspect of the society and the economy,' Teller wrote. 'My father could no longer practise law. In fact we became social outcasts ... a lawyer was a thoroughly worthless person in a "good" society. Two soldiers moved into our "extra space" ... my father's office in our home.' In fact, Edward's main memory of these two was their self-consciousness at being present in the apartment. They slept on the couches, urinated in the rubber plant and hunted for any sign of the old 'blue' currency, which though illegal, was valued more than the communist replacement. The Tellers' nanny, Magda Hesz, used her skill in bookbinding to hide their reserves in the covers of the books in Max Teller's office. They were never found.

At school the children learnt the Internationale – Edward could still remember the truly bloodthirsty Hungarian version more than eighty years later – and also that the streets and subways were plastered with threatening posters. 'On one of them a stern man, with his arm extended and his fingertip as large as if it were half an inch from my nose, said: "You lurking in the shadows, spreading horror stories, you counter-revolutionary, TREMBLE." The finger seemed to follow me wherever I went.'

The Teller family were not to experience directly the brutality for which the Kun regime became legendary, though it was a continuing fear. So, too, was the possibility of 'betrayal' by a neighbour or one of their staff. As with most of the population, the family's problems were ones mainly of sustaining themselves and obtaining food. The markets had ceased to function and so each weekend Max, accompanied by Emmi and Edward, hiked off into the country to buy what they could. 'But there was not much to buy,' Teller wrote. 'As I recall, cabbage was often all we could find. I still dislike cabbage.' On one occasion no peasants were to be found, so they took corn from the fields, leaving 'blue' money tied to the stalks.

However, this situation could not last. In the midst of that summer Max Teller predicted that the Communists would fall and that anti-Semitism would follow. 'Too many of the Communist leaders are Jews,' he explained 'and all the Jews will be blamed for their excesses.'

The Communists' support in the country had been tenuous from the start and depended not on their new ideology but on their promise to restore Hungary's old territories. This they signally failed to do. Their army was beaten off by the Serbs, and their final thrust into Romania in August 1919 resulted in a counter-attack in which the Romanian army overran and looted Budapest itself. After only 133 days in power, Bela Kun fled the country, leaving behind total chaos.

What followed was to be much more damaging and disturbing. Admiral Miklos Horthy, a war hero who had commanded the Hungarian navy, a landowner and a member of Hungary's old aristocracy, had arrived in Budapest with the invading Romanian army. During the autumn of 1919, he established a government there. They, in turn, propagated their own reign of terror, during which some 5000 people, most of them Jews, were executed.

Even though only a minority of Jews had been directly involved in the Kun revolution, they were, as Max Teller had foreseen, blamed for the chaos and excesses of those few months. Families such as the Tellers found themselves alienated and victimised, targets for an anti-Semitism that was sustained by repeated accounts of the atrocities perpetrated by the 'Jewish Bolsheviks'. There were, for instance, stories of the 'red train' used by Szamuelly – self-styled Hungarian Soviet Commissar of Agriculture – to frighten the recalcitrant peasants into submission. From its windows victims were thrown after their executioners had grown tired of torturing them: a woman who refused to reveal details of an alleged counter-revolutionary plot had her teeth dug out with a chisel; another had a nail hammered into her skull, and yet another had her tongue sewn to the end of her nose for refusing to submit to violation. Over the following decades this kind of material was to be recycled time and again, continually used to enhance the Jews' alienation and to justify the restrictions imposed on them. It was an alienation that Jews, like the Tellers, knew had its true roots in Russian Bolshevism. 'During my first eleven years,' Teller wrote, 'I had known war, patriotism, communism, revolution, anti-Semitism, fascism, and peace. I wish the peace had been more complete.'

By the end of 1919 the Horthy regime had already imposed restrictions on the categories of government jobs that Jews could apply for, and they also placed a limiting quota on the number of university places available to them. It was becoming clear that, if a Jewish child of Edward Teller's ability was to achieve his full potential, he would have to leave the country of his birth. The worry and the resentment this anti-Semitism – triggered by communist extremists – created for an over-protected eleven-year-old must have been considerable.

Up to the time they went to senior school – the gymnasium – both Edward and Emmi had been largely educated in the close environment of home. Their pursuits together, apart from hiking, were mainly intellectual ones. Max Teller was Edward's first chess partner, teaching him the moves when he was only six. For the next three or four years Max was the consistent victor, but there came the moment when Edward beat him. 'With thoughtless honesty, I told him that one of his moves was stupid,' Teller recalled. 'His reaction shocked me: he was hurt. And I was ashamed. And I do not believe we ever played chess together again.'

The same happened some years later over music. Max was an indifferent violinist who played with his increasingly proficient son until the latter actually refused to play with him any more. Theirs was never to be a particularly close relationship, and Edward believed that his father actually favoured his sister Emmi. 'I was never able to talk to him easily, nor he to me. I doubt that we had more than a dozen good talks together. To all appearances my mother made the decisions in our family, but I do not believe that was the case. When I was a little older, my father, not my mother, made the decisions about my life.'

Certainly it was his mother who taught him to read and write and in this she was assisted by the one person who seemed to leaven the atmosphere in the Teller household. Magda Hesz came to the Tellers as a nanny when Edward was seven. She had been born in Hungary, but when she was still a baby her family had moved to Chicago. Her parents had died when she was still in her teens, and she had been sent back to Hungary to live with relatives. She was no more than ten years older than Edward and, for him, she was more of a friend than part of the management . . . Missy, as we called her, never was angry, nor disliked anyone . . . She knew all the omens of good and bad luck and could tell the past and future by looking at the palm of your hand. And she told wonderful stories, most of them about Chicago . . . My fear of the dark had increased when . . . I was moved from the room I had shared with Emmi into a room of my own. Magda brought reassurance, humour, and fortitude with her. She put me to bed each night with the ditty:

> Good night. Sleep tight.
> Don't fight with yourself.

Arguably good advice for anyone, but particularly for a young boy riven with self-doubt and anxiety. Over the seven years Missy was to remain with the family, Edward's 'point of view was slowly, very slowly, coloured by her optimism'.

For her part, Magda was struck by the contrast between Edward's intellectual maturity and his social immaturity. Helping him to dress on her first morning, she was amazed to find that a seven-year-old still expected her to put his socks on for him. If he was scolded, he resorted to a childish form of emotional blackmail, accusing her of liking Emmi more than him.

Intellectually, however, he was on another plane. On holiday when he was eight years old, for instance, he and another boy continued running arguments about the existence of free will, and whether the surrounding mountains were getting smaller through erosion or growing bigger. A year or so later he was to be found in Parliament Park, digging a hole, trying to prove something he had just learnt: that the centre of the earth was hot. He was on the threshold of suffering the heartache, the frustration of being a gifted child.

By the age of nine it was quite obvious that, intellectually at least, Edward was ready for the challenge of the gymnasium and so, 'a little younger than my classmates and considerably younger in experience', he entered the Minta (Model) School set up by a friend of his father, Maurice von Kármán, then the Minister of Education, as a test bed for educational reforms.

Edward hated it. He did not like the mix of subjects in the curriculum, with its heavy emphasis on Hungarian and German studies and physical exercise, and he disapproved of his fellow students' lack of 'an enthusiasm for learning'. He saw most of the classes as in a state of 'semi-revolt'. 'I reached adolescence still a serious child with no sense of humour,' he wrote. 'My classmates laughed about our teachers: that was wrong. They also laughed at me: that was intolerable.' The years spent in the unrealistically overprotective and approving environment at home had prepared him badly for school and Edward found that it did 'little for one's social skills. My first year in gymnasium, the equivalent of fifth grade, was the beginning of a miserable time. I had no friends amongst my classmates. In fact during my first few years at the Minta, I was practically a social outcast.' Deeply unhappy, Edward would return home, to the solitary pleasures of his favourite authors, Jules Verne and H. G. Wells, and to his continuing obsession with numbers. 'Please don't talk to me – I have a problem,' he would announce at the dinner table. The problem, his family knew, was one of the calculations he performed in his head for fun. His request was respected and he would be left out of the conversation for the rest of the meal.

On one occasion Emmi, who had just started algebra, had asked her father why $(10+1)^2$ is not the same as 10^2+1^2. While Max Teller was struggling to shape a reply, Edward came up with the answer. Max was sufficiently impressed to take his son to see a friend, Leopold Klug, a retired professor of mathematics. Over the next year Edward was to have some half-dozen sessions with Klug, devouring Euler's two-hundred-year-old textbook Algebra and falling 'in love with the underlying simplicity of what seems at first complex. Klug could do mathematics all day long,' Teller wrote. 'Indeed he had done so all his life. He was the only adult (other than Magda Hesz, who hardly seemed an adult) I did not feel sorry for. Almost all the others had complaints about their jobs. I became determined to have a job that allowed me to do something that I wanted to do for its own sake.'

At school his misery was to continue. His exceptional brilliance rankled, not just with his fellow students but with his teachers as well. On one occasion the headmaster, Karl Oberle, who also taught maths, called on him in class to answer a question. The answer he gave involved knowledge not yet taught to his year, but instead of praising him Oberle put him down heavily. 'What are you? A repeater?' Teller recalled him asking – meaning: had he taken the class the year before and failed? It was meant as an insult and taken as such. It was the beginning of a long period when the teacher never called on one of his brightest pupils, 'even when I was the only student to raise a hand. Everyone noticed, but for once it was not the sort of thing that my classmates teased me about.'

There were, however, numerous things about which they did tease him. Although the school was only a matter of a thirty-minute walk from home across the centre of Budapest, his mother insisted that Magda accompany him to school. This she did, every day, from when he was nine to the age of nearly fifteen. At one point they came to an arrangement that she should walk on the other side of the street, and from there she frequently watched incidents in which he was either bullied or teased.

For a time he had to carry his sister's old lunch box with her name written across it. It gave rise to his first nickname: Emmi; but that, in time, was changed to Coco. Asked by his biographers, in the 1970s, to explain the origin of his nickname, Teller demurred, its significance perhaps re-opening a wound half a century old.* It was, of course, as he admitted in his later autobiography, a 'common name for simple-minded clowns'.

So unhappy was he at the Minta that he tried to change schools. The most promising possibility was the Catholic Piarist gymnasium. It had a good reputation and Edward had gone hiking with its scout troop – without being teased – on several occasions; but in the end the fact that he was not Catholic counted against him: he was turned down. The misery was to continue.

When Edward was nearly fifteen, Magda Hesz returned to the United States. Shortly afterwards both he and Emmi became seriously ill with scarlet fever. Edward developed the additional complication of a kidney infection and was kept away from school for several months. When he returned, things began, by a circuitous route, to take a turn for the better.

Edward found he was sharing a desk with a new student called Forgacs. Forgacs was, according to Edward, 'mentally retarded perhaps because of an iodine deficiency'; and his life at the school was 'living proof that mine could be worse ... His tormentors would dance around him in a circle, making faces and taunting noises, wiggling their hands in their ears. Forgacs would become upset, then frantic, and finally furious. That only delighted his tormentors and attracted more students to the circle. I couldn't help him ... But that full view of how teasing worked hinted at what had been happening to me.'

It taught him the basic lesson of not rising to the taunt. It seemed to work, but the more significant development arose from Edward's attempts to tutor the unfortunate Forgacs. He obviously met with some success because two students, both Gentiles and 'socially popular', came and sought similar assistance. Max Teller made his office available for the tutorials and the two students' performance in mathematics improved. After some seven years of being an outcast, the kind of experience that so often leaves an indelible mark and an accompanying vulnerability, Edward Teller began to be accepted.

His nickname was modified from the Hungarian TsoTsoh to the more Latinised KoKo – an in-joke referring to a lesson the class had had on the Roman pronunciation of Cs and Ks, but even a small change signified to Teller some kind of affection, some kind of friendship. He wrote: 'KoKo came to be seen, by the end of that year, as a badge I had earned through a long, unconscious struggle.'

Several of his previous tormentors even began to treat him as a friend. Nándor Keszthelyi, one of those Magda had earlier seen bullying Edward in the street, sometimes joined him on the walk home and 'found occasions to talk pleasantly with me', Edward modestly recalled. Another, Ede Harkányi-Schütz, whose nickname was Suki, had a younger sister whom Edward had helped with her maths. In time these three, along with the reserved but popular Tibor László, formed a quartet of the brightest pupils, who became increasingly close during the last two academically taxing years at the gymnasium. The group had a sports club in which Edward was also included. He was markedly inferior at everything, except the standing jump. 'A cosine occurs in the formula for calculating work,' Teller recalled. 'Whenever I was about to jump, they would cheer, "Go, cosine energy!" That kind of teasing I liked. The other sport in which I had a little ability was ping-pong. But I took sports as seriously as the rest of life; for me, they were not a joke. I played, but I seldom laughed.'

In spite of his serious-mindedness he was accepted outside his family for the first time. This thirst for acceptance – with the hurt and anger he felt when it was denied – was to become a defining feature of his life.

By now science, for Edward, had become 'the most exciting thing in the world'. In those last two years at the gymnasium, Max Teller, if not close to his son's private thoughts and dreams, was still a dutiful parent, and arranged for Edward to meet three young men from the Jewish community who were already working as scientists in Germany, and whose fathers he knew. Two were in their early twenties: Eugene Wigner, the future Nobel Laureate and John von Neumann, later one of the geniuses behind modern computing. The other, approaching thirty, was the physicist Leo Szilard. All three were to play a significant role in Edward's future career.

At this time, however, the teenage Edward relished the few hours he was able to spend with them when they returned home. They talked physics, Edward's growing obsession, and Szilard even narrowed the discussion further. Always a visionary, he already believed that only the theory of atoms and the theory of gravity were at the heart of the new physics. It is a view still prevalent today. As a consequence, Edward grappled as best he could with Einstein's book on relativity, coming close to convincing himself that he was not equipped for the challenge. His physics teacher actually took the book away from him, only returning it when he finished his final high-school exam, the Matura.

As this critical career-defining test loomed, Edward became increasingly concerned about his future. He wanted to pursue an academic career, but his father's foreboding seven years earlier – that anti-Semitism would severely curtail the opportunities in Hungary – was proving justified; he had therefore warned Edward that Hungary had no place for him. But his son was frightened by the prospect of leaving the only world he had known.

There was also a dispute between father and son about the course he should take. Edward wanted to study maths, but to Max this seemed to offer few job opportunities beyond teaching. To further complicate matters, Ilona Teller insisted that, at seventeen, her son was too young to leave home. A compromise was reached. In the autumn of 1925 Edward began reading chemistry at Budapest University. At the time, however, he also entered for the prestigious Eötvös competition, open to gymnasium graduates from throughout Hungary. He entered in two categories, mathematics and physics, and for the time being his preparation preoccupied him.

It was on his way to the examination that Edward recalls stopping off at his friend Suki's house, where he was met by Suki's younger sister Mici. At the time, a year or so earlier, when he had tutored her in maths, his own romantic imaginings had been fired by a young cellist, Magda Radó. After finding out which operas Magda was likely to attend, Edward had then gone himself and sought her out during the intermission. The outcome had been an increasing knowledge of the operas of Wagner and Verdi, but not much else. He had arranged for them to play chamber music together on one occasion, but it had been musically disastrous. Although the relationship, touching in its tentativeness, had never gone any further, it had preoccupied him at the time he was helping Mici. It was only now that he realised her request for tuition had had an ulterior motive. 'When Mici sent me off to do battle,' he recalled, 'I understood for the first time that she had a real interest in me.' This moment was to prove the beginning of a lifelong relationship.

When the results of the Eötvös competition were announced, Edward found he had achieved an extraordinary result: he had won the prize in physics outright and had also shared the mathematics prize with two other students. It was a major success and the impetus needed for his parents to accept that he was sufficiently exceptional to go and study abroad.

On 2 January 1926, two weeks before his eighteenth birthday, he set out for Germany to continue his chemistry at Karlsruhe. The evening before he left, he and his three friends Suki, Tibor and Nándi spent almost the whole night talking in his father's office. Edward knew how fortunate he was to be going, but that night, having to leave his friends, he said, 'the price of my good fortune felt high'.* Little could he have realised at that time just how seldom he was to see either home, close friends or parents again.

2
In the Company of Gods

Edward's parents accompanied him to Karlsruhe. His mother even had thoughts of moving in to look after him there. True to form, she immediately went off in search of grocers and butchers suitable for her son. He had then reinforced his mother's concern for his well-being when he queried whether, when buying eggs, you had to ask for hard-boiled or soft. However, much to Edward's relief, his father insisted that she leave her son to fend for himself and return to Budapest.

In spite of Karlsruhe's being the headquarters for the giant chemical company I. G. Farben, Edward remembered it as 'a sleepy little town'. A fellow Hungarian student in his third year took him under his wing for a while and helped him move from the modest lodging he had found with his parents to something more imposing. In what spare time he had he played the grand piano he found in his room, and read a great deal. That apart, he had little social life and, for much of his two years there, he was 'lonely and terribly homesick'.

At every opportunity he travelled back home to Budapest, a twenty-hour train journey, but back among his friends and family there was, initially, surprise and then irritation at the amount of time he spent in study. Just before the Christmas of his second year at Karlsruhe, for instance, he had been introduced to the mathematics of set theory and spent much of the festive season in re-inventing a solution to one of its crucial problems. 'My family was not pleased, and my friends thought I was crazy. But I remember the exploration with happiness.'

Quite extraordinarily for someone embarking on a university course in chemistry, Teller's only experience of practical work had been watching the kidney-function tests performed by his doctor at the time of his scarlet fever three years previously. From the start he thoroughly relished the theoretical aspect of the subject, but had not bargained for the repetitive procedures of the practical work, which he found tedious and boring. Within a short time he had shoes spotted with acid, a stained and ravaged lab coat, and permanent scars from shattered glassware. With neither inclination nor aptitude, he very soon looked to break out of this experimental straitjacket, and quickly discovered that the flexibility in the German academic system allowed him to add mathematics to the courses he studied. This he enjoyed greatly, but it was in his second year that he was introduced to the subject that was to change the direction of his whole life. He referred to it as 'a spectacular adventure'.

In 1926 I. G. Farben had brought a brilliant young chemist, Herman Mark, to Karlsruhe to explore the potential uses of the new quantum mechanics. When he was also invited to teach a course at the Institute, Teller became one of his most enthusiastic students and, later, a firm friend. 'Teller really wasn't a good-looking man,' Mark remembered. 'He was stubby and fat, and always a little pale. But if he wasn't handsome, he was awfully pleasant, he was always kind. Teller was popular, too; the other students respected him.'

It is possible to speculate that this popularity was more apparent to the teacher than a reality among fellow students, particularly as Mark described Teller's frequent intercessions at the end of a lecture: 'Well I think that was very interesting, but if you don't mind, I presume what you really wanted to tell us was this . . .' It was the same behaviour that had infuriated teachers and pupils alike at the Minta, but whatever the balance between popularity, irritation and envy, there was no doubting his commitment and brilliance. Very soon Mark had Teller make searches of the literature for him, reporting back to the seminars on the latest research. Mark was also one of those who, at Teller's urging, spoke to his father about the possibility of changing courses from chemistry to physics. Max Teller listened carefully, then sought the advice of a distant relative who was also a physics professor. This relative, after the briefest of exchanges with Edward, told Max that his son should go into physics. Edward was given permission to do as he pleased, and in the spring of 1928 he left Karlsruhe, and chemistry, for good. A few weeks later he enrolled at the University of Munich to study under one of the pioneers of the new physics, Arnold Sommerfeld.

The physicists who studied and taught in Germany during the 1920s formed an extraordinary galaxy of talent that, since the turn of the twentieth century, had powered one of the great intellectual revolutions – a golden age of physics that lasted until 1933, the year Hitler assumed power and that galaxy of talent was dispersed for ever. (See Appendix 1, pp. 403–5, for a summary of the path that led to quantum mechanics).

In 1928, when Edward Teller went to Munich, the discoveries and theories of the 'new physics' were truly revolutionary. Even now, some seventy-five years on, they continue to perplex. Only by comprehending the deeper meaning of the mathematics from which they have been created can they be fully understood. To a young student with a real mathematical talent, the new insights and discoveries that they promised must have seemed almost magical. Yet what he found on arrival was a real disappointment.

Sommerfeld, his new professor, had a reputation for having modified Bohr's early atomic model and had written a classic textbook on the subject.* However, at sixty, he was at least thirty years older than most of those working in the field, and still ran his department as a classic German academic hierarchy. He relished the title 'Herr Geheimrat' (Privy Councillor), with all its connotations of authoritarian inflexibility, and his whole style was antipathetic to the open discussion Teller so enjoyed.

However, he had only been in Munich for a few months when fortune took a hand in redirecting his career. On 14 July, he was heading off to the mountains, hiking with some friends. Burdened down with his hiking gear, he was riding near the front of a three-car tram when he failed to notice that the tram had reached his destination outside the railway station and had started to move off again. In a rush he pushed to the door and jumped.

'By this time the tram went too fast and I fell,' Teller remembered, 'and I saw the tram – I think three cars coupled to each other – go by. And somehow, I don't know for what reason, I remember looking back . . . and there I saw my boot lying at a distance. "My God, how will I go hiking without my boot?" And then, several seconds later it had started to hurt badly. My foot was still attached to me, but barely.' The tram wheels had almost completely severed his right foot above the ankle. He was rushed to hospital, where he was operated on immediately.

The young surgeon who operated on him used a technique of fusing what remained of his heel with the two long bones of his lower leg.* This preserved as much of the leg as possible and also meant that the stump would bear his body weight. However, it committed him to several months in hospital. Within hours of the operation his family arrived from Budapest and his mother, much to his annoyance, stayed over until he was well enough to return for convalescence to Hungary. 'She did not want to go. I wish she had,' he wrote. 'At that time I was not in need of being sustained.'

His convalescence, miserable as it must have been, was brightened by the presence of his old nanny, Magda Hesz, returning on a visit with her American husband. Edward remembers her good cheer, her stories of America and, in particular, the fact that she had a car – then such a rarity, even in Germany, that it spoke of another kind of existence across the Atlantic. Another visitor to his bedside, though perhaps a less enjoyable one, was Hans Bethe, whom Teller had met for the first time in Munich.

Bethe was some eighteen months older than Teller and, at twenty-two, was seen as indefatigable rather than brilliant, with a mind like a diesel engine, ploughing steadily through any difficulty. He was amiable, an accomplished skier and mountaineer but, while he exuded confidence in physics, socially diffident. Teller was impressed that this relatively senior figure should take time to come and see him, but recalled that his visitor knew neither what to do, nor what to say, making his escape as soon as he felt it polite to do so.

One day, after several weeks in hospital, Edward's young surgeon, with whom he had formed a close relationship, disappeared suddenly overnight without anyone knowing where he had gone. There were rumours that the surgeon, whose name was von Lossow, had gone to South America. It was thought this was because his father was the general who had thwarted Hitler's early attempt to gain power in Munich in 1923. Now, some six years later, the Nazi Party was in resurgence and von Lossow was thought to have fled in fear for his life.

With von Lossow had gone Edward's main reason for staying on in Munich and he returned home for a final two months' further recuperation. Each day he was visited by Mici Harkányi-Schütz, a cheerful antidote to his family's dolefulness. The couple had remained in touch throughout Edward's two years in Karlsruhe and were obviously devoted to one another. 'He never cared for another girl,' his cousin Illi reported. 'He never went out with another girl. He didn't care for being with other girls, for many years all he wanted was her.'

Mici – her real names, Augusta Maria, were hardly ever used – was a year younger than Edward. Small and vivacious as Mici was, Edward made a telling comparison with his sister Emmi: 'Emmi was a Jew; Mici a Calvinist.' [Her parents were Jewish but, like many other Hungarian Jews, had converted to Christianity.] 'Emmi was obedient; Mici iconoclastic. Emmi was old-fashioned in dress and manners; Mici objected to stockings and found decorum dull. Emmi was organised and careful (a trait we did not share); Mici was a free spirit. Yet they liked each other, and more surprising yet, Mici seemed to find me quite acceptable.' In spite of their obvious closeness, however, the progress of their relationship was to be hindered by their youth, Edward's absence abroad, and Mici's authoritarian stepfather.

Edward's leg healed only slowly, but in October, still barely able to walk on his new prosthesis, he returned to university. He could have returned to Munich, but Sommerfeld as well as proving an initial disappointment to Edward, had also gone on leave of absence. So Edward decided to aim for the heights: to seek enrolment at Leipzig in the department where Werner von Heisenberg was professor. To his delight, he was accepted and went off to become 'acquainted with knowledge in a way that I had dreamed of as a child. Heisenberg was only six years older than I, an enormous difference in scientific development, but not so great a hurdle in human relations. He would never play the role of a Geheimrat.'

It was the extreme youth of the twenty or so physicists gathered around Heisenberg that impressed Teller. At twenty he was the youngest in the group but there were only one or two over the age of thirty; and for the first time he met people whose interest in science was more consuming than his own. This was clearly an elite and he was intimidated. His background in chemistry and maths meant there were yawning gaps in his knowledge and he seriously wondered whether he would be able to cope. Also, he had been lonely for much of his two years in Karlsruhe and he had not enjoyed the few months that he had spent in Munich. He still feared the social ostracism that had blighted his teenage years. 'Would my fellow students accept me socially, or would I find myself on the outside looking in?' he wrote. His tactic was to make himself useful in whatever way he could. The group had a weekly social evening when they met to talk physics, drink tea, and play ping-pong. Edward offered to make the tea, and he went on making it for the whole of his two-year stay.

But his fears of being a social outcast were unfounded. This was a truly cosmopolitan group, united in their dedication to science, and there was an extraordinary absence of intolerance and snobbery. They were to become a brotherhood, and their lives and careers were to remain entwined for decades to come. They were nearly all, like Teller, living on a tight budget, lodging with families, eating in cheap cafés near the university, and spending what spare time they had in outside pursuits and in conversation. Nationality, religion, politics were all of little import. It was here that Teller first met Carl von Weizsäcker, a Prussian aristocrat whose father was to become a minister in Hitler's government and who was, himself, to work on the German bomb.

It was also here that he met and befriended the Russian physicist Lev Landau, a young man exactly the same age as himself, but already a scientist with an international reputation who would later win a Nobel Prize. Landau was a devoted communist who never failed to point out any social shortcoming he observed in the capitalist world. He also rarely failed to point out shortcomings in the communist world. There was a story told about him which, whether apocryphal or not, catches the flavour of both his independence of thought and his acerbic wit. The notorious agronomist and founder of 'Creative Darwinism', T. D. Lysenko, gave a talk to the Russian Academy on the inheritance of acquired characteristics. When it was over, Landau asked: 'So you argue that if we cut the ear off a cow and the ear off its offspring, and so on, then sooner or later earless cows will be born?'

'Yes, that's right,' Lysenko replied.

'Then how do you explain that virgins are still being born?' Landau asked.

He was eventually to be arrested and imprisoned during the Stalinist purges of the late 1930s and his suffering during this period would become one of many factors that moulded Teller's view of the Russian system.

The Swiss scientist Felix Bloch, who was to work closely with Teller at Los Alamos, was also a member of the group, as were the Americans John van Vleck and Robert Mulliken, and the unassuming Japanese Yoshigo Fujioka. When Teller's mother and sister were eventually allowed to leave Hungary in 1959, it was Fujioka, then working in Vienna, who met them on Edward's behalf. Theirs was a friendship that would bridge the horrors of nuclear war.

At the apex of this group was Heisenberg himself, who had only recently arrived at Leipzig. Like Teller, he was a mathematical prodigy and a musician who enjoyed walking and hiking. He was highly competitive, certainly, but all ideas were discussed openly and in a manner that appeared half-serious, half-joking. Teller remembered his trademark question: '*Wo is der Witz?*', translatable as 'What is the point?' but more precisely as 'What is the joke?' – the unexpected, that which goes to the heart of the matter.

It was just the kind of open challenge Teller rose to: he was always excellent at brainstorming. For Heisenberg's part, he was stimulated by his new student from the time he read early samples of his work. 'They were excellent, excellent,' Heisenberg remembered. 'From that time on he was one of the most interested as well as most interesting members of my seminary.' And Heisenberg was quick to create an opportunity for him, one that Teller regarded as his proper start in science. 'A brilliant student like Teller,' Heisenberg recalled, 'would not just sit in the lecture room and listen to what the professor says. The "start" in such a case is that the professor tells a young man "There's a problem I can't solve. Can you solve it?" '

The problem at hand concerned the structure of the simplest molecule there is, the hydrogen molecule ion. It consists of an electron and two nuclei, and a Danish and an American scientist had separately calculated the lowest possible energy levels for the three particles. Their answers were different, and Heisenberg set Teller the task of deciding who was right. He rapidly showed up the flaw in the American's calculations, and so Heisenberg set him another problem: to identify some of the possible higher-energy states of the hydrogen molecule ion.

'The assignment became my doctoral project,' Teller recalled. 'The problem involved lots of busy work, a little diligence, and no originality. I needed only to imitate Burrau [the Danish scientist] and do what he had done – again, and again, and again, and again . . . There was no end to the work.'

For this endless task Edward had the use of an old and noisy calculator. He had already established his lifelong habit of rising late and working late, and so, night after night, he was to be found pounding out his results in the department common room. This was in the same building and just below Heisenberg's bachelor apartment, and sometimes the professor would stop by late at night for a chat. 'I particularly remember one such conversation,' Teller recalled. 'He complained that physics had ceased to develop; that there was hardly anything interesting left to do. The year was 1929! I felt more than a little indignant. I thought to myself . . . "You turned physics upside down. What more do you want?" I don't recall my response, but it was considerably more polite.'

To Teller, this exchange was a challenge to his Holy Grail, to the value of physics as a field of study and of his future in it. Heisenberg was aware, to some degree, of just how insecure his young student was, and he was impressed at how, for example, Teller was coping with his new physical disability. 'I could see in the beginning that he suffered from it,' Heisenberg recalled, 'not just bodily but also mentally; but I think he overcame it rather soon. I think, in a year or so, he was quite stabilised in his mind.' In time Teller returned to playing ping-pong and Heisenberg again observed how he coped. 'He became an excellent player just because he wanted to become one. There I could see the force of the man: when he was hampered by some outer facts which he couldn't change, he really would try with all his strength to make up for it – and he did make up for it.'

Someone else who was at Leipzig at the time and watched Teller's battles at the ping-pong table with a more critical eye was the young American physicist Isidor Rabi. On one occasion he watched him play a dispassionate Chinese physicist who was regarded as the best of the players. He watched as determination, when frustrated, became aggression. 'Teller would slam and slash the ball,' Rabi said, 'and the Chinese would just stand there hitting them all back.' In the years to come, Rabi was to become one of Teller's severest and most important critics.

Teller worked on his thesis for little more than a year. Then, one night, when he was working on his calculator, Heisenberg made the point of coming down from his apartment to see him. When did he think he would finish his project? he asked. Within the next two years was the answer. Didn't he think he had done enough and that the results he had already would make a good thesis? It was advice that Teller followed.

However, the achievement of his doctorate so quickly and at such a young age – he was twenty-two – still did not silence the insecurities preying on Edward. He was mortified when, after travelling all the way to Berlin to hear Einstein lecturing on his vision of a unified field theory, he found he had not understood a word. He was also deeply embarrassed by a put-down he received from Niels Bohr during one of the regular visits Heisenberg's team made to Bohr's Institute of Theoretical Physics in Copenhagen. Seventy-five years later he could still feel the 'discouragement of that moment'. It was as if, around every corner, he expected to confront the real truth: that he had swum out of his depth and was now about to drown. He was therefore truly delighted when, not long after completing his doctorate, Heisenberg asked to him to stay on as one of his postdoctoral assistants. It meant taking on many of his professor's chores – marking students' essays, lecturing and so on – but it was still an honour; and Edward had been asked to dinner. He was sufficiently overwhelmed by the import of the occasion to 'mistake the dessert for cheese and smear it on a piece of bread. Heisenberg took his revenge by not enlightening me until it was too late.'

His assistant's duties still allowed him time to begin searching for a new direction for his research. He looked deliberately for a virgin field where, even though he saw it as much less important work than other projects in the department, he felt he could make a real contribution. He picked up on work he had done on quantum mechanics under Herman Mark in Karlsruhe. He began applying quantum mechanics to understanding the subatomic interactions within increasingly large molecules, with the aim of explaining their physical and chemical behaviour. In pursuing this field of work, Teller was following in the footsteps of someone whom, at that time, he knew only by name: Robert Oppenheimer. Some five years earlier, Oppenheimer had moved from Cambridge to study with Max Born in Göttingen. While there he had worked out a mathematical approximation that allowed some sense to be made of the complex interaction between the constituents of a molecule.

A molecule does not sit motionless in space. It rotates around its axis like a spinning top; and, at the same time, the nuclei and the surrounding electrons within the molecule's constituent atoms are also in continuous motion. The forces binding the electrons to their orbits around the nuclei are not rigid, so the electrons are vibrating in and out in relation to the nucleus. These vibrations and rotations can be observed, when a molecule is excited, to produce a spectrum, and the particular mix of vibration and rotation gives each different molecule its special spectral fingerprint. From those fingerprints can be deduced the size and shape of the molecule and how it is likely to behave; but with dozens of particles milling around, the interior of an atom is a mayhem of activity, so to simplify things Born and Oppenheimer had assumed that the nucleus, which is thousands of times heavier than the surrounding electrons, is motionless. They had also discounted the vibrations because they were thought to influence the size of a molecule by only a percentage point or two.

For Teller however, these factors were important, and he was particularly intrigued by a molecule – a compound of methane and iodine – in which excited vibration seemed to double the size of the molecule. He wanted to understand what was going on.

To help with this mammoth task, he managed to arrange for Lázló Tisza, one of the two other students who had shared the Eotvos maths prize with him five years earlier, to come and join him in Leipzig. It was work that Teller 'loved' – his first discovery, the first time he had posed, and answered, a brand-new question. The collaboration continued even when Lázló went back to Budapest, but shortly after his return something happened that stopped their work in its tracks. With the growing presence of Nazism in Germany, the two had talked often about the impending battle between communism and fascism. Edward was undecided about which he saw as the greater of two evils, but Lázló was a determined communist. Under the Horthy government in Hungary, however, the Communist Party was outlawed, and shortly after his return Lázló was caught carrying a message for some communists. He was arrested, beaten and jailed.

Edward travelled to Budapest to visit him in his prison cell, and they even made an attempt to continue their research. Edward certainly had little sympathy with communism, but he was outraged by the injustice of his friend's fate. It was first-hand experience of oppression, from whatever political quarter it had originated, and it was to colour his reaction to acts of tyranny in the years to come. Even after a short stay in jail, Lázló's future in Hungary was ruined. Edward felt powerless and angry.

That summer saw a significant change in Edward's somewhat monastic private life. Mici's strict stepfather agreed to her taking a holiday with a girl friend at the resort of Mátrafüred. She invited Edward to join them and, under the pretence of taking a solo hiking trip in the nearby hills, he spent the whole week in her company. By this time Edward had clearly decided in his own mind that he and Mici would spend their lives together. In the eyes of their friends, too, Mici and Edward were now a well-established couple; but he had said nothing, either to her stepfather or to herself. He was, after all, still living in lodgings and his parents still paid much of his upkeep. With the political uncertainties in both Germany and Hungary, he had little idea of where he was going or what he was going to do. So, after his week-long summer idyll, he returned to his monastic ways in Leipzig, working on molecular structures, lecturing and correcting Heisenberg's students' essays.

Then, out of the blue, came an offer from yet another centre of excellence in the physics firmament: Göttingen. The chemist Arnold Eucken invited him to come as his assistant. There would be no lecturing and no correcting of students' work. He would receive an assistant professor's salary and for the first time he would be an independent practising scientist. At the age of twenty-three his student days were at an end.

3
Twilight of a Golden Age

During his three years in Göttingen, Edward Teller must have been a strange sight. He may have been an assistant professor, but during his first year he went around with his head shaved – 'spared me having to waste time in the barber shop' – and wearing an oddly tailored raincoat with a large monastic hood. Mici hated his shaved look and his family denounced it as pure and simple craziness. But he persisted. His life was so tightly focused on his work that the conventionality of his appearance – or otherwise – mattered little.

When he arrived in Göttingen he was again deeply anxious about how he would be received. His acceptance by his peers and by the students at Leipzig had both 'amazed me and given me intense pleasure. But I had made a very conscious effort to be pleasant and to fit in Leipzig.' Again, however, his fears proved groundless. This, in part, was due to the presence of old friends like Carl von Weizsäcker, who had moved on there as well, but it was also because he was building a reputation. Indeed, he had not long been there when he received a phone call from James Franck. Director of the Physics Institute, Franck had also shared the 1925 Nobel Prize for his work showing that atoms can absorb and emit energy in discreet amounts, as postulated in Bohr's theory. Franck courted Teller.

Franck came from a banking family and he had a car. He took Teller on a drive showing him the beauties of the medieval town. He also described his developing interest in applying quantum mechanics to the understanding of the behaviour of molecules – precisely the area of Teller's interest and growing expertise. Within a short time he had taken Teller to meet his own research group, and had then negotiated an arrangement with Eucken by which Teller split his time between the two of them. It marked the beginning of another seminal friendship, one that was to last until Franck's death in 1964.

Franck was using spectroscopy – the detailed examination of light emitted or absorbed by a molecule – to analyse the details of the interaction between the nuclei and electrons within a molecule; but he was unusual in putting the experimental cart before the theoretical horse, coming up with results that then needed explaining. So Edward spent his first year providing that theoretical explanation for Franck's results.

In the summer he returned to Budapest, where he hoped, yet again, to spend time with Mici. Out of the blue, however, she announced that she was going to study in America for a few years and that she was leaving that autumn. 'But far worse than that,' Teller recalled, 'she broke off our relationship. The summer before, I had truly believed that Mici was as fond of me as I was of her. But now, it seemed, our romance was over. I had no idea why. I was depressed and hurt. Fortunately, science seemed not so completely incomprehensible as love or war . . .'

It is easy to speculate that Teller's obsession with his work to the exclusion of a proper social life, and his long absences in Germany, may have contributed over time to Mici's apparently sudden decision that she needed pastures new. Certainly, he had not seen the split coming and, once back in Göttingen, his work took over again. The couple were not to communicate with each other at all for the next two years.

It was at this time that he began working with, among others, the Czech physicist George Placzek. They began a collaboration using a particular kind of spectroscopy, known as Raman spectroscopy, to see what it could tell them about the rotation of an individual molecule and the effect this rotation had on the molecule's chemical behaviour. It was to turn into a long and complicated partnership and one that was to bring back a refined version of the bullying that Teller had suffered in his teens and of which he was so afraid. Placzek was some six years older than Teller and a much more experienced physicist. Hans Bethe knew Placzek at this time. 'He was a man of the world, much more sophisticated than Teller or I. He had dated girls in large numbers and he spoke eight or ten languages, all very well, and was well versed in literature and all sorts of things, so he was very impressive to me, at least.'

Teller's own view of him was sharper, comparing him to his acerbic Russian friend Lev Landau:

> He was also a communist, and he was wickedly sarcastic about the stupiditites of the social system in Central Europe. When I told Carl Friedrich drich [von Weizsäcker] that I found the two men similar, he disagreed. Landau, he pointed out, was much nicer to children ... Most of all, George Placzek reminded me of my cousin George,* who, bitter and unpleasant as he was, had a fine mind ... Our joint effort was one of the few times I did not enjoy a friendship with a collaborator.

Placzek recognised a vulnerability in Teller and derided him mercilessly. He referred to him as 'Herr Molekular Inspektor', inferring that Teller's role in their project was a mundane and inferior one. It was Placzek who nicknamed him 'Il Pellegrino' – the Pilgrim – because of the hooded raincoat he wore. According to Bethe, he eventually reduced Teller to the point where 'we used to say he apologised for being alive'.

Thus, in a short space of time his girlfriend of some four years had left him and he was inextricably involved with a collaborator who was making him very unhappy. The year 1932 could not have been the best of times for Edward Teller. However, it was during the late spring of that year that, hanging on to Placzek's coat-tails, he had travelled to Rome to meet one of the most respected and loved of physicists, the experimentalist Enrico Fermi. Placzek had an assignment in Fermi's lab but wanted to continue the project with Teller. Teller did not have the necessary funds for the intended three-week stay, so, with what Teller described as 'a really generous gesture', Placzek engineered a letter of introduction from Fermi who, until then, had not met Teller. In the letter, Fermi described him as a 'great physicist'. It was enough to gain him accommodation in the Palazzo Falconieri, run by the Hungarian government to house academics visiting Rome.

Although Teller did not work with Fermi, the Italian impressed him, as he did almost everyone who met him, with his simple ease of manner and his intellectual clarity. They played ping-pong together. Teller noted that Fermi was an enthusiast but took the game much less seriously than Heisenberg. According to Teller, when they played, Fermi won only once. 'At the game's end, when he began to crow, I raised my left arm, paddle in hand, thus pointing out that I had played left-handed.' It was a ploy indicative of a twenty-five-year-old who still had some distance to go before shaking off an adolescent smugness. It is only one anecdote, but it is not difficult to make a connection between such behaviour and his victimisation at the hands of Placzek.

It was also while Teller was in Rome, at Fermi's laboratory, that the old German general von Hindenberg narrowly defeated Hitler for the German presidency. Everyone at the laboratory heaved a sigh of relief, and returned to their science. Up to that time the physicists had, with some determination, managed to ignore the festering aftermath of the First World War, the turmoil in world finances and the prophets of racial supremacy. Even now, only one or two of those more politically aware showed signs of alarm. The most extreme of these was Placzek, who insisted that Teller buy a copy of the book *Public Faces* by the British writer Harold Nicolson.* Teller was intrigued by its story of a new, destructive weapon made from a raw material found in Iran and called an 'atomic bomb', though for Placzek the interest was not in the content, but rather in the book's potential to act as the basis of a code. He already saw good reason why he, an avowed communist, and Teller, a Jew, were soon going to have to leave Germany and would actually need to communicate secretly if they were to be successful.

However bizarre this must have seemed at the time, the seriousness of the political situation was underlined when, on 30 January 1933, while at a party at James Franck's house, the news broke that Hindenberg had just made Hitler his new Chancellor. Within a week the future became abundantly clear. 'Throughout the city, people pasted their advertisements on to innumerable bulletin-board posts (Litfassäulen),' Teller wrote. 'Under Hitler's Chancellorship, all the posts were stripped bare: everything was scratched off them. From then on, nothing could be advertised without government permission. From my worm's-eye view, that was the first sign of what was to come.'

Almost immediately the Nazi propaganda machine was directed against Jewish scientists and in particular against Einstein. His theories of relativity were scorned. A German Nobel Laureate, Philip von Lenard, was able to write: 'The most important example of the dangerous influence of Jewish circles on the study of nature has been provided by Herr Einstein with his mathematically botched-up theories, consisting of some ancient knowledge and a few arbitrary additions.'

Overnight Edward found himself isolated and his world turned upside down. In their different ways the two professors for whom he had been working over the past three years made it clear that he had no future in Germany. Arnold Eucken was an old-style German nationalist, and his decency was offended by the predicament of his Jewish colleagues. However, he made it clear to Edward that he would soon be unemployable, and that there would be nothing he could do to help him.

James Franck was, in fact, Jewish but did gain some protection from having fought at the front during the war. However, not only had he already organised a post for himself in the United States, but he was also coordinating efforts to find places abroad for as many of his Jewish colleagues as he could, including Teller.

In this he was acting as agent for one of the most remarkable rescue operations of the time. Within three months of Hitler's rise to power, the British scientific community had established the Academic Assistance Council – the AAC – aimed at helping scientists whose politics or race placed them at risk from the new regime. Anyone of ability, whether the British had a use for him or not, was welcomed in the UK. The committee was funded by Imperial Chemical Industries, and its leading lights were the Oxford professor Frederick Lindemann, soon to be Churchill's scientific adviser, and the London biochemist George Donnan. In Göttingen, James Franck's house was the focus of its activities.

Teller knew that there could be no retreat to Hungary, where Horthy remained in power and which, in terms of anti-Semitism, was little different from Germany. In an effort to find at least a temporary refuge, he had applied for a Rockefeller Foundation Fellowship to study under Niels Bohr in Copenhagen, but had been turned down as not having an assured post to which to return. Thus he was delighted when Franck arranged for him to meet both Lindemann and Donnan.

Teller did not warm to Lindemann, though he nevertheless received an offer to join his department in Oxford. However, he found George Donnan engagingly direct and was soon visiting London. Here he spent a day with Donnan, being put through his paces – everything from science through table manners to talking about his personal life. When he explained the catch-22 situation over his Copenhagen fellowship, Donnan offered him just what he needed: a post from which he could come and go as he liked. Teller was deeply indebted to a generosity that went far beyond what was required to help people in trouble; but Donnan added one small course of anglicisation that Teller had to complete before he could return: he had to read both *Alice's Adventures in Wonderland* and *Through the Looking Glass*.

In spite of all his good fortune and the prospect of a year in Copenhagen, Teller was deeply distressed by what was now happening in Germany. He wrote: '. . . a period of beauty and excitement was ending, a refuge for mind and spirit was being destroyed. I was full of anger and anguish.'

He wrote poems to express that anguish, as he often did at times of stress. In a poem entitled 'October, 1933', he wrote:

> These times are for passion, not for doubt.
> (Magnificent Mr Heroes have no concern
> For what is lost: they just destroy it.)
> Take the plane: with fervour, smooth the plank.
> Small loss if most planks break in two.
> (Small loss, if war – barbarity most rank –
> Should come with miseries old and new.)
> The times are great. No one can change the fact.

The awkwardness of the verse, certainly in translation, works to convey the depth of his emotion at the loss of the 'unique and wonderful community that was German physics in its golden years'. His sense of loss was compounded by the fact that 'for two long years . . . Mici – the one person who still tied me to the many things that were dear to me from my youth – had been in America, we had neither spoken nor written to each other'. However as he was preparing to leave Germany for London, he heard that she was back in Budapest after finishing her American course of studies. He went there immediately to see her.

'As I climbed the stairs in her apartment building, a completely inappropriate quote from Goethe popped into my head,' he wrote. 'Mephistopheles says to Faust: "You go to the chamber of your sweetheart, not to your death." I suspected that if I were ever to marry, I would marry Mici. I also knew that if I were to marry her, I should marry her now.'

In spite of the long gap in communication between them, the couple were clearly delighted to see each other again. Mici had been thrilled with America and with her switch from mathematics to her new work in personnel. It seems that the couple narrowly avoided falling out during this first meeting because of Edward's scepticism towards the newly developed technique of IQ testing, which Mici thought was so exciting. But they arranged to go walking the next day.

'In the hills of Buda, a little less than twenty-four hours later, we stopped for lunch on a meadow. About a dozen geese gathered to greet us, and there, to the loud approval of the birds, I proposed. Mici is and always has been decisive. Her answer was yes.'

The wedding was set for that Christmas and Edward continued on to London to take up his new post in Donnan's department. He also re-applied for the Rockefeller Foundation grant, telling them about his new position and also about his impending marriage. Back came a telegram from their Paris office: 'NEED INFORMATION IMMEDIATELY ON WHETHER YOU INTEND TO GET MARRIED.' It was clear that his marital status was of more than perfunctory importance and, as the meeting to award a fellowship was the following day, Donnan advised Edward to travel to Paris immediately and explain the situation in person. There, in an interview, the problem emerged: a previous Hungarian newly-wed had used his grant to go off on an extended honeymoon and the Foundation was not about to let the same thing happen again. If he wanted the grant, he would have to postpone his wedding.

That evening there followed a deeply unhappy phone call to Mici – made worse by the fact that Edward muddled the time differences between Paris and Budapest. Mici had to endure two hours of teasing from her two younger brothers before the call eventually came through; but they agreed that the fellowship must have priority and that their wedding would have to be delayed. So, a month later, Edward moved to his new lodgings in Copenhagen, which he was to share with his fellow itinerant Carl von Weizsäcker.

The two were to see a great deal of each other, both on the tram journey to and from Bohr's Institute and after hours in each other's rooms. It was here that they discussed literature, re-reading Goethe and Schiller and conducting formal debates in the Socratic style, during one of which von Weizsäcker was to defend, successfully, the thesis that 'standing to attention is ecstasy'.

On another occasion, while discussing the various human dispositions identified by Aristotle (melancholic, phlegmatic, choleric, sanguine), von Weizsäcker identified Edward's disposition in a way that he believed few of his friends or family would have done at the time, correctly naming him 'not as sanguine (confident), as some of my critics have claimed, nor as melancholic, as I sometimes feel, but as choleric (irascible), a flaw I struggled against in my youth . . .'

Edward obviously believed at the time that his struggle had been successful, that his efforts to be 'agreeable', to be 'useful', had cloaked his inner turmoil; but the roots of that turmoil ran deep. He was a highly emotional person, yet he had difficulty in winning the friendship, the acceptance he wanted so much without keeping himself in check, living within a mask of affability, trading on his talents, his usefulness. It is not difficult to see this nexus of conflicting emotions as fuelling the will to succeed that Heisenberg had identified and the aggression that Rabi had seen. It is also not difficult to see how vulnerable he remained to disapproval or enmity. It was at this time in his life, before his success as a scientist provided the crutch and the camouflage for such inner turmoil, that it would have been most evident to someone as sensitive as the philosophical von Weizsäcker.

Once settled at Bohr's Institute, Edward very quickly became caught up in his fellowship work. 'My life in Copenhagen was stimulating and challenging,' he wrote. 'I missed Mici a great deal, but I was far too busy to brood over our change of plans.' However, he had written to his old mentor James Franck, now in the US, and had mentioned the problem of the delay to their marriage. Franck was so incensed by the Foundation's actions that he launched a potent and ultimately effective protest. After some weeks of manoeuvring, mainly to allow the Foundation to save face, Teller eventually received a letter. It asked him simply to inform them of the date of his marriage.

Edward Teller and Mici Harkányi-Schütz were married in Budapest on 24 February 1934. Because Edward was Jewish and Mici a Calvinist, it was a civil ceremony, conducted in the city office just across the street from his parents' apartment. 'The official who married us weighed well over 200 pounds,' Teller wrote. 'He wore a wide sash of red, white, and green that practically encased his body, and he fulfilled his duties with few words.

Only our closest family members and friends were present, but Nándi [Teller's school friend] said the ceremony was touching.'

Their honeymoon, such as it was, consisted of the train journey back to Copenhagen, with a stopover to see friends in Leipzig. Within three days Teller was back at work in the Institute. For the first time in his life he was able to devote himself to his own research and had no obligation to satisfy anyone else. There was Niels Bohr, of course, but, as Teller put it, 'satisfying him was beyond my imagination'.

Whether because of age or temperament, their relationship remained a distant one. Unlike Heisenberg, who was warm and direct and had a sense of humour, Bohr had a dissociated air. He loved paradoxes, and was at his brilliant best explaining them in long and convoluted sentences, which somehow made sense in an inspirational but ambiguous way; but even close friends found his long rambles, delivered in a monotonous mumble, difficult to take. To Teller they were alien:

> Bohr invented paradoxes because he loved them. I imagine I understand those paradoxes, but I failed to understand Bohr. In human terms, understanding means being able to put yourself in the place of a fellow being. In those terms, I can understand Heisenberg ... In no way can I imagine myself in Bohr's place. Even though I spent a year working under him, Bohr remains an unknown and inexplicable person to me. I became acquainted with the amazing nature of his brilliance as a physicist, but we never met as human beings.

The Tellers returned to London in the autumn of 1934 in time for Edward to begin lecturing in the chemistry department of University College. Marriage had already altered Edward's lifestyle. As a bachelor arriving in Copenhagen at the beginning of the year he had been able to carry the majority of his worldly possessions. When he and Mici came back to London, they had accumulated enough to fill seventeen pieces of luggage.

At Christmas the couple returned home to Budapest. Just before they left they found a flat in Gower Street and, in anticipation of making London their permanent home, they signed a six-year lease on it. On their return after the New Year, however, they found two offers of teaching positions in the US. One was from Princeton, where Edward's friend and fellow Hungarian Eugene Wigner had found a position. The other was from George Washington University in Washington, DC, and came from a scientist who, more than any other, was to change the course of his life.

Teller had first met the Russian George Gamow during his first visit to Copenhagen in 1930, but it was during a second visit, in 1931, that the Russian had taken him under his wing. Gamow was four years older than

Teller and, at that time, had already done significant research on the nature of radioactive decay with Ernest Rutherford at Cambridge. As a Rockefeller fellow, he was spending a year with Bohr before returning to Russia.

Although always penniless and repeatedly touching his friends, Gamow owned another rarity of the time, a motorcycle. That Easter he had invited Teller to join him as his passenger on a tour of the Danish islands. Teller recalled the draughty miles clinging to Gamow, the long, rambling conversations about physics as they lazed on the beach at Fynf Hoved, and he recalled them watching a total eclipse of the moon together.

Shortly afterwards, Teller had returned to Göttingen and Gamow had gone back to Leningrad to become Master of Research at the Academy of Sciences there; but although no political animal, Gamow was not at all happy back in the Soviet Union. When his movements and contacts outside Russia had been restricted, he was sufficiently incensed to try to escape, along with his new wife Lyubov, by rowing across the Black Sea to Turkey. Exhaustion had forced them to turn back.

In early 1934, however, he had been a member of the Russian contingent at the Solvay international conference in Belgium and, by some oversight, the Soviet government had allowed Lyubov to travel with him as his secretary. This time the two had seized the opportunity to make good their escape to the West, first to Paris, where they had met Teller on his marital mission to the Rockefeller Foundation, and then to Copenhagen where their paths had crossed once more.

Following Teller's arrival on his Rockefeller grant, the two scientists and their wives had spent time together. Teller had observed that Gamow's marriage to the beautiful Lyubov (nicknamed Rho – short for Roxanne – by Gamow) was already something of a switchback affair and that George himself was drinking heavily. But the sheer excitement and the value of their discussions on physics had indelibly impressed them both. When the two had parted, Teller had returned to London while Gamow had gone off to his new post as chairman of the physics department at George Washington University.

From this new post he offered the twenty-six-year-old Edward a full professorship at three times the salary he was receiving in London. It was an offer too good to refuse. Edward, however, was very conscious of his debt to Donnan and went to see him. Donnan released him from any sense of obligation, explaining that his move was very much in the spirit of the committee's objectives. Any doubts Edward might still have had were swept aside by Mici, who was not comfortable in London and was itching to return to the US. The offer was accepted and, after the inevitable last-minute hitches over visas, the couple sailed for New York in August 1935.

Teller never forgot the charity he received from the Academic Assistance

Council. In his autobiography he wrote: 'London in 1934 made the lot of the refugee tolerable ... In London I had been a guest, and I was extremely fortunate to have been there ... The English are truly among the most hospitable and ethical people in the world.'

4

America the Beautiful

Given his alienation from so many of his scientific friends and colleagues in his later years and the public demonisation resulting from his single-minded pursuit of the arms race, it is all too easy to overlook just how close Edward Teller's friendships were with his contemporaries at this stage in his career. Also, it is easy to forget how highly regarded he was to become in those years before the Second World War.

A fellow passenger on the voyage across from England was Hans Bethe, who had also taken up a professorial appointment – at Cornell, in upper New York State. They had already crossed paths on several occasions as they had moved around the European centres of excellence, but over the next few years they were to form a close friendship.

Within a few days of their arrival, Edward was engaging with the new academic year. At that time, George Washington University was a typical city campus, similar to London's University College, which Teller had just left. It was a block or so of buildings just a short distance west of the White House. It had a well-respected medical school and law faculty, but in the pure sciences its reputation was distinctly second-rate. Over at neighbouring Johns Hopkins it was known disrespectfully as 'Washington Tech'. Its ambitious president, Cloyd Marvin, wanted to put that right by appointing a new chairman of the physics department, but he had a limited budget. He had sought advice from Merle Tuve, director of the Department of Terrestrial Magnetism at the nearby Carnegie Institute and America's future 'father of radar'. He had advised Marvin to look for a good theorist. An experimentalist could well blow his limited budget simply on equipping his laboratory, while a theoretician only needed paper, pencil and some travel money. And he suggested a name: Gamow's. Tuve had been greatly impressed by Gamow's discovery of the key to one of the forms of radioactive decay, alpha decay, and thought him a scientist of Nobel Prize stature.

Gamow had been made an enticing offer and also given the opportunity to bring on board another theoretician. 'There's this one man who knows everything, just everything,' Tuve recalled Gamow saying of Teller. 'Not just a physicist but a physical chemist who could broaden the expertise of the Department.'

Initially the relatively stellar salaries he and Gamow received looked likely to cause considerable resentment among the existing staff members, but their reputations carried the day. While in Europe, Gamow and Teller might have lingered for many years in the shadow of the giants of physics, but here in the US they represented a direct link with those giants. Both knew Bohr, Fermi, Heisenberg, Franck – all personally; and they had both, incontestably, done research of the first order. The resentment never gathered momentum.

As to the two stars themselves, they were both forceful and ambitious, and yet the chemistry between them supported a close collaboration that lasted for six years. Teller commented:

> Gamow chose me to fill the second slot at George Washington because I was a good second fiddle: I listened carefully to his craziest ideas. Bethe was disdainful; Placzek would interrupt and tell him to shut up; I was a more flexible foil. I appreciated Gamow. He generated one new theory every day, which made him a sort of force of nature. But if the theory turned out to be nonsense, as most of them did, one could tell Gamow this in a straightforward way, with no need to say 'You are right because...'
>
> On the rare occasions when I couldn't fault his notion, we wrote a joint paper. It was usually a good one, because Gamow had excellent taste in selecting questions.

Gamow was irrepressible. Well over six feet tall, blond, crumpled, ebullient, he squinted in quizzical amusement at the world through small round spectacles.* Life was fun, life was funny, and he enjoyed it to the full – his work, his food and his drink, particularly the last. Just before he died at the relatively early age of sixty-four, he joked that this was his liver now calling him to account.

There was gusto, too, in the way Teller approached his life, though tinged with moments of introspection and even depression. His bearing belied his age. Though still in his twenties, he seemed much older, with his heavy eyebrows and his deep Bela Lugosi-like voice; but in company where he felt comfortable he was gregarious and witty, punctuating his own stories and responding to those of others with an incongruously high, girlish giggle. Both men were regarded as excellent lecturers, but in very different ways. Gamow was a wonderfully clear and amusing orator, and in time, with his 'Mr Tomkins' books, he became one of the best of the early popularisers of science.

Teller's style was altogether more personal and unorthodox, and somewhat lacking in organisation. One of his students at the time was Harold Argo:

> Edward as a teacher was very exciting and stimulating. He moved incredibly fast, and he didn't often start at the beginning. But when asked, he answered our questions carefully and fully. Fermi, who taught me also, was the sort of instructor who painstakingly builds a house from the foundations up. Edward adds the roof and points out the esoteric applications off in the distance. We students would often find we had to get together after the lecture to figure out what we had learned.

It was not just the students who benefited from Teller's enthusiasm. It was a time when the physics community in the US was tiny and a specialist might be quite alone with his problems. Alfred Sklar, one of the Johns Hopkins students who had started coming to Teller's lectures, recalls how Teller made himself a kindred spirit for the whole scientific community of the area.

'All relished Teller's "compenetracion", his quickness in grasping the essence of your problem, and his uncanny ability to make good his halfjest – "I don't understand it, but I will explain it to you." This was Teller's way of saying that, without completely understanding your problem, he could nonetheless offer some ideas, which would help you understand the problem yourself. This was precisely Teller's forte . . .'

A number of these discussions led to collaborations. A majority of the sixty or so papers Teller published while at George Washington University were written with other people and they were wide-ranging and inventive. Teller's own research projects still belonged to the interface between the new physics and chemistry. To some extent he was still what Placzek had spitefully called a 'molecule inspector', but the work he now did was central to defining the nature of molecular structure, and of reaction mechanisms. It was to forge a crucial link between the high theory of quantum mechanics and its practical implementation in chemistry and metallurgy.

Back in London, for instance, Teller had collaborated with Herman Jahn in trying to explain an anomaly in the interaction between the electrons and nuclei in molecules, which had been spotted by Lev Landau. The results of this collaboration led to their defining the Jahn–Teller effect which, nearly seventy years later, has come to underpin research in fields as diverse as superconductivity and lasers.*

In Washington one of his first collaborators had been a Hungarian student, Stephen Brunauer. Together they published a fundamental paper on adsorption, which became widely applied to certain new kinds of reaction catalyst, of which the automotive catalytic converter is one. In yet another collaboration, he worked with three students to explain the hexagonal ring structure of benzene in terms of the wave-like behaviour of the electrons oscillating around the ring.

This was ground-breaking, certainly, but it was in his collaborations with Gamow that Teller ventured into the more fundamental areas of the new physics. Between them they advanced Gamow's earlier research, in which he had explained the mechanisms of radioactive decay. Following Gamow's explanation of one form of alpha decay, which takes place in seconds, he and Teller were to offer a quantum-mechanical explanation of another form, which takes place over thousands of years and which they called beta decay. 'Gamow was important for Teller because he had ideas which, maybe, Teller wouldn't have dared,' said Hans Bethe, 'but Edward took these ideas and made them honest. The beta-decay work is another credit for Teller. After Gamow had the initial idea, it was he who figured out that there were many ways that two nucleons can interact.'

There was a generosity about Teller at this time that was widely appreciated. He only needed to be excited by an idea or a problem and he would offer his time and effort freely. He was also always generous in the credit he gave to others. For example, Teller always referred to the *Landau*–Jahn–Teller effect, because it had been Lev Landau's crucial question that had set the objective for his and Jahn's further probing. This generosity of spirit reflected the way life in the small but secure pond of American physics in the 1930s was flowing Edward Teller's way. He was to describe it as a 'wonderful quiet period' in his life. His ability to find willing collaborators allowed him to expand the range of his research, and his popularity and growing reputation must have been balm to the inner doubts that had troubled him so much to date. Furthermore, in Washington there was an annual event which was to ensure that his reputation spread well beyond his own campus.

One of the conditions George Gamow had insisted upon when he took up his post at George Washington University was that the University should fund an annual conference on one aspect or another of theoretical physics. It was Gamow's attempt to ease the loneliness and isolation affecting so many physicists, and he asked Edward and Mici to become involved in the planning and management.

The conferences were tiny by modern standards – perhaps twenty or so delegates for each – but it fell to the Tellers to organise not only the speakers but the social events and the subsidiary programme for spouses. Soon after their arrival in Washington, Edward and Mici had rented a small house in Garfield Street, little more than a stone's throw from the White House. During the conference it was to become even busier than usual, the social focus for the event. 'It was a wide open, friendly house – very bourgeois,' Rose Bethe recalled. 'By this I mean that the house was very open and informal but there was no question of Edward or Mici being involved with anyone else or things like that.'

In years to come, Mici was to find her husband's fame, and subsequent notoriety, a burden; but these were years when it was hugely enjoyable for both of them. Indeed, in their warmth and closeness, Edward's relationships with his colleagues at this time were much more than laboratory acquaintanceships. Take the year 1937, the year Teller was to describe as the 'best' of this wonderful quiet period: that spring, both he and Hans Bethe wanted to attend a physics conference on the other side of the continent at Stanford, just south of San Francisco. However, some months before, Bethe had been staying with the Tellers in Washington, and had been approached by Rose Ewald. She was the nineteen-year-old daughter of Peter Paul Ewald, who had taught Teller in Karlsruhe and Bethe in Stuttgart. Ewald himself was a Gentile but his wife was Jewish and, while they had thus far survived in Germany, Rose was hoping that Bethe would be able to help her father in finding a job in America. What Bethe recalled best of the conversation, however, was that 'in that ten minutes [he] fell desperately in love'.

Thus, having first played the role of matchmakers, the Tellers were to find themselves involved in a blossoming romance. 'As a matter of fact, on this trip we were not terribly well matched as two couples,' said Rose Bethe:

Hans and I were extremely involved with one another. We hardly knew each other and it was very much an experimental trip. Edward and Mici were spectators – almost irrelevant. There was us and there were these two people on the outside watching as our relationship developed. They were very kindly duennas because, looking back, we did behave quite selfishly.

This courtship took place as they drove through some of the most beautiful scenery America has to offer. Both couples loved the mountains. In Estes Park, Colorado, Hans offered lessons in mountaineering and, even though he found extended hikes taxing enough, Edward went with him. Bethe was enormously impressed by the way he managed. 'His bad foot caused him big problems,' Bethe said of one occasion, 'but he was a very brave – foolish almost – as the gully we were in was full of snow and he kept missing his footing. But he was very brave about it.'

On through the States they travelled, to the Tetons, Yellowstone Park and Crater Lake before driving on to their destination. 'I was sad to see our small and wonderful company dissolve into the beehive of physicists,' Edward wrote. Indeed, it was a good memory for them all and the two couples were to take many more trips together in the future. 'From my experience of him in '37,' Rose Bethe said, 'I would say he was actually a very lovable man, full of fun, full of ideas. Hans and I felt very warmly towards him and to Mici.'

This trip was to continue to weave together relationships that were to prove crucial in the years to come. For it was during the conference that Teller would meet Robert Oppenheimer for the first time.

There were definite parallels between the two men's lives: both had a protected and somewhat lonely childhood; both were gifted and had successfully negotiated the difficult transition from chemistry to physics; both had gone to Göttingen in fear and trepidation of how they would be received; and both had met with considerable success, working to apply quantum theory to better understanding chemical reactions before moving on to professorships in the US. However, so much of what made Oppenheimer different, not just from Teller but from most scientists, was his considerable wealth, derived from his father's successful cloth-importing business.

When, in 1929, after four years abroad, he had eventually returned to the US to take up his professorship at Berkeley, he had become thoroughly Europeanised. He knew about wine, about French medieval poetry, about food. He had even taken a course in Sanskrit to study Eastern philosophy. And he had the means to develop these tastes into a lifestyle and surrounded himself with friends and students with whom he felt he could share it. 'He wanted everything and everyone to be special,' his younger brother Frank recalled.

That one comment by his brother holds the key to much of the Oppenheimer phenomenon: 'He wanted everything and everyone to be special.' This statement, with its implications of someone living according to some preconceived ideal, judging potential friends, students, girlfriends, colleagues by whether they fitted that ideal, seems to explain so many things in Oppenheimer's life. Those close to him found him kind and warm, but those who did not find themselves part of the circle were often antagonised by him, his clique, and the ethos he projected.

It was not surprising, therefore, that when Edward Teller was invited by him to go up from Stanford to give a seminar at Berkeley, Oppenheimer impressed him greatly. 'I found talking with him very interesting, but dining with him was daunting. Before the seminar, he took me to an excellent Mexican restaurant with food so hot I could swallow only a few bites.' As to Oppenheimer himself, he found him 'overpowering'. There is no record of Oppenheimer's reaction to Teller.

It was inevitable that, on the same visit, Teller should also meet the impresario of 'big' science on the West Coast, Ernest Lawrence, whose atom-smashing cylotron was a wonder of science at the time, for which he would win the Nobel Prize. He took Teller out in his motor boat to look at the Golden Gate Bridge. 'I withstood the choppy waters with a little less than complete equanimity,' Teller wrote. 'Californians seemed more than I could handle comfortably.'

At the end of the conference, Hans and Rose had to return to the East Coast promptly, but they left the Tellers to drive their car back at their leisure. With them was another scientist who had no pressing reason to return east, Enrico Fermi. Fermi was still working in Rome but was on a visit to the US. He was increasingly disillusioned with Mussolini's regime and the recent ruthless invasion of Ethiopia. He also had worries of a more personal nature: his wife Laura was Jewish and Italy was steadily following Germany in its pursuit of anti-Semitism. A year later Fermi was to win the Nobel Prize in physics. He and Laura were to use the opportunity provided by their trip to Stockholm to make their escape.

One afternoon the three decided to go swimming in the Pacific and went over to the coast at Half Moon Bay. It was deceptively calm and Teller basked for a while in the warm water, until he realised he was being swept out to sea. He had been caught by a riptide and, however hard he tried, he could not make progress back to shore. He shouted for help and Fermi rushed into the water to help him. It was a struggle for both men, but eventually they made it to the beach. Teller always believed that Fermi had risked his own life to save his.

Their journey back east was wonderfully eclectic. Fermi wanted to see Hollywood. Using the Hungarian network, Edward contacted the eminent aeronautical engineer Theodore von Karman ('both Mici's and my parents knew von Karman very well'). Von Karman himself knew everybody, and arranged for the scientists to watch shooting in progress. Fermi loved it. Edward was staggered at how often the actors had to repeat everything.

On through the south-west they travelled, using a route prepared for them by Oppenheimer, who had spent much of his youth riding through the region's mountains. They got seriously lost in the Arizona desert, explored the Canyon de Chelly in Navajo country, and lunched at La Fonda, a hotel in the middle of Santa Fe, which was to feature in all their lives some five years later. In his affectionate account of this trip in his *Memoirs*,

Edward created a caricature of himself as the driver on the trip whose 'speed and attention to the road varies with my interest in the conversation inside the car ... as I was still a novice driver, Fermi was anxious to help with the driving. When he was not behind the wheel, however, he constantly admonished me. I can still hear him saying "Zlllooooh!" '

This trip was to remain a special memory for Teller, but during these years there were many occasions like these, forging the bonds of deep friendship. Teller recalled a letter that Merle Tuve wrote when the University of Chicago enquired if Teller should be invited to be a professor there: 'If you want a genius for your staff, don't take Teller, get Gamow. But geniuses are a dime a dozen. Teller is something much better. He helps everybody. He works on everybody's problem. He never gets into controversies or has trouble with anyone. He is by far your best choice.'

This was Teller's memory of the letter. It may have improved in the recollection, could even be a fiction; but it did find echoes in the memories of so many of his students and colleagues – of the person Hans Bethe called the 'gentle' Teller.

In 1936, Mici and Edward had taken their first summer break away from Garfield Street on a trip back to Budapest. At that time, travel through Nazi Germany was still safe enough for this, and their stay had been spent filled with, as he put it, 'the comfortable joys of familiar and happy surroundings'. Although they did not realise it at the time, this was to be their last visit to Hungary for half a century. They were never again to see Mici's parents or her brother Suki, who died in a Nazi concentration camp, or most of their childhood friends, or Edward's father. Max Teller was to die in 1950 and Edward's mother and sister were only allowed to leave Hungary nearly a quarter of a century after their visit, in 1959.

They had intended returning in the summer of 1938, but by then it was clearly too dangerous for Jews to travel across Europe. So work began on that year's Washington conference. The topic chosen was astrophysics and, in particular, the processes that fired the Sun.

Back in the 1920s two scientists, Houtermans and Atkinson, had suggested that the energy from the Sun and the stars was generated by thermonuclear fusion – a process where, in intense heat, the nuclei of two light atoms, such as hydrogen, will fuse to produce a heavier nucleus, in the process releasing immense amounts of energy.

Throughout much of 1937 Gamow and Teller had worked at trying to understand exactly what was going on in the fusing nuclei. They had published a paper together and now wanted to extend these ideas. Teller wanted Bethe's input and telephoned him in an effort to lure him to the conference. Bethe was not a great admirer of what he saw as Gamow's whimsical approach to research, and it took Teller at least three attempts to bring him to Washington.

When he did come, he was captivated by the subject. As a result of the conference Teller's student Charles Critchfield correctly proposed that the reaction between protons was the source of the Sun's energy and it was Bethe, along with Gamow, who joined Critchfield in a joint paper on the subject. Hard on the heels of this, Bethe was to publish a review article based on the topics that were discussed at the conference, which also delineated the role that the element carbon played in the cycle of thermo-nuclear stellar reactions. It was work that was to figure substantially in Bethe's Nobel Prize. 'Edward is responsible for my fame,' said Bethe. 'I found out that the most important reaction to obtain energy in small stars like the sun is that the carbon nucleus adds protons one by one. Once it has gathered the four of them, you get the carbon back and the protons build the helium atom. It is a continuing cycle. And if Edward hadn't been there I wouldn't have gone to the conference, where he almost posed the problem for me.'

Teller had initially proposed, Bethe had disposed. While in no way detracting from Bethe's achievement, Teller's role as an intellectual catalyst on this occasion was both a significant contribution and typical of how he worked at the time. Yet there was none of the wrangling over credit that was to blight so much of Teller's later work.

Much of Teller's time over the next few months was taken up organising the fifth of the Washington Conferences, due to start early the following year, on 26 January 1939. The topic this time was low-temperature physics and superconductivity, and both he and Gamow had been delighted when they heard that Neils Bohr, who was visiting Princeton, would be arriving early enough from Europe to take part. He arrived at Gamow's house late on the afternoon of the twenty-fifth, and that evening Teller received a phone call from an extremely agitated Gamow: 'Bohr has gone crazy. He says uranium splits,' Teller remembered Gamow saying. 'That was all of Geo's message. Within half an hour, I realised what Bohr was talking about.'

Teller's assumption was to give him an uneasy night's sleep. The following day Gamow broke with the planned programme and opened the conference by announcing to a hall of some fifty delegates that Bohr had something to say.

The experiment Bohr described had taken place only a month earlier, in Berlin, but was a direct outcome of Fermi's experiments with neutrons some four years previously. These were experiments in which he had bombarded uranium with neutrons and discovered that it produced dozens of radioactive substances. His explanation – that he had produced new 'transuranic' elements, heavier than uranium – had remained unchallenged until, in late 1938, Otto Hahn and Fritz Strassman had decided to investigate the properties of two of the products of this uranium bombardment. They established that they were not dealing with a new transuranic element at all but with radioactive forms of two much simpler elements, barium and iodine.

That was as much as Bohr had known when he sailed for New York. However, in the meantime, Otto Hahn had contacted his erstwhile collaborator, Lise Meitner, an Austrian Jew who had recently had to flee Germany, and had described the new finding to her. By chance her nephew, Otto Frisch, another émigré physicist, was visiting from Copenhagen for Christmas and, over the holiday period, the two had designed an experiment to confirm Hahn and Strassmann's result. If uranium really was split, the two fragments would move apart at high speed, losing neurons and producing a great deal of energy. That energy could be picked up by a Geiger counter. Meitner and Frisch had discussed their experiment with Bohr before he had departed and then cabled their successful positive result to him while he was still on board ship.

Even though what Bohr was describing to the conference was amazing, Teller remembered the following discussion as being remarkably subdued. Some remembered Teller, others Fermi, as the person who picked up on Bohr's rambling account and focused what was on everyone's mind: if the splitting of the nuclei released more neutrons, would it release enough to set up a chain reaction? If so, a small amount of the fissionable material could be made into a bomb of extraordinary power. The prospect was awe-inspiring. Whether the notion was in the air or actively voiced aloud, Teller remembered one of his neighbours questioning whether any further discussion should take place.

When Bohr had finished, two young scientists from Tuve's terrestrial magnetism department slipped out of the meeting and headed back to their laboratory at the Carnegie Institute to repeat the Frisch and Meitner experiment. When they observed the tremendous pulses corresponding to the additional neutron release, they called Tuve, who then invited Bohr, Fermi, Teller and several others to come and witness the uranium fission. A photographer was on hand to take the historic photograph. Only Teller failed to join the group. He had found the experiment, essentially so simple, to be an anticlimax. Rather he marvelled at the fact that the result had eluded people for so long. In one of his later experiments, for instance, Fermi had actually bombarded uranium to check for telltale alpha particles released by an unstable nucleus. However, his careful screening of the experiment had prevented him from observing these products of fission,

which had a short range but extremely high kinetic energy. Fermi only
needed to have forgotten his screening procedure once and fission would
have been discovered years earlier.

There were others too – Paul Savitch in Paris and Paul Scherrer in
Zurich – who failed to interpret the results of similar experiments as
evidence for fission. For Teller, these omissions were highly significant. 'If
fission had been discovered in 1933, work on the topic in Germany and
the Soviet Union – two nations that took the military applications of science
seriously – would have been well advanced by 1939. Under different conditions the United States probably would not have been the first nation to
possess nuclear explosives. Fermi, Scherrer and Szilard, in their different
ways, had a profound and beneficent influence on history' (see Appendix
2, pp. 406–7).

On the West Coast the scientists heard about the discovery of fission
second-hand, from newspaper reports. Robert Oppenheimer had been
initially incredulous but was quickly convinced by his colleague Luis
Alvarez, who also repeated the Frisch–Meitner experiment. 'When I invited
him over to look at the oscilloscope later, when we saw the big pulses . . .
he had decided that some of neutrons would probably boil off in the
reaction, and that you could make bombs and generate power, all inside of
a few minutes . . . It was amazing to see how rapidly his mind worked, and
he came to the right conclusions.'

Shortly after this, one of his students, the American theoretical physicist
Philip Morrison, recalled: 'Within perhaps a week there was on the blackboard in Robert Oppenheimer's office a drawing – a very bad, an execrable
drawing – of a bomb.'

One evening shortly after the Washington conference had ended, Edward
and Mici were recovering from their efforts when they received a telephone
call from Leo Szilard. He had just arrived at Washington's Union Station
and was looking for accommodation. Mici objected strongly to the surprise
visit: 'No! We are both much too tired.' Edward remembered her saying,
'He must go to a hotel.' Part of the problem was that Szilard was known to
be a difficult house guest; nevertheless, when they picked him up from the
station, Mici immediately invited him to stay.

> We drove to our home and I showed Szilard to his room. He felt the bed
> suspiciously, then turned to me suddenly and said: 'Is there a hotel
> nearby?' There was, and he continued: 'Good. I have just remembered
> sleeping in this bed before. It is much too hard.' But before he left he sat
> on the edge of the hard bed and talked excitedly: 'You heard about Bohr
> on fission?'

'Yes.' I replied.
Szilard continued: 'You know what that means!'

What it meant to Szilard, Teller recalled, was that 'Hitler's success could depend on it.' Szilard had also heard about the fission discovery second-hand, but he had long been impressed by the potential dangers – and benefits – of atomic energy. He it was who, in 1934, had angered Ernest Rutherford when he suggested the possibility of a nuclear chain reaction – 'I was thrown out of Rutherford's office' was how Szilard related it.* Now he wanted to pick up where the conference had left off. 'Nuclear energy and nuclear explosives are feasible,' he told Teller during one of their lengthy conversations the day after the night of the hard bed, 'provided only that, on average, more than one neutron is emitted in the fission process.'

However, Szilard was desperately short of funds, without either equipment or even his own laboratory. On the day of his surprise visit he had been hustling an erstwhile patron, the financier Lewis Strauss, for support. So insistent had he been that he had even pursued Strauss on to the train as he and his wife had headed south from New York for a vacation in Florida. It was when he eventually left Strauss on the train at Washington that he had rung the Tellers. During their conversations the following day, they also broached the question of secrecy. Szilard had his own solution: a voluntary association of scientists that would monitor research and decide on funding and priorities. However, it was such a minefield that they decided it could only be negotiated with other colleagues when the outcome of the chain-reaction investigation was known.

Over the next month Szilard managed to collect enough equipment and find a laboratory and a young collaborator, the Canadian Walter Zinn, to work with him. In February they began their experiments. During this time Teller stayed closely in touch, encouraging and feeding him information about further experiments in Tuve's laboratory at Carnegie. It must have been comforting for the lonely Szilard to hear of enthusiasms elsewhere, as many scientists still believed a fission chain reaction to be unlikely. Ernest Lawrence at Berkeley, even the two Germans Otto Hahn and Lise Meitner, all saw it as moonshine, a long shot; but Szilard persisted and one evening, about a month later, Edward was accompanying a friend in a Mozart violin sonata when the phone rang: 'New York. Szilard,' Teller recalled: ' "I found the neutrons." He did not need to tell me anything else. That meant the nuclear energy could be released; that meant that the possibility of nuclear explosions was no longer a possibility but a probability. And you know the next thing that happened – nothing.'

What Teller meant was 'nothing of significance', but there was plenty of activity. Szilard had told not only Teller about his finding but Eugene Wigner as well, and Wigner responded vigorously and decisively. He was no doubt influenced by the news coming out of Europe, of the fall on 14 March of the country Britain and France had chosen to ignore when they had pacified Hitler at Munich: Czechoslovakia. A day later the Nazi treachery was compounded when Horthy's fascist regime in Hungary moved to invade the remaining strip of Czechoslovakia along the Hungarian border.

On the same morning that Hitler arrived in Prague, Wigner travelled down from Princeton to New York and met with Szilard and Fermi in the office of George Pegram, the dean of the graduate school at Columbia. It was the beginning of a day of heightened activity. The issue of secrecy and the direction of atomic energy research could no longer be avoided, but Wigner made clear his view that Szilard's notion of a collaborative association among scientists themselves was impractical, and that the time for such amateurishness was over. According to Szilard, Wigner 'strongly appealed to us immediately to inform the United States government of these discoveries'.

Pegram had contacts in Washington. He knew the Under-secretary to the Navy, Charles Edison, but before calling him, Pegram wanted to sort out who would carry the news. Fermi was going to Washington the following day to give a lecture and, although Italian, he was also a recent Nobel Laureate and so his views were thought likely to carry considerable weight. A meeting was fixed not with Edison but with Admiral Stanford C. Hooper, who was technical assistant to the Chief of Naval Operations. Pegram also wrote a letter for Fermi to take with him, but while the letter did describe the possible use of uranium as 'an explosive that would liberate a million times as much energy per pound as any known explosive', he concluded with the personal caveat: 'My own feeling is that the possibility is against this.' Thus armed with this ambivalent introduction, Fermi left for Washington.

With Wigner's concern about informing the government now in hand, attention turned to the pressing matter of secrecy. Both Szilard and Fermi – who had recently confirmed Szilard's findings – had written reports on their secondary neutron experiments which were ready to send to the *Physical Review*. It was agreed that the report should be sent, to establish priorities, but that the editors should be asked to delay publication until the secrecy issue had been resolved.

Wigner and Szilard then shuttled back from New York to Princeton for a pre-arranged meeting that evening with Neils Bohr. They were to be joined by Teller, who was travelling up especially from Washington, and by two colleagues of Bohr's, John Wheeler,* and the Belgian physicist Leon

Rosenfeld. Every one of these scientists had personal experience of the quality of the physicists still working in Germany and had little doubt that they also would have picked up on the implications of the Hahn and Strassmann discovery.

'We argued that, from now on,' Teller wrote, 'we should not publish the results of fission research lest the Nazis learn from them and apply them to making a nuclear explosive. Bohr, however, took the opposite view with deep conviction. "Openness," he said, "is the basic condition necessary for science. It should not be tampered with." He thought we were unduly alarmed: separating the two forms of uranium and accumulating a sufficient quality of U-235 would require effort so huge as to be impractical. "You would need to turn the entire country into a factory," he declared.'

This revelation had resulted from a collaboration between Wheeler and Bohr on a paper that would define the fission process in precise detail. Uranium metal consists mainly of the isotope U-238 but 0.7 per cent of it consists of the isotope U-235. The U-235 was the only isotope suitable for bomb-making, but because the two isotopes were chemically and physically so similar, it was, as Bohr and Wheeler predicted, going to be extremely difficult to separate them.

Nonetheless, after more than half a century of Cold War security, it is difficult to see Bohr's conviction of the need for openness as anything but unrealistic; but he had worked for decades to shape science into an international community and to establish the rules that allowed for complete honesty in the reporting of its results – results free of either political or financial interests. It was a fragile credo, openness, but Bohr saw secrecy, however justified, as subordinating science to the political system. He, like Rutherford before him, and Oppenheimer and others who followed, also abhorred the Faustian contract into which their research was irrevocably drawing them. Yet, at the same time he was deeply troubled about the Nazi threat – 'the doom of Europe', as he called it.

However, he continued to bolster his position with yet another practical argument. It was known already that other groups were working on the secondary neutron question. Frédéric Joliot, Marie Curie's son-in-law, working in France, had also discovered the necessary extra neutron. 'Joliot will not like the idea of secrecy at all,' Bohr predicted. 'He was already disappointed over credit for the discovery of the neutron.' And also over the discovery of fission itself, which he had been close to achieving as well. He was very unlikely to take kindly to anything that blocked his future claims on the priority of discovery. The discussion dragged on past midnight, but Teller recalled that eventually a decision was made. Bohr was prepared to accept secrecy, provided that Fermi, who was not at this meeting, would also agree. Teller, as everyone's friend, and someone who was considered closer to Fermi than anyone else present, was delegated the task of persuading him:

> I drove to New York late that night. The next morning, Fermi listened patiently to my arguments. I did not expect that he would want to change the practice of complete openness. Fermi's natural inclination was to abide by conventions and not challenge generally held opinions, but I also knew that he had a truly open mind.
>
> After I had introduced my case, I smiled at him and said, 'The next time you find me in a riptide, you may give me a shove toward Japan.' But after a little more gentle arguing, Fermi agreed to secrecy, again with one condition: he would not publish if everyone else agreed not to.

This was a victory indeed for Teller, which he was able to report back to the others in Princeton. At only thirty-one years old, he was already playing a significant role in the politics of the burgeoning arms race and of national security – and he was enjoying it, enjoying using his powers of persuasion, enjoying the sense of importance.

5

The Hungarian Conspiracy

On 19 March 1939 Frédéric Joliot and his collaborators published the first of two papers in the British scientific journal *Nature*, describing their realisation of the fission chain reaction. It was just as Bohr had predicted: the French had been contacted by their North American colleagues but went ahead with the publication anyway.

Their second paper, on 22 April, actually specified the number of secondary neutrons released. It was 3.5 neutrons per fission, a clear indication that a bomb was possible. Their paper was read and reported on worldwide. In the US, the voluntary secrecy initiative, to which Teller had briefly enrolled Fermi, was stifled at birth and Fermi and Szilard both finalised their papers for publication in the *Physical Review*.

In Germany, the French publication precipitated a high-level meeting within a week of publication. The meeting took place secretly on 29 April in the Ministry of Education's headquarters in Berlin, and it resulted in a decision to commandeer all existing stocks of uranium and obtain fresh supplies from the newly captured mines in Czechoslovakia.

It was indicative of the times that one of the scientists at the meeting tried to censure his fellow Germans Hahn and Strassmann for freely publishing their research on fission and thereby making it available to the whole world. He gained practically no support for his motion. In spite of the international situation and the obvious strategic importance of the work, German scientists were just as unwilling as their colleagues in America to relinquish their tradition of open exchange of information.

No response as positive as the Germans' met Fermi when he visited the US Navy in Washington on 17 March. 'There's a WOP outside,' he heard the adjutant report to the admiral on his arrival. For an hour he lectured a collection of naval engineers on atomic physics, but according to Admiral

Hooper's notes on the meeting he left an impression that a practical weapon was still years away. One or two of those present could see the potential for a power source for submarines that did not consume oxygen, but the meeting ended with the Navy saying it would 'keep in contact', and that was all.

Any effort to combat the German threat clearly remained in the scientists' hands, but there was already dissension among some of the leading figures. Fermi had been disillusioned by his first brush with the military and had returned to Columbia intent on immersing himself in an astrophysics project that intrigued him. He had then become mired in disputes with the combative and difficult Szilard.

Their disagreement at this point centred around how to create the fission chain reaction in uranium that was now thought to be possible. In order to accomplish this in a controlled fashion, in such a way that it could be studied in detail, two main elements were necessary. The first was the source of the neutrons, the uranium itself. The second was a moderator. To control the chain reaction, to prevent it from running away – from going 'critical' – the uranium had to be surrounded by a material that absorbed neutrons. Then, by withdrawing or immersing the uranium in the moderator, the reaction could be accelerated, or dampened, or stopped altogether. It was known that 'heavy' water – water in which the hydrogen is replaced by its heavier isotope deuterium – was a moderator and this was the one favoured by Fermi. Szilard, on the other hand, thought it would prove too effective and had begun looking at pure carbon, in the form of graphite, as an alternative. Szilard had then made matters worse by refusing to do any of the experimental work himself, refusing to dirty his hands 'like a painter's assistant'. To Fermi, who always worked on his own experiments, this refusal was a mortal sin. Before things became any worse, however, Teller was brought down to Columbia, ostensibly to teach at summer school.

'I lectured graduate students,' Teller remembered, 'but I was invited primarily as a consultant – peacemaker on the Fermi–Szilard chain-reaction project. Fermi and Szilard both had asked me to work with them. They were barely speaking to each other. Temperamentally, the two men were almost opposites ... Fermi seldom said anything he couldn't demonstrate. Szilard seldom said anything that was not startling and new. Fermi was humble and self-effacing. Szilard could not talk without giving orders ... So Mici and I spent the summer in New York City as the missing link in the Fermi–Szilard communication chain.'

By this time, however, Szilard was very much on his own. Fermi had gone to the University of Michigan for the summer. Szilard's three-month laboratory privileges at Columbia had come to an end and, with Fermi away, Pegram, the dean of the department, was not prepared to sanction any further work on a project that threatened to become very expensive. He tried raising the Navy's interest again but without success. He had persuaded Wigner to contact the Army, but this approach was also rebuffed. No one, it seemed, shared his sense of urgency and even he, Szilard, with his massive ego, was becoming increasingly discouraged. However, as Teller was in Manhattan for the summer and Wigner was little more than an hour away in Princeton, the three began conspiring.

Firstly, Wigner and Teller restored some of Szilard's wounded self-esteem by approving his calculations for the full-scale reactor, preferring his proposal to use graphite to Fermi's to use heavy water. Both Teller and Wigner, Szilard wrote in 1941, 'shared the opinion that no time must be lost in following up this line of development and in the discussion that followed, the opinion crystallised that an attempt ought to be made to enlist the support of the Government rather than that of private industry. Dr Wigner, in particular, urged very strongly that the Government of the United States be advised.'

Perhaps because they had already shared so many disappointments in trying to interest the government, however, they firstly focused on another anxiety they shared. They began to worry about what would happen if the Germans monopolised the large quantities of uranium that the Belgians were mining in the Congo, '. . . So we began to think, through what channels we could approach the Belgian government and warn them against selling any uranium to Germany', Szilard recalled. It was at this point that Szilard thought of Einstein. The two had been friends and collaborators. They had even taken out a patent for a refrigerator together – work funded by Szilard's patron of the time, Lewis Strauss. Szilard knew that Einstein was a personal friend of the Queen of the Belgians. For more than a decade he had corresponded with her regularly, addressing her simply as 'Queen'.

So they tracked Einstein down to a summer retreat on Long Island – beside Old Grove Pond on Nassau Point – arranging a meeting for Sunday, 16 July. In the meantime, Szilard had tried another route to someone of influence in government, and had approached another émigré, the influential economist Gustav Stolper. Stolper was a German and a Jew who, before he was forced to leave his homeland, had been a member of the Reichstag. He promised to try to identify someone suitable.

As Teller did not know Einstein personally, it was Wigner who drove Szilard to Long Island on 16 July. The meeting, in the relaxed setting of the holiday home, began with Szilard telling Einstein about the secondary neutron experiments and the possibility of a chain reaction. Szilard was astonished to find that this was the first the great man had heard of it. 'I never thought of that!' he exclaimed, but according to Szilard he was 'very quick to see the implications and perfectly willing to do anything that needed to be done. He was willing to assume responsibility for sounding the alarm, even though it was quite possible that the alarm might prove to be a false alarm.' Einstein was reluctant to write directly to the Queen herself, but eventually they reached a compromise. The letter would be sent to the Belgian ambassador while ensuring that the US State Department knew what was happening.

On his return to New York, Szilard found a reply from Stolper. He had discussed the issue with another economist, Alexander Sachs, who was a valued presidential adviser and who wanted to meet with Szilard. Sachs was also a scientist – a biologist – and he listened intently to Szilard. Then, as Szilard wrote to Einstein, he 'took the position, and completely convinced me, that these were matters which first and foremost concerned the White House and that the best thing to do . . . was to inform Roosevelt. He said that if we gave him a statement he would make sure it reached Roosevelt in person.'

Szilard was stunned, delighted after his months of frustration. Unable to reach Wigner, who had gone on holiday, Szilard called up Teller. Teller agreed that the new plan was infinitely more promising than their first, and so Szilard prepared a draft – in German, because Einstein's English was still insecure – and sent it to Einstein for approval, asking him in a covering letter if he wanted another meeting. If he visited again, Szilard wrote, he would ask Teller to drive him, 'not only because I believe his advice is valuable but also because I think you might enjoy getting to know him. He is particularly nice.'

On Sunday, 30 July, Teller drove Szilard out to Nassau Point in his 1935 Plymouth. 'I entered history as Szilard's chauffeur,' Teller once told a television audience with his particular brand of overt modesty, though it is clear from Szilard's letter to Einstein that he had made valued contributions throughout. When they arrived they found Einstein wearing old clothes and slippers. Tea was served, and he and Szilard produced a third, longer draft with Teller in attendance and acting as scribe. 'Einstein . . . made only one comment,' Teller recalled: ' "This would be the first time that nuclear energy would be used directly instead of indirectly through the processes in the Sun." That comment struck me . . . as a peculiar comment for the letter's author to make on rereading it . . . I suspect only Szilard would have felt so free as to instruct the President of the United States in detail about what to do. Einstein signed the letter and we left.'

Over the next fortnight, letters went back and forth between Szilard and Einstein, trading changes to the text, until the letter finally reached its final form and made its way to Alexander Sachs on 15 August.

However, it took the economist two months to find the right opportunity to present the letter to Roosevelt. During that time Germany, along with Russia, had invaded Poland, and Britain and France had entered the war two days later. There was no doubt now about the seriousness of the international situation and an impatient Szilard had begun wondering seriously whether Sachs had been the correct choice as intermediary. However, late in the afternoon of Wednesday, 11 October, Sachs presented himself at the White House. He did not, as he had originally intended, read Einstein's letter aloud to the President but instead used his own 800-word summary.

'Alex,' Roosevelt summarised, 'what you are after is to see that the Nazis don't blow us up.'

'Precisely,' Sachs said.

Roosevelt called in his aide, General Edward 'Pa' Watson. 'Pa, this requires action,' he told him.

Over the summer of 1939, Teller had been an integral part of what Merle Tuve called the 'Hungarian Conspiracy', and had played a significant role in producing a letter described by one observer of American life as equivalent in importance to the Declaration of Independence.* In pursuing their advocacy, however, the three were pursuing a vision that went well beyond a deterrent to the Germans. It was a vision most clearly articulated by Wigner:

> We realised that, should atomic weapons be developed, no two nations would be able to live in peace with each other unless their military forces were controlled by a common higher authority. We expected that these controls, if they were effective enough to abolish atomic warfare, would be effective enough to abolish also all other forms of war.
>
> This hope was almost as strong a spur to our endeavours as was our fear of becoming the victims of the enemy's atomic bombings.

The three Hungarians were not alone in articulating such hopes of creating a world at peace. In Germany, fission research was already well under way, organised and funded through a new department in the War Office. Carl von Weizsäcker, Teller's friend from Leipzig and Copenhagen, had been recruited and he also recalled realising that, should an atomic weapon be possible, 'then the participating nations and ultimately mankind itself can only survive if war as an institution is abolished'.

However, this belief that their efforts could also lead ultimately to the abolition of war can be seen as the moral equivalent of an optical paradox. At first it can be read as a proud vision; blink, and it appears simply as justification for still more effort. This moral paradox was to underscore and confuse judgements over the arms race throughout the next half-century – especially those of Edward Teller.

Roosevelt's instruction to 'Pa' Watson produced an immediate response. Meeting with Sachs afterwards, Watson proposed that the committee should be set up under the auspices of the Bureau of Standards, the organisation responsible for applying technology in the national interest. Its director was the physicist Lyman Briggs, and he rapidly assembled an 'Advisory Committee on Uranium', whose membership consisted of two military ordnance experts, Szilard, Teller, Wigner, Sachs, a temporary stand-in for Merle Tuve, and Briggs and his assistant. The first meeting was called for 21 October to discuss support for the experimental 'pile' (a word first used by Szilard), but Fermi, whose project it was, refused to come. 'His experience with asking for support from the Navy had convinced him that he wanted nothing more to do with governmental meetings,' Teller explained. ' "But," he told me, "I will tell you what I'd say if I were to go. You can deliver the message." Thus, I was promoted from chauffeur to messenger boy.'

It was Szilard who, predictably, led off at this first meeting by describing the proposed uranium-graphite reactor, and what they needed to find out to establish whether fission would be of practical use as an energy source or as an explosive. In doing so he introduced the army officer Lieutenant Colonel Keith Adamson and his naval counterpart Commander Gilbert Hoover to the idea of cross sections. (For an explanation of cross sections see notes, p. 414).

He explained that they needed an accurate estimate of the capture cross section of carbon: it needed to be large for the carbon in the graphite to capture enough neutrons to moderate the uranium reaction successfully. But Szilard was still thinking in terms of using natural uranium, and not of separating out the 0.7 per cent of the active component U-235. As a consequence, he was soon talking about bombs that were enormous, almost certainly too large for use by aircraft.

It is little wonder that Teller remembered Adamson interrupting contemptuously at this point: 'At Aberdeen [the Army's ordnance testing ground near Baltimore] we're offering a $10,000 reward to anyone who can use a death ray to kill the goat we have tethered to a post. That goat is still perfectly healthy.' The situation was then made considerably worse when Richard Roberts, standing in for Merle Tuve, made clear his laboratory's belief that Szilard's natural-uranium bomb was not likely to work. He had also heavily emphasised the horrendous difficulties in separating U-235, saying that it might take several years' work to overcome them. But the situation was rescued by Sachs's staunch insistence that the project simply could not wait. Szilard recalled him arguing that 'the important thing was to be helpful because if there was something to it there was a danger of our being blown up ... we had to be ahead'.

When it came to Teller's turn, he announced that, for himself, he strongly supported what Szilard had been saying, that the reactor project was essential for investigating the potential of fission. 'I said that this needed a little support. In particular we need to acquire a good substance to slow down the neutron, therefore we needed pure graphite, and this is expensive.' He also added that new work using centrifuges to separate the U-235 isotope was most hopeful but would also need help.

'How much do you need?' Commander Hoover wanted to know.

Szilard had actually not expected to receive, or even get the chance to ask for, any money, and so was taken aback when Teller instantly came up with a specific proposition. ' "The researchers at Columbia would volunteer their labour," I noted (on Fermi's instruction), "but slowing neutrons without absorbing them requires a great amount of exceptionally pure graphite, and that would be expensive. In fact," I added, "a sufficient amount of graphite would cost about $6000."

'After the meeting, Szilard nearly murdered me for the modesty of my request.'

Szilard was right. A few days later, he was able to show that the graphite alone would cost $33,000; but at the time even Teller's modest request produced a tirade from Colonel Adamson – on how it took two wars before a weapon could be judged as useful or not, and on how, in the end, it was not weapons that won wars but the morale of the troops.

'Wigner, the most polite of us, interrupted him,' Szilard remembered. '[Wigner said] he always thought that weapons were very important and ... this is why the Army needs such a large appropriation. But he was very interested to hear that he was wrong: it's not weapons but the morale which wins the wars. And if that is correct, perhaps one should take a second look at the budget of the army...'

'All right, all right,' Adamson snapped, 'you'll get your money.'

But not immediately. It took until the following February (of 1940) before it was all agreed, and then Lyman Briggs decided to sit back and wait for results from Fermi. The months went by and nothing happened. Only Szilard and Wigner remained occupied with atomic energy and Wigner described the situation as 'swimming in syrup'. Teller had returned to Washington to his regular teaching and research. In May 1940, however, as the 'phoney war' in Europe turned into the Blitzkrieg, he received an invitation to attend the Pan American Science Congress, at which President Roosevelt was to speak. Teller decided he wanted to hear what Roosevelt would say.

Roosevelt talked about the dangers not just to Western Europe but the dangers to the whole world. He made a point of how much smaller the world had become, that from century to century, from decade to decade, politics here and there were coupled more closely. And then he concluded in a remarkable fashion. He said, 'I know that you will be told that without the new weapons of science, the dangers would hardly exist. I'm telling you that this is in a great measure true, but at the same time, it is true that if you scientists don't work on weapons, the National Socialists will conquer the world.'

I listened. I had a strange feeling that Roosevelt was talking to me. We never met. I may have been amongst a couple of thousand people in the audience, but I may have been the only one who knew of the letter that Roosevelt got. Who else knew that amongst the new weapons Roosevelt had the atomic bomb in mind?

Teller always described this as *the* moment when he decided to do all he could to fight for his adopted country. 'I was one of the fortunate helped to escape from the Nazi threat ... I had the obligation to do whatever I could to protect freedom.'

For the next few weeks, as the Nazis overran Europe, Teller was on a lecture tour across the country. When he returned to Washington, he found major changes afoot in the embryonic bomb programme. Briggs's committee had been subsumed under a larger organisation, the National Defence Research Committee (NDRC), under the chairmanship of the Carnegie President, Vannevar Bush. Working with him was the President of Harvard, James Conant. Briggs, who was seen as having been overly cautious in his direction of the Uranium Committee, was to report to him. Bush's committee now had funding of its own, separate from the military, but with that financial independence came restrictions. All work on the bomb was to be secret.

The initial effect of this was a bureaucratic absurdity. The membership of the Uranium Committee changed. Only US citizens were allowed. This automatically excluded Fermi, Szilard, Wigner and Teller. An angry Alexander Sachs, himself an émigré but now a US citizen, pointed out that the entire work of the committee was founded on the efforts of the very scientists now barred from attending. The overpowering logic of this statement won the day, but the four émigrés were still not admitted as full members. They were named as advisers.

In spite of this reorganisation, however, the fission work still moved forward unsteadily. Bush had remained highly sceptical of the new research, confused by conflicting signals from different areas of research, which he did not understand. On one side, for instance, Teller himself had been doing calculations on cross sections, using results from the Carnegie experimentalists. The minimum mass of nuclear explosive needed to ensure sufficient neutron collisions to sustain a chain reaction is known as the critical mass, and Teller had worked out that it would have to be colossal – in excess of 30 tons. Such a bomb would not only be undeliverable, but would be unlikely even to explode. The Carnegie scientists were reported as describing the quest for a practical weapon as 'a wild goose chase'.

On the other hand there was some positive news. Two young scientists at Berkeley had realised a prediction by Niels Bohr that uranium would not only fission but could absorb a neuron and transmute into heavier so-called transuranic elements. Ed McMillan and Glenn Seaborg had discovered two such elements: the first, neptunium, rapidly transmuted into the second, plutonium; and they found that plutonium fissioned as easily as U-235. It had real potential as an alternative fuel for the bomb, and one without nightmare problems of separation.

There was even some progress with this Uranium isotope separation. Harold Urey, the Nobel Prize-winning chemist at Columbia, with whom Teller was consulting, was having some success at an experimental level using gaseous diffusion. It was all too clear, however, that this process, when scaled up to an industrial level, would be extremely costly. Nothing, it appeared, was straightforward.

On 6 March 1941, Edward and Mici Teller had become US citizens, sponsored by Merle Tuve and a one-time student of Edward's, Ferdinand Brickwedde. This confirmation of their citizenship sent Mici off house-hunting and within a month she had found a new house in a development called Country Club Hills in the Washington suburbs. They were obviously intending to put down roots for the first time, but they had not even moved in when, at very short notice, Edward was invited to seek a year's leave and go to teach at Columbia. The relationship between Szilard and Fermi had not improved, and he was needed to oil the wheels on the reactor project.

'After I received the invitation, I went to see Harold Urey . . . I wanted a clearer picture of what was going on and whether there really was a way for me to make a contribution. Harold was simple and clear. "By all means, come," he told me.

'Without further concerns, I made up my mind to join the work on nuclear energy and nuclear weapons. Did I know what I was getting into? Of course not. Do I regret my decision? No.'

In late August of 1941 the Australian physicist Mark Oliphant, who was head of physics at Birmingham University, arrived in the US from Britain. He had come, primarily, to talk about radar, but he had also been given a specific task in connection with atomic weapons. While confusion and uncertainty held sway in the US, in Britain they had established a committee of eminent scientists, the oddly named MAUD Committee,* to explore, in detail, the feasibility of nuclear weapons. They had made considerable progress, both on the separation of U-235 and on the structure of a possible weapon, and they had been sending their reports to the US, to Lyman Briggs. They had, however, heard nothing back.

Asked to find out why, Oliphant had contacted Briggs. What he discovered had 'amazed and distressed' him: 'I called on Briggs in Washington, only to find that this inarticulate and unimpressive man had put the reports in his safe and had not shown them to members of his committee.' So paranoid was Briggs about security that he had not even shown the reports to his two bosses, Bush and Conant. Oliphant was determined to awaken proper enthusiasm. He contacted the most influential scientist he knew – Ernest Lawrence at Berkeley – and impressed on him the urgency of the situation and the reality of the German threat. Lawrence had already suspected that there was a problem with the Uranium Committee, and took very little persuasion to decide that he must outflank Briggs.

At a degree ceremony in Chicago in September 1941 Lawrence confronted a reluctant James Conant in the living room of Chicago's Dean of Physics, Arthur Compton. Here, according to Compton, Lawrence first reported his conversations with Oliphant on how feasible the bomb was and how, if the Germans made it, they would have in their hands the control of the world.

'Conant began to be convinced,' Compton wrote:

> He turned to Lawrence, 'Ernest, you say you are convinced of the importance of these fission bombs. Are you ready to devote the next several years of your life to getting them made?'
> . . . The question brought Lawrence up with a start. I can still recall the expression in his eyes as he sat there with his mouth half open. Here was a serious personal decision . . . He hesitated only a moment 'If you tell me this is my job, I'll do it.'

At the beginning of October 1941 the British transmitted the MAUD Committee's final report. It was, in effect, a blueprint for a practical weapon, specifying a critical mass, yield, costs and an overall schedule. It was a report that Vannevar Bush now felt able, with confidence, to take to Roosevelt.

On 9 October he led the President through its findings. It had an impressive completeness – so complete, according to Bush, that Roosevelt seemed less concerned about the immediate German threat than he did about this military development that he could see changing the political organisation of the whole world; and, with the exception of Bush and Conant, he put that new future in the hands of politicians. The scientists who had initiated and fought to sustain the project throughout its troubled infancy were excluded from any policy role.*

That summer, Edward and Mici had again travelled west on vacation with Hans and Rose Bethe. It was the last relaxing time that the scientists were to have for many years. On their return, the Tellers moved out of their new Washington home and into yet another rented property, an apartment on New York's Morningside Drive, not far from the Columbia campus.

Edward began working on the reactor project, alongside his friends Wigner, Szilard and Fermi, and he was also consulting on Harold Urey's project to separate the two isotopes of uranium. Each day this group went to lunch at the Faculty Club, and now, with the Germans launching Operation Barbarossa and marching into the heart of the Soviet Union, they were talking politics. It was while Teller was walking back from one such lunchtime discussion with Fermi that the Italian raised a question, 'out of the blue':

> Fermi asked me whether I thought that an atomic explosion might be used to produce a thermonuclear reaction. Gamow and I had often discussed the thermonuclear process occurring in stars; however, such reactions are made possible not by the momentary heat caused by a neutron chain reaction but by consistently great heat and density near the centre of a star. Fermi suggested that if deuterium (which reacts more readily than hydrogen) were used as the fuel, there might be enough heat near a fission explosion – for a very short time – to produce fusion.*

Teller was always intrigued by new territory. Fermi's thought must have been much like the notions Gamow had challenged him with over the past six years, and for a week he worried at it. 'Then, during one beautiful autumn Sunday afternoon walk with Fermi, I explained why the reaction wouldn't work: at the very high temperatures we expected, most of the energy would not be present as light but as X-rays. X-rays would be radiated away without bringing the nuclei and particles closer together; they were useless for carrying on the reaction. Fermi accepted my explanation.'

When Ernest Lawrence began to devote himself wholeheartedly to war work, he had returned to Berkeley to commit his massive cyclotron to joining the efforts in separating out U-235, and he had asked Robert Oppenheimer for his help in answering some of the theoretical problems.

In October he had then taken Oppenheimer to the meeting where Arthur Compton's National Academy of Science group were to discuss the content of their third report for Bush. In preparation for that meeting, Oppenheimer had briefed himself sufficiently on the whole fission project to be able to calculate an actual critical mass for a uranium bomb. At 100 kilograms it was close to Fermi's most recent estimate of 130 kilograms.

At the meeting he would also have heard Lawrence reading a summary of the MAUD report and then been party to all the discussions about how long the building of a weapon was likely to take. Yet, only weeks before, Oppenheimer and his new wife Kitty had entertained Steve Nelson, the organiser of the Alameda County Communist Party and someone the FBI believed was involved in espionage, at their Berkeley home. According to Kitty, who had been a communist herself, the conversation had been 'about the old days, family matters'. Later, she met Nelson alone on several occasions.

From such occasions as the National Academy meeting, Oppenheimer realised that America must surely soon join in a life-or-death race with Germany to produce the most deadly weapon ever conceived. He also realised that he could play a key role in that race and that realisation, more than anything, was to change Oppenheimer's attitude to his past political contacts and activities. Up to that time he had regularly attended meetings and made payments to Communist Party officials – though without becoming a member of the Party. Now this new project made their causes and the work they did seem much less relevant.

On the night of 6 December 1941, Oppenheimer returned from a Spanish War veterans rally feeling acutely disillusioned. 'I decided that I had had enough of the Spanish cause,' Oppenheimer said, 'and that there were other more pressing crises in the world.'

The following day, Sunday, 7 December, Edward and Mici Teller were driving out to spend the day with the Fermis at their home in Leonia, a leafy suburb of New York. They stopped to buy gasoline, but had to wait for the attendant, who seemed to be preoccupied by something he was listening to on the radio.

'When he finally came out to the car, he told us that Pearl Harbor had been bombed,' Teller recalled. ' "Where is Pearl Harbor?" I asked. So it was that we learned that the United States was at war.

'On Monday, the laboratory was a changed place. Overnight, every trace of opposition to the war had disappeared. Everyone's commitment was now wholehearted, open, and complete. The day before, I had inhabited a nightmare world of colossal dangers and uncertainties; now the world was full of problems that might have solutions.'

6

Skirmishes

Edward Teller found himself very much at a loose end during the spring months of 1942. With the United States now at war, work on the fission bomb leapt forward on a wave of optimism, excitement and additional funding. One of the results of this new impetus was the decision to bring all the main research projects together in Chicago. Enrico Fermi had decamped there with his embryonic atomic pile, as had Harold Urey with his team working on the separation of U-235. Teller, however, had been excluded, and he was desperately disappointed. He was told by Arthur Compton, who was responsible for organising the moves, that it was because the theoretical problems associated with nuclear reactions had been solved, but Teller believed the truth was otherwise. 'The real problem was that I could not be cleared for classified work. Although Mici and I were both citizens, our families were behind the enemy lines.'

It seemed an immutable situation and for three months he wondered what the future held. Then, out of the blue, he too was summoned to Chicago. Although Edward didn't know it at the time, the person responsible for this reversal was Robert Oppenheimer.

Ever since the Schenectady meeting Oppenheimer had continued to make himself useful to Lawrence and Compton, helping assess in detail the new weapon's destructive potential. He had initially been put to work with Gregory Breit, the theoretician working for Briggs on the Uranium Committee, who, with his explosive responsibilities, gloried under the title of Coordinator of Rapid Rupture. Breit was even more preoccupied with security than Briggs was. 'Breit was always frightened something would be revealed in the seminars,' said Sam Allison, a member of Breit's team. 'Oppenheimer was frightened something would not. I backed Oppenheimer and challenged Breit to cut the censorship. And he accused me of being reckless and hostile to him. I failed – the seminars became uninformative.'

For four months Oppenheimer and the difficult Breit rubbed along uneasily until Breit capitulated, resigning and leaving the bomb project entirely. Oppenheimer was left in sole charge of fast neutron research, one of the critical areas of research for the bomb, and began looking for any unexploited talent with relevant experience. In his search he picked up on Teller and several others with doubtful security credentials, and pushed through their clearances. Teller set out for Chicago in early June.

On his arrival Teller discovered that no one had yet figured out a role for him and so he and another colleague, Emil Konopinski, set about re-examining Fermi's suggestion that a fission explosion might be used to initiate a thermonuclear reaction. At a time when there were so many difficulties associated with obtaining the fuels for a fission bomb, fusion held the great attraction that the fuel, deuterium, a heavy isotope of hydrogen, was both relatively cheap and easily available. Also, once ignited, there was no apparent limit, apart from the amount of fuel used, to the size of explosion possible.

In his previous calculation, Teller had assumed that the energy from the fission explosion would be rapidly siphoned off as X-rays, but their new work now showed that the X-rays were not all emitted at once, and therefore a fusion reaction might just be initiated before too much energy was lost. With typically boisterous enthusiasm, Teller felt his and Konopinski's new findings opened up possibilities that would do nothing less than reshape the entire weapons programme. The opportunity to explore and promote them further came when he was invited by Oppenheimer to join a conference that summer in Berkeley – a meeting where a group of the best theoreticians available were to re-evaluate the whole bomb programme.

As well as Edward, Oppenheimer also approached Hans Bethe, and the two scientists, with their wives, decided to rent a house together for the duration of the conference – 'a beautiful place', Teller recalled, 'it was so luxurious that I described it in my letter to Fermi a few weeks later as a "palace"'. At this time Bethe was still deeply sceptical about the project – and in particular, whether the U-235 separation would ever be accomplished. Instead he was working on radar at MIT's Radiation Laboratory. However, when he and Rose stopped off in Chicago to pick up the Tellers – 'our best friends in the country', as Bethe described them – it provided an opportunity for Teller to show Bethe over the facility there, in particular Fermi's new experimental pile, and for Bethe, this was a revelation. 'He had set up under one of the stands in Stagg Field,' Bethe remembered, 'in a squash court – with tremendous stacks of graphite.' For the first time he realised that such a pile, producing plutonium, provided the means of bypassing the isotope-separation problems with uranium that thus far had seemed to him an insuperable problem. 'On that visit Teller explained to me where Fermi stood, and that convinced me that it would work.'

This was not to be Bethe's only revelatory moment on his journey west. 'We had a compartment on the train to California, so we could talk freely ... Teller told me that the fission bomb was all well and good and, essentially, was now a sure thing. In reality the work had hardly begun. Teller likes to jump to conclusions. He said that what we really should think about was the possibility of igniting deuterium by a fission weapon – the hydrogen bomb.'

Sharing the house with the two couples was Emil Konopinski, whom Teller had asked to be included, and, as well as these three, the group Oppenheimer had assembled included John Van Vleck, who had been at Leipzig with Teller, the aristocratic Swiss physicist Felix Bloch, and Robert Serber, a one-time student of Oppenheimer's and now his close collaborator.

The group met at regular intervals throughout the summer in two attic rooms at the top of LeConte Hall, the administrative block on the Berkeley campus where Oppenheimer had his office. They met under what, for those early days, were considered strict security arrangements. The windows were all clad in wire mesh, including the exit to the small balcony, and the door was fitted with a special lock with a single key, which was given to Oppenheimer.

At the start the group concentrated on the fission bomb. They had before them all the results from the disparate groups that had been working under Breit and also the British MAUD report. They first tried to evoke an idea of the scale and impact of the weapon that they were devising. They began by studying the effects of past explosions, like that of the fully loaded ammunition ship that had gone up in the harbour of Halifax, Nova Scotia, in 1917. On that occasion, 5000 tons of TNT had completely destroyed two and a half square miles of the centre of Halifax and killed as many as 4000 people. The new weapon could be expected to produce several times the strength of that blast, so, using scale laws, they multiplied up the effect of the Halifax incident. These calculations were to enable Oppenheimer and his colleagues to visualise in much greater detail the devastating potential of the new weapon.

Next they dealt with elementary aspects of how the bomb would work, its basic structure, its size and so on. They envisaged it, at the instant before explosion, as a sphere of uranium inside a heavy metal shell that would both contain the explosion for crucial milliseconds and reflect back the escaping neutrons into the fissioning uranium metal. They even had a rough estimate, from the British, of the average distance a neutron would have to travel before colliding with an atom and producing another fission.

The figure was 10 centimetres, meaning that a sphere 20 centimetres in diameter should contain enough of the neutrons for a chain reaction to go ahead.

They were also able to give more precise figures for the length of time in which the bomb would have to be assembled and detonated: less than a millionth of a second. The specifications for the weapon would have to be very tight indeed. They had to try to work out just how well the fission reaction would continue under the disruptive physical forces of the explosion. Hans Bethe, who had done original research on the reactions inside the Sun, was able to provide a fairly precise model of what would happen. In fact, it seemed that Teller had been close to the mark when he had told Bethe that the fission weapon was a sure thing. 'The theory of the fission bomb was well taken care of by Serber and two of his young people,' Bethe explained. They 'seemed to have it well under control so we felt we didn't need to do much.'

Within days, therefore, discussion had turned to the subject of fusion and what soon came to be called by the group, the 'Super'. It came to preoccupy them for the remaining four weeks of the conference. What they deduced was both fascinating and terrifying. They calculated that a fusion bomb using just 12 kilograms of deuterium would produce an explosion equivalent to a massive 1 million tons of TNT. This output would be 500 times as great as the 2000 tons of TNT equivalent estimated for a fission weapon at the time, and 200 times as great as the Halifax blast they had studied. In fact, it would be equivalent to a fifth the total explosive power used throughout the Second World War. For those in Berkeley making the assessments for the first time it must have been awe-inspiring. (See Appendix 3, pp. 408–9, for the sketch for the Super that emerged during the conference).

Hans Bethe recalled talking to his wife, who knew in broad terms what they were discussing, 'and on a walk in the mountains in Yosemite National Park she asked me to consider carefully whether I really wanted to continue to work on this. Finally, I decided to do it.' For Bethe, the Super was a terrible thing, but its development was inextricably linked to the German threat, and to the fission bomb. This was, after all, to be the indispensable trigger for a thermonuclear reaction and, because of the Germans, they were committed to developing a fission weapon anyway. So for the time being, any moral dilemma associated with the Super itself could be held in abeyance.

However, in working through the details of fusion reactions, Teller's fertile mind had hit upon another possibility that threatened to stifle both fission and fusion bomb programmes at birth. There are other fusion reactions possible than those between the nuclei of hydrogen isotopes such as deuterium, one of them being that between the nuclei of nitrogen. In his calculation of heat build-up in a fission reaction, Teller estimated that it could trigger the fusion of nitrogen nuclei as well as those of hydrogen. Of course, nitrogen makes up 80 per cent of the Earth's atmosphere, and hydrogen isotopes are present in the world's ocean. According to Teller's figures, a fission explosion could generate enough heat to set light to both.

Memories differ as to what happened next. Teller denied that there was any detailed discussion of the issue at all. In this he was at odds with men like Compton and Bethe and the official history of the bomb project; but from this time onwards Teller did have his own particular view of events.* There is agreement, however, that Oppenheimer left the conference suddenly. Bethe recalled: 'Oppie took it [atmospheric ignition] sufficiently seriously that he went to see Compton. I don't think I would have done it if I had been Oppie, but then Oppie was a more enthusiastic character than I was. I would have waited until we knew more.' Bethe nevertheless began checking Teller's calculations while, according to Arthur Compton, Oppenheimer made contact with him at his lakeside holiday cottage in Michigan.

'I'll never forget that morning,' Compton recalled. 'I drove Oppenheimer from the railway station down to the beach looking out over the peaceful lake. There I listened to his story . . .' Compton was appalled by the vision of atmospheric and perhaps even oceanic ignition conjured up by the calculations which Teller had made at LeConte Hall. 'This would be the ultimate catastrophe. Better to accept the slavery of the Nazis than to run a chance of drawing the final curtain on mankind! We agreed there could be only one answer. Oppenheimer's team must go ahead with their calculation. Unless they came up with a firm and reliable conclusion that our atomic bombs could not explode the air or the sea, these bombs must never be made.'

For a while, according to Compton's version of events, there looked to be a serious impasse – until Bethe reported back on the checks he had made. 'I very soon found some unjustified assumptions in Teller's calculation that made such a result extremely unlikely, to say the least. Teller was very soon persuaded by my arguments.' Teller had, in fact, produced a worst-case analysis, using maximum cross-section values and only the most susceptible of several possible thermonuclear reactions. He had also minimised the energy loss to the atmosphere, as was made clear in an official history published after the war: 'Calculation showed that no matter how high the temperature [of the fission trigger], energy loss would exceed energy production by a reasonable factor. At an assumed temperature of three million electron volts the reaction failed to be self-propagating by a factor of 60.' Even this temperature, arbitrarily chosen by Teller for his calculation, exceeded that likely to be achieved by a fission trigger by a factor of 100. Therefore, as the official 1945 history reports: 'The impossibility of igniting the atmosphere was thus assumed by science and common sense.'

Once Bethe's calculations had relegated atmospheric ignition to a remote possibility – at least for the time being – the group returned to the issue in hand and continued to make formidable progress. It was during these later sessions that Konopinski pointed out that tritium, an isotope of hydrogen that does not exist naturally and has to be created in an atomic pile, would be much easier to fuse than deuterium and would therefore ease the thermonuclear ignition process. Further, the group realised that an isotope of the light metal, lithium Li^6, would, during fusion, break down to produce tritium in situ, thus reducing the need to manufacture the element itself. Step by step, a thermonuclear weapon was, in theory at least, becoming an increasingly realistic prospect.

'My theories were strongly criticised by others in the group,' Teller recalled, 'but together with new difficulties, new solutions emerged ... A spirit of spontaneity, adventure, and surprise prevailed during those weeks in Berkeley, and each member of the group helped move the discussion toward a positive conclusion.' And Teller, along with other members of the group, attributed this progress in good measure to Oppenheimer's chairmanship. 'As chairman, Oppenheimer showed a refined, sure, informal touch. I don't know how he had acquired this facility for handling people. Those who knew him well were really surprised. I suppose it was the kind of knowledge a politician or an administrator has to pick up somewhere.' Those who had known Oppenheimer longer than Teller would recognise these skills as the ones he had honed while leading his Berkeley seminars, but now he was learning to curb his arrogance – his 'beastliness', as he had described it in a letter to his brother.

During their stay in Berkeley that summer, the Tellers were pleased to be invited to dinner at the Oppenheimers' new home on Eagle Hill. Edward brought along a recording of his favourite Mozart concerto as a hospitality gift. 'However,' he later carped, 'Oppenheimer found the recording uninteresting.' It was at this dinner party that Teller believed that he brushed against Oppenheimer's political past. There was another couple present at the meal, whom he noted as pleasant company and non-scientists. They were also 'considerably left of centre. I believe that they were Haakon Chevalier and his wife.' Teller noted that Oppenheimer talked freely, though in generalities, about harnessing atomic energy to create a bomb and said that 'such a weapon could be developed'. Teller

confessed to being 'not at all surprised or alarmed' by the exchange. But he remembered it.

In late August 1942, James Conant heard the results of the summer conference at a meeting of the S1–Executive Committee, the old Uranium Committee as reconstituted by Vannevar Bush, and his notes of that meeting record the headline findings. The fission bomb would, according to the LeConte Hall group, explode with '150 times energy of previous calculation' but would require a critical mass six times larger.

His report on the Super shows that the Executive Committee were in thrall to some of the heady estimates: 'If you use 2 or 3 tons of liquid Deuterium [D] and 30 kg U-235 this would be equivalent to 10^8 [100 million] tons of TNT.

'Estimate devastation area of 1000 sq km [or] 360 sq miles. Radioactivity lethal over same area for a few days.'

He then added, 'S1–Executive Committee thinks the above probable. Heavy water is being pushed as hard as it can. 100 kg of D will be available by fall of 1943 before 60 kg of U-235 will be ready!'

The idea that the Super, with its easily available deuterium fuel, might provide a way of bypassing the seemingly intractable problems of separating out U-235 was being taken seriously. A formal status report went off to Bush echoing these findings which concluded: 'We have become convinced that success in this programme before the enemy can succeed is necessary for victory. We also believe that the success of this programme will win the war if it has not previously been terminated.' When Bush forwarded this report to the Secretary of War, he pointed up the significance of the Super. 'The physicists of the Executive Committee are unanimous in believing that this large added factor can be obtained ... The ultimate potential possibilities are now considered to be very much greater than at the time of the (last) report.'

There was now nothing incidental in the perceived role of the Super. In fact, as far as the War Department was concerned, it represented such a frightening prospect in the hands of an enemy that it had raised the whole profile of, and given new impetus to, the atomic weapons race with the Germans. As of July 1942, it was under development and Teller had every reason to believe that it was now a crucial part of the overall strategy.

As the project gained momentum during the summer of 1942, Vannevar Bush began to recognise the enormous logistical task confronting them and proposed that the military should become involved. The month of June had seen the establishment of the Manhattan Engineering District, so named because its first military boss, General James Marshall, had offices in New York. It very quickly became clear that, while he was a good engineer, Marshall was little better at fighting the priorities war in Washington or dealing with the logistics than the scientists themselves. So, enter Leslie Groves.

At the time Groves was forty-six years old, the son of a Presbyterian army chaplain. He was a graduate of Washington University, MIT and West Point, where he had been fourth in his class. He had spent years supervising the Army's major construction projects, during which time he had been responsible for building the Pentagon. By September 1942 he was responsible for all aspects of military construction work in the US. This included the building of camps, airfields and manufacturing plants, and he was controlling the expenditure of an amazing $600 million a month. Yet, in spite of his enormous power, he was still a colonel, one of the oldest in the Army, and he was restless. He was, like every other regular officer, extremely eager for service abroad as a commander of combat troops. However, that was not to be.

Instead he was given charge of a speculative new weapons project that at the time was expected to involve expenditure of little more than $100 million altogether. It could have had little appeal, but the Army sugared the assignment with a promotion to brigadier general, and that September, Groves took up his new assignment running the embryonic Manhattan Project.

Groves did not consider leadership a popularity contest. He was blunt and ruthless, and had no patience with sloppiness of any sort. Of those who worked with him, most feared him, few liked him. Colonel Kenneth Nichols, a professional soldier and a trained engineer, had worked with him in Panama in the 1930s and now joined him on the bomb project:

> He's the biggest sonovabitch I ever met in my life, but also one of the most capable individuals. He had an ego second to none, he had tireless energy – he was a big man, a heavy man but he never seemed to tire. He had absolute confidence in his decisions and he was absolutely ruthless in how he approached a problem to get it done. But that was the beauty of working for him – that you never had to worry about the decisions being made or what it meant. In fact I've often thought that if I were to have to do my part over again, I would select Groves as boss. I hated his guts and so did everybody else, but we had our form of understanding.

When Vannevar Bush first met him his reaction was: 'God help us' – a response echoed by other 'eggheads', as Groves called them. Within days of his appointment he set out on a tour of the project's main laboratories, and cut a swathe wherever he went, immediately cancelling one of the three isotope-separation processes and coming close to closing down Harold Urey's gaseous-diffusion project. When he arrived at Chicago, he was impressed by what he saw, but irritated by the air of intellectual caution he detected, which, though quite normal amongst academics, he interpreted as indecision. By the end of his visit, the scientists were antagonised and Groves was intimidated – so intimidated, in fact, that he felt compelled to make the following statement: 'There is one last thing I want to emphasise. You may know that I don't have a PhD. Colonel Nichols has one, but I don't. But let me tell you that I had ten years of formal education after I entered college. Ten years in which I just studied. I didn't have to make a living or give time for teaching. I just studied. That would be about equivalent to two PhDs, wouldn't it?'

This statement was met with silence from the assembled scientists, but after Groves had left the meeting Leo Szilard exploded with indignation. 'How can you work with people like that?'

The two people Groves found he could work with were the energetic impresario of big science Ernest Lawrence, and Robert Oppenheimer. Groves appreciated Lawrence's enthusiasm, but was seriously depressed by the problems surrounding Lawrence's attempts to separate the uranium isotopes using his cyclotron.

Oppenheimer, however, proved to be balm for the general's dismayed confusion. He had a wonderfully clear view of all the theoretical and technical problems, and he was not attached to any particular project – was not beating any particular drum. So impressed was Groves that, a week after their first meeting in early October, he flew Oppenheimer to Chicago for a prolonged discussion as Groves travelled back to New York on the famous '20th Century Limited' passenger express. It was on this journey that Oppenheimer proposed the notion of a single laboratory isolated from the rest of the world. He had already experienced the stultifying effect of security in a normal open laboratory, where, because of the fear of leaks, researchers were kept in the dark about the final significance of their work, and Oppenheimer did not want the same situation to arise again.

It was just the sort of scheme that Groves had been considering himself, and he was delighted to find a scientist who had both recognised the problem and thought about it practically. He decided to act on the proposal and began searching for a location for the laboratory. He also decided on Oppenheimer as the scientist who would lead the project. In choosing him, however, he had to ignore protests not only from those who, as one Berkeley contemporary put it, believed Oppenheimer 'couldn't run a hamburger joint', but also from both the FBI and the Army's own security. The FBI had been watching Oppenheimer for nearly two years because of his network of left-wing contacts, a number of whom were suspected of espionage. So, too, were some of Oppenheimer's own students – students who had already become involved in his research for the new weapon.

Groves almost immediately involved him in the search for the new laboratory site. This had very quickly narrowed to a selection of locations in New Mexico, and then, even further, to two possibilities: Jemez Springs and a place Oppenheimer had suggested himself, called Los Alamos. The former, deep in a sunless valley, was very quickly dismissed as being too gloomy and depressing and lacking enough space for expansion. The latter was beautiful, high on a mountain plateau. There was accommodation already in place belonging to the Los Alamos Boys' School and there was water and electricity. True, these supplies were only just adequate for the existing population, and the road up to the school was little more than a donkey track, but Oppenheimer believed that he would need only thirty or so scientists, plus back-up staff. On that basis, Groves decided there and then that Los Alamos should be the site of the new laboratory.

Following the Berkeley conference, Mici and Edward had returned to Chicago to try and pick up the pieces of their domestic life. One new factor was that, after seven years of marriage, Mici was pregnant. For some time she had been seriously concerned about the German advances across Europe, and the plight of her family in Hungary. 'Mici had been convinced that it was wrong to bring a child into such a world,' Edward recalled. 'But when the United States entered the fight, she changed her mind. She was immediately convinced that Hitler would be defeated. A few months later, she was pregnant and buying furniture.'

They had failed to find a suitable furnished apartment to rent, so had to take an unfurnished one and then fill it. In a move typical of the ad hoc way in which they lived domestically, Mici had discovered that Chicago's Congress Hotel had been taken over by the Army and was selling its furnishings. She went along and bought up a complete home-full of heavy oak hotel furniture. It would last them their sixty years of married life together, and was part of the reason why the reception rooms in the various Teller houses never lost the somewhat impersonal air of a hotel lobby.

Within weeks of Groves's appointment it became known in Chicago that the new bomb laboratory was to be established in New Mexico. The cross-currents of fear of army discipline and a belief that the real science had already been done haunted the tea-room and faculty-club conversations. Eugene Wigner was one of those who believed that the only real problem left needing a solution was that of separating the U-235 – and that was not something for him. Despite his early optimism, Teller, on the other hand, believed that things would prove more difficult in practice than in theory.

He also expected a central role in the new laboratory's organisation, and so was seriously thinking of making the move. Oppenheimer, on one of his recruiting visits to Chicago that autumn, talked with Teller. Looking back after sixty years, Teller still believed he could identify this as the moment at which his feelings towards Oppenheimer began to change.

The two scientists shared a first-class compartment on the train to Washington. 'As we travelled,' Teller recalled, 'we talked about the prospect of the Manhattan Project, about the unnecessary obstacles presented by some aspects of secrecy, and about the new laboratory, which was in the planning stage.' They also talked about the involvement of the Army and

> Oppie complained about having to work with Groves, whom he considered awful. I remember clearly one detail of those discussions. I was trying to put a bandage on one of my fingers where a hang nail had become infected, and Oppie had come over to lend me a hand. Continuing the conversation, he said, 'We have a real job ahead. No matter what Groves demands now, we have to cooperate. But the time is coming when we will have to do things differently and resist the military.'
>
> I was shocked. The idea of resisting our military authorities sounded wrong to me. 'I don't think I would want to do that,' I said. Oppie changed the subject immediately. I believe that the relationship between us changed at that instant. Oppie continued to be friendly, and he continued to encourage me to come to Los Alamos, but the warmth of our conversations vanished and never returned.

As the declared fulcrum point in the alienation of these two men, which was to influence the progress of physics and of science in the decades to come, this exchange needs to be seen in context. On the wave of enthusiasm generated by the excitement of working on the new laboratory, Oppenheimer had tacitly assumed that there would be complete integration with the Army. However, as he criss-crossed the country on his recruiting drive, Oppenheimer had certainly picked up some of the concerns about involvement of the military that Teller would have already experienced at first hand in Chicago. At the time he met Teller, he was trying to play down the security arrangements, the perimeter fences, the pass controls, the virtual incarceration; and yet, at the same time, he was participating more or less willingly in their implementation. Early in 1943 Oppenheimer had even visited the Presidio in San Francisco to begin enrolling as a lieutenant colonel, but his plans were brought up short when a group of senior scientists, among them Robert Bacher from Cal Tech and Isidor Rabi, balked at taking an army rank.

Therefore, is it not possible to see what Oppenheimer said to Teller as simply part of his deflection tactic? Was he not simply offering reassurance that, should the Army make further impositions, they would be ready to resist them? Teller had experienced the rumbling discontent over closer military involvement at Chicago, and had witnessed the antipathy to Groves. Given this, his interpretation of Oppenheimer's comments as something tantamount to treason seems at best uncomprehending, and at worst perverse. Even Teller himself later realised that his reaction could have been seen as an extreme one, and did offer an apologia: 'Perhaps I over-reacted. Perhaps, if I had continued the conversation and probed for the point behind his remark, I might have obtained a better understanding. But Oppenheimer did not invite further questions or offer clarification. Probing would have been discourteous, so I did not ask. The incident merely contributed to my impression of the complexity of Oppenheimer's personality.'

The disingenuous nature of this justification is borne out by the way that it contradicts Teller's own earlier observation that their relationship had changed 'at that instant'. This is an important incident of where Teller's own accounts of certain crucial events appear either partial and not always credible, or significantly different from those of others.

For the time being, the change in relationship between the two men was not evident. Bethe, who observed them as they worked together at that time saw their relationship as 'very good – and that was mutual. They were quite close.' Another colleague reported seeing the beginning of a 'mental love affair'. According to yet another, Teller 'liked and respected Oppie enormously. He kept wanting to talk about him with others who knew him, kept bringing up his name in conversation.' When he found himself underemployed at Chicago, Teller made himself available to help Oppenheimer. He it was who wrote a laboratory prospectus for Oppenheimer describing the work and likely conditions at Los Alamos in glowing terms, in order to help with recruitment.

But was the 'love affair' one-sided? Teller's need for friendship, for approval, would have left him highly vulnerable to someone with the armoury of charm possessed by Oppenheimer, but whose friendship was often used as a means to an end. For the time being, however, Teller's help was invaluable.

The 2 December 1942 was a momentous day for the Manhattan Project. That morning the State Department had announced that 2 million Jews had already perished in Europe and that 5 million more were in danger. The Germans were preparing a counter-attack in North Africa; American Marines were caught up in the hell of Guadalcanal. That same morning Chicago was snowbound, and the temperature in the squash court below the grandstand in Stagg Field that housed Fermi's experimental pile was only a little above freezing. The pile, which now waited to be released to breed neutrons and create plutonium consisted of 771,000 pounds (approximately 345 tons) of graphite, 80,590 pounds of uranium oxide, and 12,400 pounds of uranium metal. It had cost about $1 million to build, a relatively modest sum but a far cry from Teller's original estimate of $6000. It was, in effect, a small power station but without coolant to carry away the energy it generated. However, Fermi only intended running it up to the point where it was producing half a watt of energy – hardly enough to run a small torch. It was slowly fired up during the morning and a crowd began to gather on what had been the spectators' balcony of the squash court. Amongst them were Teller's Hungarian 'conspirators' Szilard and Wigner, but Teller himself was nowhere to be seen. Again, he had chosen to miss the practical demonstration of an idea in whose development he had been intimately involved.

It was after lunch that Fermi wound the reactor up to the final stage. 'At first you could hear the sound of the neutron counter, clickety-clack, clickety-clack,' recalled Sam Allison. 'Then the clicks came more and more rapidly and after a while they began to merge into a roar: the counter couldn't follow any more.'

Shortly afterwards, Fermi announced that the pile had gone critical. The neutron intensity was doubling every minute. Left uncontrolled for an hour and a half it would have achieved meltdown – a mini Chernobyl in the centre of Chicago; but after four and a half minutes, run at precisely half a watt, Fermi closed it down. The famous cryptic message flew from Chicago to Washington. 'The Italian Navigator has landed. The natives are friendly.'

It was believed that, with this experiment, the release of energy from the atomic nucleus had been controlled for the first time. However, some six months before, in May 1942, Heisenberg and Dopel had performed an experiment that indicated just how well advanced they were at that time. They had already built three prototype atomic piles, but the fourth – pile LIV – was bigger than the others. These piles, although they looked nothing like Fermi's pile in Chicago, nevertheless worked on basically the same principle.

During that particular experiment the pile, encased in two hemispheres of aluminium and securely bolted together, was lowered slowly into a large tank of water. The neutron source that, it was hoped, would initiate fission within the uranium metal in the pile was introduced into the pile's centre through a sealed shaft, and measurements began. These measurements showed that the pile was producing more neutrons than were being injected by the neutron source at its centre. The experiment had achieved the beginnings of a chain reaction and they calculated that, if they increased the size of the pile by a factor of approximately fifteen, they would be able to build the first chain-reacting pile in the world.

Within two months of the successful experiment in Chicago, Mici and Edward had an event of their own to celebrate – the birth, on 10 February 1943, of their first child, Paul. None of the grandparents were present or even knew of the birth. Only one, Edward's mother, would ever see her grandchild. It was a sharp personal reminder of the threat that they were now resisting with their work on the atomic bomb. Some six weeks later Edward left his new family in Chicago for the New Mexico mountains. The construction of the new bomb laboratories at Los Alamos was sufficiently advanced for the first group of scientists to move out.

> I loaded a few belongings into my black Plymouth, said my goodbyes to Mici and baby Paul [they would follow by train in a few weeks], and set out for the South-West. A few days later, I crossed the Rio Grande on the small rickety Otawi bridge and approached a place, both peculiar and outlandish, that became dear to many who lived there. But I was moving to Los Alamos willingly, and the change seemed less dramatic than had some of my other migrations.

After all the assistance he had given Oppenheimer over the previous months, and secure in the knowledge that the Super had contributed in a major way to the impetus for building the new laboratory, Teller was looking forward to a senior position at Los Alamos. He also looked forward to continuing his close working relationship with Oppenheimer.

7

Maverick on the Mesa

If Robert Oppenheimer had been left to his own devices, it is quite probable that he would have made an interesting stab at producing a bomb with his team of thirty or so carefully chosen colleagues. Priscilla Duffield, Oppenheimer's secretary, remembers prolonged rambling conversations between him and Robert Serber on whom it would be lovely to have in the team, and how one man's brilliance might mobilise another's rather more pedestrian abilities. It was as if he were casting a Broadway play rather than organising a massive scientific and technical undertaking. He even had the rather charming idea of special quiet housing facilities for the 'unmarried couples' – by which it was assumed he meant childless couples.

His approach rapidly led to problems. One of his ex-students, Robert Wilson, visited Los Alamos to prepare for the arrival of a cyclotron he was bringing out from Harvard and was appalled by the chaos he found. He and another young physicist, John Manley, cornered Oppenheimer at a party where, according to Wilson, he was 'doing his usual magic with the martinis' and they went for him relentlessly. 'And after a certain stage,' Wilson recalled, 'Oppenheimer became extremely angry. He began to use vile language, asking us why we were telling him of all these insignificant problems, that it was none of our business and so on. Both of us were scared to death. We were frightened because, if this was the leader and if the leader was going to have a tantrum to resolve a problem, then how was anything going to be sorted?'

But Oppenheimer was to learn very quickly and by the end of March he had worked out a complete organisation chart, increasing the size of the laboratory staff from a prospective hundred to one thousand five hundred. He had also begun to look realistically at the shape of the research programme. Over the next few weeks he was to preside over decisions of priority between various avenues of research, and no one was to be affected more by these decisions than Edward Teller.

Teller had moved out to Los Alamos at almost the same time as Oppenheimer himself and had spent his first few weeks there without his family, living in one of the old school buildings. His memory of that time was for ever imprinted with the sound of bongo drums emanating from an adjacent room occupied by Richard Feynman, the brilliant twenty-five-year-old theoretical physicist from MIT. He spent those early weeks in the company of the 3000 construction workers who, using rough timber and building paper, were throwing up new buildings at a great rate. Among the earliest of these to be completed was a laboratory equipped for working at low temperatures with liquid hydrogen isotopes, constructed specially for work on the Super.

The construction workers had been living through the bitter cold of the mountain winter in trailers, totally ignorant of what they were working on. Consequently their morale was very low, and they seemed to resent the arrival of the scientists. Yet in spite of this friction there was a great atmosphere of excitement among the new arrivals, a feeling described by some of the scientists as akin to the first few days at summer camp. They arrived by road or by train from all parts of the United States – some alone, some with their families, not knowing either their final destination or the true nature of the project on which they were to be working. Among them, in April, came Mici Teller, along with her two-month-old baby, Paul.

Always someone who was direct in her reactions, Mici Teller had baulked at bringing a newborn baby to such an out-of-the-way location. She had never taken to Oppenheimer – she, like Ernest Lawrence's wife Molly, thought him a poseur – and was deeply sceptical about the whole project. Like many of the scientists and their families, she had also viewed the security arrangements, the checks, the barbed-wire fences, the censored mail, with considerable distaste. Within a short time she had crossed swords with the military when they had tried to clear the only remaining clump of trees near their apartment block. She pleaded with the bulldozer driver, but to no avail. 'So I called all the ladies to the danger,' Mici reported, 'and we brought chairs under the trees and sat on them. So what could he do? He shook his head and went away and has not come again.'

The Tellers, along with three other couples and their children, had been allotted T 124, a block of four barrack-like apartments built for senior scientists. Their thin intervening walls allowed for little privacy. Central heating was in the hands of Spanish-speaking stokers, who sang and beat drums in the furnace room while alternating the little apartments between ice boxes and Turkish baths.

On one occasion a senior physicist, Ken Bainbridge, came home after a day's work to find that the carefully tended lawn in front of his apartment had simply disappeared. It had been a removed by the Army to grace the new CO's house. However, within a short while, the irritations with the military faded into the background as just another element in the exceptional fabric of life at Los Alamos. Very soon there were a cinema, square dances, and endless parties of all shades of intellectual seriousness and innocence. In time, the Mesa was to support a nineteen-piece band, a choir and orchestra, and its own radio station.* Many of the scientists, like the British metallurgist Cyril Smith and his American wife Alice, had previously lived quiet suburban lives. The Smiths had been living in Connecticut with 'pleasant friends', as Alice put it, but she soon found that the intense social life on the Mesa amongst some of the best scientific minds in the world 'created really intimate friends, changed the nature of our friendships completely'.

The Smiths lived above the Tellers in the same apartment block. The Brode family came to live in the apartment next door, and above the Brodes were Felix Bloch and his family. They all came to be close friends. 'In those days [Edward] was a wonderful man,' Cyril Smith remembered years later. 'I was very fond of him. I used to go hiking with him frequently and, of course, the ability to hike in the mountains on Sundays was one of the things that kept one sane. Almost every weekend [we would hike with], Bethe, Teller or Fermi or some of these.'

Cyril Smith's good report came in spite of the fact that the Tellers had brought Edward's Steinway with them. It occupied a large part of their living room and Edward, who still maintained his habit of rising late and retiring late, would play on into the night. 'That never bothered us,' Alice Smith recalled. 'But they had a way of putting their little Paul to bed just below our bedroom with toys and he would wake up in the middle of the night and drop them all out on the floor with a tremendous crash. Silly little things like that, you know. But as I said, we came out of all this very good friends with the Tellers.'

In her evocative book *Tales of Los Alamos*, Bernice Brode devotes a whole chapter to T 124 and the families living there, and creates a vivid cameo of the Tellers at that time.

The Smiths kept to a regular and rigid schedule, so it was no wonder that they regarded the irregular hours kept by the Tellers, who lived beneath them, as a hardship ... Neither Edward nor Mici nor the baby Paul, called Piggily, even tried to keep to a routine. It was not in their Hungarian temperament, and apparently Edward's part in the work did not require his getting over to T [Tech Area] at any set time. Edward was always being kidded about his lack of hours at T, but he took the jests lightly. He and Mici were both individualists, which made them delightful company at parties. When Alice complained that they were kept awake at night, Mici complained tearfully that it wasn't her fault that Edward got all his bright ideas at night and wanted to discuss them with her. She wanted to sleep, too.

It was usually near noon when Edward emerged, still deep in thought, and slowly walked up the road to T. He never seemed to have a sense of urgency, which I found restful. Mici, on the other hand, was always hurrying to make sudden plans that just had to be carried out then and there, like the tea party on my first day in Los Alamos. Her enthusiasm for going places and doing things was enormous, so everybody included her in the arrangements for trips. She was excellent company and generous to a fault.

Baby Paul was already precocious and something of a handful. '[Mici's] familiar question posed to the neighbourhood was, "Would anyone care to have Paul?" My boys always said under their breath, "The answer is NO." . . . He walked early and succeeded in getting around and out of any fenced-in situation contrived for his welfare . . . Someone always found him and brought him in. Edward said, "I'm glad my son won't be fenced in, otherwise I should worry about him." '

The Blochs also had a piano, a player-piano that was operated tirelessly by their two boys. When the Blochs left Los Alamos, this piano was left out at the side of the road. 'One moonlit evening, Edward came by and gave a concert that could be heard all over the Mesa. When a crowd gathered round, Cyril Smith suggested passing a hat to take up a collection to have the piano moved.'

The life Bernice Brode describes was uncompromisingly social; but while some like the Brodes and the Tellers accepted, even relished it, others like Fermi's wife Laura disliked it intensely. Bernice Brode's account of life on the Mesa is, in common with most of the recorded memoirs, an optimistic and affectionate one, but she nevertheless makes clear the tensions that existed in this most extraordinary of scientific establishments:

There is no question about it, we lived too close together and our walls were too thin. All of our men were under strain. The job was hard, and they were working against a deadline to construct the weapons before the Germans did. None of the men was used to working under any large administrative set-up . . . If they hadn't felt the call of patriotic duty, they wouldn't have come to Los Alamos in the first place, so most of them made a great effort to adjust, and since they were reasonable men of good will, the results were tensions, usually at home.

But in spite of everything, or possibly because of everything, our house was very congenial and we had some wonderful times together. The same could be said for many other four-family greenhouses as well. They had stresses and strains with conflicting personalities.

Oppenheimer was determined that communications between those working in the isolated laboratory should be as good as possible, so when the laboratory officially opened for business on 15 April 1943, it did so with a series of introductory lectures for the new staff. They assembled in a part of the main administration complex called Main Tech, where they were greeted first by General Groves. In this gathering, marked particularly by the enthusiasm of the participants, his limp handshake and the fear of impending failure, implicit in his short opening address, struck a very sour note.

Groves was followed by Robert Serber, to whom Oppenheimer had delegated the task of giving the introductory lectures. It was a role Teller was to assume with new arrivals in the future, but he was still very much preoccupied with the development work on the Super. Certainly Serber, a wisp of a man with a hazy, uncertain voice seemed an almost obtuse choice to fill the scientists with enthusiasm. 'He wasn't much of a speaker,' said one of those present, 'but for ammunition he had everything Oppenheimer's theoretical group had uncovered during the last year. He knew it all cold, and that was all he cared about.'

The briefing, by Serber and by other experts in explosives and chemistry, went on for two days, covering various practical problems surrounding the bomb, and by the time they finished, Teller must have been itching to get started. Shortly afterwards, however, he was to receive a shattering piece of news.

By the spring of 1943 the atomic bomb had preoccupied him for nearly four years – longer than either Oppenheimer or his friend Hans Bethe. He was a pioneer, a member of the 'Hungarian Conspiracy'. Few, if any, knew more of, or understood better all the scientific and technical details of the bomb, and Teller, with reason, saw himself as the obvious candidate to run the Theoretical Division. Ever since the previous autumn, when the independent bomb laboratory had become a reality, he had been helping Oppenheimer in various ways. In particular, he it was who had kept up a gentle pressure on Bethe to ensure that he joined the project. In March of 1943 he had written to Oppenheimer delightedly reporting that Bethe was now taking 'a very optimistic view, and there was no need whatever to persuade him to come'. Thus it was a severe blow when Oppenheimer appointed Bethe to the very post Teller had thought would be his – a decision in which he had been influenced by one man, Isidor Rabi. As Bethe recalled:

> Rabi told Oppenheimer that he must not direct the theoretical physics himself, there was much too much to do. There had to be a separate theoretical division and the obvious person to direct it was Hans Bethe, he said. And Teller was very upset about it. Rabi had an extremely low opinion of Teller because he had taken Gamow as his model. Rabi felt technically he was very weak and nothing basically worked. And Rabi had that opinion because Edward and I had been successively visiting professors at Columbia [where Rabi worked] and I apparently did a good job but Teller had no systematic approach.

Bethe recognised this as a crucial point in Teller's life. In retrospect, Bethe referred to the man he had come to know so well over the previous eight years as the 'gentle Teller'. 'This was the moment he began to change,' Bethe said; and his wife Rose agreed: 'Look, it repeats his experience of rejection both at school and university. It's from this point on you get those extreme reactions to people and situations which cause so many problems. And there were those at Los Alamos who egged him on in those reactions.'

Teller was certainly deeply hurt. 'I suspected that Hans thought I spent too much time in impractical pursuits, a failing that I could correct if someone disciplined me.' This plaintive reaction is understandable, though it was not only Bethe and Rabi who felt that about Teller. Many of his colleagues thought his fertile imagination was the most wonderful of assets in any discussion, but almost because of this, and because of his highly charged, at times almost childish responses to life, they thought of him as erratic and irresponsible. Still bitter sixty years later, Teller drew a critical parallel between himself and Bethe:

> Bethe enjoys turning out what Fermi called 'little bricks', work that is methodical, meticulous, thorough and detailed. He is, as one of my friends characterised him, the Dean of German physics, almost a Geheimrat. Although I have made a few tiny little bricks, I much prefer (and am much better at) exploring the various structures that can be made from brick, and seeing how the bricks stack up. Oppenheimer, in my view, also approached physics in a manner more like a bricklayer than a brick maker.

This blow, which, as Rose Bethe pointed out, repeated his experience of rejection at both school and university, was followed, a matter of weeks later, by another. Teller had been looking forward to the challenge of developing the Super, which, with its theoretical complexities, offered just the kind of project he might lead. Indeed, his expectation that the Super would form a major part of the work at Los Alamos was not wishful thinking, but something embodied in the laboratory's early schedules. However, the previous autumn Groves had set up a five-man review committee under the MIT chemist W. K. Lewis, who had a reputation for assessing the likely success of moving a scientific idea on to an industrial scale. The committee reported its findings on 10 May 1943. It approved the laboratory's nuclear physics research programme, except for one strand. It recommended that thermonuclear research – the development of the Super – should be restricted largely to theoretical investigation and was therefore to take a distant second place to the work on fission. The main reason for this decision was that as work had begun in earnest on the fission bomb, it had rapidly become clear that it was going to be far more difficult than had been expected. 'If our work was to make any contribution to victory in World War Two,' Hans Bethe recalled, 'it was essential that the whole laboratory agree on one or a very few major lines of development and that all else be considered of low priority. Teller took an active part in the decision on what were to be the major lines . . . A distribution of work among the members of the theoretical division was agreed upon in a meeting of all scientists of the division, and Teller again had a major voice.'

Playing a part, however major, in discussions where the final decision about your own fate lies in the hands of others, does not necessarily reduce any sense of disappointment – or even of betrayal. Teller was to find it extremely difficult to reconcile the decision to relegate the priority of work on the Super with the way it had acted as a spur in launching the bomb project only some seven months earlier. Then it had been considered a threat real enough to have acted as a significant political lever. Why the sudden change of heart? Why, he asked, had Oppenheimer not fought the decision of the review committee? Over the summer of 1943 he was to treat the committee's decision as a postponement. For four months he worked with Hans Bethe, helping him with the immediate problems of calculating the critical mass and nuclear efficiency of the various fission designs.

Taken individually, and objectively, these setbacks and incidents can be seen as simply the compromises needed to keep an embryonic yet complex mission on the road. To Teller, however, always sensitive to any feeling of rejection, they were deeply personal, and behind them he could always see the hand of Oppenheimer. In what he thought was a friendship with Oppenheimer, he had expected loyalty, but had experienced only expediency – and expediency at his expense.

◆

During the summer of 1943, Oppenheimer had been attracting a great deal of attention from both the Army's security officers and the FBI. On 12 June he had travelled to San Francisco, ostensibly to recruit a personal assistant, but had also visited his former lover, Jean Tatlock. Jean had been a Communist Party member and she it was who had introduced a politically naïve Robert to left-wing activities. Theirs had been a stormy relationship lasting several years; at one point Robert had nearly proposed but eventually her increasing psychiatric problems had driven them apart and she had gone east to medical school. Now she was being treated for depression and Oppenheimer had responded to a plea from her to see him. Surveillance officers had watched as the communist activist and the leader of the bomb project spent the night and the following day together. Then, on 23 August, he had paid a surprise visit to the Army's Berkeley security office. Here he had given the security officer, Lyall Johnson, the name of someone he believed they should be watching – an English engineer working for Shell: George Eltenton. Johnson had then reported the conversation to his senior officer, Colonel Boris Pash.

It is a mystery why Oppenheimer did this. He certainly knew that security were concerned about the activities of some of his students, the very students he had hoped to bring into the bomb project. He knew also about the increasingly bold attempts at espionage by some of his old left-wing contacts, such as his wife's friend Steve Nelson. He may have thought some gesture of cooperation would deflect interest from himself – or someone else whose identity he was concealing. If so, he could not have misjudged things worse.

It led first to an extensive interview with Pash, whom he surprised with a long and complicated story. He told Pash that he had been approached by an intermediary who had talked about passing on information he had about the work at Berkeley to the Russians. George Eltenton was part of the chain but not the intermediary. He also reported that he knew of subsequent approaches made to three other scientific colleagues that had troubled them. Pash had asked who the intermediary was. Oppenheimer had demurred, offering titbits of information – that he had been a Berkeley faculty member but had now moved on – but refusing to name him.

A few weeks later he faced a much sharper interview, posing the same questions, with General Groves's own security adviser, Colonel John Lansdale. To him Oppenheimer explained his refusal to name the intermediary as 'a question of some past loyalties', but the interview again ended without Oppenheimer divulging the name of his contact.

In December, Groves himself finally ordered him to name the intermediary. Without hesitation he named Haakon Chevalier, the friend and fellow Berkeley academic who Teller believes he met at dinner with the

Oppenheimers at the time of the Berkeley conference. Groves then asked him to identify the three colleagues whom Chevalier had approached. After extracting a promise from Groves that he would not pass the information on to the FBI, Oppenheimer admitted that there had only been one approach and that had been to his brother Frank. Groves was as good as his word, but the 'Chevalier incident' was to come back to haunt Oppenheimer in the years to come.

During the late summer of 1943 experiments at Purdue University optimistically reported that the fusion-reaction cross section for deuterium was much larger than expected. Teller promptly used the result to urge the renewal of the Super investigation, but while he was awaiting a decision, the mathematician John von Neumann came to the Mesa.

Von Neumann had been a close school friend of Eugene Wigner and, like Edward, was both Hungarian and Jewish. His father, a banker, had been ennobled in the 1890s, at a time when the Jewish professional classes were the engines of reform in Hungary. Some four years the elder, von Neumann, like Teller, had been a child prodigy – at six he had joked with his father in classical Greek – and had a truly photographic memory. The two had met while Edward was still at the gymnasium, where he had been much in awe of the older boy, who was a mathematical genius. Edward recalled a joke about him to the effect that a) Johnny can prove anything, and b) anything Johnny proves is correct. Von Neumann had come to the US in 1930 and in 1933, at the age of twenty-nine, become the youngest member of Princeton's Institute of Advanced Study. There it was said that he was a demigod, but that he had learned to imitate humans perfectly. He conducted himself with a bonhomie that matched his familiar of 'Johnny', but Wigner and others sensed a certain calculating coldness behind that mask, and thought that his friendships lacked intimacy.

Von Neumann was one of a handful of scientists who had agreed to act as occasional consultants for Oppenheimer and, therefore, came and went from the Mesa. He also consulted elsewhere, and had become involved in understanding the complex hydrodynamics of shock waves in the shaped explosive charges being developed for the new anti-tank weapon, the bazooka. His arrival at Los Alamos in the late summer of 1943 chimed precisely with a critical phase in the development of the implosion method for achieving the bomb's critical mass. The most straightforward way of assembling this was to force two subcritical pieces together, and a favoured configuration for this was the so-called gun method (see figure 1): a modified artillery gun fires a subcritical uranium shy into a subcritical spherical uranium target; when the two pieces come together, they exceed the critical mass and the nuclear explosion takes place.

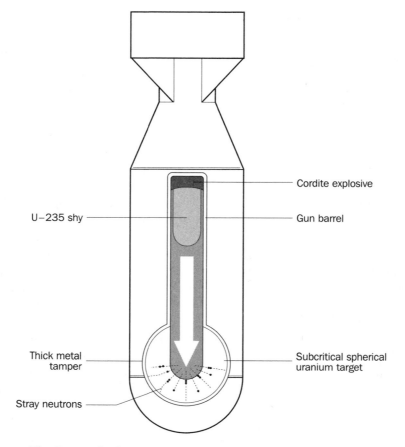

Figure 1: The Gun Method
This principle of assembling the critical mass was employed in the 'Little Boy' bomb used on Hiroshima. Inside the bomb's casing was a modified artillery gun that fired a subcritical uranium shy into a subcritical spherical uranium target. When the two subcritical pieces came together, they exceeded the critical mass and the nuclear explosion took place. Although the two pieces came together almost instantaneously, there was always the problem of surface nuclear reactions caused by stray neutrons. These could blow the bomb apart before it went fully critical, and became a serious problem when plutonium was used as the fission material.

Although this happens almost instantaneously, there was always a problem with surface nuclear reactions caused by stray neutrons, which might blow the bomb apart before it went fully critical. In an attempt to solve this problem – known as 'pre-detonation' – a physicist from the National Bureau of Standards, Seth Neddermeyer, had proposed an entirely new approach, one where the key word was 'implosion'. The fissionable material was to be shaped into a hollow sphere, which would then be surrounded by conventional explosive. When the explosive was detonated, it forced the subcritical sphere to implode, producing a solid sphere that now exceeded the critical mass, and the nuclear explosion took place. It was a method of assembly that, if it could be achieved, would be a great deal faster than the gun method.

However, Neddermeyer and his small unit had been making slow progress. Their crucial problem was how to produce an absolutely even imploding pressure wave around the core material. Any imbalance, and the core would be distorted and would spurt outwards. They were still working on imploding metal pipes rather than spheres – in effect simplifying the problem to two dimensions rather than three – but it was still proving horrendously difficult. The new head of ordnance at Los Alamos, naval captain William 'Deke' Parsons, was openly contemptuous of their efforts, believing they were attempting the impossible. It was, he said, as if they were trying to 'blow in a beer can without spattering the beer'.

Parsons's contempt for Neddermeyer's work coincided with the increasing realisation that the gun method for assembling the bomb's critical mass was unlikely to be speedy enough to overcome the spontaneous-fission problem in plutonium. Plutonium generated so many stray neutrons that with anything less than almost instant assembly of the critical mass it would pre-detonate, blowing itself apart before going fully critical. In their attempt to provide an answer, Parsons's division were battling to produce a gun with the necessary muzzle velocity of 3000 ft/sec and were reporting that it would need a barrel seventeen feet long – giving a bomb only just deliverable by one of the brand-new B29 bombers or a Lancaster.

On his first night at Los Alamos, Teller had invited von Neumann home to dinner, and there they had discussed the complexities of implosion and Neddermeyer's approach. Von Neumann's familiarity with what could be achieved with shaped charges meant that he was not fazed by the notion of a finely controlled implosion, as Parsons was. Within a short while he had calculated that, using Neddermeyer's approach, the central core could be subjected to pressures of millions of atmospheres.

It was at this point in their ad hoc seminar that Teller was able to add something crucial: 'At that point, I was reminded of a fact I had learned at the George Washington University theoretical physics conference on geophysics: iron in the core of the earth under an estimated pressure of 5 million atmospheres, is compressed by 30 per cent. I pointed out that at the high pressures Johnny was talking about, the fissionable materials would be compressed. In that case the critical mass would become much smaller.' So, not only did implosion offer the speed of assembly necessary to overcome pre-detonation problems with plutonium, but it also offered sizeable economies in the use of the scarce core materials.

A)

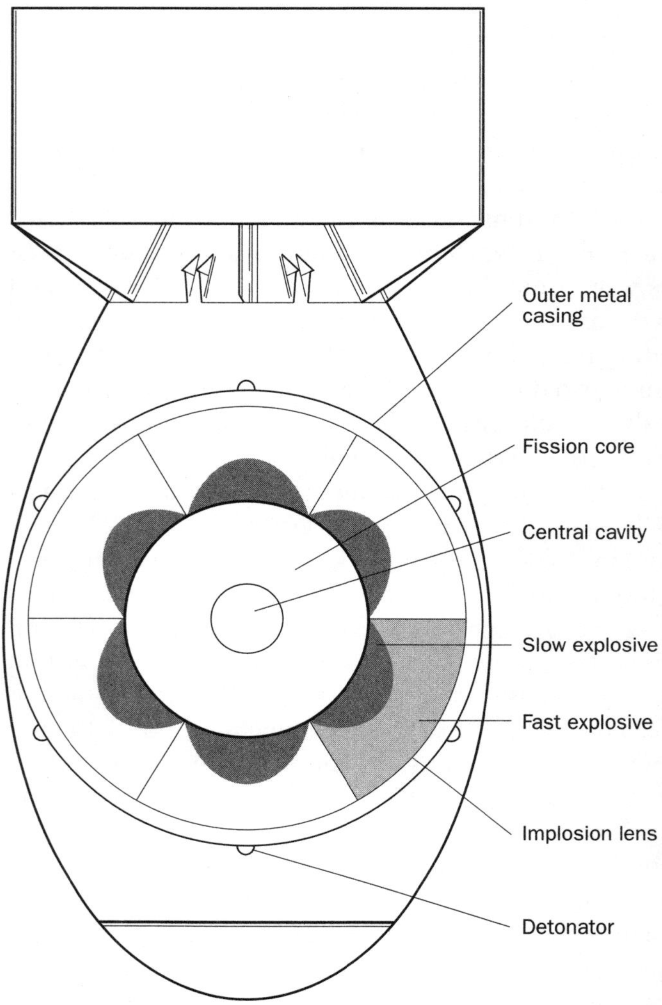

Outer metal casing

Fission core

Central cavity

Slow explosive

Fast explosive

Implosion lens

Detonator

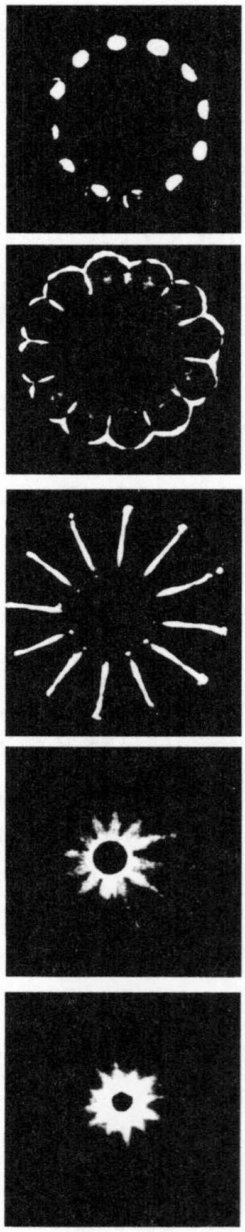

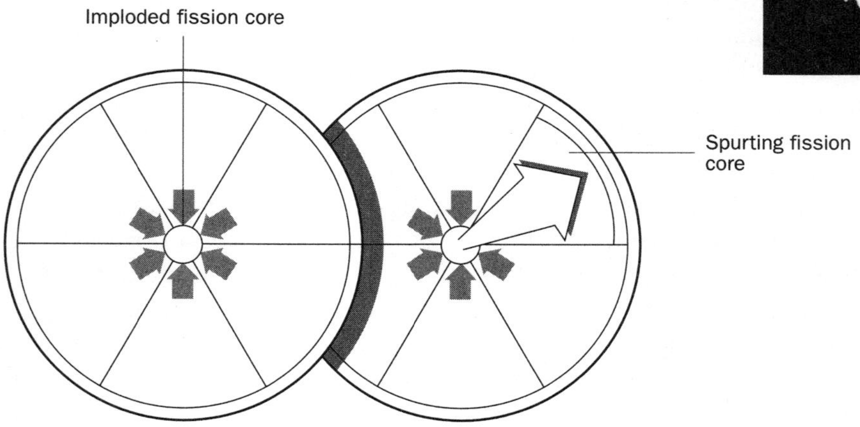

Imploded fission core

Spurting fission core

B)

C)

'The next morning, Johnny and I presented our findings to Oppenheimer. He immediately grasped their implication. Within a week, magnificent administrator that he was, he had turned the direction of research around.'

This original conceptual work, reflecting the imagination for which both men were famed, could not have come at a more opportune moment. Within a short time it had been estimated that an implosion bomb – consisting of a spherical plutonium core surrounded by a thicker layer of high explosives and encased in a heavy metal shell – would be just under five feet in diameter and could be fitted into a bomb casing only nine feet long (see figure 2). By November, work had already begun modifying a B29 to take the new weapon.*

While von Neumann moved on to territories new, Teller remained for several months enthusiastically involved with the new challenges posed by the implosion method. In particular he worked on trying to predict what would happen to the plutonium core at such incredibly high pressures. It was valuable work, but, having to some extent accepted the earlier setbacks as a postponement, he still hoped to pick up with the Super again. These hopes were, however, once more dashed when the project was re-evaluated by the laboratory's governing board early in 1944. In spite of the findings from Purdue that deuterium had a larger cross section than originally thought, and therefore would fuse more easily, they assessed that it would still be extremely difficult to ignite. Thus the Super fuel would have to contain at least some tritium, of which only a tiny amount had so far been made by bombarding lithium in a cyclotron. The large-scale production that would be needed for a bomb would require production reactors of the kind being used to produce plutonium. However, the reactors at the giant plant at Hanford were not only fully committed to producing plutonium, but not yet even in production.

'Both because of the theoretical problems still to be solved and because of the possibility that the Super would have to be made with tritium,' the

Figure 2: The Implosion Bomb
'Fat Man', the bomb used on Nagasaki, had a hollow spherical core around which were placed a number of explosive lenses. The cross section (A) shows how shock waves from detonators on the surface of the bomb are propagated through fast and slow explosives in each implosion lens. The cone of slow explosive in the centre slows the leading edge of the shockwave while the outer ends continue to be propagated through the fast explosive. If the explosives have been correctly shaped, a symmetrical and focused shockwave will result at the surface of the core as shown in diagram (B) and photographs (right).

The diagram (C) shows what happens if a detonator fails or one of the lenses is significantly weaker than the others. The core spurts under uneven pressure resulting in asymmetrical implosion, which produces a greatly weakened explosion.

official history of Los Alamos reports, 'it appeared that the development would require much longer than originally anticipated.' Work would continue, but the project had now been firmly marginalised. This time Teller had to accept the board's verdict, but he was deeply disappointed and in the wrong state of mind for what happened next. It was to prove a watershed.

Teller recalled a meeting with Hans Bethe in Bethe's office at about the time he heard of the board's decision about the Super. ' "Well," he said, "you proposed the idea of implosion. Now I think that you should work on it. I want you to take charge of solving the equations that will be needed to calculate implosion." ' In early 1944 Los Alamos had taken delivery of some IBM punch-card machines and they had promptly been pressed into service to try to clarify how to configure and shape the high explosive around the subcritical plutonium core to produce an even shock wave. It was soon clear that the velocity of the converging shock wave, travelling through each of the several shaped explosive lenses, could not be allowed to vary by more than 5 per cent. Teller baulked at the marathon of trial-and-error calculations that would be involved:

> The task Bethe was discussing seemed far too difficult. Not only were other people more capable than I of providing such work, but I also suspected that a job that formidable might not be completed in time to have any influence on a bomb that could be used during this war. Although I began explaining all those reasons to Bethe, he was convinced that I needed to tackle the job; I was just as convinced that if I did, I would make no contribution to the war effort. We talked for almost an hour without coming to an agreement. At that point Bethe excused himself to keep another appointment, and I went back to my office. Fortunately for me, we never had to resume that conversation. Although Hans did not criticise me directly, I knew he was angry. Although I hoped that he would come to understand my position, he never did, and the incident marked the beginning of the end of our friendship.

Bethe was indeed angry. There were, in fact, others who refused to involve themselves in these marathon calculations for just the same reasons as Teller, but he and Teller had been close friends ever since their arrival in the US. Bethe took his refusal to cooperate as a personal affront. 'Only after two failures to accomplish the expected and necessary work,' Bethe wrote, 'and only on Teller's own request, was he, together with his group, relieved of further responsibility for work on the wartime development of the atomic bomb.'

Teller's refusal to comply with Bethe's request caused not only a serious problem in finding a good replacement to lead a critical, if less than original, part of the Theoretical Division work; it also created a great deal of ill feeling. Oppenheimer became involved and, in May 1944, wrote to Groves about this question of 'greatest urgency', asking him to agree to Rudolf Peierls, a member of the British Mission newly arrived on the Mesa, taking over Teller's role.

Teller was driven on to the defensive. To compound his feeling of vulnerability at this critical time, he and von Neumann had spotted what they thought might be a problem with the fusion reaction that would make the Super impossible to achieve. From the beginning they had worried about the amount of energy that would be transferred away from the reaction by radiation, and the possibility that this would drain so much away that the reaction would not be able to sustain itself. Now they began to question the role of the low-energy light quanta, which were part of the wide spectrum of radiation emitted by the explosion. These quanta, they thought, might collide with the far more energetic electrons, and therefore carry away a great deal more energy than had originally been estimated. With this issue still unresolved, Teller had attended one of Oppenheimer's meetings with his section heads. He was to recall that, when asked by Oppenheimer what was new in his group's activity, he mentioned that there was a possible problem with the losses from radiation, but would rather discuss it later. Oppenheimer, however, insisted that he talked about it then.

'That arbitrary order particularly upset me,' Teller recalled, 'because I was afraid that, in light of the new difficulty, work on fusion would be scrapped completely. I stood up and left the room.' As he did so, he threatened to leave the project. He remembered regretting his petulance only minutes later. However, after the meeting, Oppenheimer came to see him. To Teller's surprise, he did not confront him but, instead, offered to meet with him on a weekly basis to discuss his group's problems. At a time when the laboratory was working overtime six days a week, it was a measure of the importance Oppenheimer himself attached to Teller, both as a troubleshooter and, no doubt, as a possible disruptive influence. Certainly, there were others who felt Teller had gone too far, and that he was behaving selfishly. Bethe recalled 'his wish to spend long hours discussing alternative schemes which he had invented for assembling an atomic bomb, or to argue about some remote possibilities why our chief design might fail. He ... spent a great deal of time talking and very little time doing solid work on the main line of the laboratory. To the rest of us who felt we had a vital job to do, this type of diversion was irksome.'

For numbers of his colleagues, Teller's idiosyncrasies – his late rising, his late-night piano-playing, his impetuosity in scientific discussion – until now accepted with good grace as eccentricities, began to grate. People also began to notice a petulance, an egotism in Teller's behaviour that had not been obvious before. By the time Lord Cherwell came to visit the laboratory in 1944, Teller's sensitivity to slight was well known. On that occasion Oppenheimer held a party, but omitted to invite Rudolf Peierls, by then the deputy head of the British Mission. In Peierls's office the next day, Oppenheimer apologised for this terrible mistake. 'But,' he added, 'there is an element of relief in this situation. It might have happened with Edward Teller.'

In spite of Oppenheimer's concession, Teller was more than ever convinced he was being manipulated. The Bethes, Hans and Rose, believed he was egged on in this view by Felix Bloch, who had his own dissatisfactions with the regime at Los Alamos. Bloch, who had a traditional European academic background, had taken exception to Oppenheimer's democratic system of open discussion, and also to the fact that he felt he was being poorly used. He eventually asked to be released from the project, a request that was granted, making him one of the very few scientists to leave Los Alamos. It was arranged that Teller would drive him to the station, but then Teller received his first invitation to join one of Oppenheimer's poker parties. He eventually saw this as a deliberate political manoeuvre by someone fearful that Bloch might have persuaded him to leave too: 'I may be unjust,' he said, 'but the whole thing looked like too much of a coincidence. He used friendships, he exploited friendships. Granted, he did not want me to leave Los Alamos, but obviously he manipulated people. In the physics community as I had known it, this just was not done.'

The fact that this was the only time he was ever invited to one of Oppenheimer's fairly regular poker evenings may have rankled too. The incident had a satisfactory conclusion, however. Teller saw his friends off – and joined his game of poker with Oppenheimer late.

There is something disingenuous in this anecdote, often retold by Teller. Oppenheimer was in charge of a three-ring circus. It was inevitable that he would have to act politically, to keep so many conflicting interests and diverse characters focused on their prime objective; but Teller took personal exception to the way the situation had been handled. Was it that being a victim of apparent manipulation defined him as an outsider? Certainly, in his last telling of the story, he referred to the store he set by friendships rather than being part of the right group; and it was from this time that he said he noted how Oppenheimer was forming a clique around him of like-thinking, liberal-minded people. Yet, by his own admission, it was Teller himself who was developing a perspective on events that, even while he was at Los Alamos, had begun to separate him off from many of his colleagues there:

The scientific community was remarkably homogeneous in its addiction to science,' he has written, 'and in its liberal political judgements, which were those commonly held on American university campuses. I slowly came to realise that in two important respects, my views differed from those held by the majority. Perhaps because I had spent the previous four years working for and on weapons research, I saw it as continuing to have great importance for the future. Unlike many in the scientific community at Los Alamos, I had become convinced that the problems connected with Communist Russia were very great.

These retrospective justifications, often given a certain political spin, often prefaced with an apology, were to become a pattern in the way Teller interpreted critical events and relationships in his past. They have been continually questioned, and with reason, and yet his position at this time, his feelings about these crucial relationships, can be explained and even justified without resort to calculated retelling.

He should, perhaps, have risen above such provocations, taken a more philosophical view of the reasons for the setbacks he had experienced rather than taken them so personally. But psychological wounds gained in childhood and adolescence do leave sensitivities. The pain of his adolescent years was still there. To himself, at least, Edward Teller was still Koko.

8

The Little Toe of the Ghost

In late 1944, Russia was an ally, one that had been fighting against the Germans for more than two years – an ally, furthermore, for whom there was a great deal of unalloyed admiration in the United States. General Douglas MacArthur had described the effort of the Red Army as 'the greatest military achievement in all history'. Throughout the winter of 1942, Americans had read daily accounts of the amazing battle for Stalingrad, and *Life* magazine, echoing General MacArthur's sentiments, had placed the Russian Army 'in the top class of fighters'. The prominent New York lawyer Herbert Brownell agreed to serve on a 'Thanks to Russia' committee: a decade later, he would become the ferocious anti-Communist Attorney General of the United States.

Throughout the war, this tide of public goodwill was to provide support for a massive Lend–Lease programme. It included thousands of B29 bombers and tanks, more than 400,000 trucks, and 2000 locomotives. The Red Army marched in 13 million pairs of American boots and ate their way through five million tons of American food – half a pound of concentrated rations for every Soviet soldier for every day of the war. Most of this was to travel by ship, but for special assignments, the US and the Russians opened up a trans-Siberian air route to avoid the predations of German U-boats out in the Atlantic.

The staging post for flights in and out of the US was Gore Field at Great Falls, Montana. Here the American officer in charge, George Racy Jordan, was witness to just what travelled back and forth. In March 1943 he had made a forced entry into the hold of an aircraft departing for the Soviet Union and discovered fifty black suitcases. As protesting Soviet couriers stood by, he noted thousands of documents – maps, naval and shipping intelligence, hundreds of commercial catalogues and scientific journals and State Department documents. Amongst them he found a large map which he recorded as 'Oak Ridge–Manhattan Engineering Dept or District, I think it was'. He also copied down words he had never heard before: 'uranium 92 – neutron – proton ... Heavy-water hydrogen or deuterons'.

As well as this outflow of information, the route was apparently being used to bring new agents into the country. 'Planes were arriving regularly from Moscow with unidentified Russians aboard. I would see them jump off planes, hop over fences, and run for taxi cabs. They seemed to know in advance where they were headed, and how to get there.'

Other evidence, both from US and Soviet sources, confirmed this influx, and also the fact that Soviet representatives were allowed to move around arsenals, factories and proving grounds without restraint – 'they had only to make known their desires'. The Soviet espionage system was one of 'mass production', with thousands of agents in both the US and in Britain.

However, there was one man who resisted this pro-Russian sentiment: the military boss of the Manhattan Project, Leslie Groves. 'The only spot I know of that was distinctly anti-Russian at an early period was the Manhattan Project ... There was never any doubt about (our attitude) from some time along about October 1942.'

At Los Alamos, the day-to-day impact of this – the security checks, the passes, the censored mail – was infuriating. Edward Teller, however, also experienced the idiosyncracies, not to say idiocies, of the attempts to restrict all dealings with scientists in outside laboratories to a 'need-to-know' basis. One of the tasks he had been given on leaving the Theoretical Division, was to investigate an issue he himself had raised at the 1942 Berkeley conference: the way energy would be transported within the exploding bomb. It was crucial information when dealing with problems such as pre-detonation and Teller had focused on the so-called 'opacity' of the exploding bomb, a measure of how easily radiation such as X-rays could move through it. The calculations involved were repetitive, tedious but important, and Teller had farmed them out to Maria Mayer at Columbia.

At their early meetings, Teller and Mayer were accompanied by an army colonel who desperately tried to employ the 'need-to-know' principle. Instead of using the word 'uranium' would it not suffice to refer to element 92, he asked? But element 92 *was* uranium, he was told. In an attempt to justify his position, the colonel explained how bits of apparently harmless unclassified information, when juxtaposed, could give away a much bigger secret. In doing this, he described how a reporter, after first hearing that the Army was looking for a place to build reactors, had then learnt that a new construction site was opening up at Hanford – and put two and two together. After the meeting, Maria told Teller how interesting she had found the colonel's comments on Hanford – the only security breach at the meeting.

Although the colonel eventually relented over the use of the word 'uranium', Teller was prevented from telling Mayer what the opacity measurements were for. However, as soon as he mentioned the temperature at which he wanted the calculations made, he recalled, Maria caught her breath. For discretion's sake, however, she remained silent.

Edward Teller had first encountered Maria in Göttingen, where her father was a professor of paediatrics. She was to become the seventh generation of her family to achieve professorial status, and in Göttingen the family had become inextricably entwined with the university community. Like Oppenheimer, she had studied under Max Born, and he, along with James Franck, for whom Teller had worked, was a close family friend. At the age of twenty-four, she had received her doctorate, then married one of her fellow students, the American Joe Mayer. The couple had then moved to the US, to Baltimore, where Joe had a post in the Johns Hopkins chemistry department.

Throughout the 1930s, nepotism rules at Johns Hopkins had prevented Maria from working in anything other than a voluntary capacity but, as well as raising two children, she had produced excellent research, applying quantum mechanics to chemistry, as Teller had done. When he had arrived at nearby George Washington University, they had begun a loose scientific collaboration and the two couples had met socially. In addition to being an extremely able physicist, Teller had found her 'also very beautiful. Slender and blonde, she had a natural delicacy and grace as well as considerable strength of mind.' The pair had begun corresponding in 1939, when Joe Mayer had taken a post at Columbia and Maria had moved with him to New York, and continued to do so for nearly twenty years; but although Maria kept most, if not all, of Edward's letters to her, only two of hers to him survive.

'I had always liked Maria Mayer, but our friendship during the war years grew deeper,' Teller recalled. 'Maria and I not only shared common concerns for friends and families about whom we had no news but also felt a common nostalgia for the music and literature of Germany . . . Maria hated Nazis but loved Germany, a feeling that I shared.' At this time her husband, Joe, was posted to the South Pacific, while Edward's role in liaising with outside laboratories often brought him to New York, and the two were consequently able to see each other frequently. 'On one early occasion, during a visit to the Ureys, Maria wistfully mentioned that she had not seen an opera for a long time. As it happens, I, too, had missed attending the opera because Mici does not enjoy it. So I invited Maria to a performance of a light opera, *Die Fledermaus*. On other occasions, we attended some of Wagner's long, involved operas.'

So just what was the nature of their relationship? The fact that they were able to meet freely away from their families and obviously cared a great deal for one another, suggests an affair. However, Edward's son, Paul, is certain that this was not the case. 'Maria and my father were clearly very important emotionally to one another, but very recently he told me, very explicitly, that he never had a physical relationship with her. This is what he told me at ninety-three going on ninety-four. I didn't solicit it. He's not the kind of person who would have told me that if it's not true.'

This view is shared by others who knew Edward and Mici well. While their marriage was often a turbulent one, Judith Shoolery, Edward's book editor for more than twenty years, believes he was entirely faithful to her. 'Edward didn't like womanisers. Both Oppenheimer and Kennedy were womanisers,* and I'm not sure what the quality is, but so many of the people at the laboratory he was not in tune with were womanisers. It's a symptom of something, an untrustworthiness, a lack of commitment to one's word which he just hated – instinctively. I saw in him a deep respect, a deep observation of Judaic law.'

Indeed, whether or not people believe the pair had a physical affair, there is a consensus questioning whether it matters: Maria was so important to Edward, with an intellect that matched his, with absolutely parallel scientific interests, as well as cultural ones, and he always turned to her in difficult times, pouring out his self doubt and fears to her.

Something, it was thought, he would have found much more difficult with Mici. The letters Edward wrote are open and honest, exhibiting a complete trust in the recipient. They speak for themselves and I shall quote from them at appropriate times later.

Teller may have responded with irritation to the everyday impact of the security measures at Los Alamos, but his feelings about the Russians as a possible future danger were hardening. Up to this time, those feelings had been accumulating piecemeal, influenced by everything from his early-childhood experiences to recent philosophical discussions with colleagues.

These feelings were further focused by the hardships his Russian colleagues experienced. The fate of Lev Landau, in particular, shocked him. Teller had experienced Landau's sharply critical tongue when they were students together at Leipzig and Göttingen and on his return to Russia this had landed him in serious trouble. During Stalin's Great Purge in 1937 and 1938, somewhere between 7 and 8 million people were arrested and imprisoned in camps, and an unknown number executed. The scientific community suffered as much as any, and many of the country's leading scientists were arrested and accused of fantastic plots against the state. At Landau's own laboratory, the Ukrainian Physics Technical Institute, where he was head of theoretical physics, the impact was disastrous. Alexander Weissberg, in his classic book on the Great Purge, *The Accused*, described the situation in the Institute's eight departments by 1938: 'The head of the laboratory for Crystallography, Obremov, is under arrest, and so is the head of the low temperatures laboratory, Shubnikov . . . The head of the laboratory for atom splitting, Leipunsky, is under arrest, and so is the head of the X-ray department, Gorsky, the head of the department for theoretical physics, Landau, and . . . myself.'

Of those arrested, Shubnikov and Gorsky were shot, Leipinsky was released after a month, Obremov after three years. Weissberg himself was handed over to the Gestapo in 1940. Landau had been arrested in April 1938, but the country's foremost physicist, Peter Kapitza, had immediately written directly to Stalin, carefully pleading Landau's case as a uniquely talented if difficult man – 'a tease and a bully', who 'enjoys looking for mistakes in others'.

Because of Kapitza's efforts, Landau was released after exactly a year, but during his imprisonment his health had been broken. As to the institute itself, it was weakened immeasurably, by fear as much as by the loss of individual talent. On the eve of the nuclear age, Stalin's regime had wrecked one of the country's most important institutes, the wellspring of the nation's technical renaissance.

Landau's treatment added a personal dimension to Teller's view of Stalin's administration, and of communist Russia – one more element to be wedded into a grand impression of corruption, opportunism and ruthlessness. Then, during his first summer at Los Alamos, he read Arthur Koestler's *Darkness at Noon*. Koestler was also Hungarian – though the two men had never met – and in the late twenties and early thirties, had been a Soviet organiser-agitator, working across Europe. He experienced the full horror of Stalin's rule at first hand, and by the end of the Spanish Civil War had become a committed anti-communist. His novel takes place during the Great Purge itself – indeed Koestler used the trial of Weissberg's wife as one of his sources. The central character, N. A. Rubashov, is a synthesis of the lives of a number of people who were victims of the Purge and who were known personally to the author. The narrative follows Rubashov, the remnant of an 'old guard', as he is arrested, imprisoned, interrogated, broken to make a confession and then tried. Koestler's psychological insight, particularly into the intricacies of the struggle between Rubashov and his interrogators, Ivanov and Gletkin, is remarkable, and he develops the arguments for communism as effectively as those against it.

For a considerable part of the book [Teller wrote], I was not sure whether the accused or the interrogator was the traitor to the cause. All that was clear was that no compromise could exist ... By the end of the book, no ambiguity remains. The accused's self-accusations are a lie, a sacrifice to the cause of communism.

Darkness at Noon brought together and crystallised the objections to the methods of control used by Russian communism, which had been forming and accumulating in my mind for fifteen years.

This extract from near the end of the book conveys something of its flavour. At the end of the accused, Rubashov's, protracted interrogation, his interrogator, Gletkin, closes his dossier and speaks directly about the mission of the Party.

'Not to perish,' sounded Gletkin's voice. 'The bulwark must be held, at any price and with any sacrifice. The leader of the Party recognised this principle with unrivalled clearsightedness, and has consistently applied it. The policy of the International had to be subordinated to our national policy. Whoever did not understand this necessity had to be destroyed. Whole sets of our best functionaries in Europe had to be physically liquidated. We did not recoil from crushing our organisations abroad when the interest of the Party required it. We did not recoil from cooperation with the police of reactionary countries in order to suppress revolutionary movements which came at the wrong moment. We did not recoil from betraying our friends and compromising with our enemies, in order to preserve the Bastion. That was the task which history had given us, the representatives of the first victorious revolution. The short sighted, the aesthetes, the moralists did not understand. But the leader of the Revolution understood that all depended on one thing: be the better stayer.'

No compromise. This was, of course, Teller's assessment of the book's message looking back over some sixty years, and his retrospectively applied statement of intent must carry at least an element of self-justification for his own behaviour over the intervening years. However, even at this early stage, he was a relative veteran who was developing an international perspective on the arms race and saw it as 'continuing to have great importance for the future. Unlike many in the scientific community at Los Alamos,' he added, 'I had become convinced that the problems connected with Communist Russia were very great.'

Teller made this pessimistic assessment without any knowledge of what the Russians were doing to prepare for the post-war situation. In fact, the information obtained from espionage, which was then flowing into Moscow, was so copious that it threatened to swamp those receiving it. Igor Kurchatov, the head of the embryonic Russian bomb programme, reported that the information so far outstripped Soviet knowledge that it was 'impossible to formulate pertinent questions that would require additional information'.

What they were receiving was, however, of the best quality and this fact enabled the Russian scientists to bypass a number of the problems that had consumed so much of the effort at Los Alamos and also to avoid some of the blind alleys down which they had wandered. Exactly how many spies the Russians had at the New Mexico laboratory is not known. Three have been uncovered, but there are thought to be others. The most important, however, was Klaus Fuchs.

Fuchs was German, the son of a Lutheran pastor who, as a student, had been active against fascism and the Nazis. He had only narrowly avoided arrest after the Reichstag fire in 1933, escaping first to Paris and eventually arriving in Britain. At the start of the war he had been interned, first on the Isle of Man and then in Canada. In late 1940 he had been brought back to Britain to work with Rudolf Peierls at Birmingham. Some six months later, he had been cleared for work on the atomic bomb, and had immediately decided to inform on the project for the Russians. For the next two and a half years he had kept up a continuous and comprehensive flow of information on all aspects of the Allied bomb programme – one that, as early as 1942, had allowed the Russians to know the contents of the MAUD report as soon as a draft was available, and whose value, according to Kurchatov, could not be overestimated.

On his arrival in the US in late 1943, he had been given a new contact, Harry Gold, to whom he was to pass on his material. Gold had offered Fuchs money for his efforts, but had received the frostiest of refusals. To Gold, Fuchs's dedication was always noble. 'When the Nazis had put a price on his head he barely managed to escape with his life to England . . . For a man of such convictions who fought the horror of Fascism at the risk of his life, I cannot help but express my admiration.'

When Teller had refused to do further work on the implosion method, Hans Bethe had requested Peierls's transfer to Los Alamos to take over Teller's role. Peierls had agreed to come – so long as he was allowed to bring along his two assistants, one of whom was Klaus Fuchs.

With D-day and the opening of the second front in Normandy, in June 1944, the Germans were in full retreat. Yet even at this late stage it was still not known how far they had progressed with their bomb project. Thus it was that Colonel Boris Pash, the security officer who, a year earlier, had interviewed Oppenheimer about the 'Chevalier incident', found himself in charge of the Alsos Mission,* a mixed unit of security men and scientists chasing across Europe in the vanguard of the advancing army, looking for evidence of the Germans' progress. It was an indication of how little real knowledge they had that just before D-day Pash and his colleagues were speculating that the great bunkers the Germans had built along the northern French coast might have been constructed for rockets with nuclear warheads.

It was not until November 1944, when the Allies took Strasbourg, that they came upon definitive evidence of the state of German research. This was where Teller's old university friend Carl von Weizsäcker had been working. He had made good his escape, but from the evidence left behind, and from that of his colleagues who were captured, it became clear that, ever since 1942, German research had stagnated. They were not even certain that an explosive chain reaction was possible.

Shortly after making these discoveries, the scientific head of Alsos, the Dutch émigré Sam Goudsmit, found himself discussing their findings with one of his military colleagues: 'If the Germans don't have the bomb then we won't need to use ours,' Goudsmit said. 'You don't know Groves,' was the reply; 'if we have such a weapon, then we'll use it.'

By early 1945 the designs of the two Los Alamos bombs were well advanced. The gun-method design, nicknamed Little Boy, was very much dependent on the separation of a sufficient quantity of U-235, but it looked as if there would be enough for at least one bomb. As to the technically more complex and innovative implosion bomb, nicknamed Fat Man, there were still significant design problems and it was becoming increasingly clear that there would need to be a test before it was used. Early in the new year Oppenheimer had worked out the project's progress day by day and the test was scheduled for mid-July.

The threat of a German bomb had been the prime motive for the work many of the scientists, particularly the leaders and pioneers, were doing at Los Alamos, but few of them, even those as senior as Hans Bethe, knew anything of the German failure until March 1945 – nine months after the Alsos Mission's discovery. Furthermore, even fewer of them knew that, ever since May 1943, the high-level Military Policy Committee, whose members included Groves and Vannevar Bush, had been considering Japan as a possible target for nuclear attack. Therefore, when the changes in strategy began to emerge in the spring of 1945, there was shock and there was anger. The moral justification for what they had been doing had been whisked away without their knowing, and the fruits of what many saw as a Faustian bargain were being taken out of their hands.

The scientists at Los Alamos were mostly too busy to give the issue much thought, but at Chicago's Met Lab, their work was winding down and they did have the time. Their feelings were fomented in large part by Szilard, who lobbied to prevent the direct use of the weapon on Japan, certainly without a warning. In this he had again co-opted the support of Einstein and, while his efforts to meet the new President had been deflected, he had met with James Byrnes, his Secretary-of-State-in-waiting. They had not got on, Byrnes complaining that Szilard had told him precisely what to do.

In the meantime Szilard had also drawn up a petition he hoped to pass to the President and begun circulating it amongst all the laboratories on the project. The recipient he chose at Los Alamos was Edward Teller. In the letter he sent, Szilard sidestepped all the complex practical and political issues and homed in on his central moral concern: 'I personally feel,' he wrote, 'that it could be a matter of importance if a large number of scientists who have worked in this field went clearly and unmistakably on record as to their opposition on moral grounds to the use of these bombs in the present phase of the war.' Up to that time Teller had thought little about the alternatives to using the bomb, but the document, drawn up by two of his friends, Szilard and his old teacher James Franck, made 'good sense' to him. Before distributing it, however, he felt he had to talk to Oppenheimer.

For the past three months Oppenheimer had been centrally involved not only in the final stages of preparing for the Trinity test but also in detailed discussions about the use of the bomb. On 12 April 1945, a weak and exhausted President Roosevelt died at Warm Springs, Georgia, and the new President, Harry S. Truman, although he had been vice-president for several months, did not even know of the existence of the atomic bomb.

It was Groves, together with the seventy-seven-year-old Secretary of War Henry Stimson, who inducted him into the secret of the new weapon. Groves, whom one journalist described as a 'blackbelt bureaucrat', went straight for the central strategic issue: that if the weapon could be used to avoid an invasion of the main Japanese islands, more than a million lives could be saved.

It was an impressive but questionable justification for the use of the weapon, which took no account of the disagreement over how many lives would be saved. For instance, General Marshall, the Army Chief of Staff, estimated that the number of lives lost during a complete invasion might be as few, relatively, as forty thousand. During the meeting, Groves and Stimson proposed an 'Interim Committee' to consider in detail all those issues surrounding the country's immediate and future weapons policy and Truman, no doubt with some relief that the immediate decision was taken out of his hands, agreed. From that point onwards, the main decision about the future of the bomb was moved one step away from a President who, in many ways, believed its use to be a fait accompli.

As only two of the committee's members were scientists, it was decided to set up a scientific panel. Chaired by Oppenheimer, its membership included the familiar names of Fermi, Lawrence and Compton, and on 31 May the panel and the Interim Committee held their first joint meeting.

The President was represented at the meeting by James Byrnes and the discussion initially centred on various technical and practical aspects of the atomic weapon programme. When they turned to best estimates as to yields, Byrnes, at this his first meeting, was shocked to hear them talking blithely about a Super bomb with a staggering potential yield of 100 million tons of TNT. Furthermore, Oppenheimer and the Chief of Staff, General George C. Marshall, had suggested that they might approach the Russians with a view to cooperation. It was a suggestion that represented a significant step towards fostering international cooperation and control, but to Byrnes the thought of giving such weapons away was unthinkable. In his first major intervention he vetoed the proposal. He was sufficiently forceful for the meeting notes to record that, after his intervention, it was agreed that the US 'stay ahead and at the same time make every effort to better our political relations with Russia'. Teller's Super had exhibited this galvanising property in the past and would do so many times in the future.

It was over lunch that some of the group discussed how the bomb was to be used against Japan – whether a demonstration was possible rather than its direct use involving great loss of life. Byrnes's notes recorded the salient points: the possibility that, if forewarned, the Japanese would use Allied prisoners of war as a human shield, and that 'the experts' warned that a static test on a tower 'would not be conclusive proof that a bomb would explode when dropped from an airplane'. Moreover, such a test might not be impressive enough for the war-hardened Japanese.

The discussion lasted probably no more than ten minutes – ten minutes to discuss moral issues that were plaguing the conscience, then as ever since, of many who had lent their skills to creating the world's first weapon of mass destruction. Those ten minutes led to a resolution that the bomb should be used without warning, and on a military target surrounded by civilian population so as to 'make a profound psychological impression on as many of the inhabitants as possible'. For anyone attempting either to introduce a new philosophy into the debate on post-war international relations, or to alter the plans for an attack on Japan without warning, this meeting was a sadly missed opportunity.

When, after the meeting, Arthur Compton returned to the Met Lab in Chicago, he found it in uproar – uproar largely fomented by Szilard. In an attempt to ease the growing tensions Compton set up a series of committees to study and report on the implications of the new weapon. The most important of these was chaired by James Franck, Teller's professor at Göttingen. Within a week Franck produced a thoughtful study clearly stating that an unannounced attack on Japan was inadvisable on many grounds – moral, political and diplomatic – and would 'precipitate the race for armaments'.

Oppenheimer had attended another meeting of the panel, held as a direct response to the uproar in Chicago. The panel had carefully considered the reports Compton had arranged but still concluded that they could propose 'no acceptable alternative to military use'. They had ended by stating that they had no claim to special competence in solving the political, social and military problems that were presented by the advent of atomic power.

It was at this time that Teller brought Szilard's petition to Oppenheimer. His accounts of their meeting varied but according to an early version Oppenheimer told him 'in a polite and convincing way, that he thought it improper for a scientist to use his prestige as a platform for political announcements . . . I did not circulate Szilard's petition. Today I regret that I did not,' said Teller.

However, his last account coloured Oppenheimer's role in a much more antipathetic fashion:

> Oppie's reaction when I asked about circulating the petition took me completely by surprise. He began talking about Szilard and Papa Franck in a way that, until then, he had reserved for General Groves. Then he asked, 'What do they know about Japanese psychology? How can they judge the way to end the war?' I had no ready answers. He went on, 'The leaders in Washington, not individuals who happened to work on the bomb project, our political leaders – who include men like George Marshall, a man of great humanity as well as intellect – those are the people who should make such decisions.'
>
> Our conversation was brief. His talking so harshly about my close friends and his impatience and vehemence greatly distressed me. But I readily accepted his decision and felt relief at not having to participate in the difficult judgements to be made.

As I have pointed out earlier, Teller's most recent accounts, written long after the event and, in this case, long after the death of the only other participant, do, with their additional detail, beg the question of how much revision and embellishment has taken place. This is particularly true in the case of Oppenheimer, where the need to justify his later actions was obviously still of considerable concern to him half a century later when he wrote his memoirs. Immediately after the meeting Teller wrote to Szilard explaining his reasons for not circulating the petition and he then sent a copy of this to Oppenheimer with a covering letter. At the core of the letter Teller wrote: 'The more decisive a weapon is the more surely it will be used in any real conflict and no agreements will help.

'Our only hope is in getting the facts of our results before the people. This might help to convince everybody that the next war could be fatal. For this purpose actual combat use might even be the best thing.'

Teller often expressed anger and regret that he took Oppenheimer's advice. He argued that Oppenheimer misled him. Yet by proposing that 'combat use might be the best thing' he took Oppenheimer's position at the science panel discussions a step further. At those meetings Oppenheimer was certainly one of those who failed to come up with a reasonable alternative to combat use, but he never actually advocated it. Even though he was clearly out of sympathy with the views of his friends in Chicago, however, Teller ended the letter to Szilard with a plea for support: 'All this may seem to you quite wrong. I would be glad if you showed this letter to Eugene [Wigner] and to Franck who seem to agree with you rather than with me. I should like to have the advice of all of you whether you think it is a crime to continue to work. But I feel that I should do the wrong thing if I tried to say how to tie the little toe of the ghost to the bottle from which we just helped it to escape.'

It was a letter written by someone who was afraid that his belief in the inevitability of progress, and his own personal ambitions for progress in this area, would isolate him. The views he expressed to Szilard can easily be seen as a symptom of the way his position was diverging from that of most other scientists who had joined the project simply to meet the German threat.

'For them, or most of them,' Teller said, 'the point that it would be wrong to continue [further nuclear weapon research] turned into a motivation, which did not take place in my case. What you read here [the letter to Szilard] is the very beginning of that controversy.'

It had been decided some months previously that Fat Man, the plutonium-fuelled implosion bomb, was so revolutionary that it must be tested before use. The test, code-named Trinity by Oppenheimer in memory of Jean Tatlock, was set for 16 July. This timing was linked to the final Allied leaders' conference at Potsdam, where the announcement of the successful test, Truman hoped, would provide a really invaluable lever in negotiations with the Russians. It was crucial that there should be no delay.

In the final weeks leading up to the test Teller's group were drawn into the immediate preparations when the possibility of atmospheric ignition was revived by Enrico Fermi. His team went to work on the calculations, but, as with all such projects before the introduction of computers, these involved simplifying assumptions. Time after time they came up with negative results, but Fermi remained unhappy about their assumptions. He also worried whether there were undiscovered phenomena that, under the novel conditions of extreme heat, might lead to unexpected disaster. In the last days before the test, Teller searched for and tested out hypotheses for such phenomena on anyone who would listen. Nothing sinister emerged, but still this did not end the stories that circulated in the mess halls of the Trinity base camp on hot evenings as the men sat around reading, drinking and playing poker. The tales passed from the scientists, to technicians, to eavesdropping MPs. The authorities were so concerned that they arranged for psychiatrists at Oak Ridge to be on standby ready to fly down to Trinity should any form of panic break out. Some men, identified as either rumour-mongers or as being particularly susceptible, were returned to Los Alamos.

On the evening of 15 July Edward Teller joined the other scientists at the laboratory who were to be bused out to Compania Hill, some twenty miles from Ground Zero, to witness the test explosion. The lab generator had failed and he stood out in the dark waiting for the bus with Bob Serber, Oppenheimer's devoted aide. They discussed the warning they had all received: to be aware of rattlesnakes. When Teller asked how he was dealing with the danger Serber said he was intending to 'take along a bottle of whiskey'. Their discussion moved on to the possibilities for atmospheric ignition and Teller tested out the arguments and counter-arguments he had considered. He ended by asking Serber what he should do about such possibilities. 'Take a second bottle of whiskey', was the reply.

The weather that night was terrible and, in the early hours, a delay and even postponement seemed inevitable. Again the rumour mill around the test site had picked up on a conversation between senior scientists about atmospheric ignition – how quickly the reaction would go and how far it would spread. The situation was reported back to Oppenheimer by an angry Ken Bainbridge, the test director. The teams and their leaders were tired and close to breaking point. A further delay had to be avoided at all costs. Around the site, the scientists had been passing the time organising a betting pool on the magnitude of the explosion. Hans Bethe's estimate of 5000 tons of TNT was the official one sent to Washington, but on site they varied from zero upwards. Oppenheimer had selected a modest 300 tons, Isidor Rabi 18,000 tons and Teller had plumped for an optimistic 40,000 tons.

As the weather worsened, Enrico Fermi sought out Oppenheimer to urge for a postponement. He had done some calculations on fallout and he was fearful that a sudden shift in the wind along with a rain shower could deluge the test site itself. It was at this moment that Groves decided to take Oppenheimer to the main control bunker, away from the highly charged atmosphere at base camp. 'There were pools of water everywhere around the bunker,' recalled Dick Watts, one of the technicians, 'and I can remember watching Oppenheimer and Groves striding up and down, dodging these pools, talking to each other intensely … trying to decide, well did we, or didn't we, shall we, or shan't we, fire this thing off?'

In the pouring rain at Compania Hill, Edward Teller was passing round the sun lotion he had brought with him:

> We were all lying on the ground, supposedly with our backs turned to the explosion. But I had decided to disobey that instruction and instead looked straight at the bomb. I was wearing the welder's glasses that we had been given so that the light from the bomb would not damage our eyes … I put on dark glasses under the welder's glasses, rubbed some ointment on my face to prevent sunburn from the radiation, and pulled on thick gloves to press the welding glasses to my face to prevent light from entering at the sides.

At 4 a.m. the rain had stopped. A short time later, after a further forecast, it was agreed the test should go ahead. At 5.10 a.m. at Compania Hill a voice echoed across the desert from the loudspeakers set up on poles above the scientists' heads: 'It is now zero minus twenty minutes.' The countdown, conducted by Sam Allison, had begun. As his voice continued its monotonous tally, a shade of pink appeared over the mountains in the east.

'We all listened anxiously as the broadcast of the final countdown started; but, for whatever reason, the transmission ended at minus five seconds,' Teller recalled:

> For the last five seconds, we all lay there, quietly, waiting for what seemed an eternity, wondering whether the bomb had failed or had been delayed once again. Then at last I saw a faint point of light that appeared to divide into three horizontal points … As the question 'Is this all?' flashed through my mind, I remembered my extra protection. As the luminous points faded, I lifted my right hand to admit a little light under the welder's glasses. It was as if I had pulled open the curtain in a dark room and broad daylight streamed in.

A few seconds later all the scientists were standing, gazing open-mouthed at the brilliance. A loud crack some two minutes later, the sound wave rolling across the desert from twenty miles away, caught them by surprise, but minute after minute they watched, with hardly a word spoken. 'The condensation cloud produced by the fireball changed shape as it was blown in several directions by the winds. Eventually it became a many-mile-long question mark. We returned to the bases with hardly a word. We knew that the next nuclear explosion would not be an experiment.'

Some thirty miles away, in the control bunker to the south of Ground Zero, they had been aware of the dazzling brightness coming from the bunker entrance facing away from the explosion. 'A few people laughed, a few people cried. Most people were silent,' Oppenheimer recalled. 'I remembered the line from the Hindu scripture, the *Bhagavad-Ghita:* Vishnu is trying to persuade the Prince that he should do his duty: "Now I am become death, the shatterer of worlds." '

As he stood outside the bunker, Ken Bainbridge, flushed with the success of his test, came up to grasp his hand. 'Oppie,' he said, 'now we're all sons of bitches.'

News of the successful test reached President Truman in stages between 16 and 21 July, each new document filling in more details of that extraordinary event. Then, on 21 July, Groves's full report of the test arrived and it made an enormous impact not only on Truman but on Churchill as well. Truman now knew that he no longer needed the Russians to help in finishing the war in the Far East. They might decide to join in the war against Japan of their own accord, as Stalin had promised, but Truman was 'still hoping for time, believing after the atomic bomb Japan [would] surrender and Russia [would] not get in so much on the kill ...' He was determined to keep Stalin's imperialist ambitions in the Far East in check, and avoid the conflicts already brewing around the Red Army's advances in Eastern Europe. A speedy and successful outcome to the first use of the bomb was now of the greatest importance.

Eventually Truman did decide to tell Stalin about the bomb and casually mentioned to him after the planning session on 24 July, 'that we had a new weapon of unusual destructive force'. Stalin's reply was unexpectedly brief. He was glad to hear it and hoped that Truman would make 'good use of it against the Japanese'. Truman was puzzled by this cool reception from Stalin. However, by that time Klaus Fuchs and the Soviet Union's other spies had already provided an accurate and detailed picture of the weapon that had so recently been tested.

9
The Legacy of Hiroshima

At 02.45 hours on 6 August, the *Enola Gay*, a B29 bomber specially adapted to carry the uranium bomb known as Little Boy, lined up for take-off from the air base on Tinian Island, near Guam in the Pacific. It took almost the entire length of the runway for the heavily laden aircraft to lumber into the air to begin its flight to Hiroshima.

Shortly after 8 a.m. local time, air-defence spotters in Hiroshima observed a flight of two or three B29s approach, but, as this was the fourth occasion on which enemy planes had been observed over the city already that day, no further alert was sounded. The flight was mentioned on the local radio as a likely reconnaissance mission, but most people continued on their way to work. At 8.15 a.m. Major James Ferebee dropped the bomb. With its release the *Enola Gay* lurched up into the air and at the same time the pilot, Colonel Paul Tibbetts, threw the plane into a steep banking dive away from the drop point.

Forty-three seconds later there was a brilliant flash, lighting up the interior of the plane, and a few seconds after that the plane was severely buffeted by two shock waves. However, it was undamaged and remained on course back to Tinian. As it left the scene, the tail gunner, Robert Caron, had the best view of the city below and gave a running commentary on the intercom:

The mushroom itself was a spectacular sight, a bubbling mass of purple-gray smoke and you could see it had a red core in it and everything was burning inside ... I saw fires springing up in different places, like flames shooting up on a bed of coals. I was asked to count them. I said, 'Count them?' Hell, I gave up when there were about fifteen, they were coming up too fast to count. I can still see it – that mushroom and that turbulent mass – it looked like lava or molasses covering the whole city

... with fires starting up all over, so pretty soon it was hard to see anything because of the smoke.

Although the crew were all deeply shocked by this awesome spectacle below them and the mushroom cloud stretching up to 30,000 feet, it is difficult to conceive of any imagining that could have matched what was happening on the ground:

> I just could not understand why our surroundings had changed so greatly in one instant ... I thought it might have been something which had nothing to do with the war – the collapse of the earth which it is said would take place at the end of the world, and which I had read about as a child ...

> There were dead bodies everywhere. There was practically no room for me to put my feet on the floor. At that time I couldn't figure out the reason why all these people were suffering or what illness it was that had struck them down ... There was no light at all, and we were just like sleep walkers ...

> My immediate thought was that this was like the hell I had always read about. I had never seen anything which resembled it before. But I thought that should there be a hell, this was it.

The number of people killed at Hiroshima either immediately or over a period of time will probably never be known accurately. The extreme confusion at the time made all assessments difficult, and each assessment could not help but be affected by emotion. The official figure is usually given as 78,000 but there are other figures. The city of Hiroshima estimates 200,000, a figure representing between 25 and 50 per cent of the whole daytime population. The centre of the city had been levelled and more than 60,000 buildings destroyed. Shortly after the bombing, great fires swept across the city.

Back at Los Alamos, Oppenheimer had received a message – bland and formal that had been flashed back from the *Enola Gay* by Captain 'Deke' Parsons, some fifteen minutes after the explosion: 'Clear cut results exceeding T R tests in visible effects,' it read, 'and in all respects successful.' Oppenheimer called the whole staff of the laboratory together in one of the auditoria.

'He entered that meeting like a prize fighter,' recalled one scientist. 'As he walked through the hall there were cheers and shouts and applause all round and he acknowledged them with the fighters salute – clasping his hands together above his head as he came to the podium.'

That same day, 6 August, President Truman warned Japan: surrender unconditionally or 'expect a reign of ruin from the air, the like of which has never been seen on this earth'. But Japan no longer possessed the ability to take quick decisions. The news that a single bomb had obliterated Hiroshima only reached Tokyo the following morning and then it seemed so incredible that an investigating team was sent to find out what had happened. The next day, 8 August, Russia declared war on Japan. On 9 August unfavourable weather forecasts prompted the local commander on Tinian to bring forward the second raid planned on the city of Kokura by fully two days. But with cloud obscuring the target, the flight was diverted and nine hours after take off, at 12.02 p.m., the second nuclear weapon, a duplicate of the plutonium implosion bomb tested at Trinity, was dropped on Nagasaki.

In the early hours of the following day, 10 August, the Japanese Premier Suzuki requested the Emperor's decision 'to accept the Allied Proclamation on the basis outlined by the Foreign Minister'. They still looked to the Americans to fulfil the one condition that had blighted previous negotiations – that the survival of the Emperor and his dynasty be assured – and this time they received it. If this guarantee had been given earlier, then there is the possibility, which even Secretary of War Stimson recognised, that neither of the bombs dropped on Hiroshima and Nagasaki would have been necessary. With the future of its imperial dynasty assured, Japan might have surrendered anyway.

In Russia, when news of the bombings broke, Stalin summoned the head of the Soviet bomb project, Igor Kurchatov, and castigated him for not demanding enough in the way of resources to ensure the maximum of progress. 'If the baby doesn't cry, the mother doesn't know what he needs,' he said irritably. 'Ask for anything you need. There will be no refusals.'

On the evening of 14 August, Edward and Mici were at home at Los Alamos when, as Edward described it,

> the mountain quiet was shattered suddenly . . . A wild racket broke upon the serenity of Jemez Mesa in a single instant, as if by a pre-arranged signal. Sirens whined, bells rang. Dozens of automobile horns blasted. I thought a giant traffic jam somehow had developed in the quiet streets. The cacophony was completed by the sounds of people running and shouting at each other. I rushed from my apartment to investigate the commotion, and soon discovered its cause: Japan had surrendered. The war was over.

The jubilant mood at the laboratory was buoyed by the reaction of a grateful nation, which believed it had been saved from the continuation of a grim and costly war. The President himself praised the laboratory publicly. 'What has been done,' he declared, 'is the greatest achievement of organised science in history. It was done under high pressure and without failure. We have spent two billion dollars on the greatest scientific gamble in history – and won.'

But the happy atmosphere could not last. More than anything, the celebrations at Los Alamos were an expression of profound relief at the achievement of a monumental task. At the same time, however, in some quarters there was a realisation of the awfulness of what they had done. Oppenheimer was one of the first to voice doubts about what had been, in isolation, such a magnificent achievement. He admitted to one of the reporters milling around the laboratory that he was 'a little scared of what I have made'. However, he then went on to say: 'A scientist cannot hold back progress because of fears of what the world will do with his discoveries.'

Since 1944 about half the laboratory had been staffed by members of the Special Engineering Detachment, technicians who were enlisted men, and by the Navy and other military personnel, and all of these just wanted to go home as soon as possible. That attitude was shared not only by junior scientists but by some of the senior ones as well.

'We all felt, like the soldiers, we had done our duty,' Hans Bethe recalls, 'and that we deserved to return to the type of work that we had chosen as our life's career, the pursuit of popular science and teaching ... it is in no way surprising that most of the [scientists] preferred the free interchange of ideas ... which goes with pure, non-secret research. However, it was not obvious in 1946 that there was any need for a large effort on Atomic weapons in peacetime.' Within months, Bethe would return to Cornell, Fermi to Chicago, and Oppenheimer himself would continue his associations with Berkeley and take a new post at the California Institute of Technology.

Teller could also have returned to Chicago, but since July he had been able to involve some of his colleagues, now freed from work on the fission bomb, in work on the Super project, and he believed he was making good progress. For him, the Super was still, as it long had been, both a strategic scientific goal and the focus of his own personal ambition. He was both intimidated by and jealous of Oppenheimer and his success at Los Alamos, and he saw the Super as his way to achieve parallel success. Thus he was deeply disappointed when, the day after VJ-day, 'Oppenheimer came to my office to tell me that "with the war over, there is no reason to continue the work on the hydrogen bomb". His statement was unexpected. It was also final. There was no way I could argue; no way I could change Oppenheimer's mind.'

Oppenheimer would later echo these sentiments in Washington, where he communicated to senior officials and politicians his strong belief that scientists as a group opposed 'not merely a super bomb but any bomb' as being 'against the dictates of their hearts and spirits'. When he then met Truman he had offended the President greatly when he declared that he 'had blood on his hands'. Truman commented afterwards that all Oppenheimer had done was make the bomb – it was he who fired it off.

For Teller, though, a continuation of the programme was vital – now more than ever. Hans Bethe recalled that it was at this time, soon after VJ-day, that he first heard him express his fears for the future. 'In this conversation,' Bethe remembered, 'for the first time in my recollection, he expressed himself as terribly pessimistic about relations with Russia. He was terribly anti-communist, terribly anti-Russian ... Teller said we had to continue research on nuclear weapons ... it was really wrong of all of us to want to leave. The war was not over and Russia was just as dangerous an enemy as Germany had been.'

At this point Teller was still very much the exception in looking beyond the immediate conflict, something amply demonstrated by the report he was asked to prepare for the Navy by his one-time student Stephen Brunauer. In it he already foresaw new concepts in strategic thinking:

> I assume that atomic bombs will be delivered by long range missiles. The natural type of defense against such missiles seems to be jet-propelled homing projectiles...
>
> It seems to me that underwater units of not too big a size will be the appropriate ships of the future. Using small piles [reactors] of power producing units, these submarines can carry with them the energy needed for an indefinite period...
>
> Incoming missiles could be detected a thousand miles away at sea. Either home bases could be notified or defensive missiles could be launched from the naval units...
>
> Such developments are likely to be so great that in the end they will probably thwart any method of defense. This, however, we cannot know for sure. Our only hope of facing future dangers in a realistic way is to explore the possibilities of improving the atomic bombs...*

This, in essence, was already a blueprint for an arms race, an arms race in which Teller had identified the main competitor. Yet back at Los Alamos, by mid-September 1945, almost all work on weapons projects had ground to a halt and more than half the staff had already left. Indeed there was a school of thought which held that Los Alamos should become a monument, a ghost laboratory, and that all work on the military use of atomic energy should now cease. Teller actually went about telling colleagues that Oppenheimer had voiced the opinion to him that the Mesa was best left to the Indians.

For those who remained it was a miserable time. The weather was terrible, the water supply froze up and families were reduced to using freezing water brought in by tanker. There was also great uncertainty about the future. Norris Bradbury, the Berkeley-trained physicist who had worked on the assembly of the atomic bomb and who was taking over from Oppenheimer as director, immediately set about trying to arrest the rot. He met with Teller and offered him the post of Head of the Theoretical Division, the post that Teller believed he should have been offered when the laboratory had first opened. He was in no mood for compromise and so set quite specific conditions for his taking the post, demanding 'a great effort to build a hydrogen bomb in the shortest possible time or develop new methods of fission explosives and speed progress by at least a dozen tests a year.'

These were conditions completely at odds with the mood and the political reality of the time. At that moment Los Alamos was floating in a political limbo, no longer the responsibility of the Army and with its future hanging on the outcome of a Congressional vote. Whatever Bradbury might have wanted personally, he simply could not have made the commitment Teller was demanding. His refusal to meet those demands, however, was to mark the beginning of a slow-burning feud between the two men that lasted for a quarter of a century. 'Unquestionably, had Bradbury indicated even the slightest inclination of being willing to fight for a thorough and ongoing programme of development,' Teller wrote, 'I would have stayed on at Los Alamos – not because I particularly wanted to, but because no one else was willing. But Bradbury maintained a cautious approach, then and throughout his career.'

However, before he made his final decision, Teller decided to talk to Oppenheimer. He offered to stay on at Los Alamos provided that Oppenheimer himself used his influence in support of further work on the Super or on new fission designs. Oppenheimer refused: 'I neither can nor will do so.' Teller then told him that he would be leaving the laboratory.

That evening William 'Deke' Parsons was giving a large party and, according to Teller, during the evening Oppenheimer 'drifted over to ask with a smile, "Now that you've decided to go to Chicago, don't you feel better?" I didn't feel better, and I said so, and added that I felt our wartime work was only a beginning. Oppie closed the subject. "We have done a wonderful job here. It will be many years before anyone can improve on our work in any way." '

Oppenheimer was in a strange state of mind, on the one hand justly proud of the laboratory's achievements, on the other darkly introspective about the consequences of their efforts and apparently confused about the future. For Teller, however, this was another remark to be added to his growing dossier of resentment, mistrust – and jealousy – towards Oppenheimer. He saw it as implying that Oppenheimer's team – the 'first team', as Oppenheimer was now fond of calling them – had done such a fine job with fission weapons that it was unlikely to be matched in the near future, either by the Russians or by anyone who might try to follow, with fission or fusion weapons.

One of the last leaving parties to be held at Los Alamos was the one organised by the British Mission, on Saturday, 22 September. Its centrepiece was a satirical revue, 'a beautiful pantomime', as Teller described it to Maria Mayer, with the physicist James Tuck wearing 'a red luminous tail and very impressive mustaches [sic]. He looked most sinister. I shall leave it to you to guess whom or what he represented.'

Three days before the party, on 19 September, Tuck had arranged for Klaus Fuchs to drive him in his Buick down to Santa Fe to pick up liquor for the festivities. While Tuck went shopping, Fuchs was left to his own devices. He then drove out to the edge of town, where he met with his contact, Harry Gold, and handed him one of the most significant tranches of information about the research at Los Alamos. It included a number of the vital practical discoveries that were the real 'secrets' of the bomb project. These included details of the properties of plutonium – the fact that it contracted rather than expanded on heating, and that it was as hard as glass one moment but as pliable as lead the next, making it very difficult to work. However, Fuchs was able to report the crucial fact that it could be alloyed with the metal gallium, and that it could then be worked to make the bomb core.

One of the documents that Harry Gold carried away from that meeting was not technical. It was a statement from the newly formed Association of Los Alamos Scientists, formed in response to the widening public debate about atomic weapons. During that first post-war autumn, Teller had become a member and had helped Robert Wilson draft a statement, adding their voice to the call for international control of atomic weapons. When it was done, they looked to Oppenheimer to take it to Washington, to circulate it there and gain the permissions necessary to make it public.

Oppenheimer agreed to do so and the petition eventually found its way to Henry Stimson and thence to the President himself, though in all probability the Russians may have seen it first. A summary of the copy Fuchs had given Gold arrived in Moscow on 29 October and noted that the scientists' 'feelings of distrust towards the government are very strong'.

Perhaps too strong for those in Washington: Oppenheimer had initially reported back optimistically on how their document had been received; then, suddenly, he wrote saying that the government had decided to make the association paper a state document, thus preventing its public release. He added, to everyone's surprise, that he agreed with this action, saying that he felt that a group statement was strongly inadvisable.

A short while later there was yet further cause for dismay, when Oppenheimer gave his support to the May–Johnson bill, a bill fostered by Groves, which proposed that nuclear power should remain under military control. It seemed an action running counter to everything he believed in. Frank, his brother, was to defend his action as targeting his influence on those issues he thought most important: 'Right from the start he felt we hadn't much time to work towards the thing he felt most important, an international agreement on arms control, and that he must therefore concentrate on influencing the system from within.'

There were those, however, who could no longer clearly discern his agenda, who felt he was flouting their trust, and Teller was certainly one of these. When, as a young senator, Brien McMahon became involved in the issues surrounding the bill and initiated a series of hearings, Teller was proposed as someone to give evidence. Teller was enormously flattered to be asked to this, his first Senate hearing. He even gave up a trip he and Mici were taking to Mexico with Rudolph Peierls and his wife Genia, leaving Mici to make the trip in the company of Klaus Fuchs. At the hearing on 1 February 1946 Teller presented much the same technical vision as his report for the Navy, but he used it on this occasion to raise the more political issues surrounding the use of atomic energy: 'Only three and a half years of intensive effort were needed to make an atomic bomb,' he said. 'Unless the possibility of a future war can be eliminated we are going to live in a world in which safety no longer exists. I do not know whether the international developments will make further work on atomic bombs necessary. I share the hope that the atomic bomb, together with other weapons of aggression, will be eliminated.'

The 'international developments' Teller was referring to included an agreement between the governments of the UK, the US and Canada to exchange information on nuclear matters with any nation that would reciprocate. They had also proposed a United Nations commission and everything had seemed highly optimistic when, that January, the Russians had voted along with everyone else to set up the United Nations Atomic Energy Commission.

To formulate American policy a special committee was appointed, headed by the Under-Secretary of State, Dean Acheson, with Groves, Bush and Conant amongst its members. Its advisory committee, however, was chaired by David Lilienthal, the boss of the Tennessee Valley Authority, with Oppenheimer as one of its members. Lilienthal and Oppenheimer took to each other immediately, though Groves, who often found himself at odds with them, viewed their relationship through jaundiced eyes. 'Everybody genuflected,' he complained. 'Lilienthal got so bad he would consult Oppie on what tie to wear in the morning.'

The report they produced, known as the Acheson–Lilienthal report, was in large part Oppenheimer's work. This extract conveys something of its essence: 'International control implies an acceptance from the outset of the fact that our monopoly cannot last ... It is essential that a workable system of safeguards remove from individual nations or their citizens the legal right to engage in certain well-defined activities in respect to atomic energy ... because they are or could be made steps in a production of atomic bombs.'

There was a real feeling that something major could be achieved, and Isidor Rabi recalls sitting in his living room on Riverside Drive, watching the winter sun set over the Hudson as he and Oppenheimer talked airily of their international schemes: 'There was a moment when our plans could have been realised, I am sure, and Oppenheimer was right to rush things, to try to push through a programme as quickly as possible. After all, think about the Union of the various states of America 200 years ago ... the call for a unification came just at the right moment to overcome [the vested interests of each state]. That was what we were looking towards with international control.'

Teller also expressed his support – and admiration – for the Acheson–Lilienthal plan. He wrote an article for the newly founded *Bulletin of Atomic Scientists of Chicago*. Its title, 'A Ray of Hope', reflected the balance in his views between international control and weapons development. In it he wrote, 'Nothing that we can plan as a defence for the next generation is likely to be satisfactory; that is, nothing but world union.'

Teller was absolutely consistent in his support for international control, even for world government, but if these aspirations failed, then his view on what needed to be done remained uncompromising. 'There is among my scientific colleagues,' he wrote in a summary of his views that he prepared for Fermi, 'some hesitancy as to the advisability of this development [the Super], on the grounds that it might make the international problems even more difficult than they are now. My opinion is that this is a fallacy. If the development is possible it is out of our powers to prevent it.' This left only one alternative to international control: the achievement of technical superiority and the maintenance of peace through strength – a strength it was hoped would never have to be used. Thus while others of his colleagues were retreating from what they had done at Los Alamos, expressing fears and regrets, Edward Teller had a clear – and to many an alarming – principle with which to steer the course ahead.

His small group at Los Alamos continued to work on the Super through-out the winter. But the task that faced them was immense, dwarfing the theoretical work and calculations that had been needed for the fission weapon. At its most basic level the Super initially consisted of a gun-type fission device and its heat and radiation were to be used to trigger the thermonuclear burning, the fusion, of a tank of liquid deuterium. In an intermediate chamber there was a mixture of tritium and deuterium which, because of the presence of tritium, fused much more readily than deu-terium alone, and could be used to help in priming the main reaction.

Because the temperatures required to initiate fusion were so high, the main problem was to prevent energy loss both from the initial fission explosion – the trigger – and from the deuterium fuel once it had begun fusing; and while there had been subcritical experiments – albeit dangerous ones – to help in understanding the fission process, no such experiments were possible at the high temperatures of thermonuclear reactions. The understanding of fusion depended on hideously complex theoretical calculations.

'The magnitude of the problem was staggering,' said the Polish math-ematician Stan Ulam, who had worked on the early Super calculations during the war. 'In addition to all problems of fission ... neutronics, thermodynamics, hydrodynamics, new ones appeared vitally in the thermonuclear problems: the ... interplay of all the geometrical and phys-ical factors became even more crucial for the success of the plan. It was apparent that numerical work had to be undertaken on a vast scale.'

George Gamow once caricatured the energy flows in what became known as the 'Los Alamos Problem' in a cartoon that neatly demonstrates the complexities they faced.

Following the letters on his cartoon:

A: The initial fusion reaction involving the deuterons, deuterium nuclei, takes place producing fast charged particles.

B: These fast charged particles either go to heat and fuse more deuterium OR:

C: they go to heat electrons.

These electrons are not part of the fusion process. The energy they absorb from the fast particles can do a whole series of things, most of them wasteful of the energy needed for fusion:

D: They can both give back and/or absorb heat from the fusion reaction.

E: They can lose their energy by straightforward conduction heat loss.

F: They can lose energy by what Gamow described as 'Bremsstrahlung', a process where the electrons brake, slow down and lose their energy in radiation.

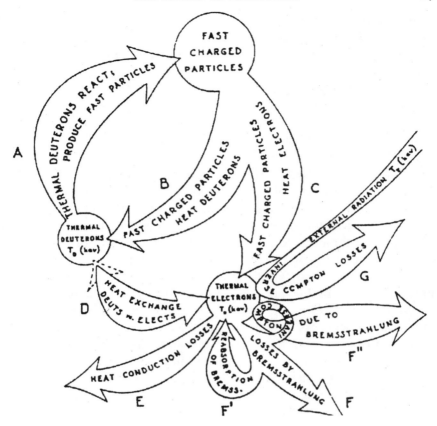

Figure 3: The Gamow Cartoon
This cartoon illustrates the various ways in which energy is transferred and lost during the fusion process.

Fi: Some of this 'braking radiation' can also be re-absorbed.

They can also lose energy (Fii and G) by a process known as the Inverse Compton Effect. This process involves the collision between fast electrons and light photons, during which some of the energy of the electron is transferred to the photon and carried away. The electrons, having lost energy, would extract more from the fusing deuterium nuclei (the deuterons) and could slow the thermonuclear reaction to the point where it stalled.

This critical process of energy exchange, even as simplified by Gamow, had so many variables, and yet in order to control it – it could after all run out of control with disastrous effect – it had to be fully understood. It was achieving this understanding that involved the 'numerical work on a vast scale' that Ulam referred to.

Yet before Christmas 1945 all that Teller and his group had on which to perform their calculations was primitive table-top calculating machines. It was then that, through the good services of John von Neumann, they gained access to ENIAC, the Electrical Numerical Integrator and Computer, developed at the University of Pennsylvania's Moore School of Engineering. With its 18,000 vacuum tubes, 1500 relays, 70,000 resistors and 10,000 capacitors, it filled an entire room at the laboratory, yet it had only 1000 words of memory.*

Nevertheless, during the months of December 1945 and January 1946, Teller's group used ENIAC for calculations that they estimated would have kept a hundred of their old machines occupied for a year. According to Ulam it raised 'hopes for a successful solution to the problem and the eventual construction of an H-bomb. One could hardly exaggerate the psychological importance of this work and the influence of these results on Teller himself and on people in the Los Alamos laboratory in general.'

In spite of this new-found optimism, however, there was still great uncertainty surrounding both the political future of Los Alamos and, with all the moves towards international control, of nuclear-weapons development in general. So, as he prepared to leave for Chicago, Teller felt he must summarise their work on the Super so far. In October 1945 he produced the *Super Handbook*, a compendium of all the computations and technical data assembled to date, and in December he filed a disclosure of invention for what was to become known as the 'booster'. This was a device that increased the yield of a fission bomb by priming the centre of its uranium core with a deuterium and tritium fusion mixture. It was also decided that, the following April, there should be a conference both reviewing the existing results and proposing lines of research for the future.

On 1 February 1946, Edward, Mici and Paul drove down from the Mesa at the beginning of their journey to Chicago. Edward left with a complex of mixed emotions, which he described to Maria Mayer: homesickness for the place where he had spent the past three extraordinary years but also both disappointment – tinged, he had to admit, with relief – at Bradley's decision not to actively pursue the Super. These feelings were further complicated by their return to Chicago. Neither Edward nor Mici enjoyed the harsh winters there and, just as they were leaving Los Alamos, Mici discovered she was pregnant again.

They also found themselves looking for somewhere to live in the midst of a severe post-war housing shortage. Eventually they found a large old house in a declining neighbourhood that was just a brisk fifteen-minute walk from the university. They rented it jointly with Bob Christy, another scientist who, with his family, had also just returned from Los Alamos. The

Christys lived on the ground floor while the Tellers spread out over the top two floors. The two families lived relatively separate lives, but Teller recalled: 'What I saw of [Christy] and his family, I enjoyed'. During their six months in the house they were to return to a lifestyle very similar to the one they had lived at their Garfield Street house in Washington before the war. The Fermis stayed for some weeks while they were house-hunting. Szilard sent a friend – whom they had never met. The wife of another colleague, still at Los Alamos, moved in with them. There were so many guests, in fact, that the house became known as the 'Teller Hotel'. 'It was fun to be back in an academic setting, to think about and discuss scientific questions with friends and students,' Teller recalled, 'but it was not the same. The world of 1946 was different in every way from the world of 1939.'

It was a measure of just how depleted of talent Los Alamos had become that, of the thirty-one scientists who gathered at the laboratory on 18 April for the three-day conference on the Super, only seven were still on the staff there. Those who had left to return to academia or elsewhere were now fast realising that the genie they had helped to escape was not to be returned easily to the bottle.

Over the previous weeks, Teller and six of his colleagues had prepared a fifty-nine-page report entitled 'A Prima Facie Proof of the Feasibility of the Super'. Right from the start, Teller was making clear his view that on prima facie evidence – the evidence to hand – he could prove the weapon's feasibility and that, as the report stated, 'a large scale theoretical and experimental programme for the development of a thermonuclear bomb is justified...'

Behind this optimism was the evidence from the ENIAC calculations and, on the second day of the conference, Nick Metropolis and Anthony Turkevich reported on their findings. They described how ENIAC had been used to the limits of its capacity but that, even so, their calculations still had to be simplified drastically, with the exclusion of some very significant processes carrying energy away from the reaction. One of these was the Inverse Compton Effect, the process in which fast-moving electrons collide with the particles of light – photons – and transfer some of their energy to them. However, they had judged that this would not significantly alter the final outcome of their calculations. On balance, the fission trigger would still generate enough energy to both initiate and sustain the fusion reaction.

There were others in the group who took issue with this. A fission bomb, they argued, produced radiation – primarily light photons – in vast quantities, and to omit any consideration of the way this flood of photons might, via the Inverse Compton Effect, carry energy away from the initiating reaction could well be seriously distorting, even invalidating, their conclusion. Certainly they queried whether the ENIAC calculations so far had actually proved the viability of anything.

On the afternoon of the second day the group discussed another possible effect of the vast quantity of radiation given off by the fission trigger. They were discussing the use of Teller's 'booster' principle, the injection of a small quantity of a deuterium–tritium mixture right into the centre of the U-235 core to enhance the intensity of the fission-trigger reaction. During this discussion Klaus Fuchs had the idea of transferring this mixture out of the uranium core and into a separate container of its own that was still open to the radiation from the fission trigger. He calculated that, under these conditions, the mixture would be subject not just to the heat of the fission explosion but to the light photons that were present in such quantities that they would actually compress the deuterium–tritium mixture, thus increasing the chance of its ignition.

It was an important idea, but the calculations needed to confirm whether it would happen in the way Fuchs proposed were simply beyond the capabilities of ENIAC to solve. Norris Bradbury, who was present at the conference, recalled 'exploring in great detail this original idea with members of my technical staff. At the time we saw absolutely no way to make it into a usable system.' Fuchs's concept of radiation compression had to remain just that, a concept. Nevertheless, he and John von Neumann thought it was of sufficient interest to develop further and on 28 May they jointly filed a patent for the invention of what they described as a new scheme for initiating Teller's classical Super. However, it was to be another five years before the full significance of their concept was realised.

During their three days together the group's discussions ranged widely from reconsidering the possibility of the more powerful Super igniting the atmosphere, as they had feared would happen with the first fission bomb, to considering how else thermonuclear power might be used other than for weaponry.

With the conference over, Teller and his group spent the next month drafting a report that took a very optimistic view. In spite of the doubts expressed over the ENIAC calculations, it concluded that, given a deuterium–tritium mixture sufficiently rich in tritium, 'a reaction can take place and propagate'. If successful, such a reaction 'can deliver energies equivalent to 1000 fission bombs and more, from a device weighing not much more than an ordinary fission bomb and containing mainly cheap materials'. However, the report did add one caveat: that the initial heating of the tritium–deuterium mixture appeared the most difficult task because 'it is questionable whether fission bombs of sufficient yield [to trigger the fusion reaction] can be constructed'. The final paragraph of the report argued: 'Further decisions in a matter so filled with the most serious implications as is this one can properly be taken only as part of the highest national policy.' But it concluded, optimistically, first with the prediction: 'It is likely that a Super bomb can be constructed and will work', and secondly with the estimate that they could be ready for the first test 'within one or two years'.

The draft of the report was circulated to the participants in the conference, some of whom were surprised, even shocked, by its optimistic tone. Robert Serber, one of Oppenheimer's close colleagues, had been persuaded to attend the conference by Philip Morrison, someone else close to Oppenheimer, who predicted that Teller would use the conference as a platform for promoting the thermonuclear programme. 'I found the report really incredible,' he recalled. 'The conclusion was that it was almost certain that it would work. I didn't want to discourage Edward from pursuing what he wanted to do, but I thought he should tell what was more close to the truth in the report, so we went over it and modified some of the more extreme statements . . . a couple of months later when the report came out, none of the changes were made that we had agreed on.'

The fact that no major push to the Super programme followed the conference left Teller deeply disappointed – and he held Norris Bradbury mainly responsible: 'He maintained that a thermonuclear weapon could not be built within the foreseeable future. The director of a laboratory has a great deal to do with whether a programme is continued. It was clear to me that all meaningful work on the project had ended for the foreseeable future.'*

In fact, the stalemate on the Super was more apparent than real. There may not have been the major push Teller had hoped for, but Los Alamos was already preparing for a series of tests at Bikini atoll in the South Pacific which had as their general aim the improvement of the crude weapons used on Japan. One of the objectives was to produce the hotter fission trigger necessary to fire a thermonuclear reaction. As well as these practical measures the Theoretical Division, now under the leadership of the Canadian physicist Carson Mark, was facing up to the complex calculations needed with the best computing equipment they could muster. At that time the division consisted of only some dozen or so scientists, but over the next four years half of them would be committed to work on the Super.

Elsewhere there were dramatic changes in the political mood of the time. On 19 February 1946 Stalin had addressed some 4000 Party members in the Bolshoi Theatre. He gave a speech aggressively denouncing capitalism as the creator of crises and armed conflict in competition for resources and markets. In the hypersensitive city of Washington the speech was seen by some as nothing less than 'the declaration of World War Three'. This was followed a month later by a speech Winston Churchill gave in the presence of President Truman, at Fulton, Missouri. In fact, at one point the old warrior declared his belief that the Soviet Union did not want war, but the speech registered most clearly for its recognition that an 'iron curtain ha[d] descended across the Continent'.

The following morning, 6 March, Under-Secretary of State Dean Acheson received a telegram, 5450 words long, which was to colour US policy for decades to come. Its author was George Kennan, a forty-two-year-old diplomat deputising for the ambassador in Moscow. His telegram offered an analysis of Soviet psychology – of how deeply suspicious the Kremlin was of the outside world, of how they had never known a friendly neighbour and always feared the West, a fear now expressed as 'capitalist encirclement'. The Russians, he said, were impervious to logic, but highly sensitive to the logic of force. They would withdraw when they encountered strong resistance from an adversary who made clear his readiness to reply.

It was a document that was picked up for the hawkish overtones that Kennan had never intended. It was in the air when Bernard Baruch, the seventy-five-year-old financier, led the US delegation to the first meeting of the UN Atomic Energy Commission in the hastily converted gymnasium of Hunter College, in the Bronx. Isidor Rabi followed the proceedings:

> Because we realised what a terrible state the world was going to get into if something like we were proposing didn't happen, I'm afraid we assumed the predicament was obvious to others, and it was to most – even the military. Then the mood began to change, helped a bit by Mr Churchill and his Iron Curtain speech and that sort of thing. The Russians helped a great deal. They were suspicious of everything, though our manoeuvring was not very apt. Yet when I think of the last 30 years or so, I wish they'd get some board of psychiatrists or physicians and find out what's wrong with the statesmen of the world. It's simply real madness what has happened.

Throughout the summer, both Dean Acheson and Oppenheimer argued with Baruch, fearful that his stance would be too dogmatic to achieve agreement. They were proved to be correct. On the one hand Baruch had insisted that any international-control agreement should have teeth and that there should be proper sanctions and punishment for those who broke it; on the other the Russians had insisted on the destruction of both the US capability for manufacturing nuclear weapons and its existing stockpile, even though they were already at work on their own weapon.

Oppenheimer watched as the battle lines of self-interest were drawn up. The two sides were never to reach any rapprochement. That special 'moment' that Isidor Rabi described was lost – for ever; and although Edward Teller was among those who continued to campaign for and write on international arms control, the world was becoming irretrievably locked into an arms race. Even though he did not recognise it, the world situation was beckoning Edward Teller on.

10
Wilderness Years

Looking back across fifty years, Edward Teller described his return to Chicago with great affection. It was to be the last substantial academic interlude in his life and it was spent 'among friends from home'. He was working closely with Maria Mayer, debating with Leo Szilard, and enjoying a close personal and family friendship with the scientists he perhaps admired most of all, Enrico Fermi and John von Neumann. It was a period that another of his friends, Eugene Wigner, was to describe as, 'perhaps Teller's most fruitful years scientifically'.

Over the three years he was there, he co-authored thirteen scientific papers and fostered a clutch of students that included one future Nobel Laureate and a number of others who were to achieve international reputations. Even though he had now left Los Alamos, he and his family often returned to join his old friends and their families for increasingly nostalgic reunions. They had come to love the place, relishing the return – for short periods, anyway – to the tacky housing, uncertain plumbing and muddy streets. 'A place which I have lived for a year or two is home for me,' he wrote to Maria Mayer. 'In Los Alamos, I know where every house has been built, who lived in it. I know, for instance, that the dog of Bob Davis is a much more permanent institution than the movie theater and considerably more ancient than the water tower. I think, I have a right to feel at home and I do.'

During their stays the family lived on 49th Street. They would take hikes into the surrounding mountains in the summer and go skiing in the winter. 'It was a wonderful place for a small child,' Paul Teller recalled. 'At the end of our street the city ended, and there was a little canyon about a hundred yards into the wood and, during the summer, we kids would play there all day long. There was a huge contrast with what it was like in Chicago. In Chicago my mother was fearful for my safety the whole time, I had fewer friends, it was a bit miserable.'

On those visits Teller consulted on problems associated with the increasing sophistication of fission weapons. He may have been irritated by the continuing paucity, even absence, of work on the Super, but he nevertheless took it upon himself to recruit others and it was as a result of this that the reunion group grew larger over the years.

His direct involvement in policy making was minimal at this time, but he wrote and lectured regularly on the world government he believed to be the only long-term answer to the nuclear threat. Some of the articles he produced were surprisingly optimistic in tone, but he remained adamant over the foolhardiness of trying in the present situation to ignore the new technology. In an article for the *Bulletin of Atomic Scientists* he wrote: 'If we should give way to fear and if we should fail to explore the limit of human power we shall surely be lost.' This was a further reiteration of the guiding principle that was to remain consistent throughout the next half-century, the principle that, he said, was eventually to bring him back to Los Alamos.

In the meantime, though, as he himself portrayed it, the life Edward Teller settled into at the University of Chicago was a satisfying one. He kept up his contacts with Los Alamos, anxious that it was understaffed, even directionless, and was prepared to do what he could to keep it moving forward. This upbeat account of his Chicago interlude, however, matches poorly with another that emerges – largely from his correspondence at the time with Maria Mayer.

In fact, things had been tricky from the start. Teller might have enjoyed the comings and goings of the 'Teller Hotel', but Mici did not. By the summer of 1946 she was heavily pregnant with their second child, and found the house hot and frantic. She had found the house on the Stanford campus that they had rented from Norris Bradbury while Edward taught summer school there a blissfully quiet retreat. She even relished the apartment they were given at Los Alamos. So when the time came and they were due to return to Chicago after their summer sojourn at Los Alamos, she point-blank refused to return until their housing had been sorted out. However, with the serious shortage, finding somewhere became one of Edward's major preoccupations and it was not until Christmas that the family was to be finally reunited in Chicago.

By that time the new baby had arrived. At about 2.30 a.m. on the morning of 31 August Edward was awakened by a light from the corridor of their Los Alamos apartment. 'I saw that Mici was not quite happy,' he wrote to Maria on the same day. 'I asked her whether I could do anything. She said, "Better get dressed." '

The baby, a girl, was born with relative ease three hours later. 'She does not have a name. She has black hair and she is, for the time being, no beauty. Mici believes that she is intelligent because she can press Mici's finger. I am not informed how rigorous this IQ test is. Anyway, I am happy that her ladyship arrived.' The remaining two-thirds of the letter is about superconductivity.

Within two weeks 'her ladyship' had a name – in fact two: Susan Wendy. This somewhat British combination arose in part from Mici and Edward having recently seen the film version of *Pygmalion*. They had been so impressed by the performance of the British actress Wendy Hiller as Liza Dolittle that they had decided to name their new daughter after her. In a very little while 'Wendy' had usurped 'Susan' as the name of choice.

The Tellers were still at Los Alamos when, in late 1946, the McMahon Bill was passed, transferring the laboratory from military to civilian control under the newly formed Atomic Energy Commission, the AEC. Edward noted an immediate effect. A real sidewalk had appeared in front of the post office, together with a sign which, when it first appeared, read 'Pedestrians shall use the side walk'. Shortly afterwards it was changed to 'Pedestrians please use side walk'. The first chairman of the commission was David Lilienthal, erstwhile leader of the Tennessee Valley hydro-electric project.

There was general approval at his appointment and also with the transfer away from the military, but Teller was one of a number initially concerned that there was only one working scientist on the commission: Robert Bacher. Teller had little time for Bacher but his fears were, in part, answered when the commission appointed a General Advisory Committee consisting largely of scientists. It was headed by James Conant and included Oppenheimer, Fermi, Rabi, Glenn Seaborg (the discoverer of plutonium) and Cyril Smith, the metallurgist who had been Teller's near neighbour at Los Alamos during the war. Initially Teller thought they were an excellent and diverse group, but in time he was to see them very differently.

On Mici's return home with Wendy, Edward was pleased to note that she was not only well but cheerful. He was always vulnerable to her mood swings, her strong likes and dislikes and her sharp tongue. However, he was affected in a different way by what seems to have been the only occasion that Maria Mayer became angry with him during their thirty-year relationship.

Edward had been remiss over a number of things – everything from telling her about Wendy's birth to providing comments on a piece of her research she had sent him. He reeled from a broadside that one can only imagine. 'But I can tell when you one thing,' he wrote in reply:

If you will sound often as you did on the phone – this is something that I cannot stand for a long time. You know why I like to be in Chicago and

I shall do everything I can to continue to have a reason for Chicago. But if the only strong reason for my being there should change I do not know how much longer I shall stay . . . When I left Chicago you told me quite seriously that now I should have a good time. As usual I tried to do as I was told and I was quite successful. For the last few days you can be quite sure, I did not have a good time. Nor do I care, if only you will stop being mad at me . . . I am really convinced that whoever is really nice to me must be crazy and you are much too reasonable . . . it would be good of you to write to me. The quality does not matter so much at present but the quality might be different. Even very little would help.

The disagreement between the two was soon patched up, but the almost childlike tone of the letter to Maria illustrates his vulnerability, his apparent low self-esteem. 'Since 5.30 p.m. yesterday I am sane again (i.e. as sane as you can expect me to be),' he wrote two weeks later after their reconciliation, 'but for the previous two weeks I thought of few other things than letters and phone calls and which parts of the world I might have to emigrate to if I get chased away from Chicago.'

The letters also provide support for Teller's own denial to me that the two were ever lovers. For one thing, alongside the intimacies are references to the letter of congratulation that Maria wrote to Mici on the birth of Wendy, to the necklace she sent as a present, and also to Joe, Maria's husband – and yes, they are personal and unguarded, but it is the minimal concern he expresses for her well-being that seems so at odds with an affair. He is happy to expose his social shortcomings, his lack of organisation, his impetuosity, his vulnerability to others' opinions, his fears for the future; but are these not the kind of things that might be confessed to a sympathetic brother or sister, or perhaps to an understanding mother figure? They are confessionals, and intimate ones, but they lack the interplay of lovers' letters. And they do contain a high proportion of science.

Initially at least, all the open-mindedness and enthusiasm so admired in Washington before the war was still evident in Edward's teaching and research in Chicago. When the British physicist Freeman Dyson had met him back in early 1946, he had found him bubbling over with ideas and jokes. 'Teller had done many interesting things in physics, but never the same thing for long,' Dyson wrote. 'He seemed to do physics for fun rather than glory. I took an instant liking to him.' His enthusiasm was legendary. Colleagues coined the term 'Teller' as a unit of enthusiasm. Finer degrees of positive thinking were measured in 'micro-Tellers' or millionths of a Teller. 'That same enthusiasm is what I experienced as a wonderful childlike quality in my father,' said Paul Teller. 'I have a boyish excitement about things, something I value, and I believe I learnt that from my father.'

But he was equally legendary for the flightiness of his interests. 'If only he could find one thing to concentrate on,' Enrico Fermi was heard to comment. 'While his ideas are always original and often brilliant, they're not always practical or timely,' Rudolf Peierls remarked. 'He pursues his ideas with great insistence, and this makes him act at times like a prima donna.'

As before the war, he was collaborating widely and providing a creative impetus to the work of others. Among these was Maria herself – Teller having originally dismissed her observations on 'magic numbers' as a detail, her steamy response caused him to reconsider, and the two discussed her research further.* The work she began that summer of 1947 represented the beginning of a detailed understanding of the stability of nuclear structure, and it was the work that was to lead to her 1963 Nobel Prize.

To begin with, Teller made progress at Los Alamos too. Arriving with his family for his first period of consultancy, he boasted to Maria that he was there in 'the proper (or rather not proper but usual) crusader spirit', but was obviously aware of the sensitivities surrounding his reappearance so soon after resigning. 'Do you think there is any chance,' he wrote, 'that I shall be somewhat less foolish than I have been?'

During this stay, he began working with Robert Richtmyer, then head of the Theoretical Division. According to Teller, Richtmyer was concerned 'about the absence of any effort on thermonuclear research', and so between them they developed another version of a hydrogen bomb that they hoped would be a workable alternative to the Super.

It was derived from the implosion mechanism of the fission bomb, with a spherical array of explosive lenses impacting on a central core in which fission and fusion material were organised in alternating spherical layers, like a bullseye sweet. Teller and Richtmyer chose to call it the 'Alarm Clock' and optimistically hoped it would re-awaken interest in thermonuclear possibilities. To some extent this optimism was well placed, for there was no question that this device would work; the issue was, how well. The intimate mixing of fission and fusion processes, rather than using the one to trigger the other, could mean that the two processes interfered with, and possibly inhibited, one another.

That summer Teller also proposed a solution to one of the many practical problems that continued to dog the thermonuclear weapons programme: the scarcity of tritium. This was produced in the reactors at Hanford and production in quantity could only be accomplished at the expense of the plutonium required for the fission weapon stockpile. Now, however, Teller suggested using lithium deuteride (a compound of the light metal lithium and deuterium) as an alternative. This offered two huge advantages: First, it was a solid at room temperature, while deuterium, a gas at room temperature, required a complicated refrigeration plant even to maintain it as a liquid. Secondly, in the extreme temperatures of a nuclear explosion, the lithium was expected to interact with the deuterium to produce tritium in situ.

However, that autumn, when Teller urged Bradbury to schedule tests of prototype fusion devices for the next year, Bradbury simply ignored him. In fact, when he sent a report on Los Alamos's various projects to the Army in November 1946, the Super was not even mentioned. This was enough of a setback in itself, but that summer Teller also heard from Fermi that the GAC had reservations about the Super. He felt driven to consult with Oppenheimer. 'We have been extremely friendly,' he wrote to Maria. 'He is a clever man. If he would really like me he would not have acted very differently. But mistakes are not corrected easily.' Teller was already expecting the worst from those he was beginning to see as adversaries.

In 1947, he spent his second summer consulting at Los Alamos and at the end he had assessed progress on the Super in a technical report entitled 'On the Development of a Thermonuclear Bomb' (LA.643). Teller's mood, as reflected by the report, had become decidedly more pessimistic since the meeting at Los Alamos eighteen months earlier.

When the GAC had reviewed the report they pronounced it 'admirable' but endorsed neither an increase in tritium production nor the thermonuclear tests that Teller had asked for. Only the fission Booster had been recommended for additional work. Then in July 1948, the Panel on Long Range Objectives, another group led by Oppenheimer, had met to consider the future of atomic warfare. They had ranged over a wide variety of topics, from small fission weapons to the world's supply of uranium, but the Super was virtually ignored and given no priority.

Teller decided not to let such a series of knockbacks pass unchallenged. A week after Oppenheimer's report, he forwarded to Bradbury his own very different version of the future. In 'the Russian Atomic Plan' he speculated that the Russians might have taken a very different course from the Americans and may have decided to build and test their bomb in secret. He also suggested that they might already have begun to manufacture the ingredients for their own thermonuclear device in decentralised sites around the Soviet Union. 'One may feel less certain about a continued superiority in atomic warfare,' he concluded. With this document, Teller expressed his realisation that the political battle for a project was at least as important as the technical. And with his report he had touched on an issue already creating anxiety – just when would the Russians produce their first weapon?

In fact it had now been around a year since the Soviet government authorised the Physics Institute of the Soviet Academy of Sciences (FIAN) to begin research on a thermonuclear weapon. In making this decision they had before them Klaus Fuchs's report of the 1946 Super conference, including the details of the patent Fuchs himself had taken out with von Neumann. Igor Tam was to be the senior physicist in charge and he immediately recruited the brilliant twenty-seven-year-old Andrei Sakharov. 'Despite summer's distractions,' Sakharov wrote, 'we worked with a fierce intensity . . . We were possessed by a true war psychology.' By the end of the summer of 1948 he had a radical new design, which he called the 'First Idea'. He had re-invented Teller's 'Alarm Clock', a concept Sakharov described as a 'layer cake' of alternating layers of deuterium, tritium and then uranium. It was to be followed by a 'Second Idea'. For this, one of Sakharov's colleagues, Vitaly Ginsberg, suggested replacing the deuterium and tritium with lithium deuteride. Within eighteen months of beginning research, the Russians had independently taken precisely parallel steps to the Americans. As Edward Teller himself had said years before, atom 'secrets' boiled down to matters of technical detail. The principles were there for anyone to discover.

By early 1948, Edward Teller was increasingly concerned that things were not working out in Chicago as he had expected. As he wrote to Maria later that year,

> I did not start things rightly in Chicago. I did not start things rightly with Enrico, or with the students or with you . . . For some reason I have to think of the fact that work is force times displacement. In Chicago, somehow, the two are never in phase as far as I am concerned.
>
> Another thing you know better than I: it's clear that I must not mix in politics again – wherever I am – not for some time to come. My mind is made up as to what I think is right. You were allways [sic] surprised why I am so bitterly disappointed about the small fact that we disagree. I do not understand it either. But I believe that I can do more about politics by talking about it less.

His bitter disappointment when he disagreed with Maria was a symptom of something that increasingly concerned others of his colleagues: his growing obsessiveness. In spite of what he had written, and continued to write, about the possibilities of a rapprochement with the Russians, numbers of colleagues noticed Teller's growing concern about their activities. 'Talk about physics and Edward was inventive and enthusiastic,' recalled one scientist. 'However, talk about politics and it turned to one thing. The Russians are coming.' Freeman Dyson, who had so enjoyed talking science with Edward, was sufficiently struck by his obsession about the world government movement to write to his family: 'He is a good example of the saying that no man is so dangerous as an idealist.' Enrico Fermi told him to his face that he was the only monomaniac he knew with a number of manias.

As the knockbacks he received accumulated, the open-minded enthusiast was slowly being transformed. While he may have wanted desperately to recapture the excitement of research, he was ten years older and his eyes had been opened to other ambitions; and always in the background must have lurked the belief that, with the Super, he could transcend his present existence. He was ambitious – 'would just as soon be the President of the United States,' was Harold Agnew's acerbic comment – and he was jealous, of Oppenheimer, who had become a national figure as the 'father' of the A-bomb, and of others too.

Throughout his time at Chicago, the Tellers and the Fermis saw a great deal of each other. Fermi thought very highly of Teller – according to his wife, Laura, he thought him 'scientifically the most congenial of the physicists and the most stimulating'. His wife, Laura, and Mici also formed a close friendship. In those post-war years the two men shared their reminiscences and it was a measure of how highly Edward regarded Fermi that he had been astonished to find that, in his youth, he too had experienced rejection by an 'in-group'. However, that admiration was sliding into envy. On 20 August 1947 he wrote to Maria about this: 'You know what my present trouble really is: I do envy Enrico – partly for his brains but much more for his energy. I wish I had done as much in three months as he did in six weeks. Not that it matters very much *what* he did. But the amount and speed with which he gets hold of things. He really makes me feel that I am a lazy bum.'

This envy served to exacerbate Edward's own sense of frustration, and was to come to a head a little later when he was polishing his and Maria's theory on the origin of the elements. When he showed the paper to Fermi, he was heavily critical, and over several months kept pointing out that they had miscalculated the strength of the nuclear forces holding their postulated clumps of neutronic matter together. These forces, according to Fermi, were simply not strong enough. Teller resented the criticism bitterly, but eventually, a year after Fermi had made his original observations, Teller had to admit that he was right: 'I know I am crazy to take Fermi's criticism so hard,' he told Maria. 'I know even more that I am crazy to take this paper so seriously. I do not know what has gotten into me. The only thing I can do is not to explode and try not to let Enrico know how I feel. I cannot fool him completely.'

In many ways, Teller's situation has all the hallmarks of a mid-life crisis.

His taste for pure research was jaded, but there was nothing obvious on the horizon to take its place; and he was increasingly unsettled by the friends, who did have the sense of purpose, or importance – or both – which were eluding him.

Then, during the terrible weather of that Chicago winter, Mici developed viral pneumonia. The demands of a sick wife and two young children served to compound his growing frustration with work and he sank steadily into a depressed state. 'Your letter just came before my second seminar talk on the origin of elements,' he wrote to Maria in early 1948. 'If anything could have helped me it would have been your letter ... I have lost the feeling of wanting to do things, I can only keep going by doing the next thing. By the time you get this letter I will probably have told you on the phone that Mici has virus pneumonia.'

Maria obviously responded sympathetically, advising him to take a break by himself. As a result, some time that April Edward checked into Hotel Governor Clinton on 7th Avenue and 31st Street, New York, '... and so I start my vacation ...' he wrote to Maria. 'Perhaps I should call up Rabi or take the next train to Washington and see the Hendricks. But you said I shall come to New York and have a vacation by myself. So I am. But I do not think that I shall go to a theatre. Edward.'

It was an innocuous enough letter reporting from a trip Maria had recommended, but some four months later he wrote to Maria describing what had really happened: 'You know that in April I had two and a half days vacation in New York. It was terrible. The ghosts from a dead existence. I stayed in the top floor of a hotel. I could not sleep. My window just looked out on the Empire State Building. I felt like getting out of that window. Then the sun came up in a street that looked like a canyon. After that I slept.' It was one of the lowest points in his life.

On Saturday, 8 March 1947, a special messenger from the FBI appeared at the Washington offices of the AEC carrying a letter and a document for the commission's chairman, David Lilienthal. The document was from the FBI's director, J. Edgar Hoover, and was a summary of the files 'relative to Julius Robert Oppenheimer ... and his brother Frank Friedman Oppenheimer.'

It contained a summary of evidence on the scientists' left-wing activities from before the war up to the present, including references to his contacts with well-known communists and his 1943 visit to Jean Tatlock. The main item, however, was the Chevalier incident.

Lilienthal read the summary over the weekend and was sufficiently shocked to call a special meeting of his fellow commissioners on the Monday morning. That afternoon they were joined by Vannevar Bush and

James Conant, who gave assurances that they had heard, and discussed, FBI charges as far back as 1942. They also warned that denying Oppenheimer clearance 'would have serious consequences on the attitude of his fellow scientists towards the project'.

But this didn't end the matter. On 12 July the *Washington Times Herald* carried the front-page banner: 'US ATOM SCIENTIST'S BROTHER EXPOSED AS COMMUNIST WHO WORKED ON A-BOMB'. The article was written by James Walters, a journalist known to have links with the FBI. It also carried a specific disclaimer emphasising that their revelation in no way reflected on the 'loyalty or ability' of his brother Robert; but in the increasingly paranoid anti-communist atmosphere of the Cold War, such innuendos were taking on the currency of hard fact.

Teller had reacted with some scepticism to the appointment of the Cornell physicist Robert Bacher as an AEC commissioner. Most of the others were unknown to him personally, but some time soon after their appointments he had an unusual introduction to one of them when he gave a talk on world government at Temple Emanu-El in New York. 'Afterwards, a sprightly grey haired woman, who I now believe to have been his aunt, introduced herself and asked me (in a somewhat remarkable fashion) whether I thought her boy's decision to take a job with the newly formed Atomic Energy Commission was a good one. I assured her that I thought so, but at that time I would certainly have underestimated the number of beneficial programmes Strauss would initiate.'

Teller may have met Lewis Strauss back in 1939 when Szilard, whose patron Strauss was, was attempting to drum up interest in the German discovery of fission. Strauss had had an extraordinary career. As a boy, he had dreamed of becoming a physicist, but a crisis in the family's wholesale shoe business had resulted in him spending his late teens as a shoe salesman. At the age of twenty-one, and against powerful competition from wealthy Ivy Leaguers, he applied for, and won, the post of private secretary to the wealthy Herbert Hoover, helping to organise post-war food relief to twenty-three countries. Strauss firmly believed that God planned his life, so God had allowed him to join the investment banking firm of Kuhn Loeb. He had rapidly risen to become a full partner and a millionaire, but he had never lost his interest in science.

Having been a Navy reservist since 1925, he had entered active service at the beginning of the war and become assistant to the Navy Secretary, James Forrestal, leaving the service at the end of the war as a rear-admiral. Strauss was regarded as clever, even original in his thinking, but he was also intellectually insecure and sensitive to criticism, attributes that, soon after his appointment as a commissioner, brought him into conflict with

Oppenheimer and his legendary hurtful arrogance. Strauss was one of the commissioners most concerned about the FBI files on the scientist and viewed him with suspicion from the start.

One situation that brewed between the two men over at least two years concerned the distribution of radioisotopes abroad for research. These were used in entirely benign ways in medicine and industry, but Strauss was seriously worried that they might find their way into 'unfriendly hands which might endanger our national security'. He made his complaint first in 1947 and he was still making it, to a Joint Committee on Atomic Energy hearing, in May 1949. A week later, and in Strauss's presence, Oppenheimer appeared before the committee and delivered the following putdown: 'No man can force me to say you cannot use these isotopes for atomic energy. You can use a shovel for atomic energy. In fact you do. You can use a bottle of beer for atomic energy. In fact you do. But to get some perspective, the fact is that during the war and after the war these materials have played no significant part and in my knowledge no part at all.'

Joe Volpe, an AEC counsel and friend of Oppenheimer, had been watching Strauss and seen his eyes narrow, and the colour rise in his cheeks. At the end, Oppenheimer turned to Volpe and said, 'Well Joe, how did I do?' With Strauss's expression clearly imprinted on his mind, Volpe replied, 'Too well, Robert, much too well.'

No such chasm of hatred was to divide Strauss from Teller. In fact, they were to become closely bonded, in part at least, by their growing mistrust of Oppenheimer.

The ennui, frustration and depression that Teller suffered from in the early months of 1948 could not have been helped by what was happening globally and the initial effect of these events on what was happening at Los Alamos. In February 1948 the Soviet armies invaded Czechoslovakia. In April of that year they imposed a temporary blockade of the railway lines into Berlin, lifting it a few days later and then re-imposing it. The US atomic-weapon stockpile at this point was so meagre that the American commander in Berlin, General Clay, was ordered to avoid confrontation and, instead, to bring in supplies by air. So began the Berlin airlift, aimed at keeping the besieged city, isolated in the Soviet-occupied sector of Germany, provided with food and fuel. All these events combined to create a sense of international danger, and many Americans saw a real possibility of armed, even nuclear, conflict with the Russians.

A large part of the reason for the immediate shortage of nuclear weapons with which to defend Berlin was to be found at Los Alamos. That spring the laboratory was a hive of activity as preparations were made for a series of weapons tests, codenamed Sandstone, to be conducted on Eniwetok atoll in the South Pacific. There were to be three shots during April and May, which were the culmination of as many years' work refining and improving the efficiency of fission-weapon design. Both weapon materials and the technicians needed to prepare the weapons for use against the Russians were, at this time, involved in the tests.

Sandstone was a resounding success. The device, which used a composite core consisting of a plutonium sphere suspended by wires ('levitated' was the jargon phrase) and surrounded by a uranium shell, produced an explosion some three or four times that of the Nagasaki bomb while using half the amount of plutonium. Its immediate effect was to make possible a 63 per cent increase in the total weapons stockpile simply by re-engineering the fissionable cores. It opened up the possibilities of enormous flexibility in fission-bomb design. With both much smaller and much larger weapons becoming possible there was now no reason why half-megaton bombs – bombs equivalent to 500,000 tons of TNT and forty to fifty times the yield of the Nagasaki bomb – could not be built from fission material.

Teller had consulted on this work. He had found it quite satisfying in the light of his frustrations in Chicago. 'If I am asked here,' he wrote to Maria from Los Alamos, 'I am asked by reason of necessity. Now, being necessary is an extremely important thing for me. You may say that I am vain (sure I am).'

However, now that the laboratory's director, Norris Bradbury, knew that he had the certain means to produce, on an industrial scale, a whole range of weapons large and small to meet the growing Russian threat, it looked as though work on the giant Super might be wound up once and for all; but Bradbury took a surprising decision.

With the heightening international tension and the effort needed to follow up on the Sandstone result, Bradbury, however reluctantly, decided that Los Alamos needed Teller's assistance on a full-time basis. 'Norris was rather diffident in his approach to the scientists who had left,' the mathematician Stan Ulam recalled. 'He felt that they should recognise by themselves how important for the country and the world it was for them to come back. As a result, although he wanted to, he did not like to ask people like ... Teller to visit. It was actually left to me, with his consent, to write such invitations ... thus, in a way, I was instrumental in bringing Teller back to Los Alamos.' Oppenheimer also encouraged Teller to return and, while Teller did not respond immediately, he began thinking seriously about the move, writing to Maria that he did 'not want another year with quite so definite a feeling of frustration as the last one'.

That September he had been invited to the Solvay international physics conference in Brussels and it had provided him with the opportunity to re-establish links with old friends and colleagues in Germany. It was a chastening experience. He found Heisenberg and his department isolated behind the Iron Curtain and 'happy about my visit – too happy. It was not a personal affair.' He had eventually come away saddened by his old professor's 'weary reserve'. Von Weizsäcker he found upset and preoccupied by the fate of his father who, as a senior member of Hitler's government, was being tried at Nuremberg. The saddest had been Eucken, the professor who had invited Teller to Göttingen. Penniless and depressed by the state of his divided country, he was to commit suicide a year later.

The visit to both sides of the Iron Curtain had refreshed the European perspective and the intensity of his concern about Russia that, Teller believed, separated him from the insularity of so many American colleagues. He was also convinced that the initiative for any action would only come from the US; but he himself was, as he put it to Maria, full of 'knots and confusions' about what he should do next. However, a visit to the doctor brought everything into focus. His leg had become painful and he was, generally, not feeling well. The doctor examining him suggested he lose a few pounds to relieve the stress on his knee, but then he added, 'I suspect that you are struggling with some decision. You won't feel better until you make it.' Shortly after this he contacted Los Alamos, telling them that he would return, full-time, at the end of the following summer.

Before returning to Los Alamos, Edward Teller spent very little time in Chicago. Los Alamos became his base and he quickly settled back into the routine and social pattern there. It obviously suited him, as his letters to Maria dwelt much less on his frustration and depression and became altogether more bullish and humorous.

'Los Alamos can be beautiful in the winter,' he wrote to her on 20 January 1949.

> It was beautiful when we arrived. I always have a fine feeling when I get to another place. Even the difficulties seem different and it is not too bad to have some.
>
> Then came the blizzard. It snowed and snowed and snowed. I did not expect that this is what the weather will do in Los Alamos. Many years ago I liked to walk alone at night in the snow. I no longer like it . . .
>
> Every evening we are invited – as usual in Los Alamos with the same people. I find myself sitting in a corner very quietly. I should be even listening if there should be anything to listen to. Mostly there are only the witticisms of Ulam, 7 – times – heard stories of Geo Gamow and, once an hour, two words of Carson Mark which are to the point.

Every morning I get up – eventually. When I do get up I am tired. I am tired all over and all the time. I don't want to think. I have not yet finished the few corrections I should make in our short paper on the Cosmic rays. It is a scandal . . .

I am doing what is expected of me at this place. Even my lethargy is counted here as praiseworthy industry. I suspect that people appreciate that I repeat the same thing every second day, contradict seldom, and say a new thing never. It is all highly satisfactory.

If I ever wake up from my hibernation I shall look into cross sections.

In the early summer he had represented the laboratory at a conference with the military personnel to discuss the practical applications of atomic weaponry. 'The conference here was not a pleasure,' he wrote. 'Explicit discussion of practical applications has some of the properties of a bad dream.'

As the international tension had increased, as China had fallen to the communists and his own home country, Hungary, had been taken over by a communist government, Teller became increasingly dismayed at how limp and unfocused the response to the communist threat was. 'Most of the military assume that we are so superior that we only need to select the most reasonable one of a number of ways how to win the war.' He reported to Maria, 'I should like to shake them and wake them up. Then I hear myself saying a few nasty things: what we ought to do to survive. And then I should like to shake myself.'

At some point over this period he stepped out of the charmed circle of the scientific hierarchy and began using his periodic visits to Washington to contact and talk with certain politicians who, he believed, might be able to create a stir in the right places. He began to pay visits to Senator Brien McMahon, the sponsor of the 1946 bill that had transferred atomic energy from military to civilian control. Very much an opportunist politician of the good-tailoring, good-staffing school, there had been considerable doubt as to whether he had actually read the bill that bore his name. However, ever since he had made atomic energy one of his special interests, he had been determined to become the 'Mr Atom' of US politics. By 1948 McMahon was a member of the powerful Joint Congressional Committee on Atomic Energy, the JCAE, endeavouring to enhance its influence so that it would match that of the AEC. He must have welcomed the visit from one of the country's best-known but discontented weapons scientists, but Teller's visits were treated with the greatest suspicion by his senior colleagues. John Manley, one of Oppenheimer's closest colleagues during the war and now one of Bradbury's deputies at Los Alamos, said:

Roughly at that time, '47, early '48, I sensed from my frequent poking around in Washington that Teller was making kind of – I was going to say a fool of himself but not that – but not doing himself any good, let's say in the AEC Washington HQ. This arose primarily because of his tendency every time to come in to Washington, to not go immediately into the Commission and have some business with them, but to go off and see Brien McMahon.

Manley had admonished him gently, but to no effect. He had broken ranks with his fellow scientists and begun what was to become the habit of a lifetime.

That summer Teller returned again to Europe, to England, for an international meeting on reactor safety. For the past two years, he had been chairman of the AEC's Committee on Reactor Safety, successfully laying down safety ground rules that were to stand the test of time. He had been accompanied on the trip by a young protégé, Freddie de Hoffmann. 'Freddie is a wonderful Gentleman's Gentleman,' he wrote to Maria – 'though I do not have a firm conviction whether the word used twice describes either him or me to perfection...'

> Now I have about made up my mind that it is time for the complete back-mutation. After all, my love of war is not terribly deep rooted. Actually I am thinking of a five year plan along the following lines: I shall become: 1) Lazy 2) Taciturn 3) Punctual 4) Observant 5) Deliberately boring 6) I shall have (God damn it) some fun. 7) I shall always talk in a low voice.
>   Do you think the five year plan will be unfulfilled?

Whatever Teller's own propensity to implement a correction to his persistent social malfunctions, events that would make such corrections irrelevant were about to take a hand. At the conference there was a good deal of tension between the British and American scientists over continuing full and free exchange of information on nuclear matters. So sensitive had the Americans become to security issues that even the British, who had contributed in a major way to the Manhattan Project, were regarded with suspicion. The matter had been compounded by Isidor Rabi who had been in the UK a short time before Teller. 'Rabi came through here a month ago,' Teller reported to Maria, 'and left a very distinct trail behind him. He must have said to different people at different times: "If the English make atomic bombs this will be felt as a menace to us." Do you think such statements are popular? Three guesses.'

Given the atmosphere, Teller was surprised to be invited to dinner in Cambridge by Sir James Chadwick, who had led the British mission at Los Alamos during the war. It was a somewhat strange event, with Chadwick verging on the monosyllabic in his contributions to the conversation. At one point Teller reported with some relish that General Groves had been eased out of power, at which Chadwick delivered an impassioned lecture on the misunderstood General's virtues, claiming that without him the likes of Oppenheimer, Conant and Bush would have achieved nothing. When Teller took his leave, Chadwick accompanied him back to his hotel. His parting words were puzzling: 'Please remember what I told you tonight. You will have need of it.'*

General Groves's view that the Soviet–American arms race was 'an ox cart-versus-automobile situation' was not the one universally held in Washington, nor was his view that it would take up to twenty years from 1945 for the Soviets to produce a fission bomb. For example, a group of experts representing the Pentagon, the AEC and the CIA reported that the Russians might well have their first atomic weapon as early as mid-1950. Edward Teller shared this view, but while Groves's estimate might not have been the only one in town, his was the one that President Truman had, thus far, chosen to work with.

By late spring of 1949 the Russian scientists at the Chelnyabinsk-40 camp were near to completing an exact copy of the implosion bomb used on Nagasaki. It was a copy largely derived from information provided by Klaus Fuchs. However, such was the suspicion that he had been the channel for a giant piece of American misinformation that, on one visit to the camp, senior officials suggested that the bomb's core was made of iron. 'Feel it, it's hot!' an angry scientist had responded. 'One of them said that it did not take long to heat a piece of iron. Then I responded that he could sit and look till morning and check whether the plutonium remained hot. But I would go to bed. This apparently convinced them, and they went away.'

When the scientists met Stalin, they found him deeply concerned that the test might provoke the US and he quizzed the scientists about using the core to make two smaller devices, one of which could be held in reserve against a surprise attack. When he was told this was not possible, he delayed the test until enough plutonium had been processed for his second back-up weapon. A date for the test was confirmed for late August.

Teller travelled back to the US aboard the liner *Ile de France* and then went straight to a meeting at the Pentagon. At the end of the meeting one of the organisers said, '... and, incidentally, what the President said today is correct'. Teller did not know what the President had said and so asked. The answer was, 'A few hours ago, President Truman announced that the Soviet Union has exploded an atomic bomb.'

It was then that Teller began to understand Chadwick's parting words. Chadwick had known both that the Soviets had tested a bomb and that Fuchs had probably passed everything he had known on to them. Teller believed that Chadwick also knew of his continuing involvement with the Super and was warning him of the resolve he would need to push things through – as Groves had done during the war.

11

The Taking of Washington

Three days in September 1949 created in the US and elsewhere in the West a crisis of confidence that took decades to overcome. In Beijing, on 21 September, Mao Tse Tung declared the People's Republic of China, thus claiming communist control of one third of the world's population and a quarter of its land mass. Two days later, President Truman made his announcement: 'We have evidence an atomic explosion occurred in the USSR.' Following as closely as they did on the Berlin crisis, and the communist takeovers of both Czechoslovakia and Hungary, the events fused in the public mind as the onward march of a single conspiracy.

Robert Oppenheimer was one of those who had been involved in assessing the first evidence that came in from the 375th Weather Reconnaissance Squadron of radioactive fall-out over the Amchitka peninsula. So incredulous was everyone, so convinced of the totality of the US monopoly, that it took some three weeks, pulling in evidence from the Navy's ships all over the Pacific and from the British, before it was realised that there had not been an error. Oppenheimer was called as a witness in front of a secret session of the JCAE, where he had the task of convincing its incredulous members. At one point Senator Arthur Vandenberg asked, 'Doctor, what shall we do now?'

'Stay strong and hold on to our friends,' Oppenheimer replied. Right from the start he had feared a panic reaction, which he believed would lead to new policies being devised without proper thought. This was, no doubt, still a prevailing concern when Teller called him on the evening after the President's announcement. 'What shall we do now?' he asked. 'Keep your shirt on,' was Oppenheimer's abrupt response.

But the hysteria was inevitable. It shattered the country's bipartisan foreign policy, which had prevailed since the end of the war, as the Republican opposition adopted a powerful anti-communist stance from which to batter the ruling Democrats. There were charges that fellow-travellers in the State Department had allowed Stalin's puppet, Mao, to defeat the Chinese Nationalist Army under Chiang Kai Shek. There were charges that traitors in the US had gone unchecked, and had given Stalin the bomb. The foundations were being laid for the witch-hunts of the following decade.

Edward Teller was travelling homeward to Los Alamos from Washington on board the Chieftain express train, when he read in his newspaper a quote from General Groves: 'We shall stay ahead of them.' It was a confidence that Teller, with his belief in the quality of Russian scientists and his deep mistrust of Stalin's regime, certainly did not share – indeed, never had done. In a letter to Maria from the train, he tried to explain the deep anxiety he felt, drawing a parallel with his feeling of foreboding when he had been about to take the Matura, his school-leaving exam in Hungary: 'You know, I used to be terribly afraid of the "Matura". I was convinced that all my sins and deficiencies will become evident on that day. Before the Matura we had two weeks holydays to give us a chance to prepare ... It was beautiful weather. I liked it the more for knowing that in a few days there comes the "Matura" and the end of the world. Now I feel in the same way.'

While Teller was journeying on through the Mississippi Valley, the Joint Committee on Atomic Energy, the JCAE, had begun a series of meetings in which they tried to formulate a response. Earlier in the year, Brien McMahon, now the committee's chairman, had been joined by a young executive director, William Liscum Borden. Following Truman's announcement, Borden and his staff had settled down to produce a list of twenty-three ways in which the production of nuclear weapons might be augmented. Their list included increasing the staff at Los Alamos, bringing industrial giants such as Du Pont back into the programme and launching an all-out effort to build a hydrogen bomb. Thanks to Borden's efforts, the committee began to focus on the latter and, towards the end of the week, took evidence on just what the realities were in pursuing a thermonuclear weapon. Borden was urging things forward.

A Yale law graduate who was only twenty-eight years old in 1949, Borden had been a bomber pilot with the 8th Air Force during the Second World War, flying out of England. On returning one night from a mission, he had seen a V2 rocket 'streaming red sparks and whizzing past us', and it had made an enormous impression, as had the bombing of Hiroshima. On leaving the services, his calling card on the world of politics had first been a book entitled *There Will Be No Time*. In it he envisaged a future war beginning with another shock attack like Pearl Harbor, the enemy attacking this time, however, with rockets armed with nuclear warheads. The US had to be strong and well prepared.

When Brien McMahon, who was a near neighbour of his parents in Washington, had become a member of the JCAE, Borden had then targeted him. He sent him a document in which he advocated giving Stalin a nuclear ultimatum while the US still had its weapons monopoly. 'Let Stalin decide – atomic peace or atomic war,' he wrote. This provocative challenge worked, at least for Borden personally. He was taken on, initially as a legal assistant, then rising to his current position when McMahon was promoted. Over the next four years he was to play an increasingly important role in Cold War politics.

During their meetings the JCAE gained the impression that Los Alamos was now working steadily – but without any great sense of urgency – towards a 1951 test of a new fission implosion bomb, one that would employ Teller's thermonuclear 'booster'. They were told that the test, which had just been codenamed Greenhouse, would be a 'step towards a possible thermonuclear bomb', but that it was going to absorb all their efforts in the immediate future. They were also left in no doubt about the scale of the difficulties that were being faced.

At that time the designs being considered for the Super were still, mostly, huge. As well as the giant Alarm Clock, another model under consideration had a fission trigger weighing 30,000 pounds, while the whole bomb would be 30 feet long with a diameter in excess of 162 feet. With such obviously absurd configurations resulting from the first attempts to turn theory into practical design, there was, quite clearly, much work to do. The JCAE were also introduced to the key problem of providing tritium for the existing designs, which could only be produced in the reactors currently making plutonium for the fission-weapons stockpile. Just one Super would absorb the reactor capacity needed to produce plutonium for as many as thirty to forty of these bombs. It was not an encouraging picture.

Elsewhere, AEC Chairman David Lilienthal was finding it extremely difficult to come to terms with the brutality and the limitless scale of the thermonuclear weapon. At a meeting of the commissioners on 5 October, he had deliberately directed discussion away from the hydrogen bomb and towards increasing the production of fission weapons. In doing so, he had tried to block discussion of a memo produced by Lewis Strauss, one of his commissioners, that called for a 'quantum jump' in the efforts devoted to the Super. Strauss, however, circulated his memo anyway and then went to one of his numerous lobbying meetings around Washington. His first was with Admiral Souers, who worked with the National Security Council and was close to the President. It was a revelation. Firstly it became clear, as Strauss had suspected, that Souers himself had never heard of the Super, and so, as Souers recalled, Strauss tested his suspicions one stage further and asked what Truman knew.

'He said, "Well, I don't think [the President] has been informed because Lilienthal is opposed to it." I said, "You should certainly see that it gets to the President so that he can get the facts and make a decision." ' It emerged that Souers was right: the President, now in the midst of his second term, had indeed never heard of the Super. When he was told, he immediately instructed Souers to 'tell Strauss to go to it and fast'. Shortly afterwards, Lilienthal eventually agreed to a proposal from Strauss to call a special meeting of the GAC. He wrote to Oppenheimer saying that he would welcome the group's 'advice and assistance on as broad a basis as possible'. The meeting would be held on the weekend of 29 October, and the brief, implying a consideration of moral as well as practical issues, was to prove of considerable significance.

On the same day as Truman first heard of the hydrogen bomb, Edward Teller received a telephone call at Los Alamos from the Berkeley physicist Luis Alvarez, asking him for an update on the Super. This call, which came out of the blue and was to prove a marvellous boost for him, had been prompted by a meeting on the Berkeley campus between Alvarez and the Dean of Chemistry, Wendell Latimer. Latimer, a fervent anti-communist, was one of the minority who had believed that it was only a matter of time before the Russians had an atomic weapon of their own. Now he discussed his new fears about their likely progress on the Super and how they 'might get there ahead of us'. Alvarez wrote in his diary: 'The only thing to do seems to be to get there first – but hope that it will turn out to be impossible.'

Alvarez had also aroused Ernest Lawrence's concern, and so it was that the two men contacted Teller. There was little of any substance that could be said on the phone for security reasons, so, with customary alacrity, Lawrence arranged for himself and Alvarez to travel down to New Mexico that night. Before they arrived, however, Teller received a 'polite call from Norris Bradbury. "Are Ernest and Luis coming to Los Alamos? Could John Manley sit in on the meeting?" Of course I agreed.'

Arriving in the early hours of Friday morning, Lawrence and Alvarez found themselves meeting with Teller, Stan Ulam and George Gamow, and also with Manley – 'mouth tightly closed, eyes and ears sharpened', as Teller remembered. For two hours Teller gave a presentation that contained essentially the same information as had been available at the 1946 conference on the Super, three-and-a-half years previously.

In spite of this lack of progress, however, and in spite of the evidence that increasingly pointed to the problems with Teller's Super design, Alvarez noted the overall sentiment of the meeting as optimistic. 'They give the project good chance if there is plenty of tritium available,' he reported. Lawrence, always an optimist himself, simply said, 'In the present situation there is no question but that you must go ahead.'

From Los Alamos, Lawrence and Alvarez were travelling on to Washington, and so an excited Teller flew down with them to Albuquerque, continuing their conversation from plane to taxi and on to Lawrence's hotel room as he washed that 1949 technological novelty, the drip-dry shirt. 'What you will have to do now will involve a lot of travelling,' Lawrence said, introducing Teller to the life of the lobbyist. 'This is the means that has made that kind of travelling possible.' It was a reminder, by an expert, of the level of commitment that was going to be needed – and, as if providing an example, Lawrence and Alvarez spent the next five days in Washington in a round of meetings. Over the weekend they met senior figures in the Pentagon. On Monday they met with Senator McMahon, warning him, Borden noted: 'Russia may be ahead of us in this competition.'

This comment, one that was to become a mantra in the years to come, bears all the hallmarks of Edward Teller's tutelage. George Cowan, who was a young chemist at Los Alamos working on plutonium, recalls how, immediately following the announcement of Joe 1, the American's name for the Soviets' first bomb, Teller had

very quickly started putting out memos to the effect that the Russian bomb was probably made using plutonium made in a heavy water reactor. This had been the original emphasis in Europe, rather than graphite, and this kind of reactor produces an excess of neutrons, neutrons which could be used to make tritium. And so he took a worst case scenario immediately which was that the Russians very possibly, and even very likely, would have the capability to make tritium and to beat us to a thermonuclear weapon. So he created an enormous sense of urgency.

Having clearly impressed the receptive McMahon with the worst-case scenario, Lawrence and Alvarez then received a chill response when they visited David Lilienthal. According to Alvarez, the AEC chairman actually 'turned his chair around and looked out of the window and indicated that he did want to even discuss the matter'. Lilienthal was equally scathing about what he saw as their fanaticism. 'Ernest Lawrence and Alvarez in here drooling over the [Super bomb],' he wrote in his diary. 'Is this all we have to offer?'

Undaunted by Lilienthal's reception, Lawrence and Alvarez then moved on, meeting with other AEC commissioners and with Isidor Rabi in New

York. They had hoped to travel on to Ottawa to talk to the Canadians, but they were thwarted by a lack of flights. Nevertheless in those five days the two physicists had played so successfully on the currency of fear among scientists, the military and politicians alike that the Super had become the only show in town. Anyone who argued against a major thrust to implement Teller's plans now would be doing so at their peril.

Teller, too, had thrown himself into proselytising on the Super. Norris Bradbury had called an open meeting of the whole laboratory staff at Los Alamos for 13 October, and both Teller and John Manley had written open letters to the scientific staff in preparation. Both were in agreement that the laboratory had lost years in developing the Super and that work must now proceed as rapidly as possible. At the meeting, however, Teller had gone further and had argued emotionally for the worst-case position. Regardless of any other consideration, he said, 'if the Russians demonstrate a Super before we possess one, the situation will be hopeless.' Such a sweeping statement must have jarred with many of his colleagues, who knew only too well the background – the success of Sandstone in showing the way to fission weapons with yields from just a few kilotons up to half a megaton – and who would have known that the nation's stockpile already stood at 169 fission weapons. But Teller wanted to provoke action, and he went on a recruitment drive, travelling east to meet, among others, Hans Bethe and Enrico Fermi, to try to persuade them to return to Los Alamos.

Teller travelled first to Cornell to meet with Bethe. 'To my great pleasure, Hans was willing to return to Los Alamos to work on the hydrogen bomb,' Teller reported. There was, however, a confusion of recollections which would, in time, contribute to the growing bitterness between the old friends and become a major issue between Teller and Oppenheimer. Teller was to testify that, during the meeting, Oppenheimer called Bethe on another matter and then invited the two physicists to go down to Princeton and discuss things with him there. Bethe recalls that they found Oppenheimer 'equally undecided and equally troubled in his mind about what should be done. I did not get from him the advice I was hoping to get ... He mentioned that ... [Dr Conant] was opposed to the development of the hydrogen bomb, and he mentioned some of the reasons which Dr Conant had given.'

In a letter to Oppenheimer, Conant had been blunt. The weapon would be built 'over my dead body', he said. His letter provided an excellent way for Oppenheimer to demonstrate his own position without committing himself. Teller, however, was wary: 'I am pretty sure that I expressed to Bethe the worry that we are going to talk with Oppenheimer now, and after that you will not come. When we left the office, Bethe turned to me and smiled and he said, "You see, you can be quite satisfied. I am still coming." ' Two days later, however, Teller called up Bethe to discover that he had thought it over and changed his mind: he would not be coming after all.

Bethe, however, remembered things differently. 'I was quite impressed by his ideas,' he testified. 'On the other hand, it seemed to me that it was a very terrible undertaking to develop a still bigger bomb, and I was entirely undecided and had long discussions with my wife.'

More than fifty years later, Hans and Rose Bethe recalled that discussion. 'Yes, we talked about it,' says Rose Bethe. 'I gave it the thumbs-down, because it wasn't necessary. There was enough of a bang, I thought, and we didn't need to be bigger. That was all.' Hans Bethe agreed with his wife: 'That is an argument that is as true today as it was in '49.'

As to what happened after they had left Oppenheimer's office and gone their separate ways, Bethe recalled going over to the university at Princeton to attend a meeting of the Emergency Committee of Atomic Scientists, a group organised by Leo Szilard and chaired by Einstein. There Bethe recalled that he was pointedly joshed by Szilard as ' "Hans Bethe from Los Alamos". I protested that I was not at Los Alamos and didn't know if I wanted to go back there.'

After the meeting, Bethe and his old friend the Austrian physicist Victor Weisskopf strolled in the grounds, imagining 'that after such a [thermo-nuclear] war, even if we were to win it, the world would not be . . . like the world we would want to preserve. We would lose the things we were fighting for.' They continued the conversation as they drove up to New York in the company of George Placzek. So absorbed were they that Bethe missed his flight back to Ithaca and Placzek and Weisskopf ended up with each other's coat. The result, however, of both this meeting and his discussion with his wife had been that Bethe had changed his mind. It was, after all, a big decision, heavy with moral and ethical implications, and likely to generate doubts and second thoughts. But that was not how Teller was to see it. Teller was disappointed and angry. He strongly suspected that it was yet another knockback that had been contrived by Oppenheimer in some way.

Although Oppenheimer had been guarded in how he had expressed his views to Bethe and Teller, that very day he put them on paper in a reply to Conant's letter. The technical problems, in his view, were no closer to solution than they had been in 1942. 'I am not sure the miserable thing will work,' he wrote, 'nor that it can be gotten to a target except by ox cart.' But this, he said, was not what concerned him most. It was that 'two experienced promoters have been at work, i.e. Ernest Lawrence and Edward Teller', and they had contributed to a major change in the climate of opinion. 'What does worry me is that this thing appears to have caught the

imagination, both of the congressional and of military people, as the answer
to the problem posed by the Russian advance . . .We have always known it
had to be done: and it does have to be done . . . But that we become committed to it as the way to save the country and the peace appears to me full of
dangers.'

Within days of his meeting with Bethe, Teller had also caught up with
Enrico Fermi. The Italian had just returned from an exhausting trip to
Europe, and when Teller appealed to him to return to Los Alamos, he was
quite sharp in making it very clear that he would not do so. However, the
conversation developed between the two men and, as Teller reported at the
time to Maria Mayer, who also knew Fermi very well, it provided a vivid
illumination of the dilemma in which both men found themselves:

> Fermi did not tell me what the General Advisory Committee proposed.
> He did tell what his own ideas are. He said: 'You and I and Truman and
> Stalin would be happy if further great developments were impossible. So,
> why do we not make an agreement to refrain from such developments? It
> [a Russian H-bomb] is, of course, impossible without an ultimate test
> and when that happens we shall know about it anyway.' . . . There is only
> one difficulty in the world: Wishful thinking. It is unexpected that Enrico
> should succumb to it.

Edward then went on to describe his own conflicting feelings: 'It would
be good if no danger existed. It would be good to continue to do physics.
Even if there is danger why should I of all people be the first to do something
about it? Any argument which gives me the right to do nothing is a good
argument. Any person who reminds me of the danger is a nuisance. Enrico
does not know what I think of him. But, unfortunately – he has an
inkling . . .'

No doubt there is justification for personal ambition in this letter. Teller
genuinely saw himself as someone who had to fight indolence, certainly
over matters requiring application; and on this occasion, in the absence of
anyone else, he believed he simply had to take a lead. 'What I saw in
Washington,' he wrote to Maria, 'makes it quite clear that there are big
forces working for compromise and for a delay . . .' Just as Oppenheimer's
views at this time are difficult to interpret as those of a Communist 'sleeper',
so Teller's are difficult to interpret, given the mood of the time, as hawkish,
verging on the paranoid. Both Oppenheimer and Teller considered the
Russians a threat. It is over the response to that threat that they differed.
They both saw a role for the Super, but while Teller was all proprietary zeal
and fired by a fear of what the Russians might be doing, Oppenheimer was
exercised by the immense technical problems, and by the Super's potential

for dominating and destroying the balance of US strategy. They were rationally held positions but, in the febrile atmosphere of the time, such differences of judgement were being distorted by politics and personalities. Oppenheimer's left-wing past was again attracting interest in some quarters while, in others, it was Teller's 'loose cannon' activities in Washington.

On his trip east, Teller had not only been on a recruiting drive – that had been approved by Bradbury – he was also doing his own lobbying and had arranged to meet McMahon in Washington. While he was having the meeting described earlier with Fermi, however, he received a phone call from John Manley, who wanted to talk to him about his Washington trip. The two men met on Teller's arrival at the capital's railway station, and Manley came straight to the point. He asked him not to see McMahon – on the grounds, according to Teller, that it would be 'unfortunate if Senator McMahon would get the impression that there is a divided opinion among the scientists'.

Teller was ready with an answer. He suggested that he would be willing to call up McMahon, but he would tell him that he had been asked not to see him, and that he would therefore not be coming. At this point, Manley backed down and Teller was allowed to continue to his meeting.* Manley's account differs in significant details. 'What happened is that he called me in Washington. He got a little worried, I guess, and asked about it. I said, "Well, you know what I told you. Of course you can see McMahon, but why don't you come to the Commission building first and see what people are thinking about and want to talk to you?" But he got all mixed up. That was sort of the beginning, really, of a strain, I'd say.'

Whichever version of events is closer to the truth, the growing mistrust is evident in both men's accounts. Teller was already seen to be working outside the system, and there was a growing resentment towards him among the management at Los Alamos. As far as Teller himself was concerned, his plea for help had been turned down by two old friends and colleagues, one of them after Oppenheimer had been brought into the discussion; and now he had been harassed by John Manley, who was not only part of the Los Alamos administration but also secretary to Oppenheimer's GAC. To Teller the patterns of loyalties were clear: 'In retrospect,' he recalled, 'that meeting with Manley had a deep effect on me. At the time I thought it was a terrible thing to demand unanimity where obviously there was none. Many years later, I realised how essential it was for me to stand up to him that day . . . My refusal to accept Manley's advice may have been crucial to injecting enough facts into the political pipeline that Truman was able a few months later to make a vital decision in regard to national security.'

◆

On Friday, 28 October, Robert Serber travelled down by train with Oppenheimer to Washington for the General Advisory Council's three-day meeting on the Super that weekend. Serber actually came as Ernest Lawrence's emissary, to urge a proactive approach to the Super. However, on the train Oppenheimer had enumerated the problems he saw with the new weapon, and by the time they arrived in Washington, Lawrence's would-be advocate had joined the ranks of the sceptics.

Among the group assembled in the AEC's panelled conference room overlooking the Lincoln Memorial there was a sense of inevitability about the outcome of the next two days' meetings despite the doubts of some prominent figures. There was such a powerful political undertow to their discussions that many of those present felt the result could be nothing other than a decision in favour of the Super.

As the meeting unfolded, however, pieces of information and surprising statements began a slow shift in the group's attitude. On the Friday afternoon Hans Bethe gave a technical evaluation of the classical Super and reminded everyone of the uncertainty surrounding even its feasibility. Then George Kennan, the State Department expert on Russian affairs, in his evaluation of Soviet attitudes, gave the group a strong impression that it might be possible to negotiate a halt to the arms race. In doing so he opened a window of opportunity no one had considered.

On the Saturday, before the committee was joined by the AEC commissioners and by the Joint Chiefs of Staff, Oppenheimer took a poll of the committee's views. According to Lee DuBridge, a GAC member, attitudes had shifted to the extent that there was now an essentially pragmatic agreement that 'there were better things that the United States could do at the time than to embark on this Super programme'. Then, when the Joint Chiefs offered their views, there was another surprise: General Omar Bradley, who did most of the talking for the armed services, took a predictable line that there was no choice but to build the Super. However, when asked what military advantages he could see for a weapon a thousand times more powerful than the first A-bombs, Bradley answered 'mostly psychological' ('a useful thing to have around the house,' David Lilienthal, the commission chairman, noted ironically in his diary).

At lunchtime Oppenheimer and Serber went out to eat together and in the lobby met another of Lawrence's representatives, Luis Alvarez. He had been there all morning, watching the comings and goings of war heroes like Generals Bradley of the Army and Norstad of the Air Force, and was now desperate to know how things stood. Joining the other two, he found Oppenheimer much more open in his opinions, arguing that they believed the Russians might follow an American example and would only build the Super if the US did. Directly after lunch a depressed Alvarez left for

California, convinced that the committee would now no longer recommend a crash programme.

The group had certainly moved a long way from their starting position,
but during the afternoon matters developed even more dramatically. The
engineer Hartley Rowe, a Manhattan Project veteran, was heard to
comment: 'We built one Frankenstein.' It was a comment that pitched the
discussion on to a new moral and personal plane. James Conant was the
most passionate – looking, Lilienthal thought, 'almost translucent, so grey',
as he argued to open up the issue to public debate. 'Cyril Smith strong for
the Conant point, as was I. Lewis quite dubious evidently,' Lilienthal noted.
It was as if Conant and other A-bomb veterans were still weighed down by
guilty consciences. 'Conant says, "This whole discussion makes me feel I
was seeing the same film, and a punk one, for the second time." '

The intensity of Conant's moral concern began to sway the others.
Gordon Dean, a new commissioner who had been McMahon's law partner,
began to sense that Conant had triggered a visceral response from those
present. They were now reacting powerfully to having to decide on the
future of a weapon of mass destruction, and it was as if the absence of a
wartime imperative left their moral justification exposed. The Russian
threat, real enough in the minds of Teller and Strauss, was not justification
enough for many of them. By the evening, feelings were running so high
that it was decided the committee report should be accompanied by
annexes expressing the personal views of the participants. Oppenheimer
and Manley set to work on the main report while the others split into two
groups.

Oppenheimer and Manley worked to produce a straightforward assessment of the Super's considerable technical problems, though they did
state: '. . . an imaginative and concerted attack on the problem has a better
than even chance of producing the weapon within five years.' They also
clearly outlined the Super's role as a weapon of mass destruction, but even
so, their conclusion was a muted one: 'that it would be wrong at this present
moment to commit ourselves to an all-out effort toward its development'.

The two annexes were anything but muted. The majority annexe that
six of them signed, including Oppenheimer, was drafted principally by
Conant and took a clear moral stance: 'We believe the Super bomb should
never be produced,' its conclusion stated. 'In determining not to proceed
to develop the Super bomb, we see a unique opportunity of providing by
example some limitations on the totality of war and thus of limiting
the fear and arousing the hope of mankind.'

The minority annex was written by Rabi and Fermi; a single-page letter
which, if anything, used even stronger language. Describing the weapon as
'one of genocide', they continued: 'The fact that no limits exist to the

destructiveness of this weapon makes its very existence and the knowledge of its construction a danger to humanity as a whole. It is necessarily an evil thing considered in any light.' They had, however, taken the notion of leading by example one practical stage further, arguing that testing of a new weapon by the Russians could be detected 'by available physical means' and that this would be the protection against any attempt to break a disarmament agreement; but could they be certain a determined enemy *would* test in advance, that they wouldn't take the chance of using a device untried? Whatever the practicalities of such a proposal, they were to be eclipsed by the emotional reaction to the denunciations in the annexes. These were to dominate perceptions of the full report.

On the evening of Monday, 31 October, only a day after the meeting broke up, Lilienthal and his committee of scientists were forced to justify their opinions to an angry Brien McMahon. There followed what John Manley described as a 'rather violent discussion'. Once again Lilienthal noted the essence of the argument. 'What he says adds up to one thing: blow them off the face of the earth, quick, before they do the same to us – and we haven't much time,' he wrote in his journal. Immediately afterwards, McMahon wrote to Truman asking to see him.

Given his vigorous lobbying of the White House in favour of the Super, Commissioner Lewis Strauss had been conspicuous in how little he had contributed to the GAC's deliberations over the weekend. However, Strauss had been privy to some very special information. Two weeks before the weekend meeting he had been visited by Charles Bates, the FBI's liaison man with the AEC. Bates had told him that the Bureau now possessed information that the Soviets had indeed penetrated the Manhattan Project. Further, he was told who the main suspect was. It was Klaus Fuchs, and there could be others.*

Strauss was not the only person who had been told about Fuchs. The newly appointed Commissioner Gordon Dean had been told a full month before, in September; and the AEC had been formally told by the FBI over a week before the General Advisory Committee's meeting. It was information that must have coloured how Strauss and Dean viewed the whole meeting. If it was confirmed publicly that the Soviets had obtained crucial US atomic secrets, the reaction would render the conclusions of the GAC meeting practically irrelevant. Strauss had immediately begun his own investigations, contacting General Groves with questions about the British Mission at Los Alamos. A few days later, Strauss had spent an hour with the FBI director, J. Edgar Hoover, following which he had contacted Groves again, this time with questions about Frank and Robert Oppenheimer. This in turn had prompted Groves to write back in an attempt to pre-empt criticism of why he had never acted on the information Oppenheimer had given him about his brother's involvement in the Chevalier incident back in 1943: '[The Frank–Haakon situation] was finally revealed to me under conditions which made it impossible to do much and it was very difficult to tell just how much Frank was involved and how much Robert was involved.'

Just after Groves had written to him, Strauss had met with McMahon at the Beverly Hills Hotel where the senator was staying, and the two had spoken to Gordon Dean, urging him not to succumb to the pressures from Oppenheimer and Lilienthal to block the Super. However, the three men, well briefed about the gathering espionage storm, must surely have speculated more widely, particularly about who the other spies in the Manhattan Project might have been.

Edward Teller was not to see the GAC's report for some weeks. However, his suspicions as to the final verdict were aroused when, on 31 October, Norris Bradbury cancelled a conference on the future of the Super that had been due early in November. A short while later, he met with McMahon in Washington, where they discussed the situation. 'It makes me sick,' the Senator commented. When, eventually, John Manley showed him the report, Teller was at first morose and silent – '*Very* unusual,' Manley noted. Eventually he bet Manley that within five years he would be a prisoner of the Russians if the US did not press ahead with the Super.

It was the two annexes that had caught Teller's attention – in particular the one written by Rabi and his friend Fermi, stating that the Super was a weapon of 'genocide' and 'an evil thing considered in any light'. 'Fermi's characteristic moderation was completely missing!' he wrote. 'I was shocked. I could only marvel at the gentle way he had expressed himself when we had talked on the same subject a few weeks earlier . . .'

Teller could not understand why the GAC had not made a distinction between developing the weapon on the one hand and using it on the other. Why did they not see the value of its very existence as a deterrent? 'I was deeply upset,' he wrote, '. . . even apart from the idea of danger of Soviet superiority, by the idea of stopping at a point of partial knowledge . . . it was glaringly obvious that knowledge about fusion was ready to harvest.'

Teller was once again adhering to his credo that what there is to know has to be known. At Los Alamos there was impatience and annoyance at the new delays. Younger members of staff had been looking forward to the challenge of the Super as a relief from the more routine work of continuing modifications to fission weapons. However, Teller was to testify that after the GAC decision work on the Super slowed almost to a standstill. He himself had turned to re-examining the Alarm-Clock concept to see if it might offer other possibilities, but had begun thinking about leaving the laboratory once more. He had an enticing offer from the University of California at Los Angeles, which particularly appealed to Mici. Everything hung in the balance as they awaited a response to the report from the President. 'I wish I could tell you some news,' he wrote to Maria on 30 November, 'but all I know is what I read in the papers. So I am learning to be patient. I make sure I agree with my friends (Los Alamos seems the only spot of sanity).'

As the days passed after the meeting, the commissioners found themselves split on what action to take on the GAC report. In the end, they sent it to Truman along with two conflicting recommendations, reflecting the irreconcilable split among the commissioners themselves – Lilienthal and his supporters on the one hand, Strauss and his on the other. As a consequence, Truman reconvened a Special Committee of the National Security Council, consisting of Defense Secretary Louis Johnson, Secretary of State Acheson and Lilienthal. At the same time, in order to avoid further confusion, Truman limited debate to this small, polarised group with Lilienthal opposing the Super on one side, and the robust Johnson supporting it on the other. When they met for the first time on 22 December, their meeting was so acrimonious that they did not reconvene again until six weeks later, when they were ready to offer their final opinion to the President. This meant that there was almost no feedback from the Special Committee to the outside world, and therefore no broader debate on the future of the Super. This did not mean, however, that the lobbyists had ceased their activities.

McMahon and Borden had bombarded the President, firstly with a 5000 word report and then three follow-up letters, two of them sent on the same day. All of them reflected Teller's belief, summarised in the first report, that if the US allowed Russia to get the Super first, catastrophe would become all but certain.

Lewis Strauss had acted independently of the AEC and, on 21 November, sent Truman his own critical response to the GAC report. The Joint Chiefs were equally critical in the report they sent to Louis Johnson, arguing powerfully that, without its atomic arsenal, the US was 'helpless to aid [its] friends' by conventional means.

While Edward Teller may not have known how the decision process at the centre was shaping, he remained an energetic campaigner, a background source of information, enthusiastically fuelling the drive for the Super. He was always on the move in what John Manley called 'a frenetic campaign to obtain converts'. He was in constant touch with both Strauss and McMahon and, along with Ernest Lawrence, had briefed the chairman of the Joint Chiefs, Omar Bradley. Because so few other scientists had access to the tiny group on which Truman had made himself dependent, Teller's closeness to the project, together with his heavy bias in favour of the Super, became increasingly significant. For instance, on the technical problem of how much expensive tritium would be needed to initiate the fusion reaction, Carson Mark observed: 'Edward promised people in Washington that they'd get by with a certain amount.' But he pointed out, 'He had no particular basis for the amount he mentioned except that it didn't appal them.'

As to those defending the GAC position, only Oppenheimer had access to the Special Committee. Acheson was well disposed towards him but failed to be persuaded by his moral arguments. 'You know, I listened as carefully as I knew how,' Acheson was reported as saying, 'but I don't understand what Oppie was trying to say. How can you persuade a paranoid adversary to disarm "by example"?'

During January 1950 matters came to a head. When the JCAE met to discuss the GAC report, McMahon read it aloud, adding his own editorial comment as he did so. On the possibility of the Russians agreeing to renounce development of the Super, he exclaimed, 'That is certainly a joke. Suppose they did? Who the hell would believe them?' Ten days after this, Louis Johnson chose to ignore his colleagues on the Special Committee and sent the Joint Chiefs' opinion straight to the President. According to Admiral Souers, Truman thought it 'made a lot of sense and ... he was inclined to think that was what we should do'.

On 27 January Sir Frederick Hoyer Millar, atomic expert at the British embassy in Washington, paid an urgent visit to Under-Secretary of State Robert Murphy. There he told Murphy that Fuchs had been arrested in London and had confessed to passing information over to the Russians, throughout the war and afterwards. Thus, on 31 January, when David Lilienthal joined the two other members of the Special Committee to present their various arguments to the President, he found him already decided. 'What the hell are we waiting for?' Truman said. 'Let's get on with it.' As they left the Oval Office, Lilienthal looked at his watch: the meeting had taken seven minutes.

Without the shock effect of Fuchs's arrest, it is difficult to assess how long any decision *not* to proceed with the Super could have held. However, Admiral Sidney Souers believed that the whole process during that autumn had been 'necessary to show the country that the President used an orderly process in arriving at his decisions', but that his 'mind was made up at the very beginning'. As another close aide noted, '[Truman] said there actually was no decision to be made on the H-bomb.'

A few minutes before one o'clock the next day, the White House press corps was read the following statement from the President: 'It is a part of my responsibility as Commander-in-Chief of the Armed Forces to see to it that our country is able to defend itself against any possible aggressor. Accordingly, I have directed the Atomic Energy Commission to continue its work on all forms of atomic weapons, including the so-called hydrogen or super-bomb.'

'I never forgave Truman,' Isidor Rabi said. 'He simply did not understand what it was about . . . For him to have alerted the world that we were going to make a hydrogen bomb at a time when we didn't even know how to make one was one of the worst things he could have done.'

The day of Truman's decision, 31 January, was Lewis Strauss's fifty-fourth birthday. For him the President's decision had been a great victory. Late in the day he went to the White House and, feeling the weight of public opinion now behind him, resigned as a protest over his own commission's delaying actions. He then went to the Shoreham Hotel for a birthday celebration, which, inevitably, turned into a celebration of his victory. Among the politicians, government officials and military men who were his guests was Oppenheimer. What must have been going through his mind as he stood there watching a celebration of the decision to make the most deadly weapon ever created we can only guess at.

There is one clue, however. One of the journalists among the guests, Ernest K. Lindley, spotted him standing alone and morose on the sidelines of the celebration. 'You don't look jubilant,' Lindley said. After a pause so long that Lindley thought he had lost track of the question, Oppenheimer finally replied: 'This is the plague of Thebes.'

For Edward Teller, the President's decision had been a moment of satisfaction if not of triumph, a moment when he knew that the tide of events had finally turned in his favour. 'I simply want to do, in this one thing, what seems right to me,' Edward wrote to Maria:

> This is a big enough ambition and I am a more satisfied than I have been in the last 35 years or so. I wonder to how many people it happens that they are set back where they have been before and that they get a second chance . . . I am going to try hard not to make the same mistakes. I never loved a fight. That is why I always was so bitter when I had to fight (or thought I had to). But this time I love the job I'm going to do – I shall even love to fight if it must be.

When he wrote these words, Teller could have had little idea just how bitter his struggle was going to be. He and his supporters had sold the Super to Washington, and finally to the President. Now he, more than anyone else, had to deliver.

12

Unholy Alliances

'The roof fell in today, you might say . . .' Lilienthal wrote in his diary on 2 February 1950, referring to the announcement of Fuchs's espionage. 'It took place during the war project; but he had been here since, which drags us in. It is a world catastrophe, and a sad day for the human race.'

When Hans Bethe heard the news, he rang Los Alamos to speak to the security officer, Ralph Carlisle Smith, about the 1946 Fuchs–von Neumann patent.

'Is it all there?' Bethe asked.

'All,' Smith replied.

'Oh,' Bethe replied.

Teller also rang to ask the same question, but his reaction was more venomous. 'I don't believe it,' Smith remembers him saying. 'If it's all there, it's because you put it there.'

'Quite a few people here are furious at Fuchs,' Teller wrote to Maria Mayer in early February:

> They feel personally insulted. I do not feel that way . . . Fuchs probably decided when he was twenty years old (and when he saw Nazism coming in Germany) that the communists are the only hope . . . From that time on his whole life was built up around that idea . . . Actually the damage that Fuchs has done is great. He surely gave away a lot and by now I feel quite doubtful whether we can keep up with the Russians in the atomic race . . . And finally we can now confidently expect a witch hunt.

Teller could not have been more prescient. On 9 February 1950 Senator Joseph R. McCarthy made a speech in Wheeling, West Virginia, in which he is reported to have said: 'While I cannot take the time to name all of the men in the State Department who have been named as members of the

Communist Party and members of a spy ring, I have in my hand a list of 205 that were known to the Secretary of State as being members of the Communist Party and who nevertheless are still working and shaping the policy of the State Department.'

Within days, following a telegram from the State Department requesting information, McCarthy panicked and reduced his total to 'fifty-seven card-carrying Communists'. Even though he steadfastly refused to put names to any of the number, the government and the President were sufficiently nervous to dignify his charges with a Foreign Relations Committee investigation. McCarthy and the full-blooded persecution of 'the Reds' were on the move.

Teller had no doubt that such a witch hunt would be damaging in the extreme, but he questioned whether there was now anyone left with sufficient authority to stop it. 'But who shall do the explaining?' he wrote to Maria, 'physicists who have refrained from working and who even now behave as though nothing more were the matter? The atomic scientists who have by now a deserved reputation for lack of judgment? I do not want to criticise – but what is the right approach?'

Teller was, by now, so sensitised to any possible setbacks in the thermo-nuclear programme as to see the seeds of yet another in the presidential announcement on the Super. In it Truman directed the AEC to 'continue' its work on all forms of atomic weapons. A harmless enough phrase but, in Teller's view, it 'gave the impression that we could produce a hydrogen bomb simply by tightening a few last screws', which was perhaps not so surprising when you consider the months of lobbying that had gone on following the announcement of Joe 1, during which the H-bomb had featured largely as a given. Despite this, however, the only practical development that had occurred at Los Alamos since then was the establishment of an informal committee which, it was hoped, would help move the project along.

The committee had been organised by Teller, the mathematician Stan Ulam, and George Gamow, Edward's old colleague from George Washington University, who was spending a year at the laboratory as a consultant; but even though Gamow was an old friend, they were not a happy group and only stayed together for a matter of weeks. 'Both Gamow and I shared a lot of independence of thought in our meetings,' Ulam recalled, 'and Teller did not like this very much.' The two scientists queried, among other things, Teller's optimistic estimate of the tritium each device required – even if, regardless of the amount of tritium used, the bomb would ever ignite. Gamow and Ulam blamed Teller for the group's demise but, in all probability, the group would have been disbanded anyway the following month when Los Alamos moved to a six-day, forty-eight-hour week – essentially a wartime footing – in answer to the President's directive. It was then that Norris Bradbury established a twenty-five-man Family Committee to oversee the thermonuclear programme, with Teller as chairman. The committee obtained its name from the nicknames given to the various weapons designs they were nurturing – 'Sonny', 'Little Edward', 'Uncle'. The biggest of them all, containing 100 cubic metres of uncompressed liquid deuterium, was known as the 'daddy of them all.'

'That's what Edward had in mind and I was enormously impressed by the fact that one man could have that impact on policy,' George Cowan recalled. 'He was everywhere. I was fascinated by his command of the subject. I didn't necessarily agree with what he said but I admired enormously how he said it. He marshalled his case. If I had needed a lawyer to defend me, I would have hired Edward.'

Certainly the three months of lobbying in Washington had boosted his confidence both in his cause and also in his abilities to persuade and direct others.* So powerful had his presence become that Bradbury made specific arrangements aimed at keeping him in check. Although he was chairman of the Family Committee, he was, nevertheless, made answerable to Darol Froman, a down-to-earth administrator, a Canadian, like Carson Mark, who had successfully organised the Sandstone tests two years earlier. 'The idea ... was to get everything that Teller could give, which was a lot,' said Froman, 'but not to let him run the thing. Because he would sure push it too hard and too fast and get all kinds of people outside the laboratory up on their ears ... people like the AEC.'

Teller had become a man with a mission. Gone for ever was the accommodating physicist who had provided the substance to back up Gamow's flights of fancy, who had contributed so enthusiastically to the earliest discussions on the Super at Berkeley in 1942, who had willingly helped Oppenheimer in the very early stages of setting up Los Alamos. 'I feel that what I can do in the next year,' he wrote to Maria that February, 'is much more important than what I can do in the remainder of my life.' So motivated was he that colleagues began to perceive his single-mindedness in terms of naked ambition. Ulam, for one, prophesied 'that great troubles would follow because of Edward's obstinacy, his single-mindedness and his overwhelming ambition'. But while accepting that personal ambition was a factor, Hans Bethe did not believe it was decisive: 'He was convinced we had to have unquestioned superiority over the Russians. He was convinced that the war between Russia and the United States would be fought with atomic weapons.'

Soon after the presidential proclamation on 31 January, Teller had set out on another of his recruiting missions, corralling old students and friends to assist on the Super project. Among them was Marshall

Rosenbluth, one of his best students from Chicago, and his reasons for coming were common to many who answered Teller's call: 'There was the Russian bomb, there was Fuchs, and there was Ted Hall. It was obvious the Russians knew we were working on the Super and, given their system, I could not believe they weren't working flat out on the project. It seemed to me pretty stupid for us not to do the same, so I went back for what turned out to be seven years.'*

Among his old friends, John Wheeler from Princeton was a major coup. Wheeler was in France on a year's sabbatical when Teller contacted him, but agreed to come immediately, bringing two of his best students with him. Emil Konopinski, who had been with him throughout the war at Los Alamos, also came, while John von Neumanm continued his hydrodynamic computations on the ENIAC computer. Teller also wrote articles exhorting scientists to join his crusade, declaring in the *Bulletin of Atomic Scientists* that they had been 'on a honeymoon with mesons [the atomic particle of choice for pure research at the time]. The holiday is over.' However, given the international situation, a surprising number he approached turned him down. Some simply did not share his sense of urgency – or not sufficiently to become embroiled in the security paraphernalia of defence research. Others objected on moral grounds – Bethe was initially one of these and Teller reported to Borden at the JCAE that he was 'shocked by the icyness [*sic*]' on the part of younger colleagues and hinted that there were more sinister reasons behind his lack of success: 'A man like Conant or Oppenheimer can do a great deal in an informal manner which will hurt or further our efforts.' His complaint to Borden earned him the opportunity to address his committee's senators in executive session, but this served to anger the AEC, who, as Teller wrote to Maria, put him 'into the lowest class of SOBs together with Szilard and similar vermin'.

The AEC's irritation with Teller is understandable. While they were trying to keep their heads while all about were losing theirs, he was seen as one more addition to the increasingly alarmist activity by the promoters of the H-bomb. First, Lewis Strauss had shown Defence Secretary Louis Johnson four secret documents on the H-bomb to which Klaus Fuchs had had access. As a result, Johnson had personally telephoned Truman to press home their significance. Secondly, Robert LeBaron, the Chairman of the Military Liaison Committee, along with two military colleagues, General Kenneth Nichols, Groves's erstwhile assistant on the Manhattan Project, and General Herbert Loper, prepared a memorandum for the President that built on Teller's initial interpretation of the residue from Joe 1 – that it showed the Russians had the means to be well advanced in preparing tritium for a hydrogen bomb. Their memorandum went even further, suggesting that the Russians might have tested one before the US's Long

Range Detection Programme had become active and 'that Russia's thermo-nuclear weapon may be in actual production'.

This memo had an electrifying effect in the Pentagon. The Joint Chiefs of Staff now endorsed an actual crash programme on the H-bomb. Then Johnson sent LeBaron's memorandum, along with this endorsement, to the President. Finally, in early March, Secretary of State Dean Acheson's committee on the Super bomb met once more and again suggested action of the 'highest urgency'. On the basis of Teller's latest estimate that each weapon would need just 100 grams of tritium, they targeted an output of ten hydrogen bombs per year. On 10 March 1950 all this activity bore fruit and Truman approved an all-out programme for the hydrogen bomb.

It was just at this moment, as the laboratory moved on to its 'wartime' footing, that Teller produced another downbeat report on the Super. His problem was that so little had been achieved. The report was little more than an updated form of the material he had presented to the 1946 Super conference, but this time Teller ended much more pessimistically: 'It may be stated that the Super is probably feasible. Its complex construction gives us little hope that it can be actually made to work in the next 3 or 4 years. It requires, furthermore, considerable amounts of tritium.'

As if to soften the blow, he tried to make positive capital out of the modest progress made on the Alarm Clock, but this was to backfire. It was certainly widely believed that this concept of interleaved spherical layers of fusion and fission material would work – but only at the megaton range or below. However, with fission bombs available in the half-megaton range, Teller had always had multi-megaton ambitions for the Super. To achieve this with the Alarm-Clock design meant adding so many extra layers of fuel that the weapon became truly gigantic, way beyond a size deliverable by air. In the report, Teller was undaunted: 'We shall see, however, that delivery by boat or submarine is capable of producing disastrous effects.' But wouldn't those disastrous effects of what would be a ground-based explosion be limited by the horizon? And wouldn't the giant ship-borne device have to be sneaked, undetected, for hundreds of miles up the Volga river in order to reach the Soviet centre of power in Moscow?

Thus, even before the start of a presidentially ordained crash programme, the central figure masterminding it was, first, having to admit openly that no significant advances had been made over the past four years and, second, having to propose schemes verging on the ludicrous. 'Nobody will blame Teller because the calculations of 1946 were wrong,' wrote Hans Bethe in 1952, 'especially because adequate computing machines were not then available. But he was blamed at Los Alamos for leading the Laboratory, and indeed the whole country, into an adventurous programme on the basis of calculations, which he himself must have known to have been very incomplete.' Teller, however, was unrepentant and continued to blame others and their lack of dedication for many of the problems they were facing. So irritated did Stan Ulam become with Teller's continual carping that he was driven to retaliate: 'Teller kept insisting on certain special approaches of his own,' Ulam wrote. 'I must admit that I became irritated by his insistence; in collaboration with my friend Everett, one day [in December 1949] I decided to try a schematic pilot calculation which could give an order of magnitude, at least, a "ball park" estimate of the promise of [Teller's Super] scheme.'

It should be said at the outset that Stan Ulam and Edward Teller had never really got on.* In the last few years, Ulam's widow Françoise investigated her husband's personnel file at Los Alamos and discovered a report on him written by Teller in 1945, early in their working relationship. While characterising his theoretical ability as 'outstanding', Teller had described both his personality and his success in recent work as 'unsatisfactory' and commented that he seemed unable 'to adjust himself to our work'. Françoise Ulam believed it clearly showed Teller's dislike of her husband, and that his brilliance was one of the reasons for this.

Socially, too, they did not get on. Teller wrote jokingly about Ulam's distaste for exercise – his not being one of those who willingly joined the Los Alamos hikers like Fermi, the Bethes and the Tellers – and his reluctance to apply himself to what Teller regarded, for others at least, as real work.* He was altogether a more opaque, elusive character than Teller. He had a disassociated manner, which could annoy, and, as one colleague explained, he could easily cut someone in the street – but only through the unfortunate combination of his shortsightedness and the vanity that prevented him from wearing spectacles. Marshall Rosenbluth, one of those working with Teller, found Ulam charming, but intellectually as well as personally elusive:

> You'd be sitting in the office working, and you'd hear these footsteps along the corridor and you'd think, 'Oh, my God, here comes Stan,' and you'd have a one hour charming conversation but not about anything that would help you with your work.
>
> Whereas Teller would come in for five minutes and maybe give you three or four good ideas, other factors you might consider. Stan liked to talk. I enjoyed it but I can't remember ever having learned anything from it.

It was not a promising background for a trusting collaboration, and trust was needed. Without adequate computers, all sorts of assumptions and simplifications were required to make sense of the complex subatomic interactions involved. Ulam's and Everett's first calculations had given them an initial 50:50 assessment of the chances of success. Now they moved on to improve the accuracy of that assessment. Ulam employed a statistical approach that he had developed, the Monte Carlo method, so called because it employed the use of random numbers generated by tossing one die of a pair of dice. It involved them in six-hour calculation sessions, day after day, using just paper, pencil and slide rule. Françoise Ulam recalled they worked alone, 'then with us the data analysts. In a couple of months the calculations confirmed his feelings. In other words Stan was the first to blow the whistle; it was not going to work.' Ulam and Everett presented their first set of calculations the day before Truman's crash programme announcement on 9 March 1950, in the form of a fifty-seven-page report. The conclusion – 'that the model considered is a fizzle' – could not have been more damning. According to Carson Mark, they had assessed that the amount of scarce tritium estimated by Teller 'was not nearly enough; so the first calculation was discontinued . . . and a second calculation, with a larger amount of tritium, was started immediately'.

Teller was furious, convinced that Ulam was deliberately sabotaging his project. 'Inside the Tech Area as well as among our visitors at home,' Françoise Ulam wrote, 'I could see how angrily Teller reacted to Stan's and Everett's crude results. He seemed to take them personally, piling objections upon objections that Stan and Everett met with more numbers and examples, and cajoling everyone around into disbelieving Stan's figures.'

But was it really surprising that Teller should be so upset? After all, Ulam had selected himself as the person to explore the feasibility of the classic Super design and Carson Mark was to comment that Ulam actually "took real pleasure" in disproving the ability of the Super because he knew Edward would feel pain over the disproof. Now the two men were locked in an unholy mis-alliance that was not of Teller's making. Theirs was a dispute that would echo on into following decades. Teller began sensing conspiracies everywhere and always Oppenheimer's influence seemed not far away.

One of those he believed to be heavily influenced by Oppenheimer – 'Oppie's men', he called them – was Los Alamos director Norris Bradbury, and this was another significant relationship to go from bad to worse during the crucial years of the crash programme on the H-bomb.

'Teller thought that Bradbury was a boob,' Herb York, who worked with both men, said. 'He said that he didn't know his job, didn't understand nuclear weapons, and didn't understand the urgency of the situation. He had nothing but bad feelings about Bradbury. They just totally disliked each other.'

George Cowan saw another reason, as well as these personality clashes, why the scientific leadership at Los Alamos and elsewhere resented Teller's presence:

> He was always being accused of overstating present capability, but he was dependent on it, this exaggerating the situation, overstating the urgency, to get the machinery for obtaining his objectives one way or the other. Those who felt Edward overstated things possibly resented his capability to engineer events. Scientists don't normally spend their time politicking in Washington.
>
> Norris Bradbury did not do that, nor did Oppenheimer who was persona non grata with the military. But Edward was a hero of the movers and shakers in Washington and, I suspect, was deeply resented by people like Norris, people who thought that scientists should not be politicians.

With so many clashes of personality and such a head of political steam developing, it was clear that even the smallest technical disagreement might spark a furious row; but the problems with Teller's classic Super design were anything but small.

Right from the beginning of his work on the Super, Teller had been struggling with the problem of how to ensure that there was enough energy to sustain the thermonuclear burn. The main approach, so far, had been to lower the energy requirement by incorporating tritium, which burned easily. However, any attempt to retain and focus the energy that was generated by the fission trigger had so far failed. Early in the process the use of compression had been discussed, but Teller had come firmly to the view that, while it might increase the speed of the reaction, compression would not affect the fatal loss of energy through radiation.

Over the years many people, coming new to the problem, had tried to argue that compression might have a role, but Teller had become increasingly dogmatic in rejecting the idea. By 1950 he had even worked a categoric statement of his opinion into his standard briefing on the Super: 'Compression makes no difference,' he used to say.

In April 1950, however, an extraordinary and paradoxical event occurred that introduced a wholly new element into the compression debate. The physicist Arnold Kramish brought an intelligence report to Los Alamos. An Austrian physicist named Schintlemeister, who had been a POW of the Russians, had recently been repatriated after having worked as a labourer at one of the Russian weapons laboratories. He had reported that one of the Russians' most senior and well-respected scientists, Peter Kapitza, was experimenting with magnetic compression of deuterium cylinders, and on 18 April Kramish briefed a group of Los Alamos scientists that included Teller. He certainly kindled interest and returned to Los Alamos several times later in the year, 'to bring Teller and de Hoffmann up to date on intelligence indicators'. Kramish wrote, 'Each time Edward returned to refinements of the Schintlemeister report.'

However, this new clue to the importance of compression, from the Russians of all people, did not result in any significant changes in approach to the problem. Teller continued to argue and bully, damaging friendships and working relationships, fostering suspicions of those who disagreed with him. 'Now my only worry is how to keep from being completely exhausted each day,' Teller wrote to Maria Mayer in early spring, 'and how to get up on time for my first appointment each morning. Gradually I am doing what the Leopard and the Ethiopian cannot do (you know about spots and skins) . . . I have actuall(y) sat upon (and temporarily suppressed) other eruptive personalities. I am behaving like a Goddam Administrator.'

Obviously the pressure to achieve results was telling on him as well as on others. But if he was referring here to Ulam as one of the 'eruptive personalities', then the suppression must indeed have been temporary. During the spring Ulam had been to Princeton to brief John von Neumann so that he could mirror and check their calculations using the ENIAC computer. He had also met and discussed his calculations with both Fermi and Oppenheimer. Oppenheimer, Ulam thought, seemed rather glad to hear of the difficulties.

By mid-May Ulam was telling von Neumann that 'the thing gives me the impression of being miles away from going', and by mid-June the second report was ready: it made the classic Super look virtually impossible. Instead of the 100 grams of tritium for each weapon that Teller had esti-mated, 3–5 kilograms would be needed, thirty to fifty times Teller's esti-mates. On the optimistic basis that the AEC reactors could produce a kilogram of tritium a year, each bomb would absorb up to five years of reactor effort.

'Teller was not easily reconciled to our results,' Ulam wrote in under-statement. 'I learned that the bad news drove him once to tears of frus-tration, and he suffered great disappointment. I never saw him personally in that condition, but he certainly appeared glum in those days, and so were other enthusiasts of the H-bomb program project.'

This was not to be the end of the setbacks. Enrico Fermi had returned to Los Alamos for the summer to help with the thermonuclear research and he and Ulam had decided to tackle the next question posed by Teller's design: would a tank of liquid deuterium, successfully ignited at one end, result in a self-sustaining burn?

The two scientists investigated the process in time-step stages, with, as Ulam recalled, 'intuitive estimates and marvellous simplifications intro-duced by Fermi'. Working with Fermi was one of his postdoctoral students,

Richard Garwin, who had a desk in Fermi's small office and watched the painfully time-consuming Super calculations as they proceeded throughout the summer. He remembered 'Fermi filling in spreadsheets – down the side times in thousandths of a microsecond and across the page would be different radii for the cylinder, different temperatures, different neutron densities ... At that time they had certainly no clue of doing anything other than depositing amounts of energy in the cylinder of deuterium and seeing if it would explode.' The result was another negative. 'You can't get cylindrical containers of deuterium to burn because the energy escapes faster than it reproduces itself,' Garwin recalled. 'With this calculation, they established for their own purposes that the classical Super could not work ... All the time [on the classical Super] was wasted. There had been miscalculation, because Teller was optimistic.'

It was another serious blow. Fermi was a good friend, and could not be accused of sabotage as Ulam had been. To compound matters further, these results coincided with an event that many thought was the start of the Third World War.

At 4 a.m. on Sunday, 25 June 1950, the powerful North Korean army, equipped by Russia, attacked along the length of the 38th Parallel dividing North and South Korea. Some 14,000 of the invading troops were ethnic Koreans from China, but the invasion was supported by both Russia and China.

The largest contingent of 100,000 US troops was in Japan, but in Korea they would have to face some 200,000 North Koreans and, no doubt, millions of Chinese who could be brought in to join them. Any reluctance to commit against such overwhelming odds was swept aside, however, when, within days, the South Koreans themselves ceased to fight. On 30 June US troops went in, but already the South Korean capital, Seoul, had been taken. 'I can't believe that Korea is it,' Teller wrote to Maria. A national emergency was declared, and the military began seriously considering the use of nuclear weapons. Curtis Le May, the hawkish head of Strategic Air Command, believed it was a way to end the war as quickly as it had been started. 'Look, let us go up there,' he recalled saying at Pentagon meetings, 'and burn down five of the biggest towns in North Korea – and they're not very big – and that ought to stop it.' He was prevented from making such a move, but there was no doubt about the dependence of the US and the West on nuclear firepower. The pressure on Los Alamos to deliver was now immense.

In theory, the national emergency should have worked to Teller's and the Super's advantage. Certainly some scientists, who had so far resisted weapons work, now offered their services and, during 1950, the numbers

The Parliament Buildings, Budapest, in the early 1900s. For much of Teller's childhood, his family lived in an apartment overlooking them.

Graduating from the Minta Gymnasium, 1925. Only in his last years there did he make friends and for much of his time he had been 'practically an outcast'.

Edward Teller, in his early twenties, at Leipzig. His acceptance by his peers, he wrote, 'amazed me and gave me great pleasure. But I had made a very conscious effort to be pleasant and to fit in.'

In the company of gods. An international assembly at Copenhagen in 1933. Front row (L–R) Bohr, Dirac, Heisenberg, Ehrenfest, Delbruck and Lise Meitner of fission fame. Teller is in the second row, third from left, with von Weizsacker on his right.

A transport tableau at Copenhagen, 1931. Lev Landau sits on the trolley, Teller himself is on skis and Gamow, with Bohr's son Aage on pillion, rides the motorcycle he and Teller used to tour Denmark.

Right George Placzek. During their scientific collaboration, Teller was so intimidated by him that colleagues said, 'he apologised for being alive'.

The Russian George Gamow in 1938. Ebullient, larger than life and a brilliant physicist, he attempted to escape from the Soviet Union by rowing across the Black Sea.

Göttingen 1931. Teller (L) stands next to Maria Mayer and her new American husband, Joe Mayer. James Franck, who was to help Teller flee the Nazis, is on the right.

Maria Mayer with her daughter Marianne, c. 1935. It was during the war years that Edward and Maria's friendship 'grew deeper'. Their revealing correspondence continued throughout the 40s and 50s.

Hans Bethe (R) and Isidor Rabi (centre) with a colleague in Copenhagen in the 1930s. Bethe was to describe the 'gentle Teller' of their early friendship. Rabi was never a friend and later became an implacable opponent.

The passport Teller used to enter the US in 1935 to join Gamow as a professor in George Washington University's physics department.

Luis Alvarez (R) stands next to Robert Oppenheimer at San Diego zoo in 1938. Oppenheimer's close associate Robert Serber is on the left.

Einstein (L) with Leo Szilard in the late 1940s re-enacting the writing of the 1939 letter to Roosevelt. Teller was present at the original meeting, helping with the drafting and acting as Szilard's chauffeur.

Scientists involved with Briggs's Uranium Committee, *c.* 1942: (L–R) Major Thomas Cranshaw, Robert Oppenheimer, Harold Urey, Ernest Lawrence, James Conant, Lyman Briggs, Eger Murphree, Arthur Compton, Robert Thornton, Kenneth Nichols.

Housing at Los Alamos, each containing four family apartments of the kind occupied by the Tellers. In spite of Edward's night-time piano playing and 'Hungarian temperament', the Tellers were regarded with affection by their neighbours.

Right Edward, with son Paul astride his shoulders – 'his favorite perch' – chats with colleagues at a Los Alamos party.

Staislaw M. Ulam

Emil J. Konapinsk

Los Alamos security passes for the Polish mathematician Stanislav Ulam and Emil Konopinski, the American physicist who proposed the use of tritium to help initiate fusion.

Teller and Los Alamos director, Norris Bradbury (R), at a party in 1947. By then the two men were already sparring over the priority given to the Super.

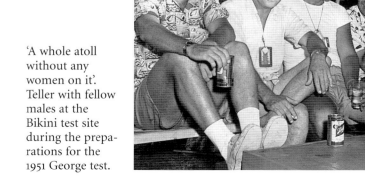

'A whole atoll without any women on it'. Teller with fellow males at the Bikini test site during the preparations for the 1951 George test.

The Mike device in situ on Elugelab. The 'sausage' is directly behind the forklift, surrounded by refrigeration equipment to cool the liquid deuterium fuel. The flanged pipes on the right carried early bomb light to distant streak cameras.

Mike, 1 November 1952. The yield of 10.4 megatons was more than twice that expected, vaporising the island of Elugelab and blasting 80 million tons of rock and coral into the atmosphere, which would fall out around the globe.

Livermore in 1952. Once a naval air station, the former medical centre served as a laboratory. Security-sensitive meetings were held in a car on the distant runways.

Teller with Ernest Lawrence (L) and Herb York *c.* 1957. Lawrence had insisted that the thirty-one-year-old York should become Livermore's first director in 1952 rather than Teller.

Oppenheimer and John von Neumann with Princeton's latest computer in 1952. Oppenheimer already knew that powerful political enemies, primed by Teller, were trying to destroy his influence on nuclear affairs.

in the Los Alamos Theoretical Division more than doubled. To add to his growing sense of frustration, however, Teller found himself fighting a priorities battle with the demand for the development of tactical weapons small enough to use on the battlefield in Korea – and it was a demand supported by Oppenheimer's General Advisory Council. In spite of the presidential order, the Super was still seen by many, and certainly by the GAC, as a long-term project that, in time of national crisis, was to be edged on to the back burner. Teller told Maria about a visit to Los Alamos by Oppenheimer, which gave him an infuriating sense of being patronised or manipulated, or both. 'It is wonderful how he can anticipate what I am going to think,' he wrote, 'and how he can get around me by agreeing with me in advance.'

Teller, however, was not about to throw in the towel. While Fermi and Ulam continued with their calculations on the classical Super design, he began working on the plans for the initial tests of the thermonuclear principles. This Greenhouse series had been proposed some nine months previously and was still scheduled for the spring of 1951. Teller felt it essential to make at least some sort of show for Washington. Thus, even though there was so much uncertainty about, work began on mounting two tests as part of the Greenhouse series, on the Pacific atoll of Eniwetok. One, codenamed Item, was a test of a fission bomb, enhanced by an addition of fusion material to its core – the booster. The other, codenamed George, was the more significant, as it would test a device known as the 'Cylinder', which was intended to demonstrate the crucial element of a fully fledged thermonuclear device. This was a capsule containing less than an ounce of a deuterium–tritium mixture, which was to be triggered by a 200 kiloton fission device, one of the largest exploded to date. As one scientist put it, 'It was like using a blast furnace to light a match.' Because the fission explosion was so large, the cylinder was used to separate off the tiny fusion element so that the team could better observe what was going on – if anything. The fusion mixture was still kept fully exposed to the heat and radiation from the fission reaction and so began to mirror an earlier configuration proposed for the Super.

'It turned out that it was very much the pattern of the Fuchs–von Neumann patent,' Carson Mark has said. 'It was not by any means a copy from that but it did about the same thing.'

The Fuchs–von Neumann patent, drawn up after the 1946 Super conference, proposed separating the fusion material from the exploding fission trigger in order that it might be compressed by the radiation energy from the trigger and the reaction thereby intensified. Seemingly by accident, Teller's configuration of the George test was moving towards a possible solution to his problem. If the test showed that fusion could be made to work, even experimentally, it would at least have some value politically.

Throughout the first months of 1950 the security services had been searching for the unknown subject 'Raymond', who had been Klaus Fuchs's contact, and on 23 May they arraigned Brooklyn-born engineer, Harry Gold. At the same time they had also been searching for the second spy they believed had operated at Los Alamos during the war, and the FBI's Albuquerque field office had begun scrutinising possible members of the scientific staff. Stanislaw Ulam was a suspect. So too was Teller's friend Victor Weisskopf, but their 'most logical suspect for Soviet agent' was Teller himself.

The web of associations they wove around him has a disturbing plausibility. He was a close associate of Fuchs at Los Alamos. At the end of 1945 Mici Teller had travelled to Mexico City on the vacation trip with Fuchs and Rudolph and Genia Peierls. Fuchs had been to dinner at the Tellers' when he had returned briefly to the US in 1947 and Teller had had considerable contact with Fuchs in England when he had passed through during the summer of 1949; but besides his association with Fuchs, he had also recommended for post-war graduate study a man with whom he had worked at Los Alamos and who had been identified as a Soviet espionage agent whilst at Los Alamos. Teller's name had appeared on the man's list of possible recruits. Unlike most of his fellow scientists, he had made frequent trips elsewhere from Los Alamos and had travelled to New York at times that matched the NKVD's cable decodes. The investigation and the accompanying freewheeling speculation might well have continued, had a conscience-stricken Harry Gold not begun cooperating fully with the FBI and named his other contact at wartime Los Alamos. So well did this man, a lowly engineer by the name of David Greenglass, fit information the FBI had that Teller was dropped from their investigations.*

In spite of the tensions surrounding the Super, Teller still found time for, and took pleasure in, life outside the laboratory. Even with the one-day weekends resulting from the laboratory's six-day week he managed to spend time with his children and their friends, telling stories, taking them swimming, going to the movies, and (along with Mici, John Wheeler and his wife) attending Indian dances; and he still played as a member of a poker school, which included both those he saw as allies and those he had described as 'eruptive personalities'. It included Stan Ulam, and Carson Mark, in whose house the school met. 'This was a fully talented group of people, good minds,' Mark recalled, 'and Teller and John von Neumann were among the best. Edward and von Neumann had a similar style; they both had a problem keeping their minds on the game. Every now and then they could make an outstanding mistake, a simply stupendous mistake. The evening's winnings were usually about $20, which was not very impressive, but winning was very important.'

As autumn approached, he heard sad news from Hungary of the death of his father. Edward had not seen him since he and Mici had departed for America fifteen years previously. This enforced fragmentation of the family, the knowledge of his powerlessness to help them during the war, the fact that grandfather and grandchildren would now never meet, all deepened the sense of regret. He was now also deeply concerned about the fate of his mother and sister, both now widows and, along with his nine-year-old nephew, in a desperate situation. 'As long as my father was alive,' he said, 'there was some hope that he would be able to marshal help from his large circle of friends and acquaintances to moderate their difficulties. Now they were alone. My attempts during the past three years to secure permission for them to emigrate had proved fruitless.' He approached the State Department for help, but they found that Teller was a person so well known and notorious on the far side of the Iron Curtain that nothing had been achieved.

It was a worrying time and various issues were coming to a head. By now, Norris Bradbury had started planning things beyond the Greenhouse test series, but had stopped any further design work on the Super until the test results were known. Given the almost universally negative findings so far, this decision was hardly surprising, but Teller took it badly: 'I was becoming truly discouraged, the more so for having been instructed not to think about what was now the most important problem of all to me; the design of the hydrogen bomb. Because I couldn't work on the design, I decided I might as well not get excited about the question until Johnny [von Neumann] had the results from the electronic computer.'

That autumn the GAC paid a visit to the facilities at Los Alamos. Although Oppenheimer was still chairman, the committee's membership had changed and it had become increasingly hawkish. Its most eminent new member was the chemist Willard Libby, a friend of Teller, who, four years earlier, had developed the technique of Carbon 14 dating, for which he was later to win the Nobel Prize. However sophisticated he may have been as a chemist, Libby's political outlook was straightforward: 'I'm an ultra-conservative,' he said. 'I believe in invention and in great new discoveries. I don't believe in mediocrity. And democracy is pretty mediocre usually – I guess one early influence was my old daddy. Tough as nails ... He had no use for hoboes and he was the origin of my belief in conservative policies.' Libby also had a straightforward view of Oppenheimer, believing him to have been 'more or less the head' of 'a strong Communist contingent at Berkeley' during the war. With the arrival of the new intake, Oppenheimer had offered to resign, but the new chairman of the AEC, Gordon Dean, had persuaded him to stay.

Now he was showing them what kind of a place Los Alamos was. The meeting took place in the shadow of rumours about the construction of a second laboratory, ostensibly to ease the load but really to provide competition for the mother lab, and for this and other reasons the tenor of Oppenheimer's meeting could not have been an optimistic one. According to Teller, Bradbury reported optimistically on small tactical fission weapons, but when it came to discussing the thermonuclear research, Teller felt Bradbury's report 'so negative that it seemed an outright attempt to squash the project'. There was, of course, little positive to report. The reviews of the Ulam–Everett and the Ulam–Fermi calculations were hardly the basis for optimism. One showed that the fusion fuel wouldn't ignite, while the other showed that, if ever it did, it would not sustain itself. Oppenheimer was able to provide a brief spark when he enthused about the plans for the Greenhouse test, in particular about the Cylinder, and its relevance to future thermonuclear models; but then Teller took the floor to summarise work on the Super. The official history takes up the story:

> In his briefing he could offer little more than determination. He saw more theoretical work was essential. He thought Los Alamos lacked people to perform the detailed calculations and to carry on imaginative thinking. More than once he stressed how much more there was to explore. He admitted to von Neumann that the practicality of the Super depended on the amount of tritium that might be needed and that the trend was unfavourable. He had no new ideas. In some way success would be grasped – how, he did not know...

Teller was damning the project from his own mouth and, furthermore, he was putting the blame squarely on his Los Alamos colleagues. 'If the [George] test showed the Super was possible – the laboratory might not be strong enough to exploit the triumph,' he wrote. The two factions were doing a wonderful job in alienating one another. So angry were Teller and John Wheeler about the presentation that they decided to write a supplementary report that discussed the current situation in a very different tone. This simply served to harden attitudes still further because in it Teller and Wheeler mentioned openly for the first time the possibility of setting up a laboratory in direct competition to Los Alamos. Teller justified this as appropriate because 'a successful test could open up a wide unexplored field of nuclear weapons . . . Los Alamos alone would not have the capability of answering all the important questions'. So here was the leading scientific figure behind the crash programme to produce the Super – indeed behind the Super itself – who, some months after the presidential declaration, had admitted to being bereft of ideas for overcoming its current problems, yet still felt able to blame his colleagues for those problems, in effect, proposing that he should now go it alone with the H-bomb.

True, he was under pressure and felt, with some justification, that he was surrounded by unsympathetic colleagues who did not appreciate the potentially disastrous international situation. True, Stan Ulam, a mathematician with whom Teller had never been friendly, had appointed himself to perform a detailed critique of his classical Super model. True, also, he was frustrated by senior scientists centred around Oppenheimer, who abhorred the thermonuclear programme and did whatever they could to change priorities. But surely his arrogant, aggressive performance at the GAC meeting, followed by his and Wheeler's report, was the worst way of dealing with the situation. He had added to those he had alienated many who were not directly opposed to him and had set a confrontational tone for the future. It was fast becoming an intolerable situation and Teller's old friend could see the effect it was having on him. 'I went as an extra consultant to Los Alamos that winter,' said Hans Bethe, 'and found they had set up a major group with Wheeler, who was a substantial scientist, and had made Edward the project's assistant director. I found Edward desperate. Desperate. He could at that time have fits of deep depression, and obviously didn't know what to do and yet still desperately wanted to do it.'

Françoise Ulam remembered:

I overheard some of his temperamental outbursts, felt the tension and bafflement they generated, as well as the political and administrative tug of war, which culminated in Oppenheimer's undoing and Teller's ascendancy. I had the feeling that nobody quite knew how to handle him and sort of caved under. All, except Stan. In short Teller turned what should have been the normal joint examination of very difficult problems into an unpleasant confrontational struggle. This, I believe, is when his thinly-veiled dislike of Stan began to smoulder.

George Gamow, who was coming to the end of his consultancy at Los Alamos, found a characteristically ironic way of showing where the Super stood. The physicist Theodore Taylor remembered an incident at a seminar: 'Gamow placed a ball of cotton next to a piece of wood. He soaked the cotton with lighter fuel. He struck a match and ignited the cotton. It flashed and burned, a little fireball. The flame failed completely to ignite the wood, which looked just as it had before – unscorched, unaffected. Gamow passed it round. It was petrified wood. He said, "That is where we are just now in the development of the hydrogen bomb." '

As 1950 drew to a close, the situation continued to sour. In December, John von Neumann reported on his machine calculations carried out on ENIAC. His results bore out all the manual calculations of the past year and showed beyond argument that the classical Super would not work. For Edward Teller, it was the end of a long, tortuous, unhappy road, from those early optimistic calculations at Berkeley in the summer of 1942, through the years of marginalisation at wartime Los Alamos, and into the barren post-war years. Finally, it had dragged on into the last miserable year, which had seen irreparably damaged relationships amongst the physics community.

The final blow was to come right at the year's end and it came from a committee chaired by Oppenheimer. This Long Range Planning Committee was charged with looking at the future of atomic weapons over the next decade. The committee included Charles Lauritsen, the highly respected President of Cal Tech, Robert Bacher, and the President of Bell Telephone Labs, Mervyn Kelly. All of these were broadly in sympathy with Oppenheimer over the possibility and need for the H-bomb, but Oppenheimer had invited Luis Alvarez to join the committee, specifically because he represented a different point of view from his own. It was a move that could be seen as either very fair, or the absolute bare necessity given the other members' views. There were by then discussions about smaller tactical weapons, and about the warheads for use on guided missiles. However, when they came to discussing the H-bomb, Alvarez recalled Oppenheimer saying 'that the hydrogen programme was going to interfere seriously with the small-weapons programme by taking away manpower at Los Alamos.' Alvarez disagreed strongly, and said so, but, nevertheless, the final report read as follows: 'In fact, we believe that only a timely recognition of the *long-range* character of the thermonuclear programme will tend to make available for the basic studies of the fission programme the resources of the Los Alamos laboratory.'

It was a clear recommendation that Los Alamos should concentrate on fission weapons. Alvarez had signed the report, so how had he missed it? Had he been duped, or hypnotised? Teller was furious, and some weeks later, when he met Alvarez, he challenged him. Alvarez's explanation was that he had been a victim of Oppenheimer's skilled politicking. He had been asked – and had agreed – to go to Pasadena, where he had considered the report in Bacher's and Lauritsen's presence. If the report had been sent to him at Berkeley, then he might have consulted Lawrence and might have had much more opportunity to view the report in a different light. As it was, he believed himself to have been outmanoeuvred. 'Oppenheimer was in effect practising Political Science 4B,' he said, 'on the postgraduate level, while I was still on political science 1A, the freshman course.'

However, while he saw nothing sinister in this, Teller did. For him and for other supporters of the thermonuclear programme, it was yet another questionable act by Oppenheimer.

Elsewhere, William Borden, Executive Director of the JCAE, had been working on the investigation into mismanagement at the AEC and, in particular, following the arrest of Fuchs, the security clearance procedures that the commission used. In November 1950 he asked the AEC to provide him with the files of ten or a dozen of the employees they considered among their most difficult 'security cases'. One of the files he was given was Robert Oppenheimer's.

13

A 'Simple, Great and Stupid' Mistake

Teller's father, he recalled, had a saying that he translated approximately as: 'Things will surely turn out somehow, because it's never happened yet that they turned out no how.'

Things were about to turn out somehow: an eleventh-hour rescue from desperation and disgrace. What is more, the rescue was to be prompted from a totally unexpected quarter, and was to generate a dispute over credit that would intensify the already deep rivalries and hostilities throughout the atomic weapons establishment.

At the same time as he had been working on his assessment of the Super, Stan Ulam had also been considering ways to improve the performance of fission weapons. Thus far, the scope of implosion had been limited by the power of high explosives needed to compress the uranium or plutonium core, and Ulam had started to look at the possibility of using shock waves from another fission bomb to produce the implosion. Only a small explosion, which nevertheless produced a flux of neutrons, would be required to provide the necessary shock to the main assembly.

In early January 1951 he went to see Carson Mark with his idea, explaining how this use of a small fission trigger would need staging. The 'primary' fission explosion would set off the main 'secondary' explosion, which would be physically separated from the first. Mark thought the idea elegant but saw no immediate use for it. However, a week or two later Ulam received a memo from Darol Froman soliciting views on how to crack the stalemate on the Super. It was this stimulus that triggered his realisation that his new 'iterative' concept, with primary and secondary stages, could be applied to the fusion weapon.

Françoise Ulam remembered the moment one day in late January:

Engraved on my memory is the day I found him at home at noon staring intensely out of a window with a very strange expression on his face. I can never forget his faraway look as, peering unseeing in the garden, he said in a thin voice – I can still hear it – 'I found a way to make it work.'

'What work?' I asked.

'The Super,' he replied. 'It is a totally different scheme, and it will change the course of history.'

I, who had rejoiced that the Super had not seemed feasible was appalled and anxiously asked him what he intended to do. His reply: He would have to tell Edward. My head spinning, and fearing in addition that Edward might pounce on him again, I ventured that maybe he ought to test his idea on someone else first.

Ulam took his wife's advice and, after lunch, went back to see Carson Mark, who, however, was preoccupied with arrangements for the upcoming George test, and did not grasp the significance of what Ulam was saying. 'I know that Stan could be oblique and pithy at times,' Françoise has written, 'especially when he was so preoccupied with a topic, that he assumed his interlocutor was on the same wavelength and already knew what he was talking about.'

Ulam also went to Bradbury and then, the following morning, to Teller. 'For the first half-hour or so during our conversation,' Ulam recalled, '[Teller] did not want to accept this new possibility,' but eventually, after a few hours, he took up Ulam's suggestions 'hesitantly at first' then 'enthusiastically'.

There is, however, another version of these events. According to Teller's last account – his account of this crucial incident in the history of the bomb programme evolved over the years – some two months earlier, in December 1950, he had been in his office when Carson Mark popped his head round the door:

> By that time [Teller wrote] Carson had joined the number of those who sided with the General Advisory Committee . . . The immediate purpose of this visit was to tell me about a conference that he and some other administrators had just had with an important visitor, an admiral. They had told the admiral that the Super was definitely not feasible and then had tried to explain some of the difficulties. Carson found it hilarious that the admiral's only answer amounted to 'Damn the torpedoes; full speed ahead.'
>
> Carson made it a practice to needle me in a subtle manner. Until that moment, I had brushed off his teasing as harmless. But his ridicule of

what he took to be the admiral's stupidity infuriated me. I knew that I would give him pleasure by showing my anger or trying to defend the admiral, so I remained silent ... I expect he has forgotten the incident but the flood of adrenaline he engendered in me had real repercussions.*

That afternoon Teller recalled beginning a review of the hydrogen bomb, reconsidering every idea that had gone into its planning, and looking for a possible mistake or a new idea. Right at the beginning they had realised the key problem that, while a fission reaction could certainly produce enough energy to heat the relevant particles to the temperatures required for fusion, a great deal of that energy was wasted as radiation.

In the classical Super design, Teller relied on the fact that energy from the fission trigger was first absorbed by the particles involved in the fusion reaction – the nuclei, electrons and other smaller subatomic particles. There is then a period of time, measured in millionths of a second, before that energy is transformed into radiation – into gamma rays, alpha rays and light photons, which then carry energy away from the fusing nuclei. Teller's aim had been to produce a situation where fusion occurred in that tiny window of time when the particle energy was at its peak. In order to achieve this, there had been developments in more efficient fission triggers, and the incorporation of tritium into the fusion reaction because it required so much less energy to burn than deuterium.

As to controlling and reducing the radiant energy loss from the fusion reaction, however, that remained a puzzle. Many people had raised the possibility of using compression, but Teller had remained adamant that this would just increase the speed of all the reaction processes, including energy loss. That afternoon for the first time, he recalled, he realised he was mistaken. What he had failed to take into account was just how light photons would behave under the extreme high temperatures of fusion. He had discounted how, in the presence of both the nuclei and the electrons of the fusing material, the light photons' energy could be reabsorbed. Furthermore, because there were three particles necessary for the process – photons, electrons and nuclei – it would be much more susceptible to the effect of pressure in bringing them closer together than other parts of the reaction that involved just two particles.

The absorption of light is usually of minor importance [Teller wrote], but it increases with the cube of compression, rather than the square, because three, rather than two, participants are involved. Thus, if the deuterium is strongly compressed, the absorption of radiation changes from a minor factor into a process crucial to a different and much simpler design, a thermonuclear reaction in equilibrium. My mistake in neglecting that process was simple, great, and stupid!

According to Teller's last account, this sudden moment of inspiration marked the birth of his new concept – some six weeks to two months earlier than Ulam's realisation of the possibilities of compression:

> Within an hour of Carson's derisive remarks, I knew how to move ahead – avoiding the torpedoes. Thus, almost at once, the new plan appeared to be ready. Many important details still had to be settled ... but the principles of the new plan would serve as the basis for the development for many years. I cannot emphasise enough that this development was unduly delayed. My single-minded focus on ideas formulated much earlier was an extreme case of mental inertia.

Given the damaging effect on relationships throughout the scientific community – of which more later – the delay caused by his 'single-minded focus' over at least five years could be described as tragic – for everyone.

His mental inertia overcome, Teller's first thought was to go to see Bradbury. However, as he put it, 'Bradbury could hardly have been more unlike Oppenheimer,' in terms of both his immediate grasp on the validity and implications of new ideas, and his flexibility of action; and, of course, Bradbury's edict – that no new ideas about the hydrogen bomb could be discussed until after the Greenhouse results were known – was still in effect. 'So the best I could do was to discuss my idea quietly with a few people' – which he did, with Johnny von Neumann, with Freddie de Hoffmann and with Darol Froman.

The first two responded positively, but Froman 'would not really listen. I must assume that he did not understand the new approach because he had not understood the old one. Or, perhaps, I made a mistake in my presentation. I probably began my discussion by telling Froman, "I have a new idea." I should have started by emphasising that I was stupid.'

On Teller's forty-third birthday, 15 January 1951, the plans for the Pacific tests were finally accepted with agreement on every essential point: 'It almost seemed like a birthday present that, at last, the conditions of Bradbury's edict had been met.' But Bradbury again refused him permission to talk about the design of the H-bomb. For nearly two months, according to the normally forceful Teller, he then sat on a revolutionary new plan which, as he put it, 'would serve as the basis for the development for many years'.

Bradbury, he recalled, knew that he had upset him and, in an effort to improve their relationship, invited Teller up to Nevada to look at the laboratory's new test facilities there. While there, he kept offering Teller advice, telling him that if he worried less he would be more effective. After the visit to the test site, Bradbury took Teller to a striptease at a nightclub in Las Vegas, 'in the hope, I suspect that it would help me follow his advice. I spent the evening as usual – worrying. What mistakes had I made in my relations with colleagues and administrators at Los Alamos?'

This brings us to the time when Ulam came to see him. In an earlier account Teller at least partially corroborated Ulam's version of events, saying that Ulam contributed 'an imaginative suggestion'. However, he later described the meeting very differently:

> He [Ulam] announced that he had an idea: Use a fission explosion to compress the deuterium and it would burn. His suggestion was far from original: Compression had been suggested by various people on innumerable occasions in the past. But this was the first time that I did not object to it. Stan then proceeded to describe how an atomic explosive should compress several enclosures of deuterium through hydrodynamic shock. His statement excluded my realisation of why compression was important, and it also included details that were impractical.

The 'several enclosures' is a reference to Ulam's 'iterative' approach: the primary fission explosion compressing the deuterium secondary – whether by shock or radiation – in a separate but interconnected container. This was a crucial element in Ulam's new approach, but Teller brushed it aside – for the moment – going on to tell Ulam how much more effective radiation would be than hydrodynamic shock in producing symmetrical compression. 'But Stan was not interested in my proposal and would not listen. I told him I would write both proposals, and we would sign it as a joint report . . . In that paper I wrote down my new plan for the first time. I explained how it would work and why it was better to compress the deuterium through radiation.'

Now with Ulam's backing, Teller said, he was in open rebellion against Bradbury's edict, and submitted the new plan. In the meantime, Ulam's 'iterative' approach had not actually gone unnoticed by Teller. 'The next day I asked Freddie de Hoffmann to write a detailed description of how the transfer of energy from the primary fission explosion would compress the deuterium and produce a secondary thermonuclear explosion.'

There, for the first time, is the iterative primary-secondary configuration referred to by Teller, its origins unspecified. At the very least Ulam's idea had added to the body of his thinking. Arnold Kramish, who had introduced Teller to the intelligence brought out of Russia on the use of compression, was involved in the calculations with de Hoffman:

Edward said that Ulam was on track but hadn't gotten it right . . . At the end of the afternoon, Edward said he had to go to dinner with family and play the piano. In his polite and forceful way, he said, 'That's it. Why don't you three work a little while and we will discuss the meaning of the solution in the morning?' The 'little while' was all night . . . Working with our Frieden and Marchand calculators we came up with an approximate solution in the morning. Teller, refreshed, became absolutely joyous, while we three were on the point of collapse.

That all-night session resulted in a paper, 'An Estimate of [deleted] Temperatures', issued on 4 February 1951. Over the following month, according to Ulam, he and Teller met several times, discussing the problem of energy transfer for about half an hour each time. 'I wrote a first sketch of the proposal,' Ulam wrote: 'Teller made some changes and additions and we wrote a joint report quickly. It contained the first engineering sketches of the new possibilities of starting thermonuclear explosions – we wrote about two parallel schemes based on these principles.'

The joint report that followed was published on 9 March, crediting work done by E. Teller and S. Ulam, and entitled 'On Heterocatalytic Detonations 1: Hydrodynamic Lenses and Radiation Mirrors'. In it they summarised their concept: 'The scheme then depends on concentrating, as much as possible, the energy released by the explosion of a fission bomb in the mass of the principal assembly and doing it so as to achieve a high compression in this mass.'

In the weeks and months following the issuing of their report, the co-authors had less and less to do with one another. Teller essentially took centre stage to move the project forward. Later, in March, he added another design feature, which, along with radiation compression, was to become part of the essential H-bomb concept. Along the axis of the cylinder of deuterium fuel, he incorporated a subcritical rod of U-235 or plutonium. This was another step towards ensuring that ignition could take place. As the cylinder imploded, it would compress this subcritical rod – the 'sparkplug', as it was called – turning it into a critical fission explosive, pushing outwards against the main implosive forces and enhancing the pressures generated.

As to Ulam, he had no stomach for political infighting. 'My impression is that, from then on, Teller pushed Stan aside and refused to deal with him any longer,' Françoise Ulam recalled. 'He never met or talked with Stan meaningfully ever again. Stan was, I felt, more wounded than he knew by this unfriendly reception, although I never heard him express ill feelings toward Teller. He rather pitied him instead. Secure in his own mind that his input had been useful, he withdrew.'

Perhaps as a form of self-preservation Ulam developed a dismissive stance to their joint discovery. 'Edward is full of enthusiasm about these possibilities,' Ulam wrote to von Neumann. 'This is perhaps an indication they will not work.' Although he was only one of a number of sceptics, Teller was to condemn his scepticism over the coming months, dubbing it disloyal. 'To me the authorship of a paper ... is a matter of responsibility.' Teller later said. 'If I sign a paper I am ready to stand up for it; if I don't want to stand up for it, I expect to have to explain why I changed my mind ... I don't know why Stan changed his mind.' It is easy to understand how the elusive, undemonstrative Ulam might have felt, increasingly 'elbowed' to one side, relieved of any credit by Teller; but just what was the balance between their contributions to an entirely new scale of weapons system, one which would refuel the arms race, and bring within reach the destruction of our own planet? And equally, how original was it?

In hindsight the discovery could be seen as a natural progression from the Fuchs–von Neumann patent, and from work already under way on the Greenhouse George test. For this test, in order to conduct basic studies of the deuterium–tritium fusion, the experimental capsule had been separated from the main fission explosion and remained linked to it only by a channel to conduct the flow of radiation. 'It's sort of obvious,' said Marshall Rosenbluth, 'once you start doing detailed calculations on radiation flowing out from a bomb to this little test experiment; then you think, why not use a radiation implosion for a secondary? ... The planning of the Greenhouse experiments, at least in my opinion, led Teller to the idea of radiation implosion.'

Carson Mark also saw an inevitability in arriving at the crucial importance of radiation as opposed to material shock waves:

> Had you sat down to design the thing, asking, now, what is the material shock doing, where is it, how fast does it move? You'd say, dear God, the radiation is going faster, it's there so let's concentrate on that. So it's hardly an important circumstance that Teller thought of radiation whereas Ulam thought of ... material [shock]. The fact that Edward thought of radiation was natural because he had been involved in much more detailed work on the George shot than had Ulam.

It was inevitable, perhaps, that these retrospective assessments would play down the originality of the discovery. With hindsight, many scientific discoveries can appear as a logical next step. However, it is always difficult to recreate the infinity of other possibilities that can engulf researchers, and it should be remembered that Carson Mark was head of the Theoretical Division at the time and one of a number of distinguished scientists, working more or less intensively on the problem, who failed to come up with the 'natural' pattern of thinking he referred to. So was Hans Bethe, but Bethe's more or less contemporary reaction, in 1952, was much more positive in assessing the achievement. He wrote that 'the new concept was to me, who had been rather closely associated with the program, about as surprising as the discovery of fission had been to physicists in 1939'.

When it comes to questions of credit and priority, public debate is blighted by the fact that the original paper is still classified. Furthermore, Teller shook off the normal trend in memory by recalling more and more detail of how the crucial ideas occurred to him as time had passed, rather than less. No one I talked to had heard the anecdotes about how Carson Mark's teasing had driven him to his new insight before he described it in his *Memoirs*, published in 2001. In the comprehensive interview he gave to George Keyworth twenty-two years earlier, in September 1979, just after a major heart attack, his memory of events was much less certain. 'I cannot tell you when I knew it,' he told Keyworth; 'I think it was in December 1950. It was not long before that January fifteenth meeting . . .' Later in the interview, referring to his realisation of the importance of the reabsorption of light energy, he said, 'All this was clear to me. I don't know whether all of it had been before the fifteenth January. I believe so.' There is no mention of the afternoon's calculations prompted by Mark's teasing and no elaborate rationalising of why he sat on his absolutely crucial realisation for nearly two months.

But Hans Bethe did believe that, early on, Teller had a notion of a solution that might have informed his thinking during the preparations for the George shot. 'The George shot was proof of an understanding,' Bethe said. 'And of course that understanding pre-dates Ulam. But if Edward meant the George shot as the way to do it, why didn't he come to me and tell me "That's the way we'll do it"? Why did he not explain to just a few people that the George shot was the key? And also, that being so, I don't understand Edward's depression at that time. There was definitely depression.'

Perhaps Teller was picking up cues from such things as the intelligence information from Russia about their interest in magnetic compression, and these cues were more or less subconsciously influencing the questions he was asking in the George test. Whether he was confident enough of a positive outcome to the test to overcome his depression, which was related as much to the antagonism fouling the atmosphere surrounding the Super project, is another question. Others close to the work at the time certainly saw Teller as leading the dance:

I see it as 99.6 per cent for Teller [said Marshall Rosenbluth]. I mean there are a lot of people, Ulam, myself, Dick Garwin, who did this and that but Teller was, first of all, the person who got the project going. He was obsessed by it. His group's work during the war was very useful for later developments. During his time at Chicago, he was probably the only major physicist spending time at Los Alamos. He was the one who kept the thing alive there. It was officially on the lab program but they didn't have the personnel or capability. The Keyworth paper is pretty fair. He was obsessively trying to pull it together in his mind. He had a better overall picture than anyone else and followed the physics quite clearly.

Herb York, who had begun his career in arms development working on the diagnostic tests for the George shot, also agreed that it was mainly Teller:

Ulam's idea, as a mathematician, was very general. That is, if you can contain the energy of a hydrogen bomb in a box for a short time – in other words achieve equilibrium – then there's all sorts of things you can imagine doing. To the extent Ulam was thinking about details, he was thinking of the energy moving around in the flow of a mass of particles. It was Teller who recognised that, no, the flow of radiation will be the main mechanism. They had in front of them the George design, and old ideas, and so Teller put them all together in a final version. And so everyone would say it's mainly Teller. Teller also agrees it was entirely him.*

Certainly the Teller–Ulam report won the political argument. 'Once it was a clear possibility, there was no longer an option,' George Cowan commented. 'Despite all the debates, no responsible government would ever voluntarily forgo developing a very powerful new weapon if it knew how to do it. That is something that you can talk about if you're not in the government, but if you are in the government it is not an option.'

Although the discovery of the role of compression gave everyone in the laboratory – metallurgists, engineers, cryogenicists – a focus for their efforts, the paper was followed by a lull in activity. Nothing very much happened. Foster Evans, a mathematician working closely with Teller, saw as the reason the fact that the Teller–Ulam report had 'outlined more than one alternative. Verification of the feasibility of Teller's proposal required some [additional] complicated theoretical work.' Teller, however, not unreasonably, saw it as yet another clear symptom of reluctance to move ahead with the new weapon and went to see the AEC chairman, Gordon Dean.

Teller was, by now, absolutely convinced that he was never going to achieve the progress necessary with the H-bomb at Los Alamos and that the only acceptable path open to him was to establish a second laboratory specialising in thermonuclear development. He and John Wheeler had already raised the possibility at the time of the GAC's visit to Los Alamos the previous September, and they knew that such a challenge would be resisted, not only by the management at Los Alamos but also by the GAC itself.

Yet again, however, there are effectively two versions of the meeting with Dean. According to Teller, he did not even raise the possibility of the second laboratory with him, but concentrated on enthusing him with his new concept for the H-bomb in the hope that he would 'help promote new efforts to move our work at Los Alamos ahead'.

Teller's memory was that, throughout the meeting, Dean was pre-occupied, 'unenthusiastic' about his description of the new concept. At the time he had put this down to the fact that, on leaving the meeting, he had found that the fly of his trousers had been unzipped throughout. Such, one might say, is the stuff of history: an administrator distracted from appreciating a major advance by a simple sartorial gaffe. Only years later, Teller reported, did he discover, on reading Dean's published diaries, that the chairman had been fully expecting complaints about the management at Los Alamos and a plea for a second laboratory. As this did not happen, Teller assumed he must have been too baffled by his long description of the new design to pay it much attention. Teller did not remember campaigning for a second laboratory at all at that time, and in retrospect stated that he 'would not [have thought] it seemly for anyone in full-time employ of an organisation' to do such a thing.

However, in a secret letter written some two weeks after their meeting, on 20 April 1951 and declassified in 1994, Teller did in fact send Dean a complete rationale for a new laboratory for thermonuclear work to be located in Boulder, Colorado. Furthermore, Norris Bradbury was on the document's circulation list.

'Following our conversation,' he wrote to Dean, 'I have given thought to the alternatives which present themselves concerning the future of the thermonuclear program ... The past two weeks have been too short to formulate such a plan but I have tried to arrive at an outline of manpower and space requirements, as well as some estimates of the cost of principal equipment.' This description belittles the scope and detail of his outline. In it he lists the divisions, from theoretical physics to photography and the staffing for each, arriving at a total of 360, and lab space of some 200,000 square feet to accommodate them. And he compared the merits of Boulder with three other locations.

It is truly difficult to reconcile this full description with Teller's own recollections. Suffice it to say that he did clearly talk to Dean about a second laboratory separate from Los Alamos, and the brief he sent the chairman shows every sign of being the product of considerable thought. Certainly, from this point on, the second laboratory became one of Teller's main preoccupations.

This, along with his later version of the controversy with Ulam over the credit for the new hydrogen-bomb concept is one more occasion on which Teller's own account of events clashed with documentary evidence, or with the memories of others, or became richer in detail over time. So is there an explanation for these differences? Certainly some do result from his worst-case analyses, his exaggerations and polemics in order to achieve a political end. As to the others, Paul Teller saw a heavy emotional component to his father's behaviour:

> He is one of the people who, more strongly than anyone else I have ever known, holds it of absolute importance to be honest ... Yet there is, of course, evidence that, at least in unimportant respects, his memories are incorrect, but nevertheless many of these are part of his mental history that I heard from the time I was a teenager. They are not something he has consciously fabricated ... Now there is a tension here. It can be resolved by questioning his honesty about his reports of how he saw things. Or it can be resolved by referring to the fact that when feelings are running high, people contrive very strong, even emotionally distorted reactions to these situations. And Edward Teller is a man of strong convictions and feels about them very strongly.

Others took a similar view: that he was essentially honest but that his strong convictions led to emotionally distorted reactions.* For others, however, the tension between truth and self-deception referred to by Paul Teller remained. One thing is certain, however: these perceptions of events did become an integral part of his judgement. As such they became a part of reality, influencing the actions he took and his attitude to others.

Shortly after the meeting with Gordon Dean, any further discussion of the second laboratory was forestalled by the build-up for the Greenhouse test series, scheduled to take place on the Pacific atoll of Eniwetok in May. 'A whole atoll without any women on it,' Edward wrote to Maria Mayer. 'In fact the nearest woman is 300 miles away ... The sadness was sincere, general and macroscopic. But on the day to day or microscopic basis, the sadness was not so great. In fact I have seldom seen so many cheerful faces. This may have had other reasons. But no women do have advantages: 1) more time 2) no need for bathing suits and 3) no racial problems.'

A team had been present on the Pacific atoll since the previous autumn, building the towers and the support facilities for the test series. On arrival they had wondered at the underwater wildlife on the surrounding reefs. They had felt the resonance of their present project in the rusty relics of the Pacific war – planes, tanks, ships – which had been bulldozed off the reef and lay half-submerged in the shallows. These sights, however, had soon palled and by the time of Teller's arrival they had been contending with months of 'no women', relieved by nightly open-air movie shows – it nearly always rained – cheap booze and endless games of poker. A joke publication was issued to all newcomers – *Sex Life on Eniwetok*. Its pages were blank.

Teller arrived some two weeks before the George test, when the race was on to meet the 9 May test deadline. There was also a realisation that George was not just an experiment but, as physicist Louis Rosen defined it, 'the vindication or non-vindication of [Teller's] major contribution'. Everyone was obviously frightened that Teller would insist on some impractical, even if important, modification at the last moment, so every effort was made to keep him occupied. He was, however, in highly optimistic mood – on his 'best behaviour', as he put it – and enjoyed an excellent few days' vacation. 'Why did one have to invent a Schnorkel for a submarine before one invented it for a man?' he wrote to Maria. 'It is a delightful invention. I never felt so fishy in my life.' As well as snorkelling, he was taken flying and given the controls. At night he joined the poker players, 'but he never had any money,' Louis Rosen recalled, 'so he was always borrowing from me'.

One of those looking after the diagnostic experiments was the young Berkeley physicist Herb York, and one sticky tropical evening he found an opportunity to meet Teller alone. 'We were in one of the slightly corroded all-purpose aluminium buildings,' he recalls, 'a blackboard up front, a few folding chairs and a simple working table made up the furnishings. Teller quickly sketched out his most recent ideas on how actually to construct a superbomb ... This was the idea that Teller and so many other brilliant minds had been groping for these last ten years. Even at the time I saw it was an opening into a new and dangerous era, and I still shudder a bit when I recall those moments.'

For Teller, George was to be the first nuclear explosion he had witnessed since the Trinity test back in 1945. There would have been a sense of déjà vu as a tropical rainstorm delayed the shot for three hours, much as had happened at Compania Hill; and then, less than an hour before firing time, a fault was found in one of the monitoring circuits. Rising early that May morning, he wrote: 'We walked through the tropical heat to the beach of Eniwetok's placid lagoon. We put on dark glasses, as had been done for the test in Alamogordo. Again we saw the brilliance of another nuclear explosion. Again we felt the heat of the blast on our faces, but still we did not know if the experiment had been a success.'

Obviously the fission explosion had worked well, but had it fired the thermonuclear fuel? The likely first indications of success would emerge from Louis Rosen's experiment to capture the image of the tell-tale fast neutrons from the fusion reaction on film plates, but they were unlikely to be ready until the following day. That afternoon, Ernest Lawrence, who had also flown out, took an increasingly pessimistic Teller for a swim. To cheer him up, Lawrence bet him five dollars that it would work. 'Early next morning, it must have been 5.30-ish,' Louis Rosen recalls, 'we were developing those plates and in came Edward':

'OK boys, what do you see from your experiment?'

'Edward, we're still washing them but you can look at them before they're dry.'

Here we were, not knowing what to expect. Failure would have meant no proton tracks [created by the fast neutrons] but what I saw were these parallel tracks in one field. I could see about 20. Out dashed Edward, moving very fast. We were near the airstrip, and I saw him go out on to the middle of the airstrip and start waving – at a plane trying to take off and he was trying to stop it. Well he stopped it and the door opened. And I could see he passed something to somebody inside. I couldn't see who. The door closed and the plane took off.

The occupant of the plane had been Ernest Lawrence. Teller was paying off his bet.

There was now no doubt about the positive results of George and the viability of the Teller–Ulam concept. George had produced a yield of 225 kilotons. Of that yield, the tiny capsule, containing less than an ounce of deuterium and tritium, had produced approaching a kiloton, 1000 tons of TNT equivalent. For the first time a thermonuclear flame had burned on the surface of the Earth.

When the task force returned home to the US, Gordon Dean decided it was 'high time that we got together all the people who had any kind of a view on H-weapons ...' A meeting was scheduled for the weekend of 16–17 June at the Institute for Advanced Study, at Princeton. Oppenheimer was to be chairman. With the success of George behind him, Teller looked forward to the meeting in pleasurable anticipation of a public vindication.

14

Technically So Sweet

There was an impressive turnout at Princeton on Saturday, 16 June 1951. The AEC commissioners were there in full force, as well as the majority of the GAC, including Fermi and Rabi, and the senior management from Los Alamos. In addition Oppenheimer had invited such luminaries as John von Neumann, John Wheeler, who was now back at Princeton, and Hans Bethe. Bethe had counted the meeting of sufficient importance to delay his departure for Europe, where he would be spending the summer. It was to turn out to be a watershed meeting, a point of no return in the already strained relationships surrounding the development of the hydrogen bomb.

In the triumphant aftermath of the George test, Teller had told the AEC chairman Gordon Dean, 'Eniwetok would not be large enough for the next one.' He may have admitted that his own 'stupidity' had resulted in considerable delay to the whole project, but, in his impatience and enthusiasm, he was already telescoping the technical and engineering difficulties lying ahead for other people, and pushing for a test as early as the following spring. 'This big frustration almost all comes from his unreasonable side,' said Herb York, who was also at Princeton that Saturday. 'He wanted everyone to do it right now. He wanted everyone to stop what they were doing and do this right now.'

In the short time since the George shot, he had become frustrated by what he saw as a lack of momentum, of enthusiasm. Behind this inertia, he yet again perceived Oppenheimer's influence and, in his mind, this had been confirmed when he had reached Princeton. Having arrived early, he had decided to go and see Oppenheimer to explain, at first hand, the new approach and his own confidence in it. Oppenheimer had listened with interest and understanding, but without, in Teller's view, the appropriate commitment. 'Although I was certain that the hydrogen bomb project needed to move forward as rapidly as possible,' he wrote, 'I couldn't guess what Oppenheimer thought, or what the GAC would decide. I had considerable misgivings; his indifference led me to suspect that he opposed taking the next step.' Such was the uncertain state of mind in which Teller entered the meeting.

In the hands of the three Los Alamos seniors, Bradbury, Mark and Froman, this began steadily, methodically. Darol Froman had prepared an agenda, and things started, as planned, with Carson Mark and Norris Bradbury working through the results not just of the George shot but of the other tests, including the 'Item' test of the 'booster'. They were then due to move on to a discussion of the problems with the classical Super, and then the Alarm-Clock configuration, now revived by the possibility of using radiation implosion. Finally, they were intending to deal with the Teller–Ulam idea, the equilibrium thermonuclear principle.

As Hans Bethe sat listening to proceedings, waiting to be called up to speak, he recalled gradually 'becoming aware that Edward couldn't contain himself'. Teller had listened to the report on the George shot, but noticed that Mark did not then mention its implications for the simplified Super design. To someone as sensitised as Teller it was an indication that his worst fears were about to be realised. There was going to be a concerted effort to suppress the true significance of George: 'I was amazed when Carson Mark, in his presentation, did not mention the hydrogen bomb report that I had handed him three months before,' Teller wrote in 1962. 'My amazement multiplied when Gordon Dean, still chairman of the AEC, spoke without mentioning the same report which I had explained to him two months earlier. My amazement approached anger as other scientists and officials who knew of the report spoke without referring to it. Finally, I could contain myself no longer. I insisted on being heard.'

'He was convinced that they were deliberately freezing him out,' Herb York said, 'that they wouldn't let him speak about the only thing that mattered, which was his ideas about "Mike", the next test. But there again, there's that difference between him and Bradbury. Bradbury wanted to start at the beginning even if the real excitement [was] coming later. I can sympathise with Teller. It's easy to see why he was frustrated and eager, but angry? He wasn't justified to be angry.'

> I asked to be heard [Teller said] but Norris Bradbury immediately opposed my being allowed to make a statement – he said (correctly) that I could not speak for the laboratory. The situation was odd. Several people present – Johnny von Neumann, Nordheim, Dean and Oppenheimer – had heard my proposal ... They said nothing. Even Bradbury, although he had never paid attention to my proposal, could

not have been completely ignorant of it. To my great and unexpected relief, a member of the GAC, Henry DeWolfe Smyth said, 'Why don't we listen to Teller?'

Teller estimated that, having moved with his uneven gait to the blackboard in front of the group, he spoke for no more than twenty minutes. He picked up and enlarged on Mark's description of the George shot, making the link that essentially showed that a much smaller fission bomb could compress and ignite much larger quantities of deuterium to produce a full-scale multi-megaton hydrogen bomb.

When he came to sit down, one of the first responses came from Oppenheimer: that his proposal was so 'sweet' that it had to be done. Teller was bewildered – but then, in Hans Bethe's view, he had, not atypically, actually been pushing against a door marked 'Pull': 'Edward may have believed they were trying to repress him but it was meant to be the opposite. He probably expected me to speak against it, whereas, in fact, we were all set to recommend that we actually pursue this. I certainly was [in favour], Nordheim was, Froman was and Mark.'

Oppenheimer himself believed the meeting established three important things. First, it established 'that the new ideas took top place', and should be pushed. Secondly, it was agreed that production should begin on lithium deuteride as the fuel for future weapons – a crucial technical step because, unlike deuterium or tritium, it was a solid at room temperature and because it produced tritium, as a by-product (see chapter 6). Finally, it showed that, although there had been those at the meeting who wished to test components of the new bomb first and had opposed Teller's wish to go straight to a full-scale test, the consensus supported Teller. This time the morality of the situation was not discussed even though many of the dissenting voices who had compiled the GAC report in October 1949 were present. It was as if everyone at least assumed Bethe's position – that if the US now found an H-bomb possible, so would the Russians. Gordon Dean remembered 'that everyone around that table without exception, and this included Dr Oppenheimer, was enthusiastic now that you had something feasible. I remember going out and in four days making a commitment to a new plant [to produce lithium]. The bickering was gone.'

Whatever the others gathered at Princeton might have felt, Teller himself greeted the outcome of the meeting with anxiety and suspicion. For him, the positive outcome ran 'so contrary to the attitudes present before the meeting that I still doubted future events. A part of me was waiting for the next disagreeable surprise to emerge.'

Earlier that summer one of Fermi's best students, Dick Garwin, had come on a visit to Los Alamos and now Teller immediately involved him in the problem. 'I told him about all these things and I asked him to put down a concrete design for the [cylinder] and I told him that I did not want a deliverable weapon, I want a proof of principle. And I want it so hard that there should be the least possible doubt about it because, as a consequence of the hydrogen-bomb controversy, it was not clear at all, if that first shot misfired, that there ever would be a second shot.'

So obsessed was Teller becoming with opposition conspiracies that he even believed that, should the Russians develop a hydrogen bomb first, there was a risk that the failure of the upcoming US test might allow this opposition to deny the existence of a Russian weapon as being extremely unlikely. Nonetheless, in a matter of weeks Garwin had liaised with engineers and come up with a design that withstood the most detailed scrutiny over the ensuing months, including that of Hans Bethe. Thus, with a concept that was metamorphosing convincingly into a hard design, Teller met with Bradbury to discuss the future of the thermonuclear programme and also what role he would play. William Borden had reported on the meeting to the Joint Committee: 'Teller made an offer to stay, if he had an administrative responsibility over that part of the program only.' However, these discussions rapidly deadlocked. It was clear that Bradbury himself was absolutely opposed to the idea of letting Teller run anything. He offered him a consultancy; he offered him an assistant directorship, but not the leadership role Teller believed should be his.

'Bradbury was entirely right,' said Herb York, who later worked closely with Teller for many years. 'He could never have done it, and it's deeper than just not being a good administrator. It is possible to be a good executive without being a good administrator. But Edward was just – they would have decided on a design course and he would have changed it. He just was erratic. It's not only a question of administration. He would have been a bad executive, which is much more serious.'

As the summer of 1951 drifted on without a decision on who would run the Mike test, Teller's threats to resign became more frequent and Gordon Dean came to believe that 'Teller would never be completely happy'. His complaints were seen at Los Alamos to be increasingly unreasonable and, what is more, to be flying in the face of the facts. For instance, he carped continually about the lack of commitment, and yet in 1951, for the second year running, the staff levels of the theoretical division had increased by a further 30 per cent. Worse still, he was also airing his dissatisfactions in Washington, where Borden made Senator McMahon's shoulder 'available for crying [on]'. As a consequence, Bradbury found himself having to cope with the backwash of these criticisms, which he knew were being generated by someone on his own staff. It was an intolerable situation, which was eventually brought to a head with Edward Teller's 'next disagreeable surprise'.

However sound Bradbury's reasons were for not appointing Edward Teller to run the Mike programme, the choice of Marshall Holloway, a Cornell-trained experimental physicist who was director of Los Alamos's weapons development division, added yet another personality clash to the developing maelstrom. Holloway had worked with Teller on the Family Committee and, in Teller's view, created difficulties over the H-bomb at every turn. 'Somewhat negative in his approach to life in general,' Teller wrote '... Bradbury could not have appointed anyone who would have slowed the work on the programme more effectively, nor anyone with whom I would have found it more frustrating to work. Norris had announced, in effect, that he did not care whether I worked on the project or not.'

In all probability Teller would have had some objection to any appointment other than his own – and yet he was by no means the only one to find Holloway difficult. While still a student at Cornell, Holloway had lost his young wife in a tragic swimming accident in which she had drowned, and some believed he had never recovered from this. Harold Agnew worked alongside him at this time: 'None of the project leaders – except one, Jay Wechsler – liked him. He never let us know what was going on. He didn't sort of fraternise with us. He was definitely non-collegial.'

These problems with Holloway would continue throughout the project. At one point, just before the Mike test, when they had been out in the Pacific labouring under Holloway's autocratic direction for some weeks, Agnew and another of the project directors were relaxing on a fishing trip when they hooked a nurse shark. At first they did not know what to do with it, until they saw an opportunity to demonstrate their dislike of Holloway. Between them – and this was a quarter century before *The Godfather* – the two men hauled the fish to Holloway's quarters and laid it in his bed. 'He never said anything,' Agnew says. 'But after that he was much more collegial.'

For Jay Wechsler, however, he was indeed the man for the job. Mike was now being driven firmly by political expectations, by the race, notional or real, for weapons supremacy; and in spite of Teller's fears, everyone did know this was work to a deadline and that the reputation of the laboratory now depended on it. 'Marshall was one of those people who force the issues on trying to get things settled,' Wechsler says. 'It may not be for the best but that's all we're going to do for now – we're moving on. And he just forced the issue on his own people – like myself, a good friend. That's very hard for people with fertile minds who keep on thinking.'

Certainly Teller could not adapt. On 17 September 1951, a week after Bradbury had announced Holloway's appointment, Teller resigned from

Los Alamos, leaving the project that had been his baby for ten years in the hands of men he neither trusted nor liked. Angry and bitterly disappointed, he was returning to Chicago, ostensibly to physics, but he was not about to let go. The hydrogen bomb had become an obsession. As he departed, he challenged Holloway to a one-dollar bet: that Mike would not work.

Eleven days later, on 28 September, the AEC's director of intelligence informed Gordon Dean that, after a gap of more than two years, the Soviets had tested their second bomb. A month later they carried out a further test. This weapon, known as Joe 3, had twice the yield of Fat Man, the Nagasaki bomb, and was dropped from a plane. The CIA analysis showed that the two devices were similar in design to those tested by the US during the 1948 Sandstone series. Their analysis provided hard evidence that the worst-case view being espoused by Teller and others was, in all probability, an exaggeration and that the Russians remained several years behind the US.

Given the hurt he understandably felt, it would, in all probability, have been wiser if Teller had stayed away from Los Alamos over the next year. However, in the build-up to the Mike test, he could not resist the temptation to return there to view progress. Inevitably, he was, more often than not, highly critical. Inevitably, he created friction and resentment. However, he also began looking beyond Mike, focusing his considerable energies on the establishment of the second laboratory. There was no doubt in his own mind that this was now the only path open to him if thermonuclear weapons were to be given the priority he believed they deserved.

As a start, he persuaded Oppenheimer to let him address the GAC personally on the need for such a laboratory. It was a bold move, needing all Teller's courage, as he knew that some of his most persistent critics, like Isidor Rabi, would be there. So would Conant and Lee DuBridge, whom, along with Oppenheimer, Teller referred to as 'three men–one soul'. The night before, on the sleeper from Chicago to Washington, he dreamt that he was in a trench fending off nine attackers but with only eight bullets. It was a despairing moment until, with the 'wonderful illogic of dreams, I felt elated instead of frightened. Nine men meant that there was a target for each of my bullets. Not one needed to go unused.'

On his arrival in Washington, he tramped through the snowbound, gridlocked streets to the AEC's office with 'a pervasive sense of cheer', where he became locked into a tense two-hour meeting. His demands were modest – too modest, to the point of being unrealistic, Rabi had argued. Rabi believed he was going to need 1500 scientists rather than the 300 he was asking for. Although Teller had left the meeting still feeling optimistic and glad that he had 'faced [his] opponents', the consensus was against him. Rabi and Oppenheimer had even gone as far as suggesting a special subsidiary unit at Los Alamos, devoted to special projects, where Teller would be working under Hans Bethe. An exasperated Thomas Murray, one of the new commissioners, claimed afterwards that they were just 'trying to juggle personalities' and, shortly afterwards, had condemned the GAC's opposition to Ernest Lawrence. More than two years after the GAC had been overruled in its decision not to pursue the Super, Oppenheimer's committee was still a force Teller had to reckon with.

Teller, however, had an increasingly powerful group of allies in Washington. As well as Lewis Strauss and William Borden and Senator Brien McMahon of the JCAE, the Air Force had recently acquired a new chief scientist. David Griggs was a geophysicist, and had been a Harvard professor before joining the Rand Corporation, the newly formed strategic defence think tank. He was now on a year's secondment to this post with the Air Force, where he had picked up on concerns that the Army was winning the inter-service struggle for atomic weapons priorities. Teller's continuing interest in large weapons was seen as broadly favouring the Air Force, and so the two men had begun a mutual cultivation.

Early in 1952 Teller had attended his first meeting as a member of the Air Force's Scientific Advisory Board. He had been invited to join by his old Hungarian family friend, the aircraft designer Theodore von Karman, and on the board was his new ally, the Air Force chief scientist Dave Griggs. Teller became convinced that without Griggs's support his efforts on the second laboratory would have foundered. Griggs introduced him to a number of highly influential figures in the armed services, including the hero of the Second World War bombing raids on Tokyo, General Jimmy Doolittle. Griggs arranged a private meeting between the two men, where Teller expanded richly on the difficulties with the hydrogen bomb. Doolittle 'listened with a sympathetic smile, but at that time I never would have guessed the far reaching consequences of that conversation'.

As a result of his conversation with Thomas Murray, Ernest Lawrence approached the young Herb York at the Radiation Laboratory's New Year's Eve party and raised the issue of the second laboratory. 'He asked me a simple question: Does the United States need a second nuclear weapons laboratory?' York remembers. Ever since Joe 1, and his initial lobbying spree on behalf of the Super, Lawrence had been searching for a role in the nuclear weapons programme. So far, however, his ambitious schemes, which involved adapting giant cyclotrons for producing the raw materials for thermonuclear reactions, had been dogged with problems. Now he dispatched Herb York on a tour of the country talking to other scientists, to the Air Force and the AEC and, of course, to Edward Teller, to find a consensus answer to his question. When the early reports from York confirmed the view that a second laboratory was needed, Lawrence contacted Teller. It should have been a terrific boost, to have a scientific impresario like Lawrence interested in backing his scheme, but Teller had very mixed feelings about him.

Nobody denied Lawrence's ability as an entrepreneur, but among scientists like Bethe and Fermi he was regarded as tight-fisted and a bully. In particular his treatment of Fermi's fellow countryman Emilio Segrè had shocked many. 'There was a time when Segrè had told Lawrence that he could no longer return to a job in Italy,' said Hans Bethe. 'And Lawrence said, "Then I don't need to pay you so much salary." In effect he was saying you're now cheap to be had. It illustrates what I found in Lawrence – his complete disregard for personal feelings.'

Lawrence's increasingly right-wing political views were also an issue with many of his colleagues, and Teller himself already had personal experience of them. Some eighteen months previously, in the summer of 1950, Teller had been seriously tempted by the offer of a professorship at the University of California, in Los Angeles, and had provisionally agreed to take the post. Then, in the growing anti-communist fever of the time, the Regents, the university's governing body, had insisted on an oath of allegiance from all its academic staff. There had been a public uproar when thirty-two of the university professors had refused to sign and had been sacked. Teller had reacted vehemently – 'the University of California is behaving like a sty-ful [sic] of pigs,' he wrote to Maria, and he had tendered his resignation. However, he had resisted his natural impulse to be outspoken in order to leave behind him 'good feelings'. In this he had been successful, with one exception. When he went to talk with Ernest Lawrence, he had been shocked by the reaction. 'Since the days of the Nazis I have seen no such thing,' he wrote to Maria. 'He did use threats and was quite unwilling to listen to any point of view except one of Nylan [Neylan, a conservative Regent].'

Lawrence's threat had apparently been to expose the names of scientists, possibly including Teller himself, who had refused the oath to Fulton Lewis, a right-wing radio commentator and supporter of McCarthy. Teller had persisted with his resignation, a gesture of principle at some personal cost and family upset as, on quitting Los Alamos, he had been forced once more to return to Chicago. So when the two men met, Teller was very wary of any permanent involvement. However, Lawrence already had a site in mind, a former naval air base at Livermore, some forty miles east of San Francisco; and over mai tais and dinner at Trader Vic's, he also talked realistically about how the new laboratory could be staffed from the pool of excellent students at Berkeley.

'So there were Teller and Lawrence, each of them independently striving to find some way to extend the nuclear weapons programmes,' says York. 'Edward had done all the political spadework for the laboratory and he had been turned down. But they couldn't turn down Lawrence. When Lawrence said, "I'll do it," that did it. It was his acceptance of the responsibility that did it.'

Everything now moved at an accelerating pace. During the spring of 1952 the Doolittle and Griggs meetings began to bear fruit, when Teller was given the opportunity of briefing Secretary of the Air Force Thomas Finletter. Initially sceptical, Finletter had warmed to Teller's account of the potential of the weapons they would be producing to wipe out whole army groups and protect Europe from a Russian invasion. There then followed a briefing for the Air Council, and then one for Defense Secretary Robert Lovett. So powerful were Teller's presentations, with their graphic details of the impact of multi-megaton explosions, that Dean Acheson and the National Security Council were soon under pressure from the armed forces to give the second laboratory serious consideration. Not that Teller's views had gone unchallenged: Finletter had decided to check out his opinions of the inadequacies at Los Alamos and had visited the laboratory. There, a poorly organised presentation by Bradbury, which included a less-than-enthusiastic account of the H-bomb by Carson Mark, had tipped the balance. The visitors had come away convinced that Teller's criticisms were valid.

This sequence was capped by a meeting at the Pentagon on 1 April, where Teller presented his critical account of the delays and misjudgements at Los Alamos to Acheson himself. 'Acheson said he thinks Teller is trying to scare the daylights out of the people in [the Department of Defense] needlessly,' wrote Gordon Dean, who was present at the briefing. 'Teller has end-runned it again ... These poor guys topside ... don't know what it is all about.' Acheson might have had reservations himself about Teller and his schemes, but he warned Dean of the political reality: that Teller's next audience could well be a susceptible President. It was then that Dean agreed to go and discuss the matter with Lawrence.

In mid-April Teller was back in Washington to give his (by now) well-honed H-bomb presentation to members of the State Department and to the National Security Council. So bullish had he become that he stopped by at the AEC to make clear to Gordon Dean that he would no longer consider any 'piecemeal' approach on the new laboratory. Convinced that he had won over the armed services (Borden certainly believed the military had 'bought Teller, hook, line and sinker') he now argued that the new laboratory should not only develop and build new weapons but should share the new Nevada test facilities with Los Alamos as well. His impulsiveness was such that it was even worrying his allies.

On three occasions over the previous winter the General Advisory Committee had been called to consider the second laboratory and on each occasion had rejected it. Now, towards the end of April, Teller's whirlwind campaign had, once again, run up against their opposition. By now, however, Strauss and Thomas Murray had conferred and come to the shared view that the committee was part of a conspiracy deliberately intent on sabotaging the H-bomb. Inevitably, both saw the 'road blocks' as yet further evidence of Oppenheimer's malign influence, and it was at this point that the factional struggle took on a much more personal and vitriolic edge.

First, Murray declared publicly that Oppenheimer was a security risk. Then the AEC's own recently retired director of research, Kenneth Pitzer, made a speech at the American Chemical Society in which he blamed delays with the H-bomb on the GAC. This was seen quite clearly as an implicit attack on Oppenheimer and William Borden arranged for Pitzer to tell what he knew to the FBI. In that interview, Pitzer declared that he was 'now doubtful as to the loyalty of Dr Oppenheimer', but added that it was Teller who could provide the more specific information. So Teller was drawn into the heart of the struggle.

During May 1952 Teller was interviewed on two occasions by the FBI. To anyone who had heard his complaints about the progress, or lack of it, in developing the thermonuclear, his theme would have been familiar. He told his interviewers about scientists whom he had approached to see if they would work for him on the Super at Los Alamos and who subsequently had not come. He explained how, in his view, Oppenheimer 'might have [had] something to do' with their declining his invitation. He referred to Henry Smyth as one of the AEC Commissioners who opposed the H-bomb (although Smyth had actually supported him at the Princeton meeting a year earlier), stating his belief that '[he] has done so through the influence of Oppenheimer'. Further, he implied that Hans Bethe had been sent by Oppenheimer to Los Alamos to see whether the H-bomb was really feasible after all.

When he moved on to discussing Oppenheimer himself, he described him as 'a very complicated person, even though an outstanding man'. According to the report, he then went into an aspect of Oppenheimer's past that had, so far, remained untouched, describing how, 'In his youth, Oppenheimer was troubled with some sort of physical or mental attacks, which may have permanently affected him' – perhaps a reference to the fits of depression and mental problems Oppenheimer had suffered from as a young man.* Teller himself seemed very concerned about the course the interrogation was taking – or amazingly devious – because, after each statement he made, he asked that they should 'not be included in a report for dissemination', because it might prove 'embarrassing' or might 'merely add fuel to an already smouldering fire'. Either way, Teller seemed anything but reluctant to share his views with the investigators.

Towards the end of the interview, the matter of Oppenheimer's loyalty came up and, according to the report, Teller said that 'in all of his dealings with Oppenheimer he [had] never had the slightest reason or indication to believe that Oppenheimer [was] in any way disloyal to the United States'. However, he did not want this view included in a report for dissemination either, because, according to the report, he felt it could be 'subject to considerable cross-examination'. In late May, Hoover was sufficiently impressed by the results of the investigation to send the transcript of Teller's interview, along with those with Pitzer and Libby, to the White House, to the Justice Department and to the Security Division of the AEC.

'There is another terrible piece of news,' Teller wrote to Maria Mayer. 'My mother and sister have been deported. I am trying to do something about it. Part of what I can do best is, it seems, not to talk about it too much.'

While his father had been alive Edward's family in Budapest had been reasonably well protected from interference by the communist regime by Max Teller's professional standing and contacts. However, since his death a year previously, Teller had been expecting the worst. At 3 a.m. on a June morning in 1951 there had been a pounding at the apartment door. When his sister Emmi had answered, she had been served with a warrant ordering the family out of the city within twenty-four hours. Piecing things together after the event, Emmi's son Janos, who was fourteen at the time, believed the authorities had used an outdated statute designed to drive prostitutes out of the city. They had applied it to the Tellers on the basis that Max Teller had at one time served on the board of a corporate client and was therefore an 'industrialist'. Since the war, Teller had studiously avoided direct contact with his family, communicating instead through Magda Hesz, the family's old governess, who was still living in Chicago and still in touch with the family. The news he had received had been frustratingly incomplete, even at the best of times; but now there was virtual silence about the family's fate.

Edward would eventually discover, years later, that his mother, along with Emmi and Janos, had been taken by train to the remote town of Tallya in north-east Hungary. There they had been billeted on a resentful peasant farmer and his wife, who were also considered 'undesirable' because they owned a 12 hectare vineyard. They had been joined by another family of four from Budapest and so, for eighteen months, these eight people had shared the farmer's tiny house with no running water. For all of them, the time had been spent working in the fields, with Janos relying on whatever material was sent from school friends to stay in touch with his studies. With so little information filtering back to America, though, Edward and Mici's imaginings embellished the severity of treatment the family were receiving. Speculation over why they had been singled out caused Edward to feel guilt and worry that his prominent role in nuclear-weapon development might be a major cause. This was borne out eighteen months later when the family was allowed back to Budapest. Shortly after their return Emmi was arrested and interrogated for three days, mainly on what she knew about her brother.

Teller applied what pressure he could through the State Department but had been advised that any direct intervention could do more harm than good. The nature of communist totalitarianism, which he had despised – and feared – from a distance for so long, had become more real and intolerable through the experiences of his own family – and that much more vivid for his having to imagine what those experiences were.

During the period in May 1952 in which Teller was being interviewed by the FBI, Oppenheimer's critics were becoming increasingly emboldened in their attacks. He was one of three members of the GAC whose term would come to an end that June, unless the President re-appointed them – and this presented his opponents with the opportunity they needed. David Griggs, for example, had openly challenged the activities of the GAC over the H-bomb and this had led to an acrimonious confrontation with Oppenheimer. At one point, Oppenheimer had asked Griggs whether he thought he was pro-Russian or merely confused. Griggs responded that he wished he knew. It was shortly after this that Oppenheimer had called Griggs 'a paranoid' and their exchange came to a heated conclusion.

For his part, Lewis Strauss managed to make his case for Oppenheimer's removal from the GAC with Truman himself. Then Brien McMahon who, that summer, had learned he was terminally ill with cancer, let it be known that he would go so far as to impeach Truman if Oppenheimer was re-elected. A lunch arranged by Air Force Secretary Thomas Finletter to see if he and Oppenheimer could come to some understanding was described by a participant as one of the most uncomfortable occasions he had ever attended. Oppenheimer was late and was then 'rude beyond belief', freezing out all of Finletter's attempts at affability.

But if Oppenheimer's behaviour was ill judged, it was understandable. He knew that his every action was under round-the-clock surveillance. He knew that his past political activities were under investigation again. A former student of his, Joseph Weinberg, was soon to be tried for having committed perjury four years earlier before the House Un-American Activities Committee. Oppenheimer knew he was a likely witness – with what consequences he could only guess at. Caught up in such an intense web of intrigue, vitriol and uncertainty, Oppenheimer decided on retreat. On 12 June he told Gordon Dean that he intended not to seek re-election to the GAC. In spite of this, his critics were to continue to encircle him.

Throughout the protracted negotiations over the new laboratory, work had been progressing on the Mike test under Marshall Holloway's leadership, and it was now set for November 1952. As the months passed and the initial design matured into a full-blown engineering project, it became increasingly clear why Bradbury had put Holloway in charge. As Jay Wechsler, one of the project directors, said:

> Edward says Dick Garwin had the design. That's like saying I've got a fission first stage and a fusion second stage and a big heavy case to put them in, so I have a design. That's almost insulting to all the people who worked on the project.
>
> There were months of arguments about the fuel to be used. Was it going to be liquid deuterium or a number of other hydrogen-rich compounds? Then there's the cryogenics, and you are not talking about liquid nitrogen but the very low temperatures needed for liquid hydrogen. Well, what company's got the capability? All these things had to be coordinated and then, how do you transport these big things to a place like Eniwetok? More and more, on and on it goes, it wasn't going to be somebody like Edward who was going to cope. It just couldn't be.

Nevertheless, because Teller's commitment to the thermonuclear device over nearly ten years was widely known, many people at Los Alamos had considerable sympathy for his position – having to stand by while his baby was taken over by others; but that sympathy was to be tested to destruction during his visits to the lab.

'Edward, being the person he was, kept coming back,' says Wechsler. 'Now people respected him for his opinion but they didn't necessarily appreciate the fact that he would say, "Oh no, don't do it this way. You've got to change this." So there were a number of people in the technical division who were pretty angry, because of his overall manner, that he didn't make good contributions, he was just critical.'

Many of his criticisms were niggles. Others involved such fundamental changes to the design – which of necessity had been frozen some ten months earlier in January 1952 – that they were impractical. On one issue, however, he did create serious concern. Just before he had resigned, Teller had persuaded Hans Bethe to return to Los Alamos for a year, and Bethe found himself drawn into this problem. 'At one meeting Edward said, "Of course it won't work. The container, the outer shell will disintegrate by Taylor instability." '

Taylor instability is a phenomenon seen at the interface between substances of different densities – between, say, water and air. 'Try to support an amount of water in air,' Bethe explained. 'You can't do it. You can't support a denser substance, water, by a less dense substance, air. And the water will fall rapidly in streams and droplets. That's Taylor instability.'

Mike was, in its simplest form, an enormous heavy metal capsule with the primary, fission bomb fitted at the top. Suspended in this capsule was a cylindrical tank of liquid deuterium, waiting to be compressed. In the first millionths of a second after the fission explosion, the gases and radiation flowed down round the cylinder walls, heating and liquefying the metal surfaces.

'The outer wall, being heated, will expand inwards,' continued Bethe, 'and then gets mixed with the air in the channel, which is supposed to let the radiation through. The denser wall will come to fill the channel and make it impossible for the radiation to flow.' This was the process that Teller believed would prevent the radiation compression essential for the ignition of the cylinder of deuterium fuel. 'This was a first time operation,' Jay Wechsler recalled, 'and every time you have a problem in your mind it's a horror story. These worries never went away.'

As with previous hypothetical problems, such as the question of atmospheric ignition with the fission bomb, the only answer would be careful and accurate calculation, which everyone hoped would prove favourable.

At the annual meeting of the American Physical Society earlier that summer, a number of Oppenheimer's friends, including John von Neumann and Enrico Fermi, had been shocked by the 'vitriolic talk' directed against him, 'notably', according to Gordon Dean, 'from some of the University of California contingent'. Both men were also friends of Teller and were concerned about his negotiations with Lawrence over the second laboratory, and they both, quite independently, visited him. Teller was disturbed by the intensity of their pleas, in particular Fermi's. Teller recalls:

> He was a reserved man, and it was unusual for him to show such emotion. But on this occasion Fermi was adamant. I could be making an immense and awful mistake if I went to California. Those people, he asserted, are not like us. We think about science as science. They are political plotters. Enrico and Johnny both believed that if I went to California, I would be excluded from the community of physicists ... I am certain that their only motivation was friendship. I was strongly

moved by their spontaneous concern for me and more than a little unsettled by it.

Whether he fully realised it at the time or not, Teller was well advanced along a path that led inexorably away from his old existence, even old values. He had already established links with men like Lewis Strauss, Willard Libby, David Griggs and powerful figures in the Air Force. A permanent association with someone as right-wing as Lawrence must have appeared to his old friends as a major and probably irrevocable step along that path. Teller himself was caught up with his own doubts and they were to be boosted by a visit he received in early summer from a somewhat embarrassed Herb York, who had brought a message from Ernest Lawrence. If the laboratory should go ahead, York told him, Lawrence expected the director to report to him. Then he went on: 'Ernest would like me to be the new director.' It was yet one more knockback, one that carried echoes of his disappointment over the theoretical-physics post at Los Alamos a decade earlier. It was made all the more unpalatable by being delivered not by Lawrence in person but by the young physicist who would take the role Teller had seen for himself. Teller, however, knew there was no realistic alternative. Chicago did not want a weapons laboratory, and a military laboratory run by the Air Force would never attract the quality of scientist Teller wanted. He had no choice but to put a brave face on things, though he only committed himself to the new laboratory for an initial year.

This apparent breach of trust by Lawrence was, however, quickly followed by a second. That June the AEC had given the go-ahead to establish the new laboratory on the Livermore site, under the aegis of the University of California, and in celebration Lawrence had arranged a big party at the Claremont Hotel in the hills above Berkeley. Just before this event, Teller remembered York bringing his draft organisation chart for his comments. On that chart there was no mention of a weapons design group. A furious Teller recalled that York's explanation was that, as the H-bomb research was well under way at Los Alamos, Livermore would concentrate on controlled fusion – power generation, not weapons. York, however, recalled things differently.

The issue was a very narrow issue of what a letter setting up Livermore should say about what Livermore was going to do. Everybody but Teller thought that Livermore was going to be working on nuclear weapons, but Teller thought 'No', because the letter sort of finessed that, saying Livermore would help Los Alamos on future developments, diagnostics, looking at the physics of things. So it didn't say Livermore was going to develop the next cycle of the H-bomb ... Teller had been so often deceived with respect to an honest vigorous hydrogen-bomb programme, however, that this, in his eyes, was one more deception.

These accounts may differ in emphasis, but this time there can be no doubt about the validity of Teller's reaction. Such a generally phrased letter offered Teller no concrete assurances and could have been interpreted by Lawrence more or less at will. Teller felt cheated and decided on a show-down, choosing to stage it on the night of the well-lubricated party at the Claremont in the presence of AEC chairman, Gordon Dean.

'He was obviously paranoid,' Herb York recalls. 'He said he thought Livermore was all a fake and that he had decided not to come. I told Lawrence and he said, "OK, we'll go ahead without him." But Captain John Hayward it was, from the AEC, said, "You can't do that. Edward has to be part of the deal." So Gordon Dean, Lawrence and Teller went off on their own and drafted an agreement which Teller was happy with.'

That new agreement contained a firm commitment on the part of Gordon Dean that thermonuclear weapons development would be part of the Livermore programme from the very beginning. There was also a renewed commitment on Teller's part to spend the next year at the labora-tory. Teller may have been 'paranoid', may have overreacted, but York, at least, realised how Teller's actions reflected the whole sequence of 'decep-tions' on the H-bomb programme.

Shortly after the Claremont party, a friend of Isidor Rabi's met Teller in a Denver street and the conversation came round to the new laboratory. 'I have quit the appeasers and joined the fascists,' Teller commented.* Fermi and von Neumann's fears were being realised, and Teller knew it.

15

Mike

Mike, the first true thermonuclear device, initially took physical form in a large galleried room at one end of the Tech Area at Los Alamos. As part of his efforts to ensure that everyone involved had as clear a view as possible of the progress on the design, Marshall Holloway had ordered a full size drawing of the bomb. According to Jay Wechsler:

> He called it a basic schematic. Everybody makes jokes about it, but it got all the necessary ideas into a picture, so we had something to talk about at each meeting. We put together plywood benches, sitting on little stands, covered with large sheets of paper. There was a balcony, and there was this full-scale drawing with these fellows in their stocking feet, on their hands and knees, crowding around the drawing, making sure there were lines where everything was. It sounds ridiculous but that was the official drawing.

Even though the plans had been frozen since January 1952, and even though these embodied the design elements Teller had asked Dick Garwin to draw up, Mike had still progressed, more or less by trial and error. Inevitably, there was a great deal more than finessing to be achieved. Central to concerns was how to maximise the flow of radiation from the primary fission explosion into and around the container for the secondary device, nicknamed the 'sausage' because of its topology (see figure 4). Just how would this cocktail of subatomic particles, heated to millions of degrees, move? Would it splash around the sides like water thrown in from a bucket? Would Taylor instability then cause the container walls to disintegrate and block the flow of radiation, as Teller had suggested? Hans Bethe's calculations indicated that it would not, but only the final test would tell.

Out at Eniwetok atoll, the facilities mothballed after the Greenhouse series had been restored and the forces required to run the test were slowly moved out there. It was logistics on a grand scale. By October more than 9000 military personnel and 2000 civilians were living aboard ships or in tents on the surrounding islands. A full task force of ships, including an aircraft carrier and four destroyers, patrolled around the atoll and the Air Force had amassed some eighty aircraft for transport and for aerial sampling. The island of Elugelab had been chosen as the first test site, and there a task force bulldozed sand and coral to raise the profile of the island before a start was made on building the six-storey-high test rig.

From the beginning, Mike had been planned as an experimental rig rather than as a prototype bomb. To assist the analysis of the fusion reaction it was using liquid deuterium and tritium as its thermonuclear fuel rather than the chemically more complex but 'dry' lithium deuteride, which would be used in an actual weapon. Both these hydrogen isotopes are gases at room temperature, so the tower was, in effect, the vast 65-ton refrigeration plant needed to keep them in their liquid state. In the midst of this rig, Mike's point zero, sat the sausage capsule, and spreading out

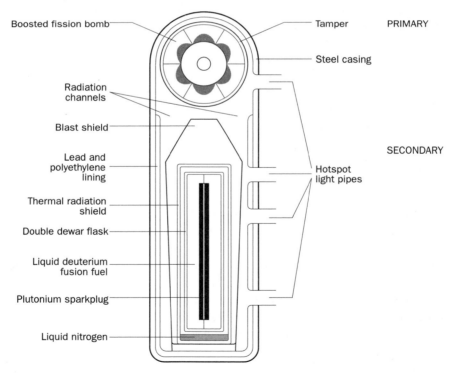

Figure 4: The Sausage
A cross section of the Mike device. The hotspot pipes were directed towards sensing cameras for observing the reaction within. Plate section shows the device in situ.

across thirty of the surrounding islands were some 500 scientific stations. The most remarkable of these was a plywood tunnel almost two miles long, running from the tower to nearby Bogon island, and packed with polythene balloons. These balloons had been filled from 20,000 two-hundred-pound bottles of helium. Through these balloons, unattenuated by the complex gaseous mix of ordinary air, neutrons and gamma rays would pass on their way to sensors that would measure the timing of the sausage's fission phase and the rise of the fusion reaction in the split second before vaporisation. It was a hugely complicated exercise on all fronts – a complexity that bore out Bradbury's wisdom in appointing an experienced administrator as director; but, even at this late stage, there were serious doubts about when, or even if, the test would take place.

With barely a month to go before the test, and with Presidential election day taking place only three days later on 4 November, there was a sharp division of opinion amongst scientists and political figures over whether to delay Mike. Vannevar Bush still harboured hopes of stopping it, as, in his view, it 'ended the possibility of the only type of agreement that I thought was possible with Russia at that time, namely, an agreement to make no more tests'. However, when he went directly to see Secretary of State Dean Acheson, he argued only for a delay in order, he said, to avoid confronting 'an incoming President with an accomplished test for which he would carry full responsibility thereafter'.

Acheson's atomic energy adviser, R. Gordon Arneson, underlined this, advising that Mike 'may well represent a point of no return', a last chance to 'avert the descent into the Maelstrom'. Oppenheimer voiced similar concerns, as did James Conant, and Bethe wrote to Gordon Dean that a test would 'undoubtedly give food to the communist propaganda machine'. Bethe also proposed that someone like Oppenheimer should be sent to brief the prospective presidential candidates, advocating a postponement.

Truman too had serious doubts, but he would only go as far as letting it be known that he would welcome a delay for technical reasons. Even he, it seems, was compromised by the mood of the times, in which there was a very real political risk in being seen to oppose progress on the H-bomb. On the other side was an intimidating line-up, including the armed services, the McCarthyite machine, the regular lobbyists and also Edward Teller, who had continued to conduct briefings and had managed to swing the opinions of important policy makers in the State Department.

On 9 October, only three weeks before the test, Truman brought things to a head. He asked Secretary of Defense Lovett 'to see to it that the situation develop[ed] in such a way that the test did not take place until after [the election]'; and he wanted it done 'without the generation of any official documents'. On 15 October Eugene Zuckert, a former assistant secretary of the Air Force and a newly appointed AEC commissioner, flew out with presidential instructions to do what he could to win a postponement. 'I ranked some Admiral out of his aeroplane in Honolulu and churned out to Kwajalein and then on to Eniwetok,' Zuckert wrote, 'and I spent many days there and tried to see if we could get that shot postponed. And finally they gave me the responsibility of determining what should happen. Well, finally I decided – I guess the day before – that because of the weather predictions, we should permit them to go ahead.'

Given Zuckert's advice, Truman backed off and the countdown was allowed to continue. Thus even a presidential intervention had been unable to slow the onward momentum into the new era of potential global annihilation. Those resisting this final step in the realisation of Teller's dream, the H-bomb, had achieved nothing except, perhaps, to heighten the bitterness and animosity between the two factions.

'I think history will show that was a turning point,' Vannevar Bush was to testify in 1954, '. . . when we entered into the grim world that we are entering right now, that those who pushed that thing through to a conclusion without making that attempt, have a great deal to answer for. That is what moved me, sir. I was very much moved at the time.'

In fact there was a last-minute problem with Mike, which came close to preventing the test. Marshall Rosenbluth had made some calculations which indicated that the fission primary core, a composite of U-235 and plutonium, contained too much plutonium and was in danger of pre-detonating. This meant that at best the primary would be compressing the secondary inefficiently, but that at worst the pre-detonation might blow the whole assembly apart before the fusion could establish itself. There were rushed meetings and all-night sessions. Should they change? Could they change? A new specification was sent back to Los Alamos for a composite core containing a greater proportion of U-235.

Mike's yield was thought likely to be around 5 megatons – 5 million tons of TNT equivalent – greater than the sum of all the conventional explosives used during the Second World War. However, such was the uncertainty surrounding the new mechanism that the arrangements for the shot took account of a yield as high as 100 megatons. As people finished their assignments, they packed up and went to sea. The whole atoll was evacuated and became seaborne – not much fun for the engineers and scientists in the bad weather leading up to the test. Even the firing was to be triggered by a radio signal from the USS *Estes*.

H-hour for Mike was 7.15 a.m. on 1 November, local time (31 October across the dateline in the US). Jay Wechsler sat up all night in the *Estes* control room, watching the dials for any blips. Their main concern at this final stage was the state of the liquid deuterium fuel, cooled to −250°C, just 23° above absolute zero. If it warmed up slightly, bubbles of gas might form, seriously affecting the yield; but as far as Wechsler could tell that night, the fuel tank in the heart of the 'sausage' remained full. As the countdown proceeded, aircraft equipped with photographic and sensing equipment took up positions circling at heights from ten to forty thousand feet. Two minutes before H-hour, two F84 jets flew into position at the highest level, ready to escort the Mike cloud and take samples. At H minus one minute, loudspeakers aboard the task-force ships ordered all personnel to don goggles or turn away, covering their eyes.

Norris Bradbury had ensured that he invited Edward Teller to witness the test of his device at Eniwetok, but Teller had decided not to go. The reasons he gave were that he was busy with his new project, the Livermore laboratory. To many, however, this decision was a clear reflection of the deep alienation, bitterness and jealousy that he felt towards all those who had, in his eyes, resisted him in progressing his concept and then attempted to cheat him of his rightful credit.

It was Dave Griggs who suggested a solution for him. One of the seismographs at Berkeley was able to detect the shock waves that a successful test would produce. So at the appropriate time Teller sat down, alone, in front of the seismograph, watching a luminous spot on its screen. Elsewhere in Berkeley was Herb York, receiving coded messages from the test site about zero hour. When it came he notified Teller by phone. 'I watched the now steady spot while the time of the shot in the Marshall Islands came and went. About a quarter of an hour later, precisely when Dave told me it should occur, I saw the dot on a seismograph screen do a little dance. The compression wave from the explosion had spent that time travelling to the coast of California.'

Shortly afterwards he was joined by Lawrence, who offered him his congratulations. 'I was glad, both for the success of the shot and for Ernest's congratulations,' Teller wrote. 'But I felt that congratulations were too much. A failure would have been a heavy blow. The success meant only that I could turn my attention to other matters.'

He decided to send a telegram via Herb York to Los Alamos, a message that made clear the result without violating security. His message read, 'It's a boy.' It was the first news that Los Alamos had received, and its import was understood immediately. It was a strangely anticlimactic and sad way of marking the final justification of a ten-year obsession.

Microseconds after the high-voltage pulse had fired the ninety-two electric detonators around the primary fission assembly, the combined fission and fusion explosion had burst through Mike's heavy metal casing,

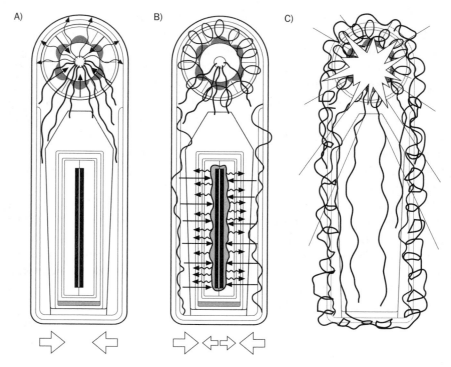

Figure 5: Mike

X-rays from the fission primary move fractionally ahead of the blast and flow down the cylindrical radiation channels, heating and vaporising the polyethylene lining to form a plasma (A). The plasma re-radiates X-rays from all sides vaporising the uranium that encases the secondary and compressing and heating the deuterium fuel to fusion temperatures. The spark plug implodes and fissions, further compressing the deuterium fuel from within (B). The arrows underneath indicate the flow of radiation inwards and the countervailing pressure from the sparkplug outwards.

Neutrons from the fusion reaction start fission reactions in the uranium casing, which produce most of the bomb's yield (C). Within millionths of a second, the entire assembly has vaporised and the fireball is spreading.

spreading almost instantly to create a fireball three miles wide. Fifteen miles out from Point H, on board one of the ships, George Cowan was meant to have told the crew in his section to turn around but forgot. 'I was surrounded by obscenities of shock and amazement,' he recalled:

> The fireball seemed to expand endlessly. I tried to set up some kind of yield scale by holding a quarter in front of me. I assumed that, if the quarter obliterated the initial fireball, then it would be within the expected yield. Actually the fireball came around the rim of the quarter. Hell, it was big. Then as you looked there was this expansion, huge,

huge, huge, with some of it disappearing up through the clouds. It's not something you have a script in your brain about, so it is breathtaking. Nowadays people have watched movies, but to see it for the first time – extraordinary.

In the sky, fifteen miles away and 40,000 feet above Point H the wings of one B36 monitoring plane heated to boiling point almost instantly. Wildlife and vegetation on the surrounding islands was vaporised, birds burnt to a cinder in mid-flight, fish stripped of their skin as if deep-fried. Within five minutes the fireball had been transformed into a purple, roiling cloud 100,000 feet high. At its full extent it rose to become a canopy of dense cloud a few hundred miles wide and thirty miles high. At ground level, the island of Elugelab ceased to exist, replaced by a crater two miles wide and half a mile deep. Eighty million tons of coral, earth and water had been vaporised and dispersed high into the atmosphere, radioactive material that would circulate and fall out around the world. Teller had been right when, a year earlier, he had commented to Gordon Dean that Eniwetok would not be big enough for the next one. It was estimated that Mike's yield had been 10.4 megatons, more than twice the estimate thought most likely.

The reactions to the test split along fairly predictable lines. Amongst the majority of those directly involved it was a project with a highly successful outcome. The team had been back at Los Alamos about a week when Marshall Holloway received a letter. 'I said "What's that?"' recalled Jay Wechsler, who was with him at the time. ' "It's a dollar." Marshall sat there with a grin on his face. "It's from Edward. I got to frame it and put it on the wall." He just loved it anytime anyone asked about it.'*

A short while after the Mike test, Teller had travelled to Princeton to see John Wheeler to show him the results and to thank him for his assistance. He had then been invited for a drink at the Oppenheimers where, as Teller recalled, Oppenheimer made an extraordinary suggestion: 'As we sat in his living room, Oppenheimer commented that now we knew the test device worked, we should find a way to use it to bring the Korean War to a successful conclusion. I was astounded and asked how that could be done. Oppie explained that we should build a duplicate device somewhere in Korea and force the Communist troops to concentrate nearby so that detonation of the device would wipe them all out . . . I did not reply to the proposal but simply said, "I don't want to think about that." '

Then, a few weeks later around Christmas, the two men spoke on the phone and Oppenheimer told Teller that he had found a means of getting the idea to President-elect Eisenhower. To Teller this was incomprehensible – an example of how Oppenheimer seemed not to be able to differentiate between possessing an extremely powerful weapon and using that weapon.*

It is an extraordinary anecdote and it begs the question of whether some tactic was being played out between the two men. At a meeting of one of the several defence advisory committees still chaired by Oppenheimer, and held at about the time he and Teller met in Princeton, the mood after Mike was reportedly 'gloomy' and 'grim'. Even though any hope of success was now severely diminished, and even though he now knew that he was under investigation and that his career was at stake, Oppenheimer still persisted with his aim of restricting the arms race. In mid-November, the Disarmament Panel he chaired forwarded its report 'Armaments and American Policy' to President Truman. By that time it was known that Dwight Eisenhower was to be his successor, but the panel hoped for one final act from Truman, which would set an agenda for the incoming administration.

Warning that the 'contest in producing weapons of mass destruction [was] proceeding grimly at an ever more rapid pace', the panel urged candour about the arms race. Very little of even the most basic information had been made available to the public, who were unaware that, while the US atomic arsenal had consisted of three fission bombs at the end of 1945, by 1952 this number had risen to over a thousand. The panel proposed, therefore, that Truman should tell the public about 'the character of major weapons, their expanding rate of production, and the enormous and important fact that they are possessed by both sides'.

On 30 December the senior members of the outgoing administration, including military chiefs, met to discuss the panel's report. They had been deeply affected by the full briefing they had received on the results of Mike, and offered the unanimous opinion that 'a weapon was in the offing which, in sufficient numbers, might have the power to destroy the world'.

After living with such a threat for half a century, we have developed an immunity that makes it difficult to appreciate the shock wave the H-bomb created with its potential for almost limitless destruction. So fearful were this group of hardened senior officers and politicians that they too endorsed the panel's call for candour on the part of the President. They hoped that public 'awareness might make it possible for statesmen, by renewed efforts, to bring about effective international control of those and other weapons'. After seven years of the arms race, the service chiefs were now among those echoing the idealism of the Acheson–Lilienthal report, drafted in large part by Robert Oppenheimer – ironically, a man some of those same military chiefs were now calling subversive.

Although they found a sympathetic ear with the President, who even went so far as to ask Finletter to draft a statement for him, Robert Lovett, the Secretary of Defense, did not share the same faith in the maturity of the

American people, nor did Dean Acheson. He believed that 'a horrendous statement such as the one proposed, without suggesting any solution to the situation, would generate a sense of utter frustration and lead to public clamor that something be done, however foolish'.

The views of these two men prevailed. An alternative statement, muted and generalised, was incorporated into Truman's parting speech to the nation. The arms race, propelled to new levels by Teller's invention, was to remain veiled in secrecy, the public lulled into apathy.*

16

'Soled' to the Californians

It was not until the first week of April 1954, a year and a half after the test, that the American public saw on film the first images of Mike, and as *Time* magazine reported, 'learned that the force and horror of atomic weapons had entered a new dimension'. That same article described how the project had been supported technologically by two 'non-conforming physicists', Ernest Lawrence and Edward Teller. When the programme to develop the H-bomb received the President's go-ahead, the article explained, 'Teller became the director of the programme and in phenomenally short time found short cuts through Oppenheimer's technical objections'. Then, with no mention of Los Alamos, the article went on to tell how, 'in the fall of 1952, Teller was put in charge of a new \$11.5 million H-bomb laboratory', and how 'a vast new taskforce began moving on the Marshall Islands for a full scale test of a complete thermonuclear "device." ' Although Teller, the article admitted, could not spare the time to be present at the test, credit for both Mike, and the upcoming test, codenamed Bravo and part of the Castle series, had been switched to Livermore.

Not surprisingly, the staff at Los Alamos were furious, their anger made more intense by the fact that security restrictions prevented anyone from putting the record straight. When the article appeared, Teller was out in the Pacific, where both Livermore and Los Alamos were sharing test facilities, and he experienced at first hand the anger the article had generated. In an attempt to defuse the situation he wrote to the editor of *Time*:

> You gave to our laboratory the kind of publicity which is most welcome to a new organisation, but you do not mention the great accomplishments with which Los Alamos is starting second decade of its existence. I should like to convey to you that the spirit on this island is a spirit of cooperation, modesty and awe in the face of the forces of nature,

which we are trying to explore for the defense of our Country to the best of our ability.

His letter aggravated the situation rather than defused it; but although he would later describe the *Time* article as 'disastrous', to an outsider, the strength of feeling, the potency of the anger felt at Los Alamos might have seemed out of proportion to the few lines of misrepresentation in just one article. However, that article reflected a bitter rivalry that had been festering ever since Livermore had opened its gates eighteen months earlier, in September 1952. And the focus of the bitterness and resentment had been Teller himself.

There was no question that, during the year Mike was being developed and built, Teller had become increasingly unpopular at Los Alamos. 'He'd been hostile,' said Jay Wechsler, 'really nasty, second-guessing people, and there were an awful lot of people, they wouldn't have worked to him for anything. So there you are. A brilliant man saying, "I'm going to put together a laboratory that's innovative, at the cutting edge, with the brightest minds." And a lot of people just said, "OK, go ahead, you do it." '

This recruitment problem had been anticipated to some extent, and resolved largely, as Lawrence had predicted, by recruiting from the ranks of the young scientists at the Radiation Lab. Teller, who at forty-four was already one of the older men of American physics, had tried recruiting from his own generation. 'They were busy,' says Herb York, 'and they didn't share his view that the world was right at the edge of the cliff. We did, but we were young and naïve. But even we didn't believe the Russians were ahead. It just wasn't true. So when it came to recruiting I recruited most, Teller recruited virtually none.'

Conceivably, this rivalry over recruiting success might well have provided yet another arena for friction, except that Teller was drawn into the excitement of Livermore's early days, days that must have potently reminded him of the beginnings at Los Alamos almost a decade earlier. For the past two years, the old naval air station at Livermore had provided a base for Lawrence's materials testing accelerator. This was what he called his 'neutron foundry', where vast quantities of power were used to generate fluxes of neutrons, which could turn spent uranium into plutonium, and deuterium into tritium. The technical problems inherent in this industrial alchemy, however, had proved overwhelming and the decision to place the new laboratory at Livermore had rescued Lawrence from an expensive embarrassment.

In the baking heat of summer, the base's former medical centre had been transformed into something resembling a laboratory: after its black, lead-lined walls had been given a coat of white paint and its floor a covering of linoleum, the X-ray room had become Herb York's office. The bathroom, the one room with running water, had become a chemical laboratory; the morgue a store for classified documents; and the old drill hall, the only place large enough to accommodate all 120 or so of the laboratory's scientists, now doubled as a machine shop and assembly hall. Particularly security-sensitive conversations were to be held in a car parked at the end of the runway. At the beginning, Herb York was the only person with a phone, two scientists shared a shower as an office and temperatures in the un-air-conditioned buildings frequently topped 100°F.

Teller had gone to Livermore on a further year's leave from Chicago, but, after a few months, Lawrence persuaded him to take a professorship on the University of California's Berkeley campus. The oath of allegiance, his main reason for having refused the post at UCLA a year earlier, had been abandoned, and the professors who had refused to take it had been invited back. Furthermore, the Teller family would, at last, be able to enjoy the Californian climate. He felt, on leaving Chicago, a slight 'I never will be missed' feeling. So he resigned, and his soul was, as he put it, 'soled [sic], sealed and delivered' to California and the Californians.

For the third time they sold yet another house – the one in Chicago next door to the Fermis, which they had owned but never occupied. Now they bought a house in Hawthorne Terrace, a quiet tree-lined street on the hill above the Berkeley campus, and only a short walk from the physics building. As he described it to Maria Mayer, it was 'a wonderful white elephant with big grounds (all weedy so it can't get any more) with columns, a huge living room, and a beautiful view'.

'No fights. No fights at all,' he wrote to Maria in the summer of 1953 as life was settling down at Livermore. 'The new existence has – apparently – started. Also a smaller and smaller number of trips. While you shall go around the world, I shall just sit and work.' At home he soon decided, after all, to do battle with the weed-infested grounds.

'I am no longer a physicist but a gardener,' he complained to Maria in the bemused terms of someone coming to grips with the unfamiliar:

Today I spent several hours cutting grass and weeding. This is most satisfying and interesting . . . The weeds are even more interesting. While pulling them out I understood for the first time the deep satisfaction that can be derived from lofty moral principles as applied to less lofty individuals. Of course the difference between a weed and a good kind of grass is merely a matter of convention but this does not diminish one's enjoyment when one succeeds to remove the evil together with its deep root . . . It is also a pity that sometimes when you remove a weed, you also remove some grass – at least I do. Finally the sharp instrument used in weeding sometimes cuts the finger. In all these respects the fight against weeds and against immorality are similar.

This summer was, for Edward Teller, the calm before the storm. It was as if, in creating a morality tale out of his experience with weeding, he had foreseen events in his own future.

To begin with, though, Teller was caught up in the bustle, organising briefing sessions just as he had done at Los Alamos. The staff, with an average age of twenty-nine, were enthusiastic and inexperienced – so inexperienced that, when Teller had finished his briefing, one recalled his colleagues sitting around saying, 'Well, what do you know? So that's how it's done.' But, in spite of the summer camp atmosphere, within the first week Teller had already crossed swords with Lawrence over the role of Livermore as a weapons laboratory. At a meeting with the AEC, Lawrence again talked loosely of Livermore's 'promising new concepts'. Teller, on the other hand, was determined not to let Lawrence soften the laboratory's brief and presented the meeting with specific projects, sketch plans for a number of radically different H-bomb designs. He was, as Herb York realised, a problem waiting to happen, and the thirty-one-year-old director made it one of his priorities to work out how best to occupy the principal inventor of the new technology that was a major part of Livermore's raison d'être. As York recalled:

We could not very well assign to a person of his stature some minor fiefdom in the laboratory, and we were not willing to give him a top-level management position. We solved the problem in a simple way. We established a small steering committee and made him a member of it without responsibility for any specific part of the laboratory or its program, but with a personal veto authority for the first year ... This arrangement enabled him to participate as he wished in whatever part of the program took his fancy.

This ingenious arrangement played both to Teller's sense of importance and to his strengths as a consultant. 'Everyone who went to see him, including myself,' York remembered, 'found the visit always fun and often inspiring. Full of intellectual energy, he brought a tone and zest to the place it would not have had otherwise.' As the astrophysicist and early recruit to Livermore Stirling Colgate said, 'Edward worked at every stage to present the lab both as intellectually dominant as well as intellectual. The game was to contribute to the Soviet perception that what you had was the best and Edward started that and represented it.'

Teller was consciously looking for a contrast with the conservatism he believed dominated the thinking at Los Alamos, and in this he was not to be disappointed. Herb York set out a working philosophy, readily embraced by the young staff, of pushing at the technological extremes. The first of the two main research divisions that he established, A division, was headed by the twenty-four-year-old Harold Brown. Their brief was to design thermonuclear weapons, based on the Teller–Ulam principle of radiation compression, but lighter, smaller and more suitable for military use than the giant devices still the focus of activities at Los Alamos. One device illustrating the new approach was the Davy Crockett, an attempt to build the smallest nuclear weapon possible. The prototype that emerged could be launched like a bazooka shell and weighed well under a hundred pounds. Herb York was able to take a full-size wooden mock-up to Washington on a regular flight in his cabin bag.

John Foster, then just twenty-nine, directed B division, which worked on fission weapons. They would also build smaller, lighter bombs using design concepts like Teller's booster principle, where a fission reaction was spiked with thermonuclear material. It was ironic that, in pursuing these smaller devices, they were actively pursuing the course advocated by Oppenheimer, for which he had been criticised by Teller's acquaintances in the Air Force. It was a double irony, then, that the laboratory's chief salesman was Edward Teller. As Herb York observed:

> Edward is charismatic in the true, simple sense. He's enthused, he does have a thorough grasp of the basic ideas, whether they're his or not, he's a teacher and he knows how to explain. I think that's the basis of his success. He has a strong superficial grace and he's always showing this false modesty that some of the other Hungarians showed. And he likes to pretend he's naïve, but he's spent a lot of time honing those arguments and he's strongly ambitious to have his way.

York was particularly impressed by the effect Teller had on Truman. The two had met shortly after the wily Truman had left office. 'Truman was just awestruck, and whatever Teller was trying to get through to Truman, Truman was for doing it – President Truman, who had seen it all.'

Nobody among the Livermore management was in any doubt that, in those early months, the laboratory's very existence was politically fragile. They needed projects with which they could prove themselves as quickly as possible, and yet their new designs were, understandably, years away from fruition. To prime the development programme at Livermore Teller had brought projects with him, but these were pet projects and were anything but new in concept. One, codenamed Ramrod, represented an attempt to realise his dream of the classical Super, but this time using a thermonuclear rather than a fission trigger to provide the necessary extra boost to burn the uncompressed deuterium fuel. Another revisited the Alarm Clock design, but this time incorporating the radiation–implosion principle. A third was a modified version of a uranium hydride bomb, using a compound of the hydrogen isotope deuterium and uranium as its nuclear fuel. Teller had originally proposed it as a concept back in 1942 – one that would make economies, much needed at the time, in the amount of U-235 used. Although the need for such economies had passed, this was chosen as the first design to be tested and Livermore pressed ahead with preparations. Two shots were scheduled, targeted to be included in the oddly named Upshot–Knothole series in the spring of 1953, alongside nine others being mounted by Los Alamos. For Teller, personally, these tests represented his first chance to hit back at his adversaries.

The first device, codenamed Ruth, was assembled at the top of its 300-foot firing tower, and early on Tuesday morning, 31 March 1953, the count-down began. It ended with a pause – 'and an explosion which has been described as a "pop".' It was quite obvious that it was a damp squib, but as the dust cleared it became apparent just how serious a failure it had been. The tower, which in all previous tests had been vaporised, still stood, its upper structure bent and broken but otherwise intact. The post-explosion silence was followed by ironic cheers and laughter fromthe Los Alamos scientists and there was a rush to find cameras. The official estimate of Ruth's yield was 0.2 kilo- tons, little more than 1 per cent of the yield of the Hiroshima bomb and less than a thousandth the force of the Mike blast. The second test, Ray, detonated a fortnight later, was a similar fizzle with much the same yield.

Teller was forced on to the defensive, arguing that the experiment, as such, had succeeded but that 'we were, indeed, embarrassed'. Hans Bethe, however, recalled experiments on hydride devices at Los Alamos during the war and the recognition then that the deuterium in the uranium hydride had had an inhibiting effect on the speed of reacting neutrons. In Bethe's view, Teller 'could have thought for an hour and reminded himself'. There was no question that the failure of both tests was a blow, both to the morale of the young Livermore team and to the laboratory's political viability.

Robert Oppenheimer still continued to frighten his enemies in the Air Force and on the Joint Committee for Atomic Energy. They might have pressured him out of the GAC, and they might have weakened his influence in the Department of Defense by closing down the development board he chaired, but he was still a major influence in Washington, a consultant to the AEC with full security clearance and the support of its chairman, Gordon Dean. In a concerted attempt to finally unseat him, William Borden at the JCAE had brought in a friend, John Walker, a fellow Yale law graduate and they began working together on two fronts.

First, Borden was one of those convinced that there was another spy still at large in the US weapons programme and that the most likely candidate was Oppenheimer. In order to provide hard evidence for this claim, he began painstakingly to compile his own case against the scientist. As a corollary to this he assigned Walker to produce an 'Atomic Program Chronology', aimed, in Gordon Dean's view, at showing that 'the Joint Committee had always been right' in supporting the case made so powerfully by Teller to the Pentagon. In essence, Teller had said that, owing to incompetence and espionage, the Russians had not only drawn level with the US in developing thermonuclear weapons but had even moved ahead of them. In order to establish this case firmly, Walker set out to find just what the Soviets had learnt from Fuchs about the American H-bomb. Fuchs had certainly been present at the Super Conference in April 1946, but it was argued by Bethe and others that this had concentrated on the classic Super model and would therefore have misled the Soviets rather than assisted them. Teller, however, claimed that the all-important radiation–implosion principle was actually discussed at the meeting, and so the issue remained unresolved.

Walker tried to gain sight of the interviews with Fuchs on his arrest, but the FBI and the AEC refused him access. So he contacted Teller's friend John Wheeler, at Princeton, who had attended the meeting, and he agreed to assist Walker. Prior to their meeting in Washington on 7 January 1953, Walker sent him a six-page document summarising details of the conference, of Bethe and Teller's disagreement over its content, and of the Mike design. On the overnight train down from Princeton, however, Wheeler somehow contrived to lose it. Borden contacted the FBI and had the coach in which Wheeler had travelled thoroughly searched and partially dismantled. Agents also walked the track back from Washington to New Jersey, but to no avail. Only the envelope was found, still in Wheeler's office.

When the new President, Eisenhower, heard what had happened, he was furious. He was still recovering from the shock of a meeting, shortly after his election the previous November, when he had been briefed on the results of the Mike test. He was reported to have visibly paled when he read that the island of Elugelab was 'missing' following the explosion. Now, confronted with the possibility that the secrets of this awesome weapon had been stolen, he angrily dressed down the AEC commissioners 'like errant schoolboys'. It was not just the security implications that concerned him but also that it was the kind of incident likely to discredit his regime and drive McCarthy into a 'feeding frenzy'. He also suspected espionage, an inside job, and he ordered that the staff of the JCAE be investigated, and Borden dismissed.

Feeling confused and at sea in dealing with an unfamiliar situation with such awesome implications, Eisenhower cast around for assistance. Early on, he met and talked with Oppenheimer. Finding they agreed on the need for greater public candour, he invited him to address the National Security Council on the subject. This served to further anger and frustrate Oppenheimer's enemies. Borden was now doubly determined to prove his involvement in espionage. He had quickly found other employment, but just before he finally left his post with the JCAE for several weeks' vacation, he managed an arrangement with his successor whereby he maintained his security clearance for another year. With access to Oppenheimer's files assured, he used his vacation to prepare what was to become an accusation that Oppenheimer was indeed a Soviet spy.

Eisenhower's own worries over Oppenheimer's past security lapses – indeed, his worries over security lapses in general – eventually persuaded him not to use him as an adviser but to turn to someone else, to Lewis Strauss. In March he appointed Strauss as his presidential adviser on nuclear matters. 'Lewis, let us be certain about this', he said to Strauss, 'my chief concern and your first assignment is to find some new approach to the disarming of atomic energy . . . The world simply must not go on living in fear of the terrible consequences of nuclear war.'

Given this objective, Ike's choice of Strauss as an adviser seems simply bizarre. Few people – not even Teller – could match his deeply held suspicions of the Russians as potential partners in any attempt to forge a disarmament programme. Few could have been less enthusiastic about communicating difficult information to the American people. Within a very short time of his appointment, he had headed off a one-to-one meeting between Oppenheimer and the President, and he had used his counsel with Eisenhower to mute the honest intent of what became known as the President's 'Operation Candor'. Within months it had become little more than a public-relations exercise.

Strauss was also one of the sources for an unsigned article that appeared in the May edition of the business magazine *Fortune*. It was bluntly entitled 'The Hidden Struggle for the H-bomb: The Story of Dr Oppenheimer's Persistent Campaign to Reverse US Military Strategy'. The author recalled events of the past four years in which, he said, the military establishment had been pitted against 'a highly influential group of American scientists' who were determined to discredit the Strategic Air Command and whose 'prime mover' was Oppenheimer.

The author was actually the magazine's editor, Charles Murphy, an Air Force reservist with contacts among senior Air Force officers.* Two months later, Murphy and Strauss were to collaborate again in mounting a response to an article Oppenheimer had written in the specialist journal *Foreign Affairs*. Because he endorsed the President's ideas on candour and openness over nuclear matters, Oppenheimer actually cleared the article with Eisenhower himself. However, that did not prevent Strauss from assisting Murphy with the *Fortune* article. Eisenhower presumably did not know about his adviser's actions, because that same month, July 1953, he appointed Strauss to replace Gordon Dean as chairman of the AEC. His rationale for the appointment of someone with aims so contrary to his own was that he believed the AEC to be both big enough and politically sensitive enough to need a manager with business experience. It was a decision that was to have heavy strategic consequences over the next decade.

On his first working day at the AEC, Strauss ordered the removal of whatever classified documents still remained in Oppenheimer's files at Princeton. The reason Strauss gave was that the files were being moved to another AEC building to save the cost of a twenty-four-hour guard.

Edward Teller was never far away from this developing situation. 'In the meantime there has been more nonsense in the local papers in connection with *Fortune*,' he wrote to Maria Mayer. 'I am doing what you said, namely nothing. Except I am asking through private channels to <u>please</u> keep my name out of it.' The private channels included Strauss, who fended off the attempts of both *Time* magazine and also Joseph Alsop, a well-known journalist close to Oppenheimer, to contact Teller. Oppenheimer was fighting back, and had good friends among influential liberal journalists. Over the next year he was to have great need of them.

Early on the morning of 12 August, the Russians exploded their fourth nuclear device. In the US it was detected seismically almost immediately and then the AFOAT-1 patrol soon picked up residues. These showed that the device was thermonuclear but that the US and not the Soviet Union was technically ahead. Joe 4 was clearly a primary device and there was no evidence of radiation compression of a thermonuclear secondary. It was believed, correctly as it turned out, that it was a device akin to Teller's Alarm Clock.

'Joe 4 was something of a fake,' said George Cowan, one of the radio-chemists who examined its fallout. 'Yes, it was a thermonuclear device, but they had used every bit of tritium they had to make it go. They could only make one at that time. They had done it as a response to Mike. Stalin needed something quickly to wave at the Americans. He knew we had a war party and was always afraid we were preparing a pre-emptive war.'

In fact the shortcomings of the Russian design were recognised by some unexpected people – by Curtis Le May, head of Strategic Air Command, and by Lewis Strauss. Strauss reported privately to Eisenhower that the Soviets had embarked on the wrong road to producing a multi-megaton weapon. The fears Teller had voiced to the Pentagon with such effect – that the Soviets were ahead – were not borne out. As far as the public and most politicians were concerned, however, the Russians had the H-bomb. Furthermore, there were rumours that, while Mike had been a 65 ton immobile monster, the Russian bomb was deliverable by aircraft. Fake or not, Joe 4 was a powerful political statement and served to foster the fear that the Russians were winning a terrifying race.

17

Bravo

On Saturday night, 7 November 1953, William Borden went to the main post office in Pittsburgh and mailed two copies of a letter containing his indictment of Oppenheimer. One went to FBI director J. Edgar Hoover, the other to the Joint Committee on Atomic Energy. To ensure that his letters did not go astray, both were sent by registered mail, return receipt requested.

At the time, Oppenheimer had gone to London to deliver the BBC's Reith Lectures – an assignment that he counted as one of the most important in his life. In one of those lectures he had expressed very clearly his view of communism: 'This is not man's fate; this is not his path; to force him on it makes him resemble not that divine image of the all-knowing and all-powerful but the helpless, iron-bound prisoner of a dying world.'

Whatever his hopes for communism had been fifteen years earlier there is little doubt that in these words he was expressing the strongest rejection of its dogma. It was a dogma that he now saw as limiting 'the open society, the unrestricted access to knowledge, the unplanned and uninhibited association of men for its furtherance ...'

When the lectures were over, Oppenheimer and Kitty went on to visit friends in Europe and, on the evening of 7 December, they dined at the home of Haakon Chevalier and his new wife Carol. That visit, the last time Oppenheimer and Chevalier were to see each other, was made under continuous FBI scrutiny. 'You know just before going to Europe, Oppie had gone out to Los Alamos,' Kenneth Nichols, the newly appointed General Manager of the AEC recalled. 'He hadn't been out there for over a year but he went and got a thorough briefing on all that was going on, then took off for Europe and made contact with Chevalier. I thought it was either arrogance or darned poor judgement, but one of our Commissioners thought it was more than that, that it was disloyalty to a trust.'

◆

The letter William Borden sent to Hoover was a remarkable document, thoroughly researched and taking a hard, not to say extreme, line in interpreting the fruits of that research. It was arranged in four phases in Oppenheimer's life. The first covered the pre-war years before he had become involved in the A-bomb project, and detailed both his connections with communist organisations and his friendships with communists, including his communist mistress (Jean Tatlock), and his 'frequent contact with Soviet espionage agents'. The second covered his entry into government service and the way he had severed those relationships, giving 'false information' to cover them up. The third covered his period at Los Alamos, where he had employed several communists, and where, once the bomb had been used on Japan, he had tried to persuade colleagues to stop work on both the A-bomb and H-bomb programmes. The final section covered his attempts since the war to retard atomic weapons development and, in particular, the H-bomb.

It was a powerful case, not a presentation of new facts but a convincing re-interpretation of well-known incidents in Oppenheimer's life into a narrative portraying him as nothing less than a Soviet agent. In his conclusions Borden stated, 'more probably than not, he has since been functioning as an espionage agent'; and, 'more probably than not, he has since acted under a Soviet directive in influencing United States military, atomic energy, intelligence and diplomatic policy'.*

The letter came at a difficult time for the new administration, increasingly threatened by the activities of McCarthy and other anti-communist witch-hunting groups that had sprung up. Only the day before Borden had posted his letter to Hoover, Herbert Brownell, the Attorney General, had felt it necessary to blame the previous Truman regime for laxness in promoting an alleged Russian spy, Harry Dexter White, to a senior position in the International Monetary Fund back in 1946. In this political climate, the charges against Oppenheimer were bound to be exploited.

The FBI spent the next three weeks substantiating the charges in Borden's letter before despatching their inch-thick report, along with the letter, to the White House at the end of November. At a meeting in the Oval Office late in the afternoon of 3 December, the President set in motion a chain of events that led to what was, effectively, a trial for treason. After consulting with Lewis Strauss, Eisenhower ordered a 'blank wall' placed between Oppenheimer and any further access to secret information until his case had been properly investigated.

Edward Teller had been due to meet Strauss that afternoon, but uncharacteristically the AEC chairman was late, 'and when he did appear he was visibly shaken', Teller recalled; 'asking me to say nothing about it to anybody, he told me that he had just come from a meeting with President

Eisenhower ... Strauss seemed as opposed to smear campaigns as I. He feared that taking action against Oppenheimer would appear to be another form of McCarthyism. He deeply hoped that some way could be found to avoid it.' Strauss certainly wanted Oppenheimer gone, but this letter containing charges sufficiently sensational to provoke the President into precipitate action was not his preferred manner of proceeding. To him the possibility of the charges creating a backlash among the scientists at the AEC was a real one. Hoover, too, was concerned about the public reaction, in particular because so much of the information built up by the FBI had been obtained illegally, through wire-tapping and surveillance. He feared that Oppenheimer might all too easily end up a martyr.

The course of action decided on was a formal one, using a personnel security board. Such boards were appointed by the AEC's general manager to inquire into an employee's suitability for one of the various grades of security clearance. The AEC held hundreds of them all over the country every year, and normally they were little more than a formality, involving an hour or two's examination of the subject by the board. Only very exceptionally, as now, would either party be represented by counsel, but even when they were, the normal rules of evidence used in a court of law were not applied. However, with a case as politically sensitive and complex as this one, the secrecy surrounding the proceedings and the absence of these rules of evidence meant that it could so easily become 'not much more than a kangaroo court'. This was the fear expressed by Joe Volpe, a former AEC counsel who had drawn up the rules of conduct for such boards and was a friend of Oppenheimer's.

The charges against Oppenheimer would normally have been framed by the AEC's General Counsel. However, William Mitchell, the present incumbent, was new in the job and so the task was handed to one of Mitchell's juniors, Harold Green.

At 4 a.m. on Saturday morning, 12 December, Green went to work in the AEC's offices. A little later he was surprised to find that the AEC's general manager, Kenneth Nichols, had come in as well and, throughout the day, kept checking on Green's progress. ' "Have you seen that yet?" he would ask, or "At last I've got the bastard." That kind of thing, firing me up to perform great deeds. I never liked him much, and I could not help being conscious of the fact that here is the guy [the General Manager] who was going to be deciding the case and I already know how he's going to decide it.'

As Green worked on alone through that December weekend, he was shocked by the amount of derogatory information that existed on such an eminent figure, but he also noted that much of it was old and had been previously investigated. 'I kept thinking to myself, "How can we be so stupid, after all these years, to do this to this kind of man?" Certainly I felt that I wouldn't have initiated these proceedings particularly as so much of the derogatory information was so old – except, that is, for the H-bomb stuff – and that it was like locking the barn door after the horse is gone.'

This 'H-bomb stuff' reflected the complaints that Teller had been voicing over the years and had made to the FBI two years earlier, and was to provide the substance for one of the most important moments in the whole Oppenheimer case. Green had been told, when he started, that he could not include this material in the charges, as it would make it appear as though Oppenheimer was being tried on his judgement and opinions rather than on whether he was a security risk. By noon on the Sunday, however, Green had finished his framing of the charges earlier than expected, and still had two hours to wait before William Mitchell was due to meet him in the office. So he began to ponder the H-bomb material, in particular those secret FBI interviews with Teller. As he did so, it occurred to him that he might be able to frame charges from it that would allow the board to measure Oppenheimer's version of events against known facts and act as a test of his truthfulness; and, with nothing better to do, Green decided to try out a draft in this vein. When Mitchell eventually arrived at the office he reviewed and accepted the charges in full, including those on the H-bomb. Now, largely on the basis of complaints made by Teller, Oppenheimer faced allegations about his attitude to the H-bomb that were to question both his 'conduct and loyalty' and his 'veracity'.

On 21 December, just a few days after his return from Europe, Oppenheimer met with Lewis Strauss and Kenneth Nichols in Strauss's office. According to the FBI's report of the meeting: 'Strauss let Oppenheimer read the statement of charges prepared by the AEC. Oppenheimer then commented that some of the charges were incorrectly stated: some he, Oppenheimer, would deny; while others were correct. According to General Manager Nichols, the meeting was amiable . . .'

Both Strauss and Nichols had hoped desperately that the matter would become a resignation issue, thus immediately defusing the situation; but while Oppenheimer raised the possibility, he then asked for time to think it over. Oppenheimer ended the meeting saying he was going to see his attorney, Herb Marks, and was given the use of Strauss's Cadillac to make the visit. He was tailed by an FBI agent and Harold Green remembers seeing reports of his meetings both with Marks and with Joe Volpe, indicating that the law offices were bugged. From the beginning, the FBI and AEC officials were breaching the bond of privacy between counsel and client. It was to set a pattern for the next four months during which Oppenheimer was under constant daily surveillance. Lewis Strauss was to comment to Charles Bates, the agent liaising between the FBI and the AEC, that 'the Bureau's technical coverage on Oppenheimer at Princeton [his home] had been most helpful to the AEC in that they were aware beforehand of the moves he was contemplating'.

Oppenheimer took two days to decide that he would not resign. He finally came to the conclusion that his resignation could too easily be leaked to the press and represented as an admission of guilt. So he felt he really had no choice but to face up to the charges. On Christmas Eve 1953, only a day after he and his wife Kitty had travelled down to New York to tell Strauss personally of his decision to face a hearing, Oppenheimer received a visit from two AEC representatives. They carried a letter instructing him that he was 'hereby directed to deliver' all remaining AEC documentation in his possession; until these had been catalogued and removed, he was refused access to his own vaults.

In late January 1954 Teller and Oppenheimer both attended a conference on High Energy Physics held in Rochester, New York. While there, Oppenheimer sought support among the attending scientists and approached Teller. 'When he saw me, he asked me, "I suppose, I hope, that you don't think that anything I did has sinister implications?" I said I did not think that – after all the word "sinister" was pretty harsh. Then he asked if I would speak to his lawyer, and I said I would.'

The lawyer Oppenheimer had finally chosen was Lloyd Garrison. Tall and Lincolnesque in appearance, Garrison was a partner in a well-established New York law firm, and a leader in the American Civil Liberties Union. He devoted what spare time he had to birdwatching, reading philosophy and Greek literature, and was a man of unimpeachable integrity and devotion to public service – even offering to forgo any fees for his work for Oppenheimer. Volpe was worried about his lack of courtroom experience, but at the time it was not thought to be of great importance, as Garrison was intending to employ leading counsel to present the case.

As agreed, Teller met with Garrison and Oppenheimer's other counsel, Herb Marks, in the former's office, where 'they devoted all their efforts to convincing me, indirectly, that Oppenheimer's actions in regard to the hydrogen bomb were innocent of disloyalty. In consonance with them, I had already made up my mind about this . . . I left unimpressed with their comments and without having changed my mind: I would testify that Oppenheimer was a loyal citizen.'

Years later, Garrison wrote his recollections of the meeting: 'He did not challenge his loyalty. He expressed lack of confidence in Robert's judgement and for that reason felt the government would be better off without him. His feelings on this subject and his dislike of Robert were so intense that I finally concluded not to call him as a witness.'

Garrison's opposite number was Roger Robb, whom the AEC had approached on the advice of the Attorney General. Robb had a reputation as one of the most robust of Washington trial lawyers. He had acted as attorney for the journalist Fulton Lewis, Jr, who was a McCarthy supporter, but he had also successfully defended the communist leader Earl Browder. At the outset he had no strong commitment to this case, but once he began reading the files, his opinion changed quickly:

> There were so many things in those files that didn't add up, unless you applied a theory to them, which was that Oppenheimer was a communist and a Russian sympathiser, and that's the only way I could add it up ... It was inconceivable to me ... that the way the Communist Party operated and having this man in their grasp today, and having this man join this top secret project tomorrow, they'd just drop him like a hot cake, but they didn't do it. I'm sure they didn't, it just doesn't make sense.

After some two months closeted in his office, chain-smoking Havana cigars and carefully sifting the material, he felt it was time to visit some of the personalities in the case. He flew out to the West Coast, where he met Teller, whom he described as 'one of the finest gentlemen I ever knew', and talked with him, in particular, about Oppenheimer's role on the development of the H-bomb programme. 'Talked' is perhaps a misnomer, because Teller had that morning just been to the dentist and, initially, was able to communicate only by signs and jottings on his office blackboard. It is worth noting that Teller's own accounts of this period do not mention this meeting. He claims only one meeting with Robb, just minutes before he was called into the hearing as a witness. He also claims he only agreed to talk to Robb then because he had also previously agreed to meet Oppenheimer's lawyer.

While Roger Robb had been immersing himself in the Oppenheimer files, Lloyd Garrison had run up against a serious problem. In mid-January he had applied for the clearances necessary for the defence to obtain access to classified material. However, while the AEC was prepared to clear Garrison, they were not prepared to clear Herb Marks, who had a record for espousing radical causes and had crossed the FBI's path on a number of occasions. When the defence team had met to decide what action to take, Marks had argued that they should not seek clearance as a matter of principle. It was argued that if they had clearance the hearing might be drawn into an exhaustive examination of technical matters and, as Garrison wrote, 'could thereby lose the main point, which is that if Dr Oppenheimer's motives were honourable, his technical recommendations were irrelevant'. As time went by, however, Garrison had become fearful that secret matters might be raised unexpectedly and that they might even be forced to leave Oppenheimer 'unrepresented and alone' in the hearing room. Thus, a fortnight or so before the hearing, Garrison had changed his mind and had written asking for clearance for himself. It was never to arrive.

As preparations for the hearing progressed, out in the Pacific, on Bikini atoll, preparations had also been continuing for the Bravo shot, the test of the first deliverable thermonuclear bomb. The device weighed some 10 tons, manageable enough to fit the bomb bay of a specially adapted B47. It was 'dry', employing lithium-6 hydride, rather than the liquefied gases used in Mike, and it was expected to yield a relatively modest 5 megatons, roughly half the yield of the previous test. Given the absence of any pressing concern about fall-out, the exclusion area surrounding Point Zero had only been increased a little on the one for Mike.

Another giant task force of some 10,000 military and scientific personnel prepared to monitor the tests and to track the dust cloud as it moved out across the surrounding ocean. By 22 February 1954 they had completed the installation of the device on its short tower on one of the surrounding reefs, and the long countdown began towards H-hour, scheduled for 6.54 a.m. on 1 March.

From the moment of ignition, Bravo had all the appearance of an explosion way in excess of expectations. The fireball expanded to nearly four miles in diameter, trapping people in experimental bunkers and menacing the task force out at sea. 'I've always said the politicians should be made to strip down and watch a multi-million-ton thermonuclear thing like Bravo,' Harold Agnew comments. 'What's scary is the heat. At 30 miles, even, it's scary. It comes on and it comes on, and it doesn't go away. You really know you're in the presence of something amazing.'

Early tests indicated a yield of 15 megatons, three times the predicted yield, but the truly frightening aspect of Bravo was invisible – or almost. Within a few minutes the firing party in the control bunker was reporting rapidly increasing levels of radiation, even with the bunker door closed. An hour later, the task force out at sea was blanketed with white flakes of vaporised coral, and was reporting radiation levels that were dangerously high. For four hours the men aboard the ships were kept sweltering below decks as their craft fled the immediate area around Bikini to re-group 50 miles out from Point Zero. It took until evening for the levels around Bikini to begin falling, but then reports of dangerous levels began trickling in from Rongelap and Ailinginae, two of the Marshall Island atolls. These lay immediately to the east of Bikini, but it had been decided to leave them outside the exclusion zone, thus avoiding the laborious task of evacuating their inhabitants before the shot.

The levels continued to climb during the night and into the following day, and so amphibious craft were sent in to rescue twenty-eight US weathermen from an atoll 133 nautical miles east of Point Zero. It was not until that evening of 2 March that it was decided to send a destroyer to Rongelap, only 100 nautical miles from Point Zero, and to neighbouring Ailinginae, to evacuate the population. The islanders were given assurances that they were only being moved for a short while, and then they were taken to Kwajalein, to hospital.

Here it was realised just how serious the situation had become. Over the day they had been under exposure, the Rongelap islanders had received over two hundred times the recognised weekly safe limit of radiation – well over the amount known to cause permanent damage. In a short time they were suffering hair loss, internal haemorrhaging and skin lesions. In terms of blood count they had suffered about the same level of damage as had the Japanese who were some 1.5 miles from Point Zero at Hiroshima and Nagasaki. Just as distressing at the time was the news that they were not going to be able to return to their home island for some time to come.

The final event in this sad saga was not discovered until a fortnight later. A Japanese fishing vessel, the *Fukuryu Maru* (*Lucky Dragon*), arrived back in Japan with all twenty-three crew members suffering from radiation exposure. The ship's log showed that, at the time of the Bravo shot, it had been some 82 nautical miles from Bikini, just beyond the eastern boundary of the exclusion area. The crew had known nothing about the test until they had seen the flash and, later, had been showered with a white powdery substance.

The reaction in Japan was intense and emotional, one newspaper commenting that the Japanese people were 'terror-stricken by the outrageous power of atomic weapons which they [had] witnessed for the third time'. Through formal diplomatic channels the US government attempted to convey deep personal concern for the injuries received. However, back at the AEC, the prime concerns of the commissioners led by Thomas Murray and Strauss was security. As a consequence, the initial goodwill of the Japanese was lost when they were refused any information about the nature of the fall-out. Strauss then created further ill will when he claimed that the *Lucky Dragon* 'must have been well within the danger area'.* As Edward Teller was to remark of this outcome to the first test of a deliverable version of his invention, 'That accident had consequences that went far beyond the lives of the unfortunate people who suffered directly.'

The date fixed for Oppenheimer's hearing was 12 April. Ever since early January, the AEC had been casting their three-man board. Harold Green had been asked his advice and had recommended one of the tough, regular boards from Chicago. He was told the commission was looking for a board with similar standing to Oppenheimer himself. However, as he later recalled, 'It was also perfectly clear to me from the conversations I'd had that they were really interested in more of a hanging jury, one susceptible at least to subtle pressures, to bring in a verdict denying clearance.'

The first member to be chosen was the chairman, Gordon Gray, a former Joint Secretary of the Army and president of the University of North Carolina. There was no doubt about his standing, but the fact that he proposed Oppenheimer's friend David Lilienthal as another possible member indicated how little he was aware of the destructive background to the hearing. It must have been clear to the commission that they had found a 'naïve' chairman. His suggestion was not followed up.

The second member of the board was Thomas Morgan, the sixty-six-year-old former chairman of the Sperry Corporation, who would distinguish himself by his almost total silence during the four weeks of the hearing. The final member had to be a scientist, and the commission needed to find someone who did not feel some commitment to Oppenheimer, one way or the other. They settled on Ward Evans, a professor of Chemistry at Chicago's Loyola University, and a regular security board member – an arch-reactionary, he almost always voted to deny clearance.

Oppenheimer did have the right to challenge the choices, but he found no fault. In the week before the hearing Roger Robb was to spend much of the time closeted with the three members, going through the case files with them. When Garrison asked if he, too, could spend some time with the board, his request was turned down. It was as if, in a court of law, the judge had met with the prosecuting counsel before the trial and then, when asked if the defence counsel might share the same privilege, refused him.

Throughout March and April of 1954, the Castle test series continued, producing still more exceptional results. Castle Romeo, mounted by Los Alamos, was another runaway, for the same reasons as Bravo, achieving a yield of 11 megatons, three times the estimate. This time, however, with the exclusion zone dramatically increased, they had persistently delayed until the weather conditions were just right and there were no possible problems with fall-out. In fact, in spite of everything, Bravo too had been counted a technical success: the cause of the excessive yield had been identified and could be corrected.* Those at Los Alamos now believed that, in the face of Teller's continuous criticism, they had mastered his invention, and were now envisioning thermonuclear weapons that could be both a great deal smaller and a great deal larger than those so far tested.

Livermore's first thermonuclear venture, codenamed Koon, was, however, a complete contrast. It had a predicted yield of 1.5 megatons and, according to Herb York, 'was built around ideas Teller had specifically brought to Livermore'. The result, fired on 6 April, was another damp squib, yielding 110 kilotons, some 7 per cent of the estimate. 'We had experienced two failures with our fission weapons,' Teller recalled. 'Now our first thermonuclear didn't work. Three failures in a row! We were chagrined to say the least . . . The only bright spot in Livermore's misadventure was that Stirling Colgate's diagnostics for the test had worked perfectly.'

Colgate's experiments showed that Teller and his co-workers had failed to shield the fusion fuel from strong radiation, which had then caused a pre-ignition before radiation compression had been properly achieved. This was a double blow, as the same design fault had been incorporated into their fourth and final test, codenamed Echo, as well. Teller took the first opportunity to fly out to Bikini, where he met with Herb York and Harold Brown. Between them, the decision was taken to abandon the shot. This moment of total ignominy was made that much keener by the presence of the teams from Los Alamos. 'After three failures, many people in the weapons group were talking about looking for employment elsewhere, and not altogether humorously,' Teller wrote. 'To me and to the young people it seemed far from unlikely that the weapons program might be closed down.'

At the end of two years of bitter rivalry, there was absolutely no love lost between the two laboratories. In the face of Teller's continual carping about their commitment and their abilities, Los Alamos had conducted what were, in technical terms, two hugely successful test series. It was no surprise that they gave no quarter to the teams from the new laboratory, who, in their turn, smarted from the way their failures were received. Little surprise, either, that there was such a bitter outcry from Los Alamos when, a week after the failure of Koon and the abandoning of Echo, James Shepley's *Time* article made its appearance, in effect giving credit for Mike and Bravo to Livermore.

As for Teller himself, nothing seemed to be going right. Already the focus of criticism from his peers, now his own hopes of avenging himself against Oppenheimer's old laboratory had seriously backfired, leaving him little room for future optimism. It was in this frame of mind that he had boarded a plane home from Bikini, piloted by General Curtis LeMay, head of Strategic Air Command. At one point during the flight the general handed over the controls to come back and play poker with his passengers. Teller was distracted, his mind wandering, and he 'not only lost money but also a considerable amount of the general's respect'. It was during the game that Teller learnt for the first time from LeMay that the hearings against Oppenheimer were going ahead.

18
The Hearing

T3, the Atomic Energy Commission's building, was one of those depressing 'temporary' government constructions that has long since passed its intended lifespan. Room 2022 was a standard government issue executive's office once occupied by the AEC's director of research, but now prepared to act as a miniature courtroom.

It was in this bare, utilitarian setting that Robert Oppenheimer would spend four weeks listening to the details of his life being dissected and painfully torn apart. Sitting for much of the time on a leather sofa against one wall of the room, he would see only the backs of the witnesses as they sat just in front of him, addressing the three-man board ranged behind a table at the far end of the room. Branching from this table, and stretching back towards Oppenheimer like the arms of a 'U', were the tables at which the two legal teams sat facing each other.

The first morning of Monday, 12 April, started scrappily for Oppenheimer. He and his team arrived late, having been caught up in a discussion with James 'Scottie' Reston of the *New York Times* on whether, and if so when, to make the case public. So far all the preparations had been in secret and, during his opening remarks, the board's chairman, Gordon Gray, instructed that the hearing itself was to be 'regarded as strictly confidential'. However, fearing the story would be leaked anyway, Lloyd Garrison had, during the first day's lunch break, rung Reston and authorised him to go ahead and publish. In retrospect, then, the first afternoon was the calm before the storm, with Garrison leading Oppenheimer through a measured account of his career. The next morning, however, Reston's article appeared under the following headline: 'DR OPPENHEIMER SUSPENDED BY AEC IN SECURITY REVIEW Scientist defends record; Hearings started: Access to secret data denied nuclear experts – red ties alleged.'

The very full story that followed was picked up by the media worldwide.

It was just what the AEC and Lewis Strauss had been fearing for months. Now they were as much victims of an administration frightened that the Oppenheimer case would become yet more fuel for McCarthy to harangue them with incompetence. Trapped in the middle was Gordon Gray, who found that the nature of the whole inquiry he was heading had been dramatically reinterpreted by the outside world as a witch-hunt. He immediately attacked Garrison for failing to indicate to him in any way that documents had been handed to Reston – a serious setback for the lawyer, who had based his strategy on a non-aggressive approach, tuned to appeal to the urbane Gray.

At the same time as the Reston article was souring the atmosphere at the hearing, another separate journalistic storm was blowing up – one driven by the impact of the radioactive fall-out from Bravo on the Marshall Islanders, and the reaction in Japan to the return of the *Lucky Dragon*. Eisenhower had initially sidestepped any discussion, but the pressure was so great that he agreed to deal with it in his weekly press conference on 17 March. Or rather, he had Lewis Strauss deal with it. A 'calm-voiced' Strauss made a nine-minute statement in which he noted the possibility that the Russians had gone to work on a thermonuclear bomb 'substantially before we did' and announced: 'We now fully know we possess no monopoly of capability in this awesome field.'

As to reports that the Bravo test had run out of control, he gave the impression that they were untrue, stating, 'The yield was about double that of the calculated estimates – a margin of error not incompatible with a totally new weapon.' Having done his best to cool this line of inquiry, he was then asked by a journalist for some basic information about the H-bomb. Strauss replied that it could be made as 'large as you wish' and that 'an H-bomb can be made as – large enough to take out a city'. (Strauss seemed to check himself.)

'How big a city?' he was asked.

'Any city,' he replied.

'Any city? New York?'

'The Metropolitan area, yes,' he replied.

As they left the press conference, Eisenhower gently admonished Strauss: 'I wouldn't have answered that one that way, Lewis.' In checking himself, Strauss may just have avoided saying that an H-bomb could be large enough to destroy the planet. Nevertheless, the statement he had made was shocking enough to have triggered graphic representations in the newspapers of the Manhattan skyline with the fireball profile superimposed on it, as well as a number of investigative features. The injuries to the Marshall Islanders and the fishermen of the *Lucky Dragon* paled into insignificance beside the possible obliteration of New York, Washington or Chicago, as a result of just one bomb. Strauss did note that he had meant to say 'put out of commission' rather than 'take out', but his slip of the tongue touched off a debate that was long overdue. 'Let us cease all further experiments with even more horrifying weapons of destruction,' wrote the biologist Lewis Mumford, 'lest our own self-induced fears further upset our mental balance.' It was a letter that attracted nationwide attention and was reprinted widely. Within a month the White House was receiving more than a hundred letters a day protesting against the continued tests.

'Dear Mr President,' read many identical cards illustrated with a mushroom cloud, 'there must be a world-wide ban on the H-bomb.' Suspecting a communist plot, such cards were forwarded to the FBI, but the protest could not be pigeon-holed so easily. Protests from well-known figures, both from the US and from abroad, testified to the broad range of people genuinely concerned about the aftermath of Bravo. Albert Schweitzer, the world-renowned medical missionary, and the Pope both expressed moral concern over continued testing. Prime Minister Nehru of India went a stage further in proposing a moratorium. This chimed with Eisenhower's own belief that 'this terrible problem' – a world overshadowed by the hydrogen bomb – could only be solved by first establishing a test ban.

He asked his Secretary of State, John Foster Dulles, to explore the technical feasibility of such a ban, and Dulles in turn referred the problem to Strauss and the AEC. Strauss then involved both Teller and Norris Bradbury. Their assessment was that small tests could remain undetected and that such a moratorium would be impossible, practically, to police.

In the middle of the hearing's third day Garrison finished leading Oppenheimer through his version of events and Roger Robb took over with a cross-examination that would last for twenty hours spread over a further three days. From the start he began laying traps that might only bear fruit after hours of careful questioning.

It was after lunch that Robb turned to the sensitive matter of the 'Chevalier incident', and lurking behind the pattern of his questioning were the tapes of the interview between Oppenheimer and Boris Pash some eleven years earlier. At the start of Robb's cross-examination, Oppenheimer was unaware both of the tapes' existence and of the fact that Robb knew of his brother's involvement with Chevalier. This was something that Oppenheimer had always determinedly tried to hide. During his examination, he even ensured that Garrison asked him whether there was a connection between the two, in the hope that a firm denial would close off the issue. If he really thought that this would ward off any further questioning about Frank's role, he was mistaken. It must have been a sickening experience to slowly realise that Robb knew so much more than he had expected. One by one, Robb raised leading questions – about the use of microfilm in transmitting any forthcoming information; about the possible use of contacts in the Soviet embassy – each of them based on statements Oppenheimer had made in the interview with Boris Pash. One more probe from Robb brought him a step away from raising Frank Oppenheimer's name:

ROBB: Did he tell you or indicate to you in any way that he had talked to anyone but you about this matter?
OPPENHEIMER: No . . .
ROBB: Did you learn from anybody else or hear that Chevalier had approached anybody but you about this matter?
OPPENHEIMER: No.

Oppenheimer's last negative had been almost inaudible. He had visibly slumped in the witness chair. The gaps between question and answer had grown longer and longer. There was a feeling of impending revelation.

ROBB: Now let us go back to your interview with Colonel Pash. Did you tell Pash the truth about this thing?
OPPENHEIMER: No.
ROBB: You lied to him?
OPPENHEIMER: Yes.

'His statements to Pash were completely at odds with his testimony at the hearing,' recalled Robb. 'Oppenheimer was smart enough to know that that was a serious discrepancy, and I remember him sitting there with his hands between his knees, washing his hands between his knees, head bowed, white as a sheet. I felt sick.'

The informal conversation of eleven years ago was now being inflated into a situation in which the scientist perjured himself, and Robb began to push harder:

ROBB: So that we may be clear, did you discuss with or disclose to Pash the identity of Chevalier?
OPPENHEIMER: No.
ROBB: Let us refer then, for the time being, to Chevalier as X.
OPPENHEIMER: All right.
ROBB: Did you tell Pash that X had approached three persons on the project?
OPPENHEIMER: I am not clear whether I said there were three Xs or that X approached three people.
ROBB: Didn't you say that X had approached three people?

OPPENHEIMER: Probably.
ROBB: Why did you do that, Doctor?
OPPENHEIMER: Because I was an idiot.

Oppenheimer's answer had come after a long and painful pause, but whatever concern or sympathy Robb might have felt towards him, he drove onwards, confronting Oppenheimer with section after section of the old recordings. Time after time, he made him go back over the various details that made up what Oppenheimer was eventually forced to admit was a 'cock and bull' story. Then finally Robb administered the coup de grâce.

ROBB: Isn't it a fair statement today, Dr Oppenheimer, that according to your testimony now you told not one lie to Colonel Pash, but a whole fabrication and tissue of lies?
OPPENHEIMER: Right.

Shortly after this exchange, one that would later play a significant role in shaping the evidence Teller gave, the day came to an end. Everyone in the hearing room seemed to realise that this was a turning point in the inquiry, but few could have known the question, identifying Frank, that Oppenheimer feared. In the end, Robb had not pushed Oppenheimer for the final admission of his brother's involvement. Neither he nor Strauss had wanted Oppenheimer to be seen as a hero, protecting his younger brother. In this they succeeded, while at the same time achieving their main objective of establishing the eminent scientist as a self-confessed liar.

'I came home that night, after midnight – I'd been working – and my wife was still awake,' says Robb, 'and I said "I feel bad." She said, "What's the matter?" " I've just seen a man destroy himself on the witness stand."

'That's the way Oppenheimer impressed me. You can't begin to feel the drama of the thing. Some people thought that Oppenheimer was going to commit suicide that night. I said, "Oh no, he won't." And he didn't, of course. Next morning he was back just as cheerful as ever.'

Fitted into the fourth day of the hearing was Major General Groves, now a businessman working for the Remington Rand Corporation and one of Oppenheimer's witnesses. Groves had promised his support in writing, but his testimony was affected by the fact that he had not only disregarded Oppenheimer's left-wing past during the war but had also promised to keep Frank's involvement with Chevalier secret – information that Strauss was aware of and used against him. Nevertheless, when Groves came to the stand he reaffirmed his view that he would be 'amazed' if Oppenheimer would ever commit a disloyal act. As to Oppenheimer's refusal to reveal Chevalier's name, he shrugged that off as 'the typical American schoolboy attitude that there is something wicked about telling on a friend'. He had named Eltenton, and that was the main thing.

However, when it came to Robb's question as to whether he would clear Oppenheimer at the present time, Groves backed away from the full support Oppenheimer had expected. He replied that, largely on the basis of Oppenheimer's pre-war associations, he 'would not clear Oppenheimer today' under his interpretation of the new security standards. In covering his own back in this way Groves considerably blunted the helpfulness of his testimony.

When Oppenheimer returned to the stand after Groves's departure, his bearing could not have been more different from the day before. He was his old confident, fluent self again, as can be seen in an exchange with Robb about the visit of Teller and Lawrence to Washington just before the 1949 GAC meeting. In a letter to Conant he had referred to the fact that 'two experienced promoters ha[d] been at work'.

ROBB: Would you agree, Doctor, that your references to Dr Lawrence and Dr Teller and their enthusiasm for the super bomb . . . are a little bit belittling?

OPPENHEIMER: Dr Lawrence came to Washington. He did not talk to the Commission. He went and talked to the joint congressional committee and to members of the military establishment. I think that deserves some belittling.

ROBB: So you would agree that your references to these men in this letter were belittling?

OPPENHEIMER: No, I pay my great respects to them as promoters. I don't think I did them justice . . .

ROBB: You think that their work of promotion was admirable, is that right?

OPPENHEIMER: I think they did an admirable job of promotion.

This exchange was so neat, so quick, that it underlines the level of duress Oppenheimer must have been under the afternoon before to make him act the way he did. The same thoughts may well have been going through the minds of the board as they listened to him parry questions about the hydrogen bomb that were aimed at testing his veracity. Oppenheimer demanded to see the transcript of the Joint Committee. Garrison also asked to see the transcript, but was, once again, confronted by the curtain of secrecy that was increasingly being used by the commission at its own convenience.

That night, in the Georgetown house where the Oppenheimers were staying, the talk was on what to do about the way the commission was carrying on, withholding evidence and manipulating the procedures. After listening to the descriptions of what had been going on, Joe Volpe told Oppenheimer in no uncertain terms to 'shove it, leave it, don't go on'. But the consequences of a boycott, one of which would be a legal trial, were equally fraught with dangers, so nothing was done. The following morning, Oppenheimer returned to the hearing for one last session and then, at 12.15, he stood down.

During the second week, an array of luminaries from a variety of different spheres were called into Room 2022. Among them were the Nobel Laureates Enrico Fermi, Isidor Rabi and Hans Bethe; senior figures in government and scientific administration, such as James Conant and Vannevar Bush; and both Strauss's predecessors as chairman of the AEC, who, along with three commissioners, testified on Oppenheimer's behalf.

As the days went by, a pattern began to emerge in the examinations. Each witness would describe why he was convinced of Oppenheimer's loyalty. Then Robb, or perhaps Gray, would cite the Chevalier incident, or imply the existence of some other Oppenheimer misdeed unspecified, and ask how this affected the witness's feelings. More often than not they replied that times – or Oppenheimer – had changed. When James Conant came to the stand, he lost no time in attacking any censorship of Oppenheimer's opposition to the H-bomb as 'it would apply to me, because I opposed it – as strongly as anybody else'; but the edge was taken off his evidence, as it had been with other witnesses, when Robb exploited the fact that, without security clearance, the defence counsel and their witnesses did not know what evidence the board had.

ROBB: Of course, Doctor, you don't know what the testimony before this board has been?
CONANT: No, I don't.
ROBB: Nor do you know what the record or file before the board discloses?
CONANT: No. I only know what is in the letter of General Nichols.

There were others, however, who responded robustly to this approach. Isidor Rabi, whose friendship with Oppenheimer stretched back to their student days, made the point that, whatever Oppenheimer's behaviour, it should be judged against his contributions to the national interest. Here Rabi was forceful. 'We have an A-bomb and a whole series of [them]. What more do you want, mermaids?' But when Robb suggested 'perhaps the board may be in possession of information which is not now available to you ...' Rabi's response threw into relief the obsessive triviality that had marked so much of the hearing so far: 'I think that any incident in a man's life of something of that sort you have to take it in sum,' Rabi responded.

'That is what novels are about. There is a dramatic moment in the history of man, what made him act, what he did, and what sort of person he was. That is what you're really doing here. You are writing a man's life.'

After Rabi came an angry Vannevar Bush. 'I feel that this board has made a mistake and that it is a serious one. I feel that the letter of General Nichols which I read, this bill of particulars, is quite capable of being interpreted as placing a man on trial because he held opinions, which is quite contrary to the American system, which is a terrible thing.'

Following this assault a shaken Gordon Gray tested out the possibility with Robb of bringing the hearing to an end. Robb had told him that such a move was not possible. At this midway point it was clear the hearing was turning into a bitter affair and Robb had taken a strong personal dislike to Oppenheimer: 'My feeling was that he was a brain and as cold as a fish, and he had the iciest pair of blue eyes I ever saw.' Robb was also becoming more and more convinced of Oppenheimer's guilt.

This dislike was mutual. On one occasion he and Oppenheimer were outside during a short recess when Oppenheimer developed a coughing fit. Robb, who had had TB when he was younger, was solicitous, but Oppenheimer turned on him, snapping out something Robb did not understand. The prosecutor walked away.

To Lloyd Garrison the source of the bitterness was clear: 'Of course, Robert was in the most overwrought state imaginable – so was Kitty – but Robert even more so. He would pace his bedroom floor at night, so Randolph Paul [in whose house he was staying] told me, and he was just an anguished man. His anxieties were added to our own and it was a great torture really.'

For his part, a tense and nervous Strauss took to bombarding the FBI with requests for any kind of derogatory information they might have on Oppenheimer. Agents interviewed one of his erstwhile graduate students about rumours of a homosexual affair between them. The student denied it. Others chased an army private who recalled finding copies of the *Daily Worker* among the scientist's rubbish at Los Alamos.

Now into the hearing's third week, Strauss was facing a major problem. That week saw the start of part of the proceedings without precedent in any of the AEC's regulations or previous hearings: the presentation of the 'government case' through witnesses called by a 'representative of the government'. This group was to include Teller, Lawrence, Alvarez, and Griggs, but Strauss was having to contend with their increasing nervousness and threats of desertion.

A few days before he was due to appear, Luis Alvarez had received a phone call from Ernest Lawrence, who 'announced emotionally that he wouldn't testify and that I shouldn't either. He said people had convinced him that the Radiation Laboratory would be greatly harmed if he testified and that he, Ken Pitzer, Wendell Latimer, and I were viewed as a cabal bent on destroying Robert. I had never seen Ernest intimidated before. He was suffering terribly from the ulcerative colitis that soon afterwards claimed his life.'

Lawrence had attended a gathering of laboratory directors at Oak Ridge, where he had been confronted and intimidated by, among others, an angry Isidor Rabi. Lawrence himself was able to plead illness – another serious attack of his colitis. To prove that he was not malingering, he had taken a fellow physicist into the bathroom to bear witness to the blood in the toilet bowl. However, this did not prevent an apoplectic Strauss from haranguing him about his responsibilities, and when Alvarez called Strauss to say that he was not attending either, he received more of the same. 'He prophesied that if I didn't come to Washington the next day I wouldn't be able to look myself in the mirror for the rest of my life. I had never disobeyed Ernest's direct orders, I said, and I wasn't about to start now.' Strauss's last jibe, however, had the desired effect and that night Alvarez booked on to the TWA midnight flight and flew east, disobeying Lawrence for the first time.

Lawrence and Alvarez were not the only ones from the Berkeley 'cabal' who were having second thoughts: even Kenneth Pitzer, the AEC's director of research and the only one, initially, prepared to question Oppenheimer's loyalty, was having second thoughts. Only Wendell Latimer, Berkeley's dean of chemistry, was secure, but his main qualification as a witness seemed to be simply his dislike of Oppenheimer. The strength of feeling among the whole scientific community, so powerfully expressed by the witnesses at the hearing, now threatened to destroy Strauss's case.

Edward Teller, of course, had not been immune to the emotional pressures influencing the other government witnesses. On 22 April Charter Heslep, one of the AEC's public information officers, had visited him to talk about the commission's Atoms for Peace programme, to find that he 'was interested only in discussing the Oppenheimer case'. Teller was brimming with pent-up emotion and talked non-stop for more than an hour. Heslep was sufficiently concerned about his state of mind to send a six-point memorandum of the conversation to Strauss. Some of the points from it provide a valuable insight into Teller's thinking at the time:

2. Since the case is being heard on a security basis, Teller wonders if some way can be found to 'deepen the charges' to include a documentation of the 'consistently bad advice' that Oppenheimer has given, going all the way back to the end of the war in 1945 . . .

4. Teller said that 'only about one per cent or less' of the scientists know of the real situation and that Oppie is so powerful 'politically' in scientific circles that it will be hard to 'unfrock him in his own Church'. (This last phrase is mine [wrote Heslep] and he agrees it is apt.)

5. Teller talked at length about the 'Oppie machine' running through many names, some of which he listed as 'Oppie men' and others as not being on his team but under his influence. He says the effort to make Conant head of the National Academy of Sciences is typical of the operation of the 'Oppie machine'. He adds that there is no organised faction among the scientists opposing the 'Oppie men'.

6. Teller feels deeply that this 'unfrocking' must be done or else – regardless of the outcome of the current hearing – scientists may lose their enthusiasm for the program.

It is an account that gets behind the statements Teller carefully honed over the years to rationalise his actions. His jealousy, the depth of his feeling of impotence as an outsider facing a man of great personal magnetism who, he believed, directly or through those he influenced, had blighted his life and the things he believed in for more than a decade, are all there. These were the emotions he sought to control as he journeyed towards Washington for his appointment at the hearing on 28 April.

Teller arrived in Washington to find the city full of physicists. The big annual meeting of the American Physical Society was being held there, and the gossip was serving to fan the already fevered atmosphere surrounding the hearing. It was the main topic of conversation and various groups, including one from Los Alamos, were organising petitions and protests. On the evening of Teller's arrival, Hans Bethe, who was the society's president that year, had brought Oppenheimer straight from the hearing to the society's dinner, where he had been given a standing ovation.

Wherever he went Teller saw one colleague after another whom he either knew, or suspected, supported Oppenheimer. In an attempt to avoid confrontation he latched on to a confirmed ally, John Wheeler, and spent much of the evening of 27 April with him. 'As he describes it,' Teller recalls, 'I spent the evening pacing his hotel room floor and worrying about how best to make my statement without creating doubt about Oppenheimer's loyalty.'

At some point in the evening, however, Teller met Marshall Rosenbluth and the two talked for some time. According to Rosenbluth:

He was clearly quite anguished about it, and I told him he was really going to damage himself if he did testify . . .

His attitude was that he wasn't going to say he was a spy or anything, but he said the people who had backed him and gotten the US bomb programme going and so on, felt they had to do this – get Oppenheimer out of the middle so they could have a better rapport with the scientific community. He said, 'I feel I can't refuse to go in and tell the truth.' He felt that it was his duty, at least as he portrayed it. I believe he was aware of the consequences and was quite unhappy about being put in this position, but nevertheless went ahead.

This impression is largely borne out by the recollections of Hans and Rose Bethe, who also ran into Teller and invited him back to their room: 'We were still very much on friendly terms,' Rose Bethe recalled, 'and we were anxious that he shouldn't testify, or at least not make his testimony detrimental, keeping it at least neutral. But as our conversation went on, he became stiffer and stiffer. He had to do it, he said.'

In Hans Bethe's words:

The two of us, Rose and I, talked to Edward for at least an hour and tried to make clear to him that he should not testify against Oppie . . . It was a desperate discussion. He said that he didn't like Oppenheimer making decisions. He had made so many mistakes. The fact that he was destroying a person's life by speaking out didn't seem important. He was sorry, but that was what he had to do. I don't think it was what he owed other people. I think it was entirely himself, and the strongest reason was that Oppie had opposed his pet project. As simple as that.

When Freeman Dyson, the Princeton theoretician, met Bethe in the hotel lobby later that evening, 'He was looking grimmer than I had ever seen him . . . "Are the hearings going badly?" I asked. "Yes," said Hans, "but that is not the worst. I have just now had the most unpleasant conversation of my whole life. With Edward Teller." He did not say more, but the implications were clear.'

When Teller arrived in Washington he had found a note waiting for him from Robb, requesting a meeting. At first, Teller said, he hesitated: 'I considered it not quite right. But considering that Oppenheimer had asked me to do the same thing [meet his lawyer], I felt that I could not say no to Robb.'

This justification seems to underline just how uncomfortable he continued to feel about his actions at this time. Would someone who only days before had allegedly stated that Oppenheimer must be 'unfrocked', be so coy about seeing the commission's counsel? Also, Teller had already met Robb to talk about the case. What was different about this meeting in

Washington? Anyway, at some time before Teller gave evidence, the two men did meet and Robb asked him the key question: whether he was going to testify that Oppenheimer was a security risk.

'I told him,' Teller remembered, 'that I did not consider Oppenheimer a security risk. Whereupon Robb showed me part of the testimony, showing that Oppenheimer had lied to a security person [Pash]: that he admitted so lying, but that he could not now be prosecuted for what was a criminal offence because the statute of limitations had taken effect.'

The part of the testimony Teller was shown was where Oppenheimer had been pinned down by Robb over the Chevalier incident, finally admitting that he had been 'an idiot'. 'When Robb showed this to me and gave me this argument,' admitted Teller, 'I said I guess that I cannot simply say that Oppenheimer is not a security risk. So in retrospect I am quite unhappy about this event. Had Oppenheimer not asked me to talk to his lawyer, I would not have listened to Robb. I would have gotten into less hot water, personally. On the other hand, you may say that this brought out more of the actual facts.'

Given this suggestion that he was conscious of being manoeuvred by Robb, one might wonder why Teller did not simply refuse to read the material Robb offered him; but Teller was by no means independent of Robb. Between them was a bond, thus far unspoken, that tied Teller firmly to the government case. Robb knew that, if necessary, he could invoke Teller's 1952 testimony given to the FBI, the testimony used by Harold Green to draft the charges. He could do this with or without Teller's collaboration. So Edward Teller came to the witness stand struggling to find a balance that would not draw the anger of his friends and colleagues but that, equally, would not induce Robb to introduce his FBI testimony. Very quickly Robb's questioning got to the heart of the matter:

ROBB: To simplify the issues here, perhaps, let me ask you this question: Is it your intention in anything that you are about to testify to, to suggest that Dr Oppenheimer is disloyal to the United States?

TELLER: I do not want to suggest anything of the kind. I know Oppenheimer as an intellectually most alert and a very complicated person, and I think it would be presumptuous and wrong on my part if I would try in any way to analyse his motives. But I have always assumed, and I now assume, that he is loyal to the United States. I believe this, and I shall believe it until I see very conclusive proof to the opposite.

ROBB: Now a question which is a corollary of that: Do you or do you not believe that Dr Oppenheimer is a security risk?

TELLER: In a great number of cases, I have seen Dr Oppenheimer act – I understood that Dr Oppenheimer acted – in a way which for me was exceedingly hard to understand. I thoroughly disagreed with him in numerous issues and his actions frankly appeared to me confused and complicated. To this extent I feel I would like to see the vital interests of this country in hands which I understand better and therefore trust more.

In this very limited sense I would like to express a feeling that I would feel personally more secure if public matters would rest in other hands.

These were the first of the evaluations made by Teller during his evidence that would come back to haunt him for years to come. For the next two hours Robb led Teller in re-rehearsing the complaints against Oppenheimer. It was a less definite – some thought tamer – account than ones Teller had previously given, but then again, he was clearly intimidated by the fact that he was in the presence of the man he was accusing. In his account, for instance, of trying to recruit Bethe to the H-bomb programme in 1950, he no longer firmly stated that Oppenheimer had changed Bethe's mind: it was implied but not stated. Harold Green believes that if Robb's prime objective 'was to make the case against Oppenheimer as effectively as possible, Robb should have pressed Teller to repeat under oath his earlier allegations'.

Instead the AEC's counsel was to allow Teller's muted version of his complaints to pass unchallenged, knowing, however, that the board – but not Oppenheimer's lawyers – had the original FBI reports to hand. In the final minutes of Teller's testimony, Gordon Gray did try to push him into a clear answer to the question the board itself finally had to address:

GRAY: Do you feel that it would endanger the common defense and security to grant clearance to Dr Oppenheimer?
TELLER: I believe . . . that Dr Oppenheimer's character is such that he would not knowingly and willingly do anything that is designed to endanger the safety of this country. To the extent, therefore, that your question is directed toward intent, I would say I do not see any reason to deny clearance . . .

If it is a question of wisdom and judgment, as demonstrated by actions since 1945, then I would say one would be wiser not to grant clearance.

This was the second evaluation that was to haunt Teller for the rest of his life. At ten past six that evening he was excused from the witness chair. As he left the room, he paused before the frail figure seated on the couch and held out his hand. 'I'm sorry,' he said.

'After what you've just said,' Oppenheimer replied, taking Teller's outstretched hand, 'I don't know what you mean.'

In his most recent reassessment of his answers, Teller said that his most glaring error was not revealing that Robb had shown him part of Oppenheimer's testimony. 'As a result everyone assumed that my testimony was meant to accuse Oppenheimer for his opposition to the H-bomb. My comments were only meant to question his behaviour in regard to Chevalier.'

After two hours of examination and cross-examination, concentrating on the H-bomb and not mentioning the Chevalier incident once, would it not have required the skills of a mind-reader to have discerned that preoccupation from Teller's replies? All the evidence – his statements to the FBI, to Charter Heslep and to those friends who had tried to dissuade him from testifying – points to the critical statements in his testimony being a fair reflection of what he thought about Oppenheimer's role in the development of the H-bomb. 'I think he was quite honest in what he said at the hearings,' says Hans Bethe. 'He just didn't like Oppenheimer making decisions.'

Teller's implicit challenge to Oppenheimer's loyalty, given the mood he knew existed among his fellow scientists, was an act of some courage and was to stand in marked contrast to Lewis Strauss's other witnesses for the government case. They were to disappoint in one way or the other. Kenneth Pitzer did not carry through his challenge to Oppenheimer's loyalty. Instead he charged that Oppenheimer's sin had been lack of enthusiasm, which had affected recruitment for work on the hydrogen bomb. Wendell Latimer, although a fellow academic on the Berkeley campus, had never been close to Oppenheimer either professionally or socially. This did not prevent him from advancing a case based on the physicist's 'hypnotic power'. 'Many of our boys came back from it [Los Alamos] pacifist,' Latimer said. 'I judged that was due very largely to his influence, this tremendous influence he had over those young men.'

Ernest Lawrence had used illness as his excuse not to appear, and even Alvarez, who had gone against Lawrence's wishes to appear, was not as effective as Strauss might have hoped. He testified for the best part of a day about his difficulties back in 1949 in promoting the H-bomb. It was towards the end of his testimony that, for the first time, the board member Ward Evans asked questions that seemed to take a pro-Oppenheimer turn: 'The main thing that we have gotten out of you,' he said to Alvarez, 'is that you have tried to show that Dr Oppenheimer was opposed to the development of the super weapon. Is that true?' Alvarez agreed and Evans then asked, 'What does that mean in your mind – anything?'

ALVAREZ: ... The point I was trying to bring out was that every time I have found a person who felt this way, I have seen Dr Oppenheimer's influence on that person's mind. I don't think there is anything wrong with this ... I just point out the facts as I see them.

EVANS: It doesn't mean that he was disloyal?
ALVAREZ: Absolutely not, sir.

The two witnesses from the Air Force, David Griggs and General Roscoe Wilson, both pursued much the same line. They both objected to the 'pattern' of Oppenheimer's judgements in favouring the tactical use of nuclear weapons, rather than their use in a nuclear strike force such as Strategic Air Command.

It was late on Friday, 30 April that the government's last witness appeared. William Borden, the man whose letter of charges had catalysed these proceedings, had to be subpoenaed to appear. Now, for the first time, Oppenheimer was to see both the letter and the man who had written it.

As Borden began reciting the contents to the board – that Oppenheimer 'more probably than not [was] an agent of the Soviet Union ... under a Soviet directive in influencing United States policy' – he must have sounded incongruous indeed to those who had sat through the three weeks of evidence that had done so little to back up his charges. Gordon Gray dismissed outright his central conclusion, saying there was 'no evidence' linking Oppenheimer with the Soviet Union. Roger Robb asked him only a few questions, Garrison none. His accusations were now irrelevant, having proved a catalyst for an event now fuelled by personal animosities and political neuroses.

During the final week, Oppenheimer was to return to the stand to be questioned once again about the Chevalier incident. In an effort to moderate the effect of his earlier collapse, he admitted that he should have reported the incident immediately, 'but that it was a matter of conflict for me ... I may add one or two things. Chevalier was a friend of mine.'

Was the board likely to take account of this conflict of loyalties between friendship, on the one hand, and national security on the other? The hearing finally ended on 6 May 1954. After four weeks in Room 2022, everyone involved needed some respite. The board members took a ten-day break, returning home to rest and to gather their thoughts. It was a short interlude before the backstage politics resumed in earnest again.

Immediately after the hearing, Teller spent three days recruiting at the American Physical Society conference before returning to Livermore to attempt to repair the damage to the laboratory's reputation inflicted by the total failure of its weapons tests so far. From here, in mid-May, he wrote to Lewis Strauss, reporting what he described as the 'rough time' he had received the evening before he had testified from 'people I like to consider my friends'. He then went on to the National Academy of Sciences, where he lobbied successfully to prevent them coming out in defence of

Oppenheimer. He now wrote optimistically to Strauss: 'Most of the physicists are quite level-headed about the Oppenheimer case. I fear[ed] that the scientific community will be split wide open. I'm still apprehensive but much more optimistic than I was two weeks ago [when attending the hearing]. In any case I feel very certain that Oppenheimer's continued interfering was exceedingly harmful. I hope I can speak about that interference in the past tense.'

Thus, nothing in his experiences at the hearing had yet modified Teller's determination to remove Oppenheimer from any position of influence. Indeed, he now believed his colleagues were, at least, accepting what had happened to Oppenheimer and, perhaps, even condoning it. At this point, a matter of days before the board reconvened, it seemed as if everything was going the way Teller had hoped.

On 17 May, Gordon Gray and Thomas Morgan travelled back to Washington on the same plane together and, on comparing notes, found that they agreed that Oppenheimer should be denied his security clearance. Given that Ward Evans had often shocked them with his anti-Oppenheimer views, they arrived convinced that the board would rapidly come to a unanimous decision. They were amazed, therefore, to find Evans already at work on a recommendation to restore Oppenheimer's 'Q' rating. They suspected that Evans had been 'got at' back in Chicago, but Evans denied it and was now unshakeable in his support of Oppenheimer. The news of this change of heart filtered back to Strauss and, according to an FBI report, Strauss immediately took action trying to involve J. Edgar Hoover in a direct intervention with the board: 'Commissioner Strauss . . . said that the Oppenheimer Hearing Board is in the last stages of its considerations and that things are "touch and go". Commissioner Strauss said that he felt most urgently that the three members of the board . . . together with counsel Roger Robb, should see the Director.'

Hoover wisely fended off Strauss and Robb's approach, and so Robb proposed another course of action. 'I thought it would be a tragedy if the Board went the other way. No doubt about that at all after the hearing . . . and I didn't want "Doc" Evans's opinion to be too vulnerable. If it was . . . it would look as if we put a nincompoop on the Board.' So Robb worked with Evans on his minority report, paradoxically helping to produce a document that was to be praised for its 'jabbing directness and its conciseness in showing up the charges against Oppenheimer'.

The board's findings were ready on 27 May. The majority report found no less than twenty out of the original twenty-four charges either 'true' or 'substantially true'. The board had considered all the charges together – except the single H-bomb charge – and were able to conclude on these 'that Dr Oppenheimer is a loyal citizen'. As to the H-bomb charge, they dismissed any implication of disloyalty that might be inferred from his resistance to the H-bomb. For Gray and Morgan, however, this was out-weighed by their 'conclusion that, whatever the motivation, the security interests of the United States were affected'.* Thus a man employed and used as a consultant was being condemned not for any motives he might have had but for the effect of his genuinely held opinions.

By contrast Ward Evans in his minority report stated: '[Oppenheimer] did not hinder the development of the H-bomb, and there is absolutely nothing in his evidence to show that he did ... If his opposition to the H-bomb caused any people not to work on it, it was because of his intellectual prominence and influence over scientific people and not because of any subversive tendencies.'

It was indeed a jabbingly direct document, but predictably it did not affect Kenneth Nichols's recommendations, which, along with the board's findings, he passed on to the commissioners for their final decision. In them, Nichols agreed with the board's majority opinion that Oppenhei-mer's lack of 'enthusiasm' had delayed the hydrogen bomb and he also elevated Oppenheimer's several versions of the Chevalier incident to 'crim-inal dishonest conduct'.

On the following day, Oppenheimer and his lawyer received their copy of the board's reports and, after some deliberation, decided on a course of action that mirrored the one taken at the start of the hearing: they released the report to the press along with their letter of response to Kenneth Nichols. It was a defensive move, as they anticipated, not unreasonably, that sections of the report unfavourable to Oppenheimer might be leaked at any time. However, it initiated a running battle in the press that further fanned the powerful emotions generated by the hearing.

The next weekend, Teller wrote to Maria Mayer asking urgently to see her when he was in Chicago about 'the Oppenheimer situation which was becoming more terrible by the week'. Since the board's decision had been made public, speculation had begun 'on the underlying reason that Oppenheimer had lost his clearance', said Teller. 'Every major newspaper promoted a different idea – the army was behind Oppenheimer's fall, or Lewis Strauss was, or Ernest Lawrence, or I.'

The following Sunday, 6 June, a nervous Lewis Strauss called in 'Scottie' Reston of the New York Times for an 'exclusive' on the Oppenheimer case. The following day the commissioners were sufficiently perturbed to consider the possibility of releasing the transcript of the whole hearing, even though every witness, including Teller, had been given an assurance that the AEC would not take the initiative in publishing it. Four days later an accident provided them with the excuse to break that assurance. Eugene Zuckert, one of the commissioners, left a copy on the train while travelling home to Connecticut. It was discovered almost immediately but, at an emergency meeting, the commissioners decided on release. The densely printed 993-page transcript was produced and distributed within two days. The effect, on Teller in particular, was devastating.

There for everyone to read was testimony that he had believed would never see the light of day. His colleagues read: 'If it is a question of wisdom and judgment, as demonstrated by actions since 1945, then I would say one would be wiser not to grant clearance.' For those with more intimate knowledge of the H-bomb saga there were also his often-voiced complaints crystallised in the evidence and the charges against Oppenheimer.

'The exile I was to undergo at the hand of my fellow physicists, akin to the shunning practised by some religious groups, began almost as soon as the testimony was released,' Teller wrote. 'My recollection of that painful period is general rather than specific: I was more miserable than I had ever been before in my entire life.'

A week after the publication of the transcript, Teller, accompanied by Mici, attended a scientific meeting at Los Alamos. These were normally amiable reunions, particularly when wives were present, but this one was different. On 23 June, Strauss recorded that he had received a phone call from the laboratory:

> Dr Luis Alvarez called to tell me that he felt very much concerned about Dr Teller. He said that Dr Teller was now in Los Alamos with his wife and was being given fairly rough treatment there ... I called Teller who confirmed what I had heard from Alvarez. 'I have made up my mind that I can take the gaffe. I am no longer interested in anything except truth. I have complete confidence in the fact that my friends will eventually come to the conclusion that in telling what I believed to be the truth, I have done the cause of science in the service of my country the best that I could.'

What was to bring things to a head at this summer meeting was that, as a consequence of the broadcast on television of the Mike film and articles like the one in *Time*, Teller had, over the preceding weeks, become a celebrity. The press now referred to him as 'the father of the H-bomb', or simply 'Mr H-bomb'. On arrival at Los Alamos, Teller had unwisely arranged to be interviewed by yet another journalist, Robert Coughlan, who was preparing a profile for that great reputation-maker of the time, *Life* magazine. Stirling Colgate and his wife Rosie were also attending the same meeting.

Rosie Colgate recalled:

Coughlan was there to interview Edward, and Edward asked Stirling to go to his room so that he had someone friendly there. We were Livermore, and given the atmosphere, we were really the only people that could be a friend to Edward. In the evening, the other people at the conference began gathering on the terrace for a picnic, and it became known that Edward was being interviewed in his room by *Life* magazine. This generated just so much anger, particularly as Edward was later and later for the picnic.

As Teller recalled:

One of my friends later told me that Coughlan's friendly presence there was like pouring salt in an open wound. I was resented because I was receiving media attention and because I was suspected of trying to vindicate myself. At the time I was oblivious to the implications. My sole concern was to make sure that all the statements I made to Coughlan were accurate . . .

Our interview ran later than I hoped, so by the time Coughlan and his group were leaving, many of the participants in the conference were already assembled for a picnic on the terrace outside the dining room. The first person I spotted was Bob Christy, with whom we had shared a house for a year. I hurried over, reaching out to greet him. He looked me coldly in the eye, refused my hand, and turned away. I was so stunned that for a moment I couldn't react. Then I realised that my life as I had known it was over. I took Mici by the arm, and we returned to our room upstairs. Our last exile had begun.

Back in their room Edward Teller broke down and wept. 'If a person leaves his country, leaves his Continent, leaves his relatives, leaves his friends, the only people he knows are his professional colleagues. If more than ninety per cent of them come around to consider him an enemy, an outcast, it is bound to have an effect. The truth is it had a profound effect. It affected me, it affected Mici, it even affected her health.'

19
Aftermath

Sir:

Your testimony before the review board has been most disturbing to me and many other physicists. The statement that you would feel more secure if Oppenheimer's influence were eliminated from the AEC seems to indicate that your ideal of security is a state where you would be running the show surrounded by a bunch of yes-men . . .

In fact you should feel more secure by the very fact that men with convictions dare to disagree with you. If we ever get to a state of uniform agreement, that would be cause for anxiety.

In the weeks following the publication of the hearing transcript and the AEC commissioners' confirmation by a 4:1 majority of the Gray board's findings, Edward Teller received many letters similar to this one from Joseph Jauch, the Austrian-born professor of physics at Iowa State University. Others who had criticised Oppenheimer, like David Griggs and Luis Alvarez, were also censured, but it was Teller who was seen as the fulcrum of the assault on Oppenheimer. Like Jauch, many of his critics picked up on the concern he had expressed about Oppenheimer's opinions – and in early July Teller felt driven to write to Strauss, desperate to put out some kind of corrective to what he saw as a misinterpretation of his testimony: 'In my testimony I did not imply that the right to disagree should be limited,' he wrote in the draft he sent to Strauss. 'I consider the right as essential in a free society. That my testimony in this connection should have been misinterpreted is a matter of greatest concern to me.'

Strauss passed the note on to Roger Robb, who then wrote directly to Teller: 'Unless you think it absolutely necessary to do so, it seems to me that you would be well advised to refrain from making any statement at all. No statement is necessary for your friends, and your enemies will try to misinterpret or twist anything you might say.' Nevertheless, Robb enclosed a revised version of Teller's statement, one that removed the element of apology and placed the fault for misinterpretation firmly in his critics' court. Robb ended his letter by expressing his 'great admiration' for the courage and character Teller had shown, and belief that 'the people of this country should be proud of [him]'.

In fact the press was largely sympathetic both to the final verdict and to Teller himself, but it was poor comfort to someone who felt rejected by the group that mattered to him most. He wrote to Maria Mayer:

> I came back from Los Alamos a few days ago. I was there for two weeks. I felt like Daniel in the Lyons [sic] den. After some time you learn to distinguish the Lyons (here I had to interrupt. I am now back in Los Alamos for some 18 hours) according to their growls. And as you see I am now back in the Lyons' den. The second time it takes no less courage...
>
> The worst of them is Rabi. He never was my friend but now he is terrible. Tomorrow I have to give a report. He will be there to heckle me. I hope I shall not loose [sic] my temper. This is not much but it is something.
>
> Last night I dreamed there was a Raven and I did not dare to go to sleep because he may pick out my eyes. Please translate Raven into German [The word for raven in German is Rabe]. I found this amusing because the Raven started to smile and I slept quite well.
>
> I am reading the Hearings. If it would not involve me so deeply I should find it interesting ... I do not know what people will think but they should know the facts.
>
> I am ducking newspapermen. It is not easy and not fully successful. There is one advantage: I am finding out who are my friends...
>
> Paul and Wendy are in a camp near Santa Fe. They are fine and they are wonderful. We have arrived at a point where I am missing them more than they are missing me. (mainly they don't, which is quite all right).
>
> I remember a conversation a long time ago about backbone. I seemed to get along fine without one. Now there seem to be some growing pains. I also wonder whether it is growing in the right direction.
>
> So long.
> Edward.

At this stage the relationship with those at Los Alamos was barely tolerable, but Bradbury still invited him back from time to time to give reports or to join in discussions. If things had continued as they were, maybe some level of détente might have developed, which would have brought improvement over time; but such hopes were brought up short by two publications. In September the *Life* article, for which Robert Coughlan had been interviewing Teller on the evening of the Christie incident, appeared under the title 'Dr. Teller's Magnificent Obsession: Story behind the H-bomb is one of a dedicated, patriotic man overcoming high-level opposition'. It praised its subject lavishly, crediting him with having 'kept the idea of an H-bomb from dying of pure neglect' and reminding its readers that, without this weapon, President Eisenhower had judged: 'Soviet power would today be on the march in every quarter of the globe.'

Teller wrote to Coughlan shortly after its publication, describing the article as 'wonderful', adding, 'knowing how conscientious you have been in this last article, I promise to believe anything you write about anybody.' Later, after he experienced the backwash from it, he described it as 'insulting, since the same article had blackened my friends. It was the source of the greatest possible embarrassment.' Any problems he may have experienced as a result of this article, however, were to pale into insignificance beside the consequences of a second publication that appeared a fortnight later. It was to change everything.

James Shepley of *Time* magazine had used his article of the previous March as the springboard for a collaboration with colleague Clay Blair to produce a book. Entitled *The Hydrogen Bomb, the Men, the Menace, the Mechanism*, it expanded Shepley's view of all the issues and conflicts that had caused such an upset six months earlier.

Teller had been interviewed for the book and, while it took a more extreme stance than he ever did – certainly in public – there was no doubt that in many respects the book mirrored Teller's own view of his fight for the H-bomb. Within it, Oppenheimer was portrayed very much as the Machiavellian figure of Borden's charges, his role as director at Los Alamos carefully minimised; Norris Bradbury came across as incompetent; and Hans Bethe was described as not wanting to work on the H-bomb because the pay was too low. On technical matters the book made it appear that implosion had only been invented in 1950, not at Los Alamos during the war, and that the use of a fraction of the critical mass to create an atomic explosion – a wartime Los Alamos development – had simply been 'sparked by Teller's intuition'. It certainly did not stint in applying its economies of truth to the glorification of its hero, Edward Teller.

This all served to raise the bitterness and anger among Teller's opponents to new levels, and caused many who had so far been uncommitted to look askance at his role. It ensured that he received no further invitations to Los Alamos. The Alsop brothers, who had supported Oppenheimer during the hearing, summed up the mood: 'Gordon Dean, former AEC Chairman, has called the Shepley–Blair book a "blood-stained Valentine to Edward Teller". The blood, of course, is that of Teller's fellow scientists. The scientists are the more bitter because Teller has not yet seen fit to repudiate the "blood-stained Valentine" and to expose the falsehoods and distortions of the book.'

Teller had indeed remained silent, although doing so, as his friend and former student Steve Brunauer pointed out, was 'bound to be interpreted as meaning only one thing: namely that you concur in everything the book says . . .' Teller argued that enough people were already contradicting them, so that he did not need to add his voice, but was this the real reason? What Teller said to Shepley during his interviews for the book can only be guessed at, but they had met in the period leading up to the hearing when Teller was voicing his criticisms most vehemently and widely. After publication Shepley wrote Teller an admiring letter:

> It might be worthwhile now for the sake of what I hope will be a continuing friendship to make a journalist's statement of conviction. I did not write the story of the great controversy over the thermonuclear weapons program simply to make a hero out of Edward Teller. I sincerely regret any embarrassment I have created for you by an account which showed you to be right when so many of your eminent colleagues were wrong. But I feel that in the long run it will be more than compensated for by the widespread recognition of the American people, both officially and unofficially, of the debt which they owe to you.

Teller responded a week later with a note that provides few grounds for assuming that he thought Shepley had exaggerated or misrepresented him:

> It was a great pleasure to get your letter. Kind words are always welcome, and for me they are more than usually welcome these days.
>
> There is one thing I should like to say to you with all possible emphasis. What you took for modesty in my statement to you was rather respect for the facts as I see them, and an urgent desire to continue to work in harmony with my colleagues . . .

Shepley had hoped for 'continuing friendship', regretting only 'any embarrassment' at showing his subject up for the hero he was. Teller, for his part, not only mentions his great pleasure at receiving Shepley's letter but then states that his 'modesty' was in part the result of an 'urgent desire to continue to work in harmony with my colleagues' – implying that what he had told Shepley had been necessarily muted. There were no recriminations, just good wishes. It would seem that Teller was trapped, unable to dispute an account in which he had collaborated.

This was perhaps one of the most desperate periods of Teller's life. Friends had been able to admire his courage at the hearing, whether they agreed with him or not. With the Shepley–Blair book, however, it was difficult to dissociate Teller from its egocentric and often inaccurate account. Teller was embarrassed and lonely – a loneliness about to be compounded by news from Chicago.

Some time earlier, Enrico Fermi had been diagnosed as terminally ill with cancer of the stomach. Teller had gone to visit him in hospital, taking with him a draft of an article he hoped to place in one of the popular magazines, like *Atlantic Monthly*. It was his response not to the Shepley–Blair book itself but to the general grievance among his colleagues, initiated by Shepley's *Time* article, that he had been hogging the credit for the development of the whole thermonuclear weapons system. Entitling it 'The Work of Many People', he had sent drafts of the article to a few close allies, and now he had wanted Fermi's opinion.

Fermi had been predictably philosophical about his medical condition. 'Isn't this a dirty trick on me?' he had asked rhetorically, but he had had no wish to dwell on the subject and in a short while they were talking about Teller's problems and his draft article. When Teller had confessed that, after all the criticism, he was so confused that he was no longer sure what to do, Fermi had advised him in the strongest terms to publish. He had also clearly spoken his mind.

Exactly what he said to Teller is not known, but the conversation between these two friends of twenty years went far beyond the practicalities of whether or not to publish:

> Fermi, whether he believed in God or not, was a Catholic [Teller wrote]. In most religions a man who is dying confesses his sins. This now seemed to me to be the wrong way. If a man is dying, particularly if a good man is dying, you should confess your sins to him, because maybe he is going to see God and maybe he can do something about your sins. In a completely irrational manner, that is how I felt about it. Fermi was a very wonderful man.

Not long after this, Teller had taken his leave and Fermi had talked about his visit to a close colleague, Herb Anderson, and his wife Jean, who later offered an indication of how intense it had been: 'According to Enrico, Edward talked non-stop about the effects of the hearing,' Jean remembered.

'He was so emotionally overwrought at that time that Enrico actually thought he was close to suicide.'

This was to be the last time that Teller saw the man whom he had most admired, and envied, as a scientist and as a person. Fermi had welcomed him as a gauche young researcher to his Rome laboratory, had first aired the possibility of what had become Teller's lifelong obsession, the thermonuclear weapon, and had counselled him during the fights and frustrations with 'Oppie's men'. He had been a constant example, a point of reference, and Teller was to miss him terribly.

Fermi was one of several friends who recommended publication, but the popular magazines Teller had hoped to interest turned him down. Instead it was published on 25 February 1955 in the prestigious journal *Science*. In the article he spoke glowingly of his enemies. The A-bomb was completed in time to impact on the war because of 'the leadership of our director, Oppenheimer'. Los Alamos's survival after the war he credited to 'a few determined people, among them Norris Bradbury'. As to the basic theoretical understanding underpinning his invention, 'the most remarkable part was done by Bethe'. He even credited Ulam with 'an imaginative suggestion'.

In retrospect it stands as a fair summary of the efforts of the major contributors, but at the time, after everything that had preceded it, the article was seen as a calculated manoeuvre and failed entirely to lift the cloud of resentment that Teller found so hard to bear. He began to experience wrenching pains in his lower abdomen. He went to see the same gastroenterologist as had treated Ernest Lawrence, and was diagnosed with the same condition: ulcerative colitis. In one of the few personal conversations the two men had, Lawrence explained its link to stress, and the fact that it was potentially fatal unless treated carefully.

At much the same time Mici became ill too. 'She had fainted at a dinner party,' Mici and Edward's son Paul recalled, 'and because of the way she had fainted, the doctors were concerned that it could be a symptom of a brain tumour, of her having cancer. She also felt terrible. There was a long period where she was confined to bed and everybody was extremely upset and concerned because nobody had any idea what the problem was.'

A journalist who visited Teller at home in January 1955 for an interview, found him trying to cope with the children and the house as Mici had just returned from hospital. He sensed 'a large residual strain and a deep sadness. His smiles, when discussing the [Oppenheimer] matter at hand, were forced, and halfway through the interview his eyes became moist. This description is germane ... To understand his position as he outlined it, you have to understand that he is a man struggling under a great sense of burden, sadness, and what seems almost to be despair ... One might almost imagine that his and Oppenheimer's roles had been reversed.'

It was a miserable time. 'At the present μ-second I am not sorry for myself,' Teller wrote in one letter to Maria. In another he declared, 'Life, at present, is only fun in the general sense. Somehow I believe that it might be fun again at some improbable time, in some impossible way...'

'He had the aura of someone who'd been terribly, terribly hurt and taken by surprise,' said Paul.

He was not prepared for anything like the very violent reaction he got. He described himself to me as having lost most of his friends, friends that were really important to him, that were irreplaceable because he had shared a body of important life experiences with them. And for that reason alone they were just not replaceable. If he'd just lost one or two – but there were just so many of them.

His positive response was to develop a new circle of friends and many of these were business world people. Among these people he had a certain positive notoriety and he looks like he enjoyed it, but it wasn't the same. It may have been their walk of life, or it may have been his old friends just cutting him out but I cannot tell you quite how awful it all was for him. Everything after this was a poor substitute.

Rosenbluth sums up Teller's situation in terms of a classical tragedy:

My own view of things was that events at Los Alamos at the time were like the classic tragedy *Coriolanus*. The hero performs an incredible feat – in this case Teller, saves the state. Following this, of course, hubris sets in and he decides he's so brilliant that he should be director of everything. At the same time, of course, on the other side, petty jealousies set in and people don't want to admit their hero was smarter or stronger than they were. I even think that the origin of the Ulam campaign was that people did not want to admit that Teller had cracked it.

The hero then goes into exile and joins the enemy. That's Livermore and the Air Force. The parallel is quite exact and it was, in a way, a tragedy.*

Sick, isolated and deeply, even suicidally depressed, Teller was desperate for comfort and support. His links with Los Alamos and moderates like Hans Bethe severed, Teller's closest colleagues now included hawks like Ernest Lawrence, Willard Libby, David Griggs, employees of the Rand Corporation and various military personnel. His bitter feeling of rejection was sharpened by his own conviction that he had achieved a great deal. He had ensured the US now had a new class of weapon as different in order of destructive power from fission weapons as they had been from conventional explosives, and he perceived that he had achieved it against serious and committed opposition. He also believed that what he had done was essential to the security of his adopted country. 'Look, let me put it in perhaps a not permissible way, in words of one syllable,' he summarised years later. 'I do not want a hydrogen bomb because it would kill more people. I wanted a hydrogen bomb because it was new. Because it was something that we did not know and could know. I am afraid of ignorance.'

This was the principle that he had first voiced to justify his continuation with thermonuclear research at the end of the war and that he was to adhere to throughout his life. For him, it was entirely separate from the morals and ethics of use, for, as he saw it, 'Throughout all written history the killing of people was never limited by the ability to kill people but always by the amount of intention to kill people.' He supported his argument with an event early in Hungarian history just after Genghis Khan's death, when

> the Mongols under two generals descended on Hungary and killed 90 per cent of the Hungarians in a few weeks. You know what they did was – they would attack the city, kill everybody, then leave, then come back in a week, and kill those who crawled out from the weeds ... The atomic bombs killed 150,000 people out of the 50 million that had been killed in the Second World War. The limitation is not in the ability but in the intention ... So there was reason to fear at any time.

Teller was to diverge from colleagues in years to come when it seemed that he believed that fear of the Russians was justified not just at *any* time but *all* the time; but at this moment, with Stalin's Russia well along the road in the race to create thermonuclear weapons, there were many who shared Teller's fear of the Soviets. Even though the Oppenheimer affair had alienated so many of his old friends and scientific colleagues, he was widely regarded by the public as a visionary, not perhaps of Churchillian proportions but in the same mould – as someone who perceived a serious threat that others chose to ignore. Shepley and Blair's book had contributed to this impression and so, to a large section of the public, had his role in the hearing.

There were a number of scientists too who, as Marshall Rosenbluth put it, had judged 'that with the way Stalin ran things in Russia, by God, all the top people there, the best scientists, were getting involved'. Rosenbluth, and even those as reluctant as Hans Bethe, had accepted the need to fight to 'stay ahead'. In the face of the Stalinist threat, it was hard to accept the arguments of Oppenheimer and the GAC that there was enough 'cover' in the existing weapon stockpiles to sufficiently intimidate the enemy.

This view was particularly prevalent among the young staff at Livermore who took him, as he himself put it, 'worryingly seriously'. This must have been balm indeed for the psychological wounds of the past year. 'Going to California was like going to a new country,' he wrote to Maria. '... I never worked so hard as now and, incidentally, I am establishing a reputation that I never fight and am always pleasant ... a thoroughly new existence.'

However, he knew better than anyone that, following the failure of its first three tests in the spring of 1954, Livermore was in serious danger of closure. At a GAC meeting soon after the hearing verdict, Isidor Rabi described its performance as 'amateurish' and commented that Teller's laboratory did not have responsibility for any 'necessary' part of the weapons programme. Then, that September, close on the heels of the Shepley–Blair book, Norris Bradbury wrote to the AEC suggesting that the second laboratory should be made subordinate to his own, Los Alamos. 'The brilliant new ideas have not appeared,' he pointed out.

To compound matters, the AEC also cancelled its order for Ramrod, the thermonuclear trigger that Teller had hoped would at last make feasible his old classical Super design. In response , Teller tried to impress Washington with plans for a device with a preposterous yield of no less than 10,000 megatons, a thousand times that of the Mike shot. It was dismissed by Rabi as an advertising stunt. As a consequence of these threats to Livermore's survival, Teller encouraged anything that forged bonds with the Pentagon. Thus, it was of great interest when the Department of Defense invited projects that would aim at minimising or eliminating fallout. Livermore began working not on modifications that would *reduce* fallout but on one that would seek to eliminate it entirely.

Throughout that year, the public reactions to the implications of fallout had waxed and waned as, first, the Oppenheimer affair, then the Army–McCarthy hearings (during which the Wisconsin senator finally received his come-uppance) and then the desegregation riots in the Southern States, stole the headlines. However, in the autumn the results of the first serious studies into the fall-out from Bravo began to emerge. Professor A. H. Sturtevant of Cal Tech, one of the world's foremost geneticists, had made a theoretical projection of the effect of the additional global fall-out generated by the test. Based on the assumption that one roentgen of radiation would cause an undesirable mutation in every 10,000 persons exposed, he calculated the genetic cost of Bravo as damage to 1800 children. 'And every new bomb exploded,' he concluded, '... will result in an increase in this ultimate harvest of defective individuals.' They were chilling findings, findings that rekindled public alarm. In an address to the conference of the nation's mayors in December, Willard Libby, recently elected an AEC commissioner, tried to play down the issue, but did so in full knowledge that, back at the AEC, the commission's own report was causing sufficient alarm that the State Department blocked its publication for fear of its international impact. Everything was giving added impetus to Livermore's new project for developing its so-called 'clean' weapon.

In principle, at least, a 'clean' bomb is relatively simple. The isotopes of hydrogen used as fuel in the secondary fusion stage of a hydrogen bomb do not themselves generate the isotopes that make fall-out so dangerous. Rather, the majority of these come either from the fission of uranium or plutonium in the ignition primary or, most significant of all, from the interaction between the thermonuclear secondary and the heavy uranium tamper that encases the whole bomb. When a bomb detonates, the tamper is not simply blown apart. The flood of neutrons produced in the secondary creates a giant fission reaction within it; and it was this that had generated the massive radioactive fall-out in the case of Bravo.

'Fifty per cent of the yield comes from the uranium fission in a thermonuclear bomb,' Herb York explained. 'That's the efficient way to build them – you can build them smaller and everything else. If it's already working well enough, then you can afford to replace the uranium in the tamper with tungsten or something and that's all a "clean" bomb is. That's the change you make.' The removal of the fissionable tamper, however, still leaves the problem of the fall-out that will result from the explosion of the fission primary itself. This may represent a small percentage of a multi-megaton bomb, but with the smaller bombs being developed for the battlefield it becomes an increasingly larger fraction of the whole. The crux of the 'clean' bomb problem was how to reduce this fall-out and yet still achieve the conditions necessary to fire the thermonuclear secondary. It was a problem that the scientists at Livermore were not going to find it easy to resolve.

In February 1955, after two months' delay, leaks and other alarmist reports forced the State Department's hand and the AEC's report on Bravo was published. It covered both the local and global effects of the test. It described the cigar-shaped footprint of debris from the blast, 220 miles long and 40 miles across at the widest point – 7000 square miles. The report admitted that, if such a bomb were dropped on a city, half the population living up to 160 miles downwind would be killed. However, it was the main section, which dealt with the global effects of testing, that attracted the most attention. For the first time, the name strontium 90, described in the report as one of the most dangerous of radioactive by-products from a nuclear explosion, became familiar to the public. Strontium 90's half-life, a measure of its radioactive longevity, is twenty-eight years and, because strontium belongs to the same chemical group as calcium and reacts in much the same way, it was realised that it could eventually find its way through the food chain into the bones of humans. The fear was that this radioactive source might cause cancer, particularly in growing children, who absorb more calcium than adults. To make matters worse, these isotopes would be carried high into the atmosphere by the explosion, to be spread indiscriminately around the globe, well beyond the measurable radioactivity footprint.

The AEC attempted to reassure the public, pointing out that the amount in the atmosphere would have 'to be increased many thousand times before any effect on humans would be noticeable', going on to compare the exposure to fall-out from Bravo as equivalent to 'exposure from one chest Xray'. The report then ended up with what was its essential credo: that while testing did create risks, both to present and future generations, these were minute compared to the advantages gained for 'the security of the nation and of the free world'.

The matter-of-fact tone of the report found many supporters, but there were also powerful and vocal critics. One of the country's leading civil defence experts, Val Peterson, raised the prospect of a cobalt bomb, a hydrogen bomb with a cobalt metal jacket that would produce so much fall-out that the cloud would circle the globe, killing wherever it went. Then enter, for the first time, Linus Pauling, Nobel Laureate in chemistry, who shocked an NBC television audience when he said that global radiation might be sufficient to trigger leukaemia in thousands of those susceptible to it.

Another who spoke out was Professor Joseph Rotblat, a Polish-born physicist who had worked at Los Alamos but left once the German threat was found to be non-existent. Now he was in London researching into the medical uses of radioactivity. He had become aware from the Japanese data that Bravo could have produced a thousand times more fall-out than the US authorities were making out – and then picked up on the seemingly comforting comparison between exposure to radiation from fall-out and chest X-rays. 'Now this alarmed me,' he recalled 'because a chest X-ray is quite a large dose, and with a few bombs of that type you could get quite a high dose and it would be very dangerous. This is why I published my data and caused quite a furore and I was attacked by the government and so on.'

Rotblat argued publicly that while X-rays are targeted at specific organs, fall-out affects the whole body including the reproductive organs, and therefore great damage might already have been done. His findings and the whole post-Bravo furore came at a time when pure genetics research was suggesting that the damage to human genes from radiation had been seriously underestimated. Herman Muller, father of radiation genetics, suddenly found that his thirty-year-old results on the ability of radiation to cause genetic mutations had become highly controversial. He had been invited to give a paper at the 1955 Atoms for Peace Conference in Geneva but, at the last minute, Lewis Strauss interceded to prevent his appearance. It was a clumsy move as not only was Muller present in the audience of the session he was due to address, but his paper had already been distributed and he had received a standing ovation. It served simply to add to the growing sense of mistrust, to the feeling that Strauss's Commission, and those who sided with it, had something to hide.

It was the beginning of a fierce and frightening debate that was to rage on for more than a decade and that Teller was soon to join on the side of the AEC. The public outcry and adverse publicity was influencing a President whose growing antipathy towards nuclear weapons was strengthening his resolve to introduce a test ban – with all its inherent implications for the futures of both Los Alamos and Livermore.

The 'shrimp' device, used in the Bravo test. Innocuous looking, it ran away to 15 megatons, a yield more than three times that of all the explosives used in World War II.

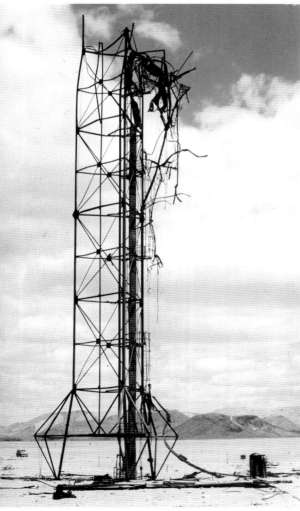

The aftermath of Ruth, Livermore's first nuclear test on 31 March 1953. The tower was hardly damaged and the post-explosion silence was broken by ironic cheers from Los Alamos rivals as they rushed for their cameras.

Haakon Chevalier in 1936. Professor of Romance languages at Berkeley he was a close friend of Oppenheimer's and major catalyst in his security problems.

Lawrence, Teller and Mark Mills (R) meet with AEC Chairman Lewis Strauss on 25 June 1957, the day after they had enthused Eisenhower with the humanitarian virtues of their 'clean' bomb and prevented a moratorium on testing.

Teller took pride of place among his contemporaries in *Time*'s November 1957 US science review. He was praised for working 'hardest and most belligerently to send the warning that the Russians are coming'.

The Teller family and home in the late 1950s. Paul (L), Mici, Edward and Wendy with the family's dog.

Right Edward and Mici in the studio preparing for the 1957 TV series on the atom. He had a formidable reputation as a lecturer and his courses were always oversubscribed.

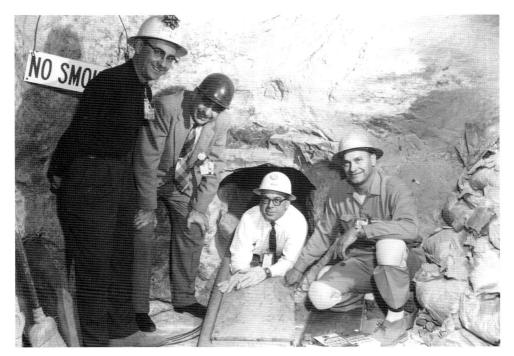

Teller on the site of the 1961 Project Gnome, the first Plowshare test. Alaskans were to describe Livermore physicists as 'firecracker boys'. With Teller are (L–R) George Cowan, Arthur Hudepris, and future Livermore director Roger Batzel.

'If your mountain is not in the right place, drop us a card.' Teller's epigram was offered to reporters as he and Mici flew out to view preparations for the 1.4-megaton Project Chariot blast to create a harbour near Point Hope, Northern Alaska in June 1959.

President Kennedy's visit to Livermore in spring 1962. With him are (L–R) Norris Bradbury, John Foster, Edwin McMillan, Glenn Seaborg, Edward Teller, Defense Secretary Robert McNamara and Harold Brown.

On 2 December 1962 Kennedy presented Edward Teller with the Fermi Award watched by AEC Chairman, Glenn Seaborg (L), and Mici (R). After the ceremony, Teller and the President once again crossed swords.

In December 1963 Oppenheimer received the Fermi Award from President Johnson. After the presentation he posed for the cameras with Teller in a gesture of reconciliation. The expression on Kitty Oppenheimer's face (L), however, speaks volumes.

On 23 November 1970, Berkeley students staged a tribunal accusing Teller of 'war-crimes'. Earlier in the day, Lowell Wood (L), in student dress, and Teller met to discuss tactics before Wood went off to keep an undercover watch on events.

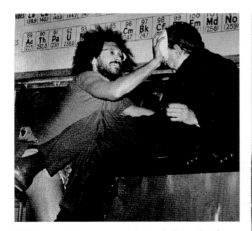

Teller receives a custard pie full in the face delivered by leading student activist, Jerry Rubin.

Right: The canister containing the five-megaton Cannikin device, begins its mile deep descent underground on Amchitka island. There were fears that this, the largest underground blast ever, would trigger earthquakes right around the Pacific 'rim of fire'.

Roy Woodruff, head of Livermore's 'A' Group, was initially a great enthusiast for the X-ray laser, but fell out with Teller over exaggerated claims for the project.

The reluctant inventor of the nuclear pumped X-ray laser. Peter Hagelstein had a strong antipathy to defence work but found himself inexorably drawn in.

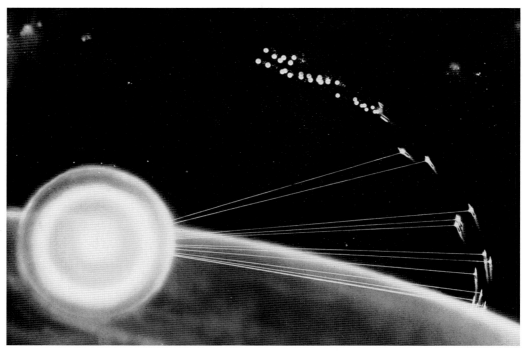

Artist's impression of Excalibur in action. Laser rods would focus the blast of a small hydrogen bomb into beams up to a trillion times as intense as the original explosion, each of which could be aimed at enemy missiles.

The moment at the White House dinner on 8 December 1987, which marked the signing of the Intermediate Nuclear Force Treaty, when the President introduced Teller to Mikhail Gorbachev. The treaty was a compromise spiked by conflict over SDI, and Gorbchev refused to shake Teller's hand.

Teller and Soviet H-bomb pioneer, Andrei Sakharov, meet for the first time at a Washington dinner on 16 November 1988. Sakharov declared his 'deepest respect' for Teller, but took the opportunity to damn SDI.

26 July 1988. Lowell Wood, on Teller's left, demonstrates Brilliant Pebbles to the National Security Council. Second to the right of Reagan is Vice-president Bush and on the table sits a shrouded model of a Pebble.

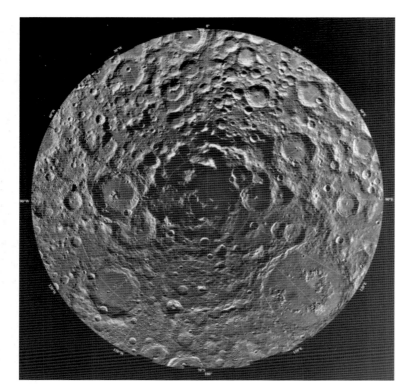

Clementine's 1996 map of the moon's surface. The satellite's sophisticated package of sensors, based on those developed for Brilliant Pebbles, produced the most detailed map so far and discovered water at the lunar south pole.

On 26 August 2003, two weeks before he died, Edward Teller received the Presidential Medal of Freedom, the United States' highest civilian award.

20

'Almost like Ivory Soap'

'Mici is becoming better quite impressively,' Edward wrote to Maria, taking stock of his life and his family in the summer of 1955:

> The diagnosis is at present: neurotropic virus infection. Clearly what one does not know of the central nervous system is quite a lot. Mici no longer keeps secret the present theory . . .
>
> In the long run this terrible trouble with Mici may have good sides. She has become for the first time in her life, a little doubtful about herself and consequently she is beginning to find out some things.
>
> It was also very good that all this did not happen in Chicago. Mici was so incredibly full of resentments and to a considerable extent still is. I do not know what would have happened.
>
> Paul and Wendy are better than ever. It is wonderful how they escaped any harmful effect; at least it seems so. Actually I can not believe they are my children: Paul has tact and Wendy has assurance.

After a traumatic year Edward could sense some equilibrium returning, both to his family and to his workplace, though, as he put it to Maria, 'only till the next chapter (I feel fit to be put into a soap opera)'. He noted that California was fine, but added: 'I am not making any friends any more.' The only exception he mentioned was Dave Griggs, whom he described as having 'wit and . . . character and I cannot predict what he will say'. He still travelled a great deal but he was at Livermore enough, he explained to Maria, to begin establishing a routine:

1) Five days a week I get up at 6:45 (in case of doubt it is AM).
2) Four days a week I arrive at Livermore around 9 AM (The average is honest).

3) Seven days a week I have no fights (but with nobody).
4) One day a week I go to Berkeley and make (somewhat unsuccessful) attempts to be a physicist.
5) I have come closer than ever before to work and, more than that, to work on a single job.

Those evenings he was at home, he spent time with the children:

He [Paul remembered] would read to me almost every night he was at home for an hour or so. When I was 11, 12, he read the whole *Horatio Hornblower* series and I only found out years later that, after the first two, he was just having a hard time because he was forced to read it. I didn't have a clue it wasn't what he wanted to read. Other books? *Winnie the Pooh, Alice in Wonderland*. A very British selection because the books from his own childhood were not available and he didn't know what was traditional American reading for a child.

The legacy of bitterness left by the last year, however, was never far away, stalking him at science conferences and meetings, and also in less predictable locations, in clubs and restaurants. On one occasion he had been taken to the Senate dining room, as the guest of Senator Henry 'Scoop' Jackson, one of his allies on Capitol Hill. By chance the young John Kennedy, senator for Massachusetts, passed by their table and Jackson introduced the two men. Kennedy, according to Teller, 'smiled a broad and beautiful smile, though perhaps not as beautiful as that of Eisenhower' and said, ' "I read so much about you in the Shepley–Blair book." '

Teller's reaction was one of irrational anger. The pain, the damage, the friendships lost came flooding back to him, rekindling the months of misery. He nevertheless managed a civil if somewhat strange response, singing, in recitative, a line from Gilbert and Sullivan's *Yeoman of the Guard*: 'The things they have related, they are much exaggerated'; whereupon Kennedy, smiling 'perhaps a little less sweetly', moved on. Teller believed he was never to dissociate the hurt he felt on that occasion from Kennedy himself, and was to harbour dislike and mistrust of him from then on.

It took until March 1955 for the ideas of the new generation of Livermore scientists to arrive at the testing stage. Recently, they had been exploring ways of achieving implosion using a cylindrical configuration rather than the usual spherical one, designing smaller, more efficient devices that would fit missiles or artillery pieces on the battlefield. Tesla and Turk, the two experiments that they included in the Teapot series of tests running from February to May of that year, were more successful than anyone had hoped, and resulted in Livermore receiving a contract from the Army for a bomb that could be fired from a cannon. However, despite this, the big contracts continued to go to Los Alamos, and in the summer of 1956 Livermore lost in a bid against the New Mexico laboratory to supply the warhead for the latest 'ultimate weapon', the hypersonic Atlas missile. While the 'clean' bomb was still a Pentagon priority, they needed a major contract, to quieten carping voices, to establish their place in the defence firmament.

It was at this point that the Navy announced they would convene a special study group at Woods Hole Oceanographic Institution, Massachusetts, at the end of July. Called Project Nobska after a nearby point of land, the group had been established to consider ways of combating the growing Soviet nuclear submarine threat.* Present were Carson Mark and a group from Los Alamos, and Teller, who, because of his young colleagues' inexperience at what was likely to become a pitching session, led the Livermore contingent. What followed was an object lesson in how he could operate.

By the time the conference had convened, the current Redwing series of tests had been running for some time, and both Los Alamos and Livermore had successfully tested their own lightweight prototype warheads. The Los Alamos device, codenamed Egg, weighed 793 pounds and produced a yield of 250 kilotons, while Livermore's Swan–Flute device delivered a larger yield of 350 kilotons but weighed in at 1,100 pounds. The yield per pound weight of the two devices could not have been much closer, but the Los Alamos device included significant innovations that promised greater scope for further miniaturisation. This is what the Navy wanted, as they were specifying a yield of a megaton, nearly three times that of the Livermore device.

Teller acted precipitately and forcefully. How much he consulted either Harold Brown or John Foster, who were with him at the meeting, is not clear, but he rapidly made his pitch – that Livermore could deliver a ballistic missile in five years that would yield a megaton.

Such a clear-cut offer brought the meeting up short. According to Teller, Carson Mark immediately claimed that it could not be done, and there followed a heated 'yes-it-can, no-it-can't' debate, which had to be stopped by the presiding naval officer, who then asked both laboratories to specify what they could do in the next five years. Without pause Teller waded in with the offer of a 600 pound warhead – the size of a conventional high-explosive torpedo warhead – with a one-megaton yield. This represented more than a fivefold improvement in yield per pound over the devices tested during Redwing. Would anyone believe that was possible?

'Then,' Teller recalled, 'Carson Mark made a tactical error':

The Navy might not have awarded a contract if Carson had stood by his earlier statement that the job could not be done, or said that I had exaggerated by a factor of ten, or said that such an undertaking required further study. But Carson offered an estimate. His cost estimate was higher, the explosive power was less, his estimates of the size and weight of the weapon were greater, and he added a couple more years to the development time. [The admiral] then said, 'All that doesn't make much difference. The important thing is that you now agree that it can be done. However, since Teller has promised us more, let him do it.'

This victory over his bitter rivals gave Teller great satisfaction. He was, he wrote, 'happy' at their embarrassment. It was becoming a classic Teller ploy: making a grand claim on little more than instinct and offering it within a five-year time frame. He had done it with the Super, he was apparently having success with the 'clean' bomb and now he was staking his reputation on what was to become the Polaris project. It was this boldness that excited those who worked with him, but on this occasion it created near-panic back at Livermore. 'I'm the person who has to make good on your estimate!' an anxious Harold Brown pointed out to Teller on their return to the laboratory.

However, there was a positive response to Teller's challenge from Livermore's young staff, which resulted in some radical thinking.* They began what was to become a Livermore tradition of looking at every warhead component as a potential element in the explosive. They fabricated them either from a fissile or a fusionable material, or from material that would convert to an explosive fuel once the explosion had started and it was irradiated by neutrons, X-rays or gamma-rays. They also reconceived the warhead's outer casing so that it would also serve as the re-entry vehicle, as well as simplifying the detonation system, reducing it from dozens of detonators to just two. Within a short time a new programme was under way that showed every promise of producing the megaton device weighing the 600 pounds that Edward Teller had promised. But just as work started the possibility of a nuclear test ban reared its head again during the Senate's hearings on fall-out, threatening the whole project.

During his testimony, General James Gavin, chief of research and development for the Army, had confirmed recent journalistic reports of estimates that a large-scale thermonuclear attack could kill or maim some 7 million persons and render hundreds of square miles uninhabitable for perhaps a generation. Even more dramatically, he had offered the prediction that a US retaliation on the Russians would spread death from radiation across Asia and into Japan. If the winds blew the other way, such an attack could, he estimated, kill hundreds of millions in Europe. Privately, another Pentagon spokesman estimated that if the wind were 'unfavourable', the death toll might rise as high as 500 million and include nearly half the population of the British Isles. Against such estimates, the whole nuclear arms race seemed to many an obscenity and the pursuit of a nightmare.

President Eisenhower was anxious to do anything to counter the impact of such evidence, and another Livermore success in May 1956 gave him what he was looking for. The Bassoon test, another part of the Redwing series, was the laboratory's first multi-megaton device, with a yield of 3.5 megatons. It was also a 'clean' device. However, the fact that the fall-out, representing half of its fission products, was scattered over 17,000 square miles of the Pacific illustrated just how relative that term was.

Lewis Strauss was charged with making political and public-relations capital from the successful test, but his press statement was cryptic – and disastrous. He concluded by saying that the Redwing test had proven 'much of importance not only from a military point of view but from a humanitarian aspect'. It was a statement that reaped a whirlwind, one that had already been gaining energy from the commission's hedging and dissembling during the past two years. The physicist Ralph Lapp wrote a devastating piece in the *Bulletin of Atomic Scientists*. After carefully assessing all information on fall-out, and making clear that 'cleanliness' was only a relative term, he observed: 'Part of the madness of our time is that an adult man can use a word like humanitarian to describe an H-bomb.'

The issue would not be wished away. Albert Schweitzer used his Nobel Peace Prize address in April 1957 to speak out, concluding that the genetic danger from global fall-out was 'a catastrophe that must be prevented under every circumstance'. Linus Pauling initiated a petition against further testing among scientists; in a short time he had gathered more than 2000 signatures, thirty-five of them from Nobel Laureates. Shoals of letters again poured into the White House. Many were friendly, urging the administration to 'go the second mile' in reaching agreement with the Russians; but others were bitter. One elderly Californian living close to the Nevada test site suggested that the President vacation there with his grandchildren so that 'you may all have the advantages of this healthful spot. In this family,' the writer added, 'an eight-year-old child is slowly dying of leukaemia.'

Eisenhower was deeply disturbed, both personally and politically, by the increasingly hostile reaction and therefore, when his special assistant for disarmament, Harold Stassen, returned from the deadlocked London peace talks reporting that the Russians genuinely wanted to end nuclear testing, he responded warmly. At a White House meeting on 25 May he held the line against strong opposition from Strauss and from the Chairman of the Joint Chiefs, Admiral Radford (Radford had publicly declared: 'We cannot trust the Russians on this or anything'), and he sent Stassen back to the talks. With him he took a brief to make a tentative offer of a short suspension in testing in return for limitations on production.

To some, this was seen as a 'magic moment' of opportunity, the moment when there was the will in the administration to change course. Teller, however, watched the administration's manoeuvres with dismay and decided that positive action was necessary. He had the considerable resources of Livermore to call upon and, on Memorial Day 1957, he and Ernest Lawrence entertained Senator Henry 'Scoop' Jackson at Livermore, showing him the work of the laboratory, emphasising the smaller weapons designs and the 'clean' bombs. At the end they took care to point out just how deleterious a test ban would be for the laboratory, and Jackson was sufficiently impressed to invite them to meet the Joint Committee on Atomic Energy subcommittee on military applications, which was holding hearings at that time. The two scientists appeared before the subcommittee on 20 June. In introducing them, Jackson commented how impressed he had been 'with the progress that they were making in low-yield weapons, the possibility of making them smaller, the possibility of making them cleaner', and said that he had noted 'the gleam in the scientists' eye of making them almost like Ivory Soap, [but] not quite'.

In their testimony, the two scientists presented carefully constructed ethical as well as practical reasons for maintaining testing. The US knew how to make 'dirty' weapons of any size, Teller explained, but smaller, cleaner weapons still had to be perfected. Lawrence then expanded their arguments, stating that Americans had to realise the crime that would be committed if 'dirty' bombs, indiscriminate in their impact on friend or foe, combatant or civilian, had to be used in war. If 'clean' weapons were not made available and 'dirty' ones had to be used, Lawrence argued, 'Well God forbid ... we will have to use weapons that will kill 50 million people that need not have been killed.'

Both wished the committee to believe that they were arguing as much from a moral position as men like Pauling or Schweitzer. Also they argued from the position that the fall-out hazards associated with testing were negligible. They therefore concluded that it would be 'misguided', even 'foolish', to stop testing and thus prevent the development of weapons that carried such a moral imperative.

This unexpected assumption of the moral high ground was sufficiently impressive for them to be taken to meet the full Joint Committee the following day, and here they elaborated on the same argument – taking the even more extreme view that 'it would be a crime against the people' to stop testing. With the vision of a 'clean' weapon backing them, they were exploiting the very same arguments their adversaries used. They must have made a formidable team, the Nobel Prize-winning impresario of Big Science and the voluble 'father of the H-bomb'.

When Teller moved on to explain that there was now evidence that the Russians could hide both underground and upper atmospheric tests during a test ban, members of the committee reacted with shock and confusion. This information was new to them, they said, and they wanted to know whether the administration was familiar with it. Here were Lawrence and Teller making a powerful, original and focused case for the continuation of nuclear testing while, in London, Harold Stassen was carrying out a presidential brief to end it. As one of Stassen's advisers, Lawrence should have kept him informed on Russian efforts to conceal their tests, but Teller side-stepped any potential embarrassment by explaining that the possibilities were known in outline but not in detail. An urgent telephone call was made to the White House. It was arranged for Teller, Lawrence and another physicist from Livermore, Mark Mills, to see the President three days later on the following Monday.

On 24 June, the three scientists spent forty minutes with the President. Lawrence repeated his assertion that the US failure to develop 'clean' weapons 'could truly be a crime against humanity'. He also boasted of how the nuclear laboratories already knew 'how to make virtually clean weapons down to small kiloton weapons'.

Teller explained how the fall-out-free 'clean' bomb would be ideal for use on the battlefield and particularly in the defence of Europe. It could, potentially, be available within six or seven years, he said – if only their work was allowed to continue. He then made claims about the future promise of 'peaceful nuclear explosions', talking in glowing terms of setting off H-bombs in deep underground cavities lined with steel and filled with water in order to mine or release deposits of oil. They might also be used to alter the course of rivers, or 'perhaps even modify the weather on a broader basis through changing the dust content of the air'. Lawrence then added a proposal that a United Nations group should be invited to verify that the Americans were testing 'clean' weapons.

Eisenhower's reaction was calm and measured. He acknowledged that 'no one could oppose the development programme they had described', but countered by referring to the worldwide concern over fall-out and testing, stating clearly that the US could not permit itself to be 'crucified on the cross of Atoms'. The test ban proposals were, he said, an integral part of what was to be a total disarmament package, an end to war. Teller had countered that a test ban could not be properly policed, but Eisenhower responded that testing was becoming too much of an issue. It was both creating opportunities for Soviet propaganda and, at the same time, dividing public opinion at home. Eisenhower was firm in his belief that it was no longer possible to ignore the effect on people of reading 'fearsome and horrible' reports of fall-out.

Then he made a surprising concession: perhaps he could say something at his next news conference to clarify things, by explaining that the US wanted to continue testing primarily 'to clean up weapons and thus protect civilians in event of war'.

The two scientists' arguments had carried the day. Indeed, they had been almost too persuasive, for, as Teller and Lawrence were leaving, Eisenhower made a further suggestion. He pointed out that, in the long run, the US might want ' "the other fellow" to have clean weapons too – and perhaps it is desirable to turn over our techniques to him'. This was a natural extension of the moral position on 'clean' weapons, ensuring vulnerable Western populations at least some protection against the fall-out from the enemy's indiscriminate 'dirty' weapons. Teller responded to this alarming proposition by pointing out: '[American weapons] incorporate other technological advantages of great value that we don't wish to give to the Soviets.' There was no doubt that Livermore was on a war footing and that its moral charity had definite limits.

Teller and Lawrence had, however, impressed the President greatly. In conversation with his Secretary of State John Foster Dulles, the following day, he surprised with his new-found enthusiasm for the 'clean' bomb. His comments to Dulles reflected how far he had been seduced as he commented: 'The real peaceful use of atomic energy depends on their developing clean weapons' in as little as four or five years. He was also as good as his word when he repeated much of this at his next two press conferences. At the same time he made bold – and largely unjustified – claims that American nuclear weapons were already 96 per cent fall-out free, with the prospect of an absolutely clean bomb only a few years away.*

The contradictions implicit in the President's statements were to be criticised, even ridiculed, both at home and abroad. 'How can you have a clean bomb to do dirty things?' Khruschchev had asked and even those friendly to Ike questioned the wisdom of his apparent change of heart. Eisenhower ignored the controversy. 'At the time he had Teller and Lawrence and all going in to brief him about how wonderful the clean bomb would be, and how humanitarian,' Herb York says, 'Eisenhower gave two press conferences, one before and one after he saw these guys. At the one before, he talked enthusiastically about getting limitations on nuclear-weapons, then, afterwards, you find him talking about new ideas, and that we have to proceed very cautiously along this road.' The seductive charms of the 'clean' bomb as presented by Teller and Lawrence had ensured that the 'magic moment' when a test-ban might have taken root had passed.

This was to be the first of several crucial moments where the work going on at Livermore had tipped the balance in shaping international disarmament decisions.

On 4 October 1957 the Soviet Union launched Sputnik. It was a technological Pearl Harbor, shaking the West like no other Soviet accomplishment in the decade. The orbiting Soviet satellite suggested that the fears of Russian scientific advancement were well founded, and that American technological dominance, which had been taken for granted since the Second World War, was beginning to crumble. More specifically, it also served to announce to the world the superior Soviet position in missile technology.

For the moment, at least, the US was defenceless against the possibility of rocket attack, and the satellite's launch generated feverish attempts to meet the challenge. Espionage, however, subsequently revealed that the Americans were not so far behind with their ballistic missiles after all, as the Russians did not, as yet, have a miniaturised warhead with which to arm their ballistic missiles. Furthermore, in a routine assessment of fallout from Russian tests, Hans Bethe discovered a defect that made their thermonuclear devices vulnerable to 'pre-initiation'. As they were transported through space on their ballistic missile before re-entering the atmosphere, they could be made to explode harmlessly by a flux of neutrons generated from another nearby atomic explosion.* This had led Isidor Rabi, now GAC chairman, to propose an emergency missile defence system for the immediate future based on exploding nuclear missiles out in space. In the long term, however, Rabi believed the only answer was a worldwide moratorium on nuclear explosions, and, on 29 October, he met with the President and Lewis Strauss. Within a short time the meeting degenerated into a slanging match between Strauss and Rabi over the test-ban issue, Rabi stating that it had been 'a great mistake for the President to accept the views of Drs Teller and Lawrence' in the controversy over clean bombs.

Quite extraordinarily, nearly four years after the Oppenheimer hearing, this was the first time Eisenhower had become aware of the deep personal and ideological split both in the scientific community and among those who advised him. 'I learned that some of the mutual antagonisms among the scientists are so bitter as to make their working together almost an impossibility,' Eisenhower wrote in his diary that day. 'I was told that Dr Rabi and some of his group are so antagonistic to Doctors Lawrence and Teller the communication between them is practically nil.'

He also became aware that he had been caught between the two factions. He had responded positively to Lawrence and Teller and had changed direction accordingly. He had then listened to Rabi, 'a friend of long-standing', and although he now more than ever saw the attraction of Rabi's proposed moratorium, he was trapped, unable to see how a complete and sudden reversal in the US position 'could be achieved in the face of our public opinion and the opinion of our allies'.

This situation convinced Eisenhower that he should listen to those who had counselled him to appoint his own scientific advisers. Some ten days after his meeting with Rabi, he announced the formation of the President's Scientific Advisory Committee (PSAC), and with it the appointment of James Killian, President of MIT, as his personal scientific adviser. For the first time since he had appointed Lewis Strauss as his special adviser, the opinions of liberal scientists were to be heard in the White House.

Derailed by Sputnik, the quest for a test ban should now have been back on track. However, on 19 September 1957 Teller had masterminded another Livermore test, the Rainier shot, a 1.7-kiloton device, which was exploded in a tunnel drilled 2000 feet into a mountainside. Teller had watched the explosion from an observation post a few miles away, and reported: 'The Mesa shivered and appeared to lighten in colour. The top of it jumped upward nine inches, throwing up some sand that cascaded down the slopes. Then the earth fell back into place, apparently unchanged except for a few fissures . . . No trace of radioactivity escaped. The experiment was complete.'

Early thoughts were linked to the peaceful uses of nuclear energy, of the kind with which Teller had enticed Eisenhower when they had met in the summer. In addition, it also demonstrated that if world opinion forced the US to stop testing in the atmosphere, it would be perfectly possible to continue doing so underground. However, it was quickly realised that if they could do this, then so too could the Russians – and herein lay a problem. The only way of detecting these explosions was by using a seismograph and, annoyingly, the readings on Rainier had been inconclusive. In spite of prior warning, many reporting stations failed even to detect the shock waves, and those that did found them difficult to distinguish from a minor earthquake that had occurred fifty minutes later. Thus, while Rainier offered the prospect of fall-out-free tests, it also offered underground testing as a route round any future test ban – unless, that is, detection techniques could be improved.

21

A Matter of Detection

In late 1957 *Time* magazine produced a special edition on the state of science in the US. It appeared on 18 November, and there, on the front cover, was the magnificently eyebrowed countenance of 'Scientist Edward Teller'. Inside there were short profiles of such Teller associates as Ernest Lawrence, Glenn Seaborg and Luis Alvarez, and also two of his adversaries, Rabi and Oppenheimer. In pride of place, however, with four pages devoted to him, was Teller himself. In the opinion of *Time*, he more than any of his colleagues had 'worked hardest and most belligerently to send the warning that the Russians were coming'.

The extreme paranoia of the early 1950s may already have seemed a bad dream, but there remained a widespread and genuine fear of a worldwide communist conspiracy, backed by Russian military technology. In many people's eyes, Sputnik had substantiated Teller's oft-repeated claim that the Russians were ahead and validated his arguments against any kind of test ban. The *Time* article followed and bolstered this line, painting a sympathetic portrait of a totally committed man. It described how his 'harried' secretary worked with him over breakfast, then took dictation on the drive out to Livermore, and how, once at the laboratory, he spent his time juggling theoretical physics with calls from the Pentagon, before making hurried departures for the airport. It was a hectic lifestyle, *Time* reported, which had prevented him from completing the 'atomic alphabet' he was working on for his children.* It described also the damage to his health, the ulcerative colitis, and his daily doses of atropine and pheno-barbital sticks, and his medically controlled diet – problems the article attributed to his workload rather than to the distress following the Oppenheimer affair.

On Capitol Hill Teller now had powerful allies on important committees, such as Senators Jackson and Strom Thurmond; and of course he maintained his close relationship with the Air Force as well as being a member of Oppenheimer's one-time stronghold, the GAC. He was also recognised as a science teacher of exceptional interest. His 'Physics 10' course at Berkeley had long been thought one of the most stimulating available, and had always been oversubscribed, even before he became famous. In 1958 the local TV station, KQED, made a twelve-part series based on it. Mici assisted him and the first programme opened with Edward at the piano, playing Bach's Prelude and Fugue in C Major.

This was a public portrait of Edward Teller at the height of his powers, a brilliant man with a total commitment to his adopted country. It was not long, however, before a combination of factors, some within Teller's control, others not, began to erode this image.

In 1958, the year after his petition, Linus Pauling returned to the fray over fall-out with a book and new claims about its damaging and potentially lethal effects. There had recently been some new and alarming assessments by Lawrence Kulp of Columbia University that, in just one year, levels of Strontium 90 in human bones had risen by a third – and in children by up to ten times as much. Taking these figures, Pauling had concluded that the current rate of testing would produce 8000 deaths a year from leukaemia, 1600 more from bone cancer and 90,000 from the shortening of life expectancy. Pauling described Strontium 90 as 'a terrible poison', one teaspoonful of which, if 'distributed equally among all the people of the world, would kill all of them within a few years'.

He had also come forward with a new danger: Carbon 14. When any thermonuclear device explodes, whether 'clean' or 'dirty', it produces a massive flux of neutrons, and these neutrons interact with the nitrogen in the atmosphere to produce radioactive Carbon 14. Compared to Strontium 90, this is a relatively weak source of radiation, but its half-life of some 5000 years is much greater. Thus its effects are much more prolonged; for while Strontium 90 might accumulate over decades, Carbon 14 will do the same over centuries, indeed over millennia. Pauling was not afraid to state what the effect of this would be. He claimed that, over the next 300 generations, Carbon 14 from the bombs already tested would cause at least one million defective births.

Supported by his avuncular charm and Nobel status, Pauling's claims were taken seriously by a nervous public. However, the hard evidence needed to back them up was, to say the least, scant and Teller was determined to show that Pauling was playing fast and loose with the facts. Unlike others of his colleagues who also resisted the test ban, he took the dangerous path of confronting Pauling. He began by admitting to some small genetic effects created by fall-out, but then attacked Pauling's figures, taking particular issue with Carbon 14's supposed impact on life expectancy. Claiming that smoking one pack of cigarettes a day could shorten a person's life by nine years and that being 10 per cent overweight would shorten life expectancy by one and a half years, he then compared this with the effect of fall-out on longevity. 'World-wide fallout,' he assessed, 'is as dangerous as being an ounce overweight or smoking one cigarette every two months.'

He also took up Pauling's claims about increases in the rates of diseases like leukaemia. His approach was to compare the disease rates in cities at sea level with those in the mountains, such as Denver, where the population was subject to much higher radiation from cosmic rays from space. Far from having a higher rate of leukaemia, Teller pointed out, Denver had a lower rate than cities like New Orleans and San Francisco. 'If such small doses of radiation really are dangerous, we had better evacuate Denver,' Teller wrote:

> Brick contains more natural radioactivity than wood ... If worldwide test fall-out really is dangerous, we should tear down all of our brick houses ... we should throw away bedside alarm clocks with dials that can be seen in the night, because they are spraying the occupants of the bed with radiation ... we know that worldwide fall-out is not as dangerous as living in Denver rather than San Francisco, that it is not as likely to induce cancer as smoking a pack of cigarettes a day, that it is not as likely to give rise to harmful effects as are many unsuspected chemicals in the food we eat or in the air we breathe, that it is not as apt to produce mutations as wearing trousers. It is, in other words, not worth worrying about.

His arguments were telling and Pauling was also criticised by, among others, Lawrence Kulp who concluded a letter to the *New York Times*: '... exaggerated statements by respected scientists only add to the public's confusion.' But the scientific evidence was really just ammunition in an increasingly violent political battle. As William Libby said in 1978:

> Linus Pauling and Bertrand Russell led the argument that we shouldn't fire any bombs in the air because it will kill five babies in the whole world. Well they knew better. They were lying. I think people who lie in public for a political purpose should be brought to task ... They said that any amount of radioactivity, no matter how small, is intolerable. [Then if there is a] one per cent increase in the natural dose in Los Angeles, then a 200 percent increase [as in Denver] ought to be damn obvious. But it isn't. Public health records show no effect. We had populations some places in the world where the increase was 10,000 per cent;

they showed no affect . . . They're just plain liars . . . For a purpose to stop
the armament of this country . . . At least Muller was, and Pauling was,
and Russell was. All very well known. All right?

So, in retrospect, what was the true situation and which of the two
parties were misleading the public? 'I would say that both of them were
misleading – but not deliberately,' said Nobel Peace Prize Laureate, Sir
Joseph Rotblat,

> To a large extent it was due to ignorance. We did not know at that time
> very much about the effects of small doses of radiation and therefore the
> conclusions which were come to depended from what angle you came; I
> mean, a scientist should not begin from a preconceived idea but often in
> practice it does happen. Pauling felt very strongly about nuclear weapons,
> therefore he took the standpoint from one extreme which indicated
> high risk of radiation; on the other hand Teller, who was interested in
> continuing the testing of nuclear weapons started from the other extreme
> and said there was a threshold.

The situation continues much the same today. While the majority view is
that there is no limit below which radiation is safe, a significant minority
argues that the attitude to radiation is over-protective. In a report written
in 2000, the General Accounting Office, the investigative arm of Congress,
said: 'The Standards administered by the EPA and NRC to protect the
public from low-level radiation exposure do not have a conclusive scientific
basis, despite decades of research.' In 1958, such a report would have been
music to Edward Teller's ears.*

At the time, however, the issue remained very much alive. Pauling's
concern over the world's future health – the health of children – put him
clearly on the side of the angels. In arguing against him, Teller was placing
himself outside that charmed circle; and whatever the strengths or weak-
nesses of his logic, Teller was already seen to be aligned with the mendacious
AEC, and with the military. He may have won the first round of the test
ban argument, but he was embracing a course that would erode the public
regard for him.

With the backing of PSAC, his scientific advisory committee, President
Eisenhower joined combat for the second round of the battle to establish
a test ban. This time he was determined not to be misled by technical
objections and on 7 January 1958 James Killian set up an inter-agency panel
chaired by Hans Bethe to investigate its feasibility. Six days later, Eisenhower
proposed to the Russians that scientists from both countries should meet

for 'technical studies of the possibilities of verification and supervision'.*

The eleven-man Bethe panel had a broad membership drawn from the AEC, the Department of Defense, the CIA and PSAC, and it included Edward Teller. It took three months to produce a report and during that time Teller's voice was one of those most often heard. Looking back over more than forty years, the radiochemist George Cowan saw a particular rationale behind Teller's interjections:

> On the Bethe panel, Edward was always talking about what the Russians were doing. He was always pulling up the props, banging the drum, developing the worst case analysis. Hans would say, 'Look, we can conjecture about this and that, but let's consider what we know.'
>
> But Edward would always stick to his theme – that they're ahead of us and this is the way they've got ahead of us. I just sense in Edward that he knew what decided policy, which, by the way, is like magic. It certainly has very little to do with the facts, but he knew how to conjure very well. He valued the perception of power. He knew that the actual use of this bomb would be evil, but believed that illusions have their own reality.

Teller had certainly become a very considerable performer on the Washington stage. He knew how to attract attention, how to impress; but by now an increasing number of people were beginning to question his actions. Was his obsessive mistrust of the Russians disturbing his judgement? Were opportunities for peaceful moves being missed as a consequence? Were the expense and the dangers of the escalation of the resulting arms race too high a price to pay?

Interestingly, however much Teller dominated the Bethe panel's discussion, its final report took a cool and optimistic technical view. It concluded that verification of a test ban was feasible, and could be achieved by setting up approximately seventy instrumented stations throughout Russia and China, which would be augmented by on-site inspections and overflights using the new high altitude spy planes. Using this mixture of detection methods, Bethe claimed, the US would be able to detect clandestine tests with yields as small as a kiloton, whether underground, underwater or in the atmosphere. It boded well for the talks that Eisenhower had proposed and that Khrushchev had now agreed should take place in Geneva during July and August of that year.

One of those who attended this so-called Conference of Experts that summer was Ernest Lawrence. Towards its end he became so ill from colitis that he had to be brought home for surgery, but his condition going into the operation was so poor that he survived for only a day or two. He died on 27 August. He and Teller had never been close personal friends. Before they had begun working on establishing Livermore, Teller had been suspicious of him, and Teller counted the one conversation they had had about their common ailment as almost their only intimate exchange. Nevertheless, theirs had become a close collaboration, and Teller wrote:

> When Ernest died, I was just 50 years old. For me, Fermi, von Neumann, and Lawrence had each filled a role similar to that of an older brother. They were guides and counselors as well as friends . . . In the subsequent years, I have had other friends and other helpers, but I have never had one who could take the place of any of these three. Losing them, I could say, was the last and most painful stage of growing up.*

This was, however, not the first ally in the fight against restrictions on nuclear testing that Edward had lost that summer. Two months earlier, at the end of June, Lewis Strauss had resigned as Chairman of the AEC, ending more than ten years' association with the commission. It was indicative of the changing times Teller now had to face that Strauss had bowed out because of political enemies who were making life impossible for him. He left behind him a commission at odds with the administration and with the PSAC, a commission viewed widely as manipulative and mendacious. It was an inheritance that was to taint attitudes towards nuclear affairs for decades to come.

Before his departure from office, Strauss had been very careful to ensure that the commission itself remained in sympathetic hands. His last appointments to the GAC were all men sympathetic to his aims – as was his successor as chairman, the businessman John McCone. Strauss had also recommended that Edward Teller be one of the representatives at the Conference of Experts, but had met such strong opposition from within the government that he was forced to withdraw his name. His attempts to block Bethe's attendance had been similarly frustrated – he eventually went as a consultant.

The conference was deliberately modest in scale with small delegations from the US, France, Britain and Russia. The US party was six strong, including the three delegates, James Fisk, the vice-president of Bell Laboratories, the physicist Robert Bacher and Ernest Lawrence. Broadly speaking, Fisk and Bacher were in favour of a ban and Lawrence opposed to one. However, because of his illness, Lawrence attended very few of the meetings and so the other two were able to act with very much a common purpose. Hans Bethe joined them from time to time in his consultancy role: 'It was a good meeting, and gradually we found we agreed more and more with the Russians. It was clear that any test in the atmosphere could be detected, any test in the ocean could be detected because even when they're very diluted the radioactive atoms are so very easy to detect. We did however have some difficulty about underground tests.'

Even on this politically delicate issue, however, there was a convergence of views. On the basis of data gathered from Teller's Rainier test and from the findings of Bethe's panel, the conference ultimately recommended an increase in control posts from Bethe's initial 72 to some 160 spread across the globe. In addition there was an agreement, in principle, over on-site inspections. As the conference drew to a close, it was hailed as a triumph. Scientists, it seemed, had succeeded where the professional negotiators had failed. At the end of August 1958, Eisenhower felt able to announce a one-year moratorium on testing to start at the end of October. He also proposed to Khrushchev that negotiations on a permanent ban should begin that autumn, and the Russians agreed.

Edward Teller reacted to news of the moratorium with determination. Some months earlier Herb York had moved on to take up a post in Washington, and after his intended successor as director, Mark Mills, was killed in a tragic helicopter accident, Teller had stepped into the breach. Under his leadership the laboratory he had at his disposal had become a hive of single-minded activity.

'Livermore was wildly exciting' – so wrote the British scientist Freeman Dyson to his parents in the autumn of 1958:

> The days I was there were the last days before the test ban went into effect, and they were throwing together everything they possibly could to give it a try before the guillotine came down. Everyone was desperate and also exhilarated. Edward Teller, who is head of the Lab talked to me quite a lot about his plans. He was in very good spirits and pressed me with invitations to come and work for him. There are so many wild ideas and enthusiastic people at this place, I almost felt sorry to come back here at the end of the week.
>
> A lot of the talk at Livermore was about cheating the test ban. We found a lot of ways to cheat which would be quite impossible for any instruments to detect. The point of this is not that the Livermore people themselves intend to cheat, but we are convinced the Russians can cheat as much as they want any time they want, without being found out.

During his visit Dyson witnessed experiments whose political significance might be enormous. If Livermore were to discover methods of cheating that undermined the detection method agreed in Geneva, rendering it unreliable, even useless, the whole test-ban initiative would be in jeopardy.

In a 1985 interview with historian Gregg Herken, Rand physicist Albert Latter has described how, at Teller's urging, scientists at the laboratory devised 'some wild schemes' for cheating the test ban and discrediting the conclusions from Geneva. They included exploding a bomb in a giant egg-shaped structure strong enough to contain the force of the nuclear blast. In order to hide tests in outer space, they proposed testing on the far side of the Moon or using a gargantuan shield to hide the telltale X-rays from earthbound observers. They even investigated the possibility of setting a bomb off beneath a mountain with parabolashaped sides such that the shock waves would be reflected straight down to the centre of the earth – an ingenious concept, sadly thwarted by the fact that parabolic-sided mountains do not exist in nature.

Nonetheless, just before the end of the year something significant did emerge. The results of one of the last tests before the moratorium took effect showed a serious problem in distinguishing between seismic signals from explosions and those from earthquakes. This was not the problem of distinguishing between a confusion of signals, as had happened after the Rainier test, but of actually knowing what the identifying features of the two signals were. Hans Bethe was involved in assessing the findings:

> Now we thought at that time that an underground nuclear explosion should have as its first [seismic] signal a compression. So that's what we proposed ... but a few months later, Edward and his crew wrote a paper on the basis of some underground test, which Livermore had done for the purpose. In it they showed that, in three places out of four, indeed the first motion was compression, but in the fourth place it was expansion. That was pretty bad for us and it essentially doomed our efforts.

This discovery of a major inconsistency in the monitoring process used to distinguish nuclear explosions from other seismic events was indeed damaging and called into question the very foundation of the detection system that had underpinned the previous summer's successful Geneva test-ban talks. On 5 January 1959 James Killian informed Eisenhower that America must now change its negotiating position, and call for some 650 inspection stations, almost four times the number agreed, along with a corresponding increase in the number of on-site inspections. Given the Soviet sensitivity to any kind of inspection, this was, at the very least, going to cause difficulties.

Then, later that January, came another blow to hopes of achieving a foolproof detection programme, and it again stemmed from work at Livermore. Teller had asked Albert Latter to explore whether it would make any difference if an underground explosion was detonated in a large cavern. Latter's tentative findings were startling. The explosion, he discovered, would be effectively decoupled from the surroundings and the strength of the resulting seismic signals would be reduced in power by as much as 300 times. It fell once more to Hans Bethe to check Latter's findings and he found them correct – which meant that the Russians could, if elaborate enough precautions were taken, disguise the detonation of a 300 kiloton bomb as one of just a kiloton. Along with experiments recently carried out by the Air Force, which had demonstrated that tests carried out in space would be difficult to detect, it was a vindication of Edward Teller's comment: 'In the competition between prohibition and bootlegging, the bootlegger will win.'

When Eisenhower was told that they would now have to go back to ask for a fourfold increase in the number of monitoring stations and on-site visits, he predicted that the Russians would treat the findings as a breach of trust, and he was correct. The results of the previous summer's patient collaboration in Geneva, and one of the most cherished ambitions of his presidency, were left in tatters.*

The apparently insuperable detection problems raised by both underground testing and testing in space left Eisenhower angry and frustrated. So, too, was his science adviser, James Killian, who had come to believe that the comprehensive ban had been deliberately sabotaged. His worst fears were confirmed when, at about that time, he met Albert Latter. 'Dr Latter said to me in casual conversation,' Killian reported, 'that whatever advances might be made in detection technology, the West Coast group led by Teller would find a technical way to circumvent or discredit them.'

That May, Killian announced his intention of resigning as Eisenhower's adviser, ostensibly because of his wife's illness; but many believed his frustration with Teller's 'West Coast group' had played the major part. Years later, he was to regret not having been more aggressive in dealing with their technical objections, but his successor, George Kistiakowsky, was determined not to make the same mistake.

Kistiakowsky was the distinguished Harvard chemist and explosives expert who had directed the design of the implosion lens at Los Alamos. A Ukrainian by birth, he had fought the Bolsheviks in Russia before emigrating and, like Teller, was fiercely anti-Soviet. However, unlike Teller, he was convinced that a comprehensive test ban represented the best chance of gaining control of a dangerous arms race. He immediately challenged the obstacles that were being erected by the Livermore scientists. 'The notion of evasion through Latter holes is completely nonsensical,' he wrote in his diary. 'These things are too uncertain and too costly for any national program of evasion to be based on them.'

Here Kistiakowsky was echoing the views of other scientists who believed that Teller's scenarios by which the Russians might avoid detection were wildly over-elaborate. He set about persuading the President himself that these should by no means be the only consideration – and he met with some success. Much to the fury of the AEC and the Department of Defense, Eisenhower extended the moratorium by a further two months, announcing at the same time that he wanted to continue working for a comprehensive test ban.

However, he soon realised that he was facing not only some extremely determined opponents, but also a great deal of political power and skill. In effect, the group opposing the test ban had spent more than a decade playing off the AEC against the Senate's JCAE, and the Senate against the White House – keeping everything in a state of political uncertainty. Kistiakowsky wrote that he was 'convinced that a comprehensive treaty would not be ratified by the Senate since AEC, DOD, and Teller will all testify in opposition'; and this was just what happened in the run-up to the summit between Khrushchev and Eisenhower that was due to take place in Paris in May 1960. John McCone, of the AEC, engineered a public hearing on the test-ban issue under the auspices of the JCAE's radiation, research and development sub-committee. It was carefully programmed to bring home to the public all the risks associated with the uncertainties of detection. Teller was one of the main witnesses, and he intrigued his audience by showing how the Russians might pattern three of their nuclear blasts together to actually mimic an earthquake. He concluded that, however small a weapon had to be to escape detection, they could still use tests on such devices to gain all the knowledge they needed to perfect their new and deadly weapons.

Watching from the sidelines, Isidor Rabi despaired of Teller's role in events: 'I've never seen him take a position where there was the slightest chance in the interest of peace. I think he is an enemy of humanity. When it came to the first steps in arms control [the test ban] Teller was brilliant in inventing excuses and ways it could be circumvented, far beyond any reaches of common sense. We spent enormous fortunes trying to meet his objections.'

Eisenhower's remaining hopes were pinned on the Paris summit, but on 1 May the Russians announced that they had shot down a U2 spy plane, piloted by Gary Powers, in Soviet airspace. The weeks following the U2 incident, and with it the cancellation of the Paris summit, saw the collapse of Eisenhower's last remaining hopes of achieving a test ban before the end of his administration. In his frustration he spread the blame widely. Kistiakowsky recalled that when he had tried to defend the scientists involved, Eisenhower had 'flared up', but then had 'ended very sadly that he saw nothing worthwhile left for him to do now until the end of his presidency'.

When he came to leave office seven months later, time had softened his anger and disappointment. However, his experiences had impressed on him the new dangers facing society from a whole governmental and industrial infrastructure that had grown up to support the arms race. In his final televised address to the nation, on 17 January 1961, he spoke of his fear of 'the acquisition of unwarranted influence, whether unsought or sought, by the military industrial complex', and of the 'danger that public policy could itself become the captive of a scientific–technological elite'.

Herb York, who, after having left Livermore and gone to work in Washington, had come to know the President well, recalled a conversation he had with Eisenhower after he had left office:

> Eisenhower spent winters out here in Palm Springs and I got close enough to him as President so that we were able to continue conversations. At one point, I asked him the question everyone is interested in about his farewell address. People forget. He said two things. He said we needed a military–industrial complex but because we need it, we have to make sure it doesn't get too much power. And then he went on to say something that's often forgotten. We need a scientific and technological elite, but because we need this we have to be aware of it gaining too much power.
>
> We then talked about it generally, and at one point I asked him who he had in mind as this elite. Without hesitation, he said 'Teller and von Braun'. That's the technological elite we were to beware of in Eisenhower's mind. They're super-salesmen, and they come in here, saying, we've got these things and if we don't adopt them and promote them, we're doomed. Von Braun and Teller.

22

Plowshare

On 14 July 1958, while the Conference of Experts was meeting in Geneva to discuss an end to nuclear testing, Edward Teller and an entourage from Livermore landed, unannounced, at Juneau, the seat of the Alaskan government. On arrival, they demanded that the local Department of Health office organise a press conference and that the Rotary Club bring their members together for a presentation. Out of the blue the reporters and businessmen of Juneau found themselves listening to the mesmerising, resonant tones of Edward Teller as he described his intention to create a deep-water harbour at Cape Thompson, one of Alaska's most northerly and inaccessible points, by instantaneously blasting 70 million cubic yards of earth and rock with several hydrogen bombs.

The project, codenamed Chariot, must have seemed bizarre, to say the least, if not incredible. 'We looked at the whole world – almost the whole world,' Teller corrected himself, as their search had not looked beyond the US. 'On the basis of a preliminary study,' he said, 'the excavation of a harbour in Alaska, to open an area to possible great development, would do the job.'

Back in 1955 Teller had titillated Eisenhower with the possibilities of using 'clean' bombs to perform amazing feats, from the gigantic in civil engineering to the global in weather transformation – ideas that had resulted in a conference at Livermore in February 1957. Predictably, the conference, also attended by scientists from Los Alamos, had generated a whole gamut of interesting possibilities. Teller had opened proceedings with one or two mind-expanding proposals, including the thought: 'one will probably not long resist the temptation to shoot at the Moon ... to observe what kind of disturbance it might cause'. The meeting had then continued in much the same adventurous, not to say cavalier, vein, with discussions on how explosives might be used for anything from creating diamonds to mining for oil. However, by far the bulk of the programme had dealt with landscaping and, in particular, the possibility of using the power of nuclear detonations to blast a new canal adjacent to the existing Panama Canal. It was planned to run at sea level, thus avoiding all the bottlenecks at the locks on the existing canal, and there were several possible alignments for it. The preferred one, the conference estimated, would require twenty-six nuclear devices totalling 16.7 megatons.*

Not long after that meeting the project had acquired a name. In July 1957 Harold Brown met with a sceptical Isidor Rabi. 'So you want to beat your own atomic bombs into plowshares?' Rabi had asked, colloquialising the famous quote from the book of Micah. Brown had a name – Project Plowshare was born.

The real impetus for Plowshare, however, came from Teller's Rainier test, fired in September of that year. Not only had the test shown that it might be possible to avoid detection by hiding atomic explosions underground; it had also shown that such an explosion would not generate atmospheric fall-out, and had not interfered with nearby mining operations. This demonstration of such an apparently controllable and fall-out-free source of massive power had turned speculation into possibility. That possibility now needed a demonstration to become a reality, and a demonstration in America's own backyard rather than someone else's – particularly one that involved diplomatic complexities like Panama. The site needed to be wild, underpopulated and near the ocean, in order to test the effect of water on the stability of crater slopes. 'And that ended up with Alaska,' said Gerald Johnson, test division leader at Livermore; but then creating a harbour would be like blasting out one of the craters they had produced in the South Seas, but with an entrance channel opened up to the sea – wouldn't it?

However, as he moved from meeting to meeting rallying support, Teller found people questioning the practical need for a harbour as far north as Cape Thompson. It was pointed out that the site would be iced up for nine months of the year, and that the coalfields he hoped to open up were some 400 miles away over the most impossible terrain. His response was to ask for alternative suggestions, assuring his audiences as he did so: the 'blast will not be performed unless it can be economically justified – it must stand on its own economic feet'. He added that they wanted 'not just a hole in the ground, but something that will be used'.

Towards the end of his tour, therefore, the engineering aims of the project had loosened considerably, to include possibilities ranging from constructing dams to creating lakes or even canals across the Alaskan peninsula – whatever, it seemed, the Alaskans themselves wanted. But regardless of which plan was adopted, one group remained sceptical. The local scientists were concerned about the use of thermonuclear devices, even 'clean' ones. They had not been convinced by Teller's assurances, his estimates that Chariot, this prototype Plowshare project, would produce less radiation than the background radiation we are all exposed to from cosmic rays. At this early stage, however, they took comfort from the fact that Chariot had, first and foremost, to make commercial sense, and judging by the loose nature of the Livermore plans, there seemed little chance that this would happen. His initial tour completed, Teller was able to fly back to California carrying with him the general approval of the local press, the politicians and the business community.

However, when Livermore representatives returned soon after, they found attitudes polarising sharply. The commercial interests in this financially starved state were still very much in favour, but scepticism among the scientists had hardened into resistance. Some of them had harried the Livermore men about fall-out, demanding more specific information about the scale of the blast, and about their indecent haste to get started, which would almost certainly exclude a proper pre-shot baseline study of the environment. In one exchange on the local radio, a fishing instructor summed up the feeling of many when he dubbed them the 'firecracker boys'.

Indeed, to those who collaborated with Livermore at this time, the laboratory seemed, as one of the Alaskan scientists described it, an 'autonomous and irresponsible bureaucracy ... They had a lot of money, it would be something fun to do, and they wanted to do it.' Even to senior insiders like Gerald Johnson the laboratory, now under Teller's directorship, seemed to be exercising a remarkable level of independence: 'The thing that people need to realise is that in those days in the Laboratory we could go directly to the top. We didn't go to the President, but we'd go to the Joint Committee ... We didn't feel we had to work with anybody in between ... It wasn't a normal bureaucratic structure. Now the structure resented that ... We had unlimited priority, unlimited money ... I was rarely asked how much money I spent and no one cared.'

In spite of the criticisms and the alternatives proposed by the Alaskans, the harbour near Cape Thompson remained the favoured scheme for Chariot. Teller had described the overall dimensions of the harbour, its key-like shape with an entrance channel more than a mile long opening on to an oval basin half a mile wide and a mile long. To date, though, he had been economic about the operational details of the blast – and he would remain so. It was only in classified documents that he had estimated the required two megaton blasts needed to create the basin and four 100 kiloton blasts to open the channel. This total of 2.4 megatons represented no less than 40 per cent of the estimated six megatons of explosive used by all combatants during the Second World War. And this massive explosion was to be detonated within thirty-two miles of the thriving Inuit community who lived at the tip of the Point Hope promontory, a thin peninsula situated between Cape Thompson and Cape Beaufort.

However, the results from one of the last tests before Eisenhower's moratorium took effect, the Neptune shot, allowed a moderation of this giant total. Neptune had been a modest tunnel shot, one of the final tests on Polaris, but it had blown the roof off its containment, showing that smaller charges would shift larger volumes of earth than expected – and that, at a modest 100 foot depth, they could do so with the release of only minimal radiation. This finding enabled Livermore to reduce the size of the explosion by more than 80 per cent; but still Chariot was going to be a 460 kiloton package – thirty times the yield of the Hiroshima bomb.

Teller returned to Alaska twice in the summer of 1959, doing so in the knowledge that measures had been taken to at least soften the resistance to Chariot. The academics at the university had been offered good research grants to perform an inventory of the likely social and environmental impact Chariot would have. The mood certainly seemed to have improved in May, when he received an honorary doctorate from the university 'in recognition of his fearless endeavours to strengthen his adopted country against the menace of tyranny'. He took the opportunity to shape his acceptance speech around the exciting possibilities of Chariot, to give press conferences and to meet influential people. He assumed much the same public-relations role when he returned a month later, this time in the company of Mici, for a tour that took them from Juneau in the south-east, to Kotzebue and Cape Thompson in the north-west. At Elmandorf Air Force Base he had some fun with the press, offering up one of his classic aphorisms. When asked what he meant by 'geographical engineering', he responded, 'If your mountain is not in the right place, just drop us a card.'

In the midst of this banter, however, he let something significant drop, when he acknowledged that Cape Thompson harbour would have no economic value but might provide a model for future projects. When asked where these might be, he replied, 'That is like a little girl asking what do I want for Christmas? It's up to you.' But elsewhere on his tour he continued to talk about Chariot as an end in itself.

The Tellers ended their Alaskan tour by flying north to Kotzebue and then on to view Chariot's Ground Zero, Ogotoruk Creek. Because Teller was an important figure, the chartered plane had to have twin engines in case one failed, and so needed more space to land and take off than provided by the short airstrip at Ogotoruk. Instead Teller flew over the site, face pressed against the window as he looked down on the inlet and the embryonic engineering camp below. It was to be the setting for his re-enactment of the third day of creation, re-ordering the land and the waters. The plane circled several times before carrying Edward and Mici Teller away south on their return journey to California.

Back in the US, Edward became involved in a heavily publicised incident that once more reopened the wounds of the Oppenheimer affair. When Lewis Strauss had resigned as the chairman of the AEC, he had done so because of his intense rivalry with the New Mexico senator Clinton Anderson, who was chairman of the JCAE. This rivalry had its roots back in the battle for a second weapons laboratory, which Anderson had seen as an attack on Los Alamos. On leaving the AEC in June 1958, Strauss had hoped the dispute was behind him. A year later, however, it was to resurface when Eisenhower proposed to re-employ him in the cabinet post of Secretary of Commerce. Normally the confirmation of such an appointment would be a formality, but in the recent mid-term elections the Democrats had gained control of the Senate, and they were looking for any issue they could use to embarrass the Republican Eisenhower administration. This had given Anderson an opportunity to orchestrate a very effective opposition to Strauss's appointment, claiming that Strauss was too conservative, that he believed nuclear testing was safe, that he opposed science.

Teller was brought in as a witness on the last two points, arguing that Strauss had helped set up the Office of Naval Research and that he had supported basic research; but the hearing had already become very sour and he found himself being attacked personally. 'My testimony for Strauss was of little help,' he wrote. 'During the following discussion period the unrelated question of my contributions to the hydrogen bomb (as compared to those of Stan Ulam) became one of the main topics.'

The past was re-visited for both men, in full public view, and paralleled what had happened to Oppenheimer four years earlier. It gave his supporters no little satisfaction. 'It's a lovely show – never thought I'd live to see my revenge,' the wife of a Berkeley physicist telegraphed to Oppenheimer. 'In unchristianly spirit, enjoy every squirm and anguish of victim. Having wonderful time – wish you were here!'

In the middle of the hearing, the American Physical Society invited Strauss to be the main speaker at an annual banquet in Washington. There was outrage amongst its members, stirred by Ed Condon, Oppenheimer's first assistant director at Los Alamos. He wrote a letter to the Society that crystallised the feelings of many of its members. He criticised Strauss for the way that, in his joint role as chairman of the AEC and the President's adviser, he had exercised 'one man dictatorial control over all atomic energy matters' and had 'misled the President and misinformed the public' over

fall-out. Finally he held Strauss responsible for 'directing the security persecution' of Oppenheimer.

Teller must have known of the currency of these powerful sentiments,
but his loyalty to a friend overrode any concerns he might have had about
his own self-interest. Once more, Teller spoke out for an unpopular cause –
or rather two causes. While in the early years, he and Strauss had ridden
on a tide of uncritical wonder at nuclear energy, in the recent past they had
become increasingly embattled in its defence. The idea of nuclear energy
had become tainted, no longer perceived as the foundation of security and
a potential source of miracles, but as a threat to both the fabric of society
and to life.

After what Eisenhower described as 'the second most shameful day in
Senate history', second only to Andrew Johnson's impeachment a century
earlier, Strauss lost his nomination battle. Immediately afterwards, Teller
wrote to him and Strauss replied: 'You were good to write to me as you did
... and I shall always treasure your letter. You and I have stood shoulder to
shoulder on many battlefields, and I think that time is abundantly vin-
dicating the causes for which we contended. As for the enemies whom we
have made, history will accord us the more respect because of them ... Be
of good cheer and let us be sure to keep in touch with each other.'

Teller then experienced a bitter attack from a long-time enemy, Drew
Pearson, the political columnist. He accused Teller, as a government
employee, of being a paid consultant with industry and demanded that he
stop. The company in question was General Atomic, established by Teller's
former factotum Freddie de Hoffmann. In fact, as an employee not of the
government but of the University of California, which was contracted to
run the weapons laboratories, Teller was allowed to consult, but following
the Pearson article he nevertheless ended his contract. It was one more
episode that enhanced an increasingly unfavourable public image and it
was stressful. It was after these last two episodes that he was hospitalised
with his ulcerative colitis for the second time.

The AEC seriously underestimated the Tikirarmuit villagers of Point Hope.
By the autumn of 1959 they had heard and read of meetings about Chariot,
a major nuclear test near their village and in the midst of their hunting
grounds, but as yet no one had come to see them. As Don Foote, the
anthropologist who was living with them, remarked, they were 'as well
read and concerned about fall-out etc. as any good New Englander'. Their
protest was led by a young Eskimo whaling captain, Daniel Lisbourne,
who, until recently, had been president of the Village Council. They had
written to the Alaskan Senator Bob Bartlett, raising their concerns, and
also petitioned the AEC directly, rejecting an explosion 'for any reason at

any time'. As a consequence, in March 1960 three AEC representatives arrived in the area, and began moving from village to village, showing a film entitled *Industrial Applications of Nuclear Explosives* and answering what queries there were. The PR officer with the group noted 'no significant questions' – until they arrived at Point Hope.

At the time, tape-recorders were very popular among the Inuit. With no written language of their own, they used them in lieu of letters, and in the absence of phones, to send messages to one another. Thus the meeting in Point Hope's Browning Hall on the afternoon of 14 March was recorded.

Three AEC representatives, the scientists Robert Rausch and Russell Ball, and the PR man Rodney Southwick, faced about a hundred people, sitting shoulder to shoulder, on the floor, against the two eighty-foot-long side walls of the narrow hall. The audience were suitably impressed by the special effects in the film – the blue sea running in to fill the entrance channel and the harbour basin as the narrative stated reassuringly, 'Activity in the region of the water would be washed into the ocean rapidly and essentially removed from the Biosphere. The activity on land near the crater would rapidly decay.' However, unlike in the other villages visited, the end of the film was greeted with a barrage of questions, which rapidly exposed the holes in the scheme.

The effect of radiation on the wildlife, the villagers' source of food, was obviously uppermost in their minds and there were questions about the effect of the blast on fishing. '[At Eniwetok] they found no evidence that fishes were destroyed,' they were told, 'or that there was any significant radiation in them.' When Lisbourne questioned the possible interference with caribou hunting, the response was: 'They have been keeping cattle in the Nevada area, in the immediate vicinity where the shots have been made ... and they have yet to find any evidence of any damage.' Equally, when concerns were expressed about the timing of the blast, likely to be in the spring, and the potential interference with their annual collection of thousands of crowbills' eggs, the villagers were reassured that if the study being conducted showed there was a problem, then they wouldn't do it.

'I hope you don't,' responded Kitty Kinneeveauk, who had already crossed swords with the Chariot engineers and, like many of the villagers, regularly read *Life* magazine. 'Once I read some news from magazines about Indians where you work on this too, blasting their town, and none of these atomic people help them.'

Russell Ball replied that he understood her fears thoroughly, but 'the testing we have done so far has had no affect on the Indian people any place'.

'I've read it on a book,' she insisted. 'It happened while they blasted their homes.'

Ball asked her whether she had in mind the natives of Eniwetok, but she was uncertain. Then Lisbourne remembered a problem with a fishing boat, 75 miles away from the blast. 'That's right,' Ball responded, 'and because they were where they were told not to be . . . They were within the danger . . . area.'

The meeting was by now becoming mutinous. A voice broke in claiming that fifteen years after the blast, children would be deformed. Ball turned to Robert Rausch for an answer, but he declined to give one. Instead Ball gave an account of the findings after the attacks on Japan: 'And we have found no evidence of any effect upon the children of those people from the radiation dose they received. No evidence of any indication all.'

But the answers had been too lacking in conviction for the audience. 'The more you talk, the more scared she gets,' the visitors' pilot, Tom Richards, translated for one of the older women. Shortly after this – and against Richards's protestations that they actually had plenty of time – Ball made excuses that they had to get back for another meeting, and the three AEC representatives left.

They had faced three hours of trenchant, well-informed objection, which had exposed the gaping holes in the experimental foundations for Chariot. The Point Hope community had been told that the Pacific tests had not contaminated the fish with radiation, that the cattle on the Nevada ranges had shown no effects, that the harmful constituents of fall-out would largely be gone within hours, that US nuclear testing had not harmed 'Indians' anywhere, and that the children of survivors from the Japanese bombs had suffered no ill effects. Some of the statements were wrong, others at best questionable, and they were all down on tape.*

At the same time as the AEC was facing opposition from the Tikirarmuit of Point Hope, another conflict was brewing between the local researchers making the environmental inventory for Chariot and the project's bio-environmental committee. The committee, which included only one Alaskan, had produced a report whose positive conclusions were not borne out by the research results. There were rumours that the committee had been bribed with further offers of research contracts to give Chariot a positive bill of health. When the local researchers protested at the mis-representation of their work, they were threatened with either termination of their contract or the loss of their university post. Yet the research was undoubtedly showing up some serious and unexpected problems. In par-ticular, it had been discovered that in Alaska the uptake of Strontium 90 in caribou, and in the Eskimos who hunted them, was higher than any-where else in the world where tests had been carried out. Yet, globally speaking, Alaska itself was still a region of low fall-out.

Bill Pruitt, a seasoned local, was one of several researchers who had focused their search for an answer on the caribou's diet, and on the lichen that was their main food. Lichens are an extraordinary example of symbiosis between algae and fungi, in which algal cells interweave with fungal filaments to produce a plant-like body that can live almost anywhere. This is because lichens do not need to draw nutrients from the soil, as many plants do, but can instead derive water and dissolved minerals from the atmosphere. This meant that, unlike grasses, with the thick protective cuticle on the surface of their leaves, lichens absorbed virtually 100 per cent of the radioisotopes that fell on them and, because they were so long-lived, accumulated the products of many years' fall-out.

It was a result that threatened to destroy the AEC's notion of evenly distributed fall-out by describing an unexpected mechanism that created dangerous 'hotspots'. The wrangles over publication continued for a year until early 1961, when the Chariot shot seemed imminent. 'We all feel', wrote one of Pruitt's colleagues, 'that there is a great deal of urgency about acting quickly.' The group went public, sending accounts of their work to several different environmental newsletters.

Their account went to Barry Commoner, among others, a plant physiologist at Washington University in St Louis, who had recently organised an action group called the Committee for Nuclear Information. Commoner found Pruitt's material a revelation. 'It was my introduction to ecology...' said the man who was to become a leading figure in the environmental movement a decade later. 'It was when I realised that the different ecosystem in Alaska deeply conditioned the outcome of this technological impact, that I realised that what we were doing in our work on radiation was really an aspect of what is now called environmentalism.'

The new material on Chariot was published in his news-sheet *Nuclear Information* and was accompanied by an article questioning the AEC's assumptions about the amount of radioactive material that would be vented into the air by the combined Chariot explosions. Would it be 5 per cent, as the AEC estimated? it asked; would it be lower at 1 per cent? or would it be as high as 25 per cent? Such uncertainty, it was argued, was simply not acceptable.

The impact of the publication was immediate and extensive. The local Alaskan press, which had been largely favourable to Chariot, began to question it. So did various conservation groups and also the main national newspapers. Later in the year the national press took up another story associated with Chariot, the David and Goliath story of the Point Hope villagers who had made waves by writing directly to the new President, John F. Kennedy, and had successfully involved the Association of American Indian Affairs in their cause. This, in turn, had drawn in the Department of the Interior, whose secretary, Stewart Udall, had been successfully lobbied by the AAIA.

There was little sympathy for Chariot in Udall's department. They questioned the AEC's 'roles as investigator, judge and jury in safety matters', and they expressed serious doubts about 'whether the land could be taken from the Eskimos and made available for this experiment'. It would, they believed, take an executive directive from the President himself or congressional approval – but such a move was becoming increasingly unlikely. The young and idealistic Kennedy had embraced the concerns of his predecessor over atmospheric testing and the need to achieve disarmament.

Early in 1962 there were rumours that Chariot might be deferred. The public reaction to the revelations about the project had triggered a re-evaluation. John Foster, who was now the director at Livermore, began to question whether his mentor's four-year-old project was becoming obsolete, whether the information it would provide on cratering techniques could be obtained just as well from experiments planned for the Nevada test site. However, in order to minimise the embarrassment of a deferral, it was decided to time the announcement to coincide with a 'convincing event' – an event which could be made to justify the deferral.

The event chosen was Project Sedan, a cratering experiment using a 100 kiloton device, to be exploded 635 feet below the surface of the Nevada desert. On detonation, at 10 a.m. on 6 July 1962, the surface of the desert domed to 300 feet before incandescent radioactive gases burst through the surface, hurling rocks and earth 2000 feet into the air. An hour later, although it was too soon for precise analysis, the AEC felt able to issue a statement that no more than the estimated five per cent of radiation from the device had escaped into the air, and that most of it had landed close to the crater.

The truth, as finally revealed some twenty-one years later, was very different. The blast, which was the largest to have occurred in North America up to that time, was substantially dirtier than admitted. It 'deposited nearly five times as much fall-out on and near the test site than had been predicted' – a fivefold error, which matched closely the higher estimate for Chariot in the *Nuclear Information* article.

Once aloft, the wind caught the radiation cloud and sent it northwards at twelve miles an hour. It was already off the test site by the time the AEC made their first optimistic press announcement. That afternoon it crossed State Highway 25 and the police had to close the road and evacuate ranchers living nearby. By four in the afternoon the radioactive dust cloud was so thick in Ely, Nevada, 200 miles from Ground Zero, that the street lights had to be turned on. Five days later, after the clouds had moved northwards across six states and into Canada, seven miles of Highway 25 had to be pressure-washed clean before it was safe to re-open it.

This was a practical measure of the risks Edward Teller's 'firecracker boys' were taking. If a shot the size of Sedan had been detonated at Ogotoruk Creek, it could have spread fall-out along the entire length of Alaska's North Slope, and if the wind had been from the east it could have penetrated 1000 miles into Soviet Siberia; but Chariot, even in its final manifestation of 280 kilotons, would have had nearly three times the yield of the Sedan shot.

Seven weeks later, the AEC, as planned, formally announced that Project Chariot was being 'held in abeyance', and that they expected data originally planned to come from Chariot to be provided by other tests. It was a tactic that did soften the impact of what was a U-turn in the face of powerful opposition from the Department of the Interior and an increasingly unsympathetic administration; but without its showpiece experiment in a real location, the various other projects in Teller's Plowshare portfolio were to suffer a severe setback.

While discussion on the Panama Canal scheme was to continue, fuelled by the knowledge that the Russians had already used nuclear explosives for such a scheme, there were to be no more Plowshare experiments for five years. By then they were accompanied by direct-action environmental protests and action in the courts. The 'clean' bomb, so much the basic essential of the Plowshare vision, was to remain an enticing chimera for ever on the horizon, one of Teller's motivations in his continuing fight against the test bans. To this day, it has never been fully realised.

One final fact that emerged from Chariot was the reason for Teller's flexibility over the form the project should take. It had seemed to matter little whether it was a canal or a harbour. However, in the classified minutes of an AEC meeting on 29 May 1959, just at the time of Teller's second visit to Alaska, the AEC's chairman, John McCone, said, 'it is clear that Dr Teller believes that detonations carried out in the near future under the Plowshare program will be related to our weapons program: therefore the Commission must exercise care in any arrangements made for conducting nuclear shots under this program.'

When Chariot was initiated in 1958, the Geneva test-ban talks were at their most fruitful. A partial test ban seemed certain, a comprehensive ban a possibility, and the moratorium was looming. Only one line of testing seemed likely to be permitted and that was for peaceful uses. Chariot was intended as a cover for military activities. Indeed, so integrated did the peaceful programme and weapons development become at this time that the AEC even abandoned the attempt to set up a separate civilian applications office. The harbour at Ogotoruk Creek was little more than incidental to Teller's prime concern: to keep military testing going. The fact that the project remained a live issue for four years is a measure of the determination to push the arms race as hard as possible.

23

Confounding Camelot

It was in August 1963, at the height of Kennedy's battle to secure a test-ban treaty, that one of his advisers described Teller to him as 'John L. Lewis and Billy Sunday all wrapped in one'. Lewis had been a frighteningly aggressive leader of the Union of Mineworkers during the 1940s, a leader who inspired an extraordinary loyalty through his determination and his grandiloquent oratory, and who, in his 1947 testimony to Congress, had spoken for five hours. Billy Sunday was a professional baseball player turned preacher, whose crusades in the twenties and thirties had drawn crowds of millions every month and whose blazing-fisted evangelism was a model for many who followed.

The juxtaposition of two such powerful archetypes is a measure of just how persuasive, how powerful, Kennedy found Teller's evangelising for the causes he espoused. Yet three years earlier, when he had come to power in 1961, the young President had believed he could at least clip the wings of such opposition. He had been impatient and self-confident. During his briefing by Eisenhower he had declined the outgoing President's perspective on disarmament talks in Geneva, saying that he had already been briefed on the subject.

Kennedy found that he had inherited a vast nuclear armoury of some 18,000 nuclear weapons. Gone were the days of 'Fat Man' and 'Little Boy', bombs that had been hand-crafted by a small army of scientists. Weapons were now mass-produced and ranged from small kiloton battlefield weapons that could be launched from Jeep-mounted mortars to 10 megaton monsters. These were mounted in B52 bombers kept in a state of constant readiness; at times of emergency, these bombers would fly to a pre-assigned line a certain distance from the Soviet Union and, unless ordered otherwise, turn around and fly back. This formed the basic scenario of *Dr Strangelove* – in the film one bomber does not turn back.

Waiting in the wings were the Inter-Continental Ballistic Missiles (ICBMs) such as the solid-fuel Minuteman, and the intermediate range Polaris, mounted on submarines that roamed, practically undetectable, beneath the seas. In the future these two were planned to provide the strategically crucial 'second strike', the weapons that would survive the enemy's 'first strike' and inflict the retribution – the 'mutually assured destruction' that kept the two forces at bay.

The Soviet armoury was smaller, perhaps containing some 2000 weapons, but this still left them with plenty of 'second strike' capability and plenty of 'overkill'. It was little wonder that Eisenhower's successor was just as keen to bring the arms race – this runaway – under control.

On gaining office, Kennedy determined to move quickly and put out feelers for an early meeting with Khrushchev. These met with a positive response and a summit was arranged for June 1961. Then, along with his new science adviser, Jerome Wiesner, an electrical engineer from MIT and a great supporter of the test ban, he set about dismantling such previously sacrosanct Pentagon projects as the nuclear-powered bomber.

On Wiesner's advice he both appointed new and more sympathetic advisers in the Pentagon and replaced the hawkish John McCone as chairman of the AEC.* The new chairman, the Nobel Laureate Glenn Seaborg made clear his personal ambivalence about the test-ban issue very early on and, with extraordinary frankness, advised Kennedy against giving 'undue weight' to the opinions of either his own agency, the AEC, or the two weapons laboratories. Nevertheless, Wiesner still invited Edward Teller to become one of four new members of his advisory committee, the PSAC. 'I agreed,' Teller recalled, 'but I was rarely asked for my opinion and cannot think of one occasion on which it was reflected in the policy adopted.'

He was, however, an important voice on the Air Force's Scientific Advisory Board, and they continued their complaints that, ever since the start of the existing moratorium begun during Eisenhower's administration, the Soviets had been cheating. They also reviewed and approved a secret study from Rand entitled 'Some New Considerations Concerning the Nuclear Test Ban'. It described all the projects suffering from the ban on testing and concluded that there were 'new aspects of the nuclear problem that [said], in sum: it does matter that we test'.

Kennedy, however, had sought the advice of another scientist and member of PSAC, Wolfgang Panofsky, who dismissed the complaints of cheating as 'unproven' and advised the President not to allow technical questions to preoccupy him as they had Eisenhower. Instead, Panofsky advised the President to concentrate on strategic questions, such as whether the US should aim modestly at deterrence or at achieving the overwhelming superiority needed to win a nuclear war at the first strike.

This is what Kennedy intended doing. Sadly, however, his plans started to go wrong from the beginning. April 1961 saw the Bay of Pigs invasion, in which Cuban refugees, backed by the CIA, attempted to overthrow Fidel Castro's new communist regime by force. The invasion was already being planned when Kennedy took office, but the CIA had given him a hopelessly optimistic view of its likely success. The result was a fiasco that damaged the President's image at home and weakened his position with the Russians. Arriving at the Vienna summit in June, Kennedy felt at a definite disadvantage.

The two leaders spent many hours together, irritable meetings in which Khrushchev tried to lever the US out of Berlin, and threatened to break up the post-war arrangements over the city. The President stood his ground, but came away, according to the British Prime Minister, Harold Macmillan, having been 'completely overwhelmed by the ruthlessness and barbarity of the Russian Chairman'. A little over a month later, Khrushchev instigated the building of the Berlin Wall.

Still Kennedy persisted with his plans for a test ban and briefed his Geneva negotiating team accordingly. His notes, jotted down during a briefing, testify to the nature of his concerns: 'We have been the leader [in peace initiatives]. Soviets have dragged their feet. Diminish danger. Fallout – radiation. Children.' At the end of August, however, before negotiations could begin, the Russians gave two days' notice that they were resuming testing. Worse, with thirty-one shots, ten of them in the megaton range, it proved to be the most ambitious series the world had yet seen. Kennedy and Macmillan together wrote to Khrushchev urging no further megaton tests in the atmosphere because of fall-out, but their plea was ignored.* Even then Kennedy resisted calls for the US itself to resume testing but then, on 30 October, the Russian series culminated in a test which, quite exceptionally, the Russians widely publicised beforehand. At 58 megatons, the equivalent of 58 million tons of TNT, this was – and still is – the largest single device ever detonated. It was straightforward intimidation, meaningless in military terms, but a major gesture in the psychological warfare that was such a major constituent of the Cold War. Given the amount of preparation needed for such a comprehensive test series, it was also clear that the Russians had, at the very least, gone against the spirit of the moratorium. It was further justification for the test-ban opponents and made it impossible for Kennedy to resist the resumption of testing underground any longer.

It was at this time that Wiesner strongly advised the President to meet with Teller and to at least hear out the opposition. The two men spent forty-five minutes together, a meeting Teller described as a strain for both men. He urged Kennedy not only to resume unrestricted testing but also to embark on a nationwide scheme to construct nuclear shelters for the whole population. After the meeting, Weisner found Kennedy obviously shaken by the extent of his plans for deep underground shelters and, no doubt, by the detail and completeness of the dark vision of which they were part.

In mid-September the US responded to Russia by beginning its own test programme. Initially this was restricted to underground testing, but following the Russians' 58 megaton explosion at the end of October, Kennedy had little choice but to order a full resumption. It then rapidly emerged that the Russians had not been the only ones active during the moratorium. According to Teller, Los Alamos and Livermore had 'taken the precaution of making some preparations in case the Soviet Union withdraw from the gentlemen's agreement'. Just three weeks after the presidential order to resume testing, the two laboratories had tested two small devices, the first in the seven-month-long Operation Nougat, during which they, too, tested thirty-one low-yield devices. Further, during that September, Livermore had once more gone on a six-day week in the build-up to Project Dominic – a further series of no less than thirty-six shots. The Livermore contribution ranged from tests of new experimental design concepts through to the final proving of the Polaris missile under simulated battle conditions.

In 1960 Edward Teller stood down as the director of Livermore. He had successfully faced the imminent cutbacks threatened during Eisenhower's last years, maintaining the morale of his staff during a difficult time; but he had wanted the freedom to campaign more. Ever since his report to the navy in 1946, Teller had been preoccupied with the idea of anti-missile defence. Now he was determined to champion that cause, and find a way to break the deadlock of 'mutually assured destruction' in which whole populations were virtually defenceless hostages in the struggle. In his 1962 book *The Legacy of Hiroshima*, which was partly autobiographical and partly an exploration of the future, he wrote, 'A retaliatory force is important. A truly effective active defense system would be even more desirable. It would be wonderful if we could shoot down approaching missiles before they could destroy a target in the United States.'

Teller saw defence as something that would change the whole balance of terror. It was Oppenheimer who had likened that balance to 'two scorpions in a bottle, each capable of killing each other, but only at the risk of his own life'. In Teller's view, however, this was simplistic – and dangerously misleading. He was convinced the Russians believed America would never launch the first attack, and so felt free to continue their steady erosion of capitalism and its influences. For them there was no deterrent. 'Step by step, nation by nation, convert by convert,' he wrote, 'it will conquer the world eventually. And thus a policy of mutual deterrence does not deter.'

In order to stem this inexorable process, Teller proposed a number of measures. He urged the use of such passive defences as nuclear shelters, as he had done to President Kennedy. However, it was pro-active defence, blocking the effectiveness of a Soviet attack, which he believed would free the West from its straitjacket and enable it to fight communism in whatever way was necessary. The problem was that, at the beginning of the 1960s, the technology necessary for anti-ballistic missiles (ABMs) – in effect the speed and accuracy necessary to hit one bullet with another – was beyond reach. Instead, Teller intended avoiding this problem of pinpoint accuracy by using ABMs topped with a nuclear device – its fireball blasting anything in space within a sizeable radius. In the 1950s, he had flown over the test of a small device at some 25,000 feet, and been surprised at how effectively its blast was absorbed – 'so slight I could barely feel it'. This had led him to believe that a blast out in space would have a negligible effect on the ground below and he used one of the shots in the Dominic series to see if this was the case.

At 23.00 on 8 July 1962 a Thor missile lifted into the air from Johnston Island in the Pacific, carrying a thermonuclear device 248 miles out into space. There the device, codenamed Starfish Prime, exploded with a force of 1.4 megatons. In Honolulu, 800 miles away, startled people watched as the night sky was turned into daylight for six minutes. On Kwajalein, 1600 miles west of Johnston, observers were treated to a truly spectacular display of multicoloured auroras. None of these effects were unexpected, but there was one that caught everyone by surprise: the huge surge of electrons that rippled out through space. As darkness returned in Honolulu, the air was shattered by the sound of hundreds of burglar alarms. Fuses blew, circuit breakers were tripped, and strings of streetlights on Oahu Island were extinguished. Days later, enough of the electron surge was still trapped within the Earth's magnetic field to begin damaging the solar panels of several weather satellites. This was the first example of what was to become known as the Electromagnetic Pulse (EMP) and it occurred with each of the five other high-altitude shots that were detonated over the next four months. In the lowest of these, the electron surge was powerful enough to disrupt the ionosphere, breaking down radio communication for several hours after the blast.

It was quickly realised that, if the blast was being used for defence, such a catastrophic disturbance would most likely occur over home territory, not only causing major disruption generally, but also potentially hindering or even paralysing other defence activities. The surprise side effect of EMP appeared an insuperable difficulty, and yet work did not stop on ABM devices based on a nuclear blast in space. It was yet another example of Teller's 'fear of ignorance' – his fear that there were possible solutions and that the Soviets would discover them.

Back in the spring of 1962, during the preparations for Project Dominic, President Kennedy had paid a visit to Berkeley and met with some of the senior staff from Livermore. While waiting for proceedings to start, Teller had taken the opportunity to remark on what he saw as the misinformation being publicised about the dangers of small doses of radiation. Radiation fall-out was still of major public concern, and a recent study on more than 750,000 pregnancies in the north-eastern United States had confirmed British work that suggested even one X-ray during pregnancy increased a child's risk of leukaemia and other cancers by 40 per cent. It was an important and well-publicised study, but others had produced contrary results, and Teller wanted to ensure that the President knew about them.

'I was about to point out that some evidence indicated that amounts of radiation in slight excess of background radiation might be beneficial,' Teller recalled. 'Before I could do so the President interrupted me to observe, "Dr Teller, if you are trying to convince me that radiation is good for me, you will fail." In this case, as in many others, Kennedy demonstrated more talent as a politician than as a scientist.'

This was a clear rebuff, one indicative of the suspicion with which the President regarded Teller; but this antipathy was mutual – one Teller explained using much the same quirky rationalisation as he had used to explain his initial alienation from Oppenheimer. With Oppenheimer it had been the incident when they had discussed keeping the military intervention at Los Alamos to a minimum. With Kennedy it had been the President's ambiguous reference to Teller's heroic depiction in the Shepley–Blair book. These explanations, however, seem to cloak a deeper, more instinctive reaction. In their different ways, both Oppenheimer and Kennedy had the qualities, the charisma, to create their own inner circle. In Oppenheimer's case, this consisted of the colleagues and students that Teller referred to as 'Oppie's men'; in Kennedy's case it was made up of the members of his 'Camelot' administration.

Teller had never gained entry to Oppenheimer's inner circle, nor would he to Kennedy's. Later that year, however, he was, at least, the recipient of official presidential recognition. In 1954 Congress had created an annual award to be presented to a scientist recommended by the General Advisory Council of the AEC for having made a significant contribution to nuclear physics. The first recipient had been Enrico Fermi. At the time he had been terminally ill and subsequently the award had been named after him. The list of winners included Lawrence, Bethe and Seaborg, and now Kennedy approved Teller's nomination 'in recognition of his contribution to chemical and nuclear physics ... his leadership in thermonuclear research ... and his efforts to strengthen national security'.

Teller was due to make a speech of acceptance when he received his award from the President at the White House and, in the weeks between his nomination and the ceremony, he brooded on what he should say. Before leaving for Washington, he sent a draft to Harold Brown, who had moved from Livermore to take over directorship of Research and Engineering at the Pentagon. En route for Washington, Edward and Mici spent the night of 1 December with Edward's friend and colleague on the Air Force Science Advisory Board, Ted Walkowicz. The following morning, the day of the ceremony, Walkowicz was woken at 6.30 a.m. by a telephone call from an anxious Harold Brown, who pleaded with Walkowicz to do what he could do to get Edward to tone down his speech. 'Let's not open up old wounds,' Brown said, a sentiment that was repeated minutes later when a second call came through – this time from one of Kennedy's aides.

Giving up any ideas of sleep, Walkowicz ventured downstairs, to find Mici and Edward with 'their backs to each other', feuding furiously. Mici had also read the speech and, like the others, felt her husband was making a big mistake, wanting 'to make a stupid speech and reopen all that argument'.

According to Walkowicz, the original draft referred to things Fermi had said on his deathbed that were critical of Oppenheimer and could have seriously exacerbated the schisms that had existed since the hearing. Fermi had not been uncritical of Teller's own role, but he had been the one person close to Oppenheimer who had understood why Teller had behaved in the way he had. The text of the draft speech no longer exists, but Fermi's criticism of Oppenheimer was apparently seen by Teller as a cue for some self-justification.

Even after eight years – years made miserable by ostracism and illness – feelings ran deep. The impassioned quarrel between Edward and Mici was only brought under some semblance of control when Walkowicz told them of the two calls he had received that morning. So he sat with Mici, editing the speech, 'with Edward fighting like a lion over every word and phrase. We finally turned it into something very innocuous.' Walkowicz's final advice to the overwrought Teller was to 'remember that the most important thing is to get invited back to the White House'. But when the time came, Teller was sufficiently disoriented not to heed his advice.

'As President Kennedy handed me the medal,' Teller remembered, 'he asked me about the proposed Plowshare plan of a second sea-level canal across the Panama isthmus. My response to the President was truthful but inexcusable. "It will take less time to complete the canal than for you to make up your mind to build it." '

This response must have been largely prompted by the fact that only two months previously the AEC had announced the deferral of Project Chariot, a precursor to the Panama Canal project. Teller believed that decision to have been the result of 'political timidity' – he continued to feel strongly about what he saw as the unnecessary concerns for minute doses of radiation. By that time President Kennedy's 'unnecessary concerns' would have been informed by news of the fall-out resulting from the Sedan test. According to Teller, the President gave a 'sour grin' and went off to talk to someone else.

One other factor must have coloured the interaction between the two men on that occasion. Back in October, the thirteen-day Cuban Missile Crisis had given the President a real and powerful sense of the Armageddon implicit in nuclear warfare. For those two weeks he had borne the burden of the millions of deaths that could so easily have ensued should the negotiations have gone wrong. Afterwards Kennedy's efforts on behalf of a test ban had therefore been redoubled. In his view, the risks resulting from the Russians cheating or mounting undetectable tests 'pale[d] in comparison to those of the spiralling arms race and a collision course toward war'. In Teller he must have seen the embodiment of all those groups, the armed services, the AEC and the Joint Committee on Atomic Energy, who stood against him, and whose efforts, like his own, were accelerating.

Throughout the spring of 1963 the JCAE and the Armed Services Committee conducted hearings aimed at casting doubt on America's ability to verify a test ban. The new director of Livermore, John Foster, took the stand to testify against any treaty, and so too did Teller, appearing for a whole day, and delivering a presentation that caused a considerable stir. On returning, tired, to his hotel, he received an urgent call to go and address the Southern Governors' Conference being held at the time in Arkansas. The weather was so stormy that Teller doubted he would find a flight, but his contact was insistent, urging him to take the overnight train. 'It seemed only a brief time before the porter roused me,' Teller wrote. 'There was J. R. [his contact Dr J. R. Maxfield] at the station to inform me that Kennedy had sent a message to the governors in protest of my speaking on the grounds that it would be a one-sided presentation ... I felt the presidential protest had a greater element of flattery in it than anything I had yet received from that quarter.'

His campaign was indeed an increasing source of concern to both Kennedy and to his senior aides. At a meeting on 14 June, both Secretary

of Defense McNamara and Secretary of State Dean Rusk had given a
warning to the President. There was, they told him, little chance of winning
the necessary support for a ban from the Joint Chiefs and other groups
while Teller and Foster continued to provide private briefings on the essential importance of testing for diagnosing and correcting defects in the US
arsenal. It was known, too, that Teller had delivered a further impassioned
plea for the rejection of the test ban to a breakfast meeting of nearly a
hundred senators.

In the hope of brokering some compromise, Wiesner again organised a
meeting between Teller, this time accompanied by Foster, and the President.
Teller reported, with some irony, that he was 'privileged to see Kennedy in
his famous rocking-chair, face to face. In a brief conversation I stated ...
that the amount of knowledge that we needed was far greater than the
knowledge we possessed about how nuclear explosives could be used in
ballistic missile defence.' Teller did nevertheless concede that 'underground
testing was far preferable to no testing' and Wiesner and Kennedy came
from the meeting believing they had achieved a consensus of sorts, albeit
on a partial ban only. However, even this limited aim would be far from
easy to achieve. Since the missile crisis, the mood among the press, public
and those on Capitol Hill had hardened towards the Soviets and there had
been growing opposition to a ban – opposition that Teller had played a
major part in mobilising. It had become a force with its own momentum,
threatening any kind of compromise. At a meeting with conservative
members of Congress, Kennedy was advised that he would have to abandon
his prime objective of a comprehensive ban if he was to have any hope of
achieving even a partial one.

It was the Soviets who finally crystallised the situation. Khrushchev, it
seems, had been under much the same pressure from conservative groups
as Kennedy, and he too pulled back from discussing a total ban. Instead,
on 2 July, the Soviet Chairman announced that Russia was prepared to
prohibit testing everywhere but underground. There were diplomatic
attempts to change his mind but he proved immovable. It was not the total
ban Kennedy had hoped for, but the final negotiations could now sidestep
all the tricky questions over monitoring stations and on-site inspections
that would have been involved in a total ban, and they moved swiftly to a
conclusion. On 25 July a draft treaty was initialled by Dean Rusk and
Khrushchev and passed on to the Senate for ratification. What followed
amounted to the final showdown between the pro-and anti-ban scientists,
as they testified before Senator William Fulbright's Foreign Relations
Committee.

When Teller appeared on 20 August he shattered any fragile consensus
that might have emerged between himself and the President. Faced with

the draft treaty, he effectively returned to square one, arguing that 'the United States research on defense hardly could be accomplished without opportunities to test in the atmosphere', and he laid particular emphasis on the requirements for progress with ballistic missile defence:

A few years ago I believed that missile defense was hopeless [Teller testified]. I am now convinced that I was wrong ... I am now convinced that we can put up a missile defence that can stop the attack of any weaker power, such as China, for at least the next two decades. In a time when we rightly worry about proliferation, we must not neglect our defences against an attack from a quarter other than Russia. In addition, I also believe that our defense can be partially effective against the Russians ... Missile defence, by deterring the Russians, may make the difference between peace and war.

This was the time when Kennedy's aide described Edward Teller as 'John L. Lewis and Billy Sunday all wrapped in one'. His testimony may have reflected hopes rather than existing technology, but it was delivered with the full force and conviction of the 'father of the H-bomb', the proponent of the successful Polaris system and of 'clean' weapons. Such a performance had, however, been expected and Wiesner had lined up a whole group of scientists, many of whom had previously expressed misgivings about a ban, but were now prepared to speak out in its favour. The group included Harold Brown, Glenn Seaborg and Norris Bradbury. Teller, along with John Foster, was very much isolated. And Wiesner did not leave it there. Four days after Teller's testimony he issued a press release featuring an endorsement of the treaty by some thirty-five Nobel Laureates. The President's advisory committee then issued a statement directly refuting the claims Teller and Foster had made, which was sent to the President.

Fulbright's committee endorsed the limited test-ban treaty, by sixteen votes to one, but still Teller continued to fight, appearing on the TV programme *Meet the Press* and before an audience at the National Press Club. It was a rearguard action, and Teller had finally to admit defeat: on 24 September the Senate ratified the treaty.

Whether viewed as dangerous obsession or deep commitment, there could never be any doubting Teller's full and personal involvement in his campaigning. When Kennedy broadcast a statement hailing the Limited Test Ban Agreement, Teller's long-time secretary Genevieve Phillips went outside to listen in her car. She hadn't wanted him to hear the speech in his Livermore office. When she returned to her desk, Teller attempted to console her, saying that they had known for weeks it was going to turn out this way. When Teller left the room, Mrs Phillips put her head down on the desk and cried.

Yet in hindsight the Limited Test Ban Agreement can be seen as an unsatisfactory compromise, once described by James Killian as a 'propaganda step' – a move that served to defuse the fall-out issue and remove it from the headlines but that did nothing to achieve Kennedy's main aim of slowing the nuclear arms race. Looking back, the agreement seems a temporary, even hollow victory for the President. Teller, on the other hand, had, by playing his part in ensuring the continuation of underground testing, contributed in a major way to maintaining a future both for his laboratory and for his anti-Soviet crusade.

Not long after he had received the Fermi award at the end of 1962, Teller, as the most recent winner, was approached for his thoughts on a nominee for the following year's award. He was not alone in thinking of Robert Oppenheimer; but while others saw the award as a timely opportunity to rehabilitate him, after nine years of political exile following the security hearing, Teller also saw things in more personal terms. He hoped it would end his differences with Oppenheimer and lead to an easing of his own miseries of the past decade.

Following the hearing in 1954, Oppenheimer had returned to the Institute for Advanced Studies at Princeton. Lewis Strauss had failed in his quest for his resignation and, with Oppenheimer's re-election as the Institute's director, the FBI at last judged it politic to withdraw their full-time surveillance of him. He became a much-travelled and much-visited celebrity, almost, it seems, a landmark on the map of intellectual tourism: a person whom everyone wanted to meet and to have spoken with.

Staying in Princeton, it was inevitable that Oppenheimer would have to come to terms with Strauss and, just after his reappointment as director, Strauss appeared in Oppenheimer's office and had himself announced. Oppenheimer emerged from his inner office, hand outstretched. 'Lewis ...' he said as if greeting a much-missed colleague. 'Robert ...' Strauss beamed back. An onlooker in the room who witnessed this reunion thought that, if a match had been lit, the whole place would have exploded.

'He missed his role in politics, moving in elevated circles, terribly,' said Marvin Goldberger, who, in the late 1950s, became professor of physics at nearby Princeton,

> and he failed really to connect with physics again. I discovered much to my surprise, that he and his wife had very few friends, having antagonised many by his arrogance. So he really latched on to me and to my wife although we didn't share the Los Alamos experience.

He did seem so lonely, so isolated, that I tried to bring him closer to the local physics community by organising weekly lunches in his office. I would take along one or two of my research staff to talk about their work and what was going on generally. I'm afraid this was all a little patronising, but it was certainly accepted.

We also saw quite a lot of Robert and Kitty socially, but we did find Kitty very difficult. Quite honestly she was a bitch, and she was an alcoholic. After a while my wife would not invite them to our house or go there, because of her behaviour.

In fact both Oppenheimer and Kitty had drunk heavily while at Los Alamos, and the tensions of the last unhappy years had taken their toll. The director's house, Olden Manor, accrued the nickname 'Bourbon Manor'. The unhappiness inevitably communicated itself to the two children. Their son Peter, twelve years old at the time of the hearing, became a painfully shy and reclusive teenager, eventually dropping out of college and leaving his family to live with his uncle Frank. Their daughter Toni was, years later, to commit suicide.

The arrival of the Kennedy administration, however, had eased Oppenheimer's exile greatly. The new men in the administration, the liberal intellectuals such as Arthur Schlesinger, Jr, McGeorge Bundy and Dean Rusk, with whom Edward Teller was intuitively at odds, had all learned from Oppenheimer and saw him as something of a champion. He was again invited to functions at the White House. It was at one of these that Glenn Seaborg suggested he might submit to another security hearing to clear his name. 'Not on your life,' Oppenheimer is said to have replied. So his friends in government had to think of other ways of helping his rehabilitation and it was at this point that the Fermi Award had been mentioned. In April 1963 his nomination, supported by Teller and others, was approved unanimously by the AEC, though not without considerable controversy. There was even a question mark over whether Kennedy would make the award, but on 22 November it was announced that he would. That afternoon the President was assassinated in Dallas.

Early the following week it was revealed that Lyndon Johnson would still make the presentation and on the date already fixed, 2 December. This was twenty-one years to the day after Fermi's pile had gone critical in Chicago. It was also ten years, all but one day, since President Eisenhower had placed a 'blank wall' around Oppenheimer, pending his security hearing. Speaking in the cabinet room of the White House, in the presence of Oppenheimer's family and friends and eminent scientists, including Teller, President Johnson said, 'One of President Kennedy's most important acts was to sign this award.' Then, 'on behalf of the people of the United

States', he presented Oppenheimer with the citation, the medal, and a cheque for $50,000.

For a few moments Oppenheimer stood, silently reflecting on the situation and the medal, and then, turning to Johnson, he said, 'I think it is just possible, Mr President, that it has taken some charity and some courage for you to make this award today. That would seem to me a good augury for all our futures.'*

After the ceremony, there was a reception and it was here that Oppenheimer and Teller were asked to pose in a handshake of reconciliation. The two men seemed to find no difficulty in obliging, but Kitty Oppenheimer's face, as she looks on in the resulting photos, tells another story.

Many people commented on both the rehabilitation and the handshake of the two rivals. *Life* magazine, in a highly emotional account of the Oppenheimer affair, saw in the handshake a hope for an end to the ten-year schism among scientists. Practically, however, the rehabilitation changed nothing. Robert Oppenheimer's security clearance was not restored. He was still considered unworthy of being trusted with his country's secrets. As to the handshake, it was Teller who so much wanted a rapprochement. Some people, such as George Cowan, believed he felt genuine contrition. Others, however, believed his concerns were largely motivated by self-pity, and the schism among scientists remained unhealed. Teller himself could still not be sure how he would be received anywhere. Nothing had changed.

24
Struggling Uphill

It was just two months after Teller and Oppenheimer had shaken hands at the Fermi Award ceremony that Columbia Films took advertising space to report their record first-week takings for Stanley Kubrick's film *Dr Strangelove*. As described in the introduction to this book, Kubrick's black comedy savagely mocked the President, the military establishments of both East and West and the Cold War rhetoric that fuelled the nuclear arms race.

Many critics, both at the time and ever since, have hailed it as a master-piece, the film of the decade, but even among those who hated it, it was seen as a cultural landmark. '*Dr Strangelove* is straight propaganda, and dangerous propaganda,' wrote one correspondent to the *New York Times*. 'It is an anti-American tract unmatched in invective by even our declared enemies.' Another correspondent added that the film 'indulges in the most insidious and highly dangerous form of public opinion tampering con-cerning a vital sector of our national life, the sector which needs public funds, public understanding and public support to do its job'.

Some of the film critics expressed similar sentiments. 'When virtually everybody turns up stupid or insane,' wrote one '– or, what is worse, psychopathic – I want to know what this picture proves.' Another, after panning it as a film, nevertheless wrote, '. . . it is a milestone. It promises a beginning to large scale consideration of the folly of American and Soviet warmongering.'

Despite the heaviness of its satire, many reviewers praised its believ-ability. The *Newsweek* reviewer described the scenario as one 'which Kubrick makes perfectly plausible,' while another commented: '. . . it contains horrors that, though outrageous, ring absolutely true.' These reactions, along with the film's universal success, were exceptional and can only be seen as springing from the deep fears and anxieties of a society living with the very real threat of a thermonuclear holocaust.

It was in this context that Edward Teller had recently fought in every way he knew, both in public and in private, to resist the Limited Test Ban Treaty, signed six month earlier. Over the previous five years, he had publicly resisted the outcry over nuclear fall-out. He had been associated with Livermore's cavalier, risky and reprehensible Plowshare activities in Alaska, which were to have been used as a front for continued weapons testing and carried out with little regard for the local population. He had lent his public support to a controversial chairman of the AEC and he had recently been publicly accused by a popular political columnist of venality, of using his position to obtain lucrative consultancies. It is difficult to think of anyone placing themselves more substantially at odds with the powerful political mood of the time than Edward Teller had done by 1964. Yet each of the stands he had taken was inspired by committed beliefs.

His mistrust of the Russians was as strong as ever, and had been bolstered by Soviet actions over Berlin and the Cuban Missile Crisis. He continued to believe that the dangers of fall-out from testing had been greatly exaggerated, and he was by no means the only expert who thought Linus Pauling had been intellectually dishonest. Even though other Plowshare projects were under increasing scrutiny in the wake of Chariot, Teller still hoped for positive returns from nuclear energy. And he had supported Lewis Strauss as a friend and ally, regardless of the personal cost. He had wanted to develop backbone, he once told Maria Mayer, and this he had done. But now the mood of the times was so determinedly against him that both his vision and his loyalties were to be tested at every turn. At the time, friends like Willard Libby were concerned about his apparent need to seek out controversy. 'I didn't talk publicly. I talked privately,' said Libby, 'very effective places. I knew where the buttons were . . . See, Edward makes speeches and I don't. If I do say so, sometimes it's better not to make speeches.'

Teller even became drawn into controversies outside the field of nuclear power. On 9 November 1965 New York City experienced a total blackout for the first time in its history. By chance, Teller was travelling into Manhattan from the airport and witnessed the famous blazing skyline suddenly go dark. That evening he walked the streets in the moonlight, enjoying the unusual sense of community and friendship that the darkness generated.

Shortly after this, Nelson Rockefeller, then Governor of New York, had asked his advice on how to prevent a recurrence.* His first suggestion of employing state-of-the-art computer technology to maintain the supply and demand situation minute-by-minute did not carry the day, but he soon found himself advising on another scheme on Rockefeller's behalf aimed at solving New York's power problem. Con Edison, the power company, were proposing a hydro-electric scheme and Teller was drawn into its defence. The scheme threatened the flooding of some of the most beautiful valleys along the Hudson River near Storm King Mountain, and ranged against Teller was a very well organised environmental protest group, Scenic Hudson, led by the folk singer, Pete Seeger. Nevertheless, together with two of his protégés Harry Sahlin and Lowell Wood, he began appearing at various public meetings.

Anyone concerned with their political image would have steered well clear, but Teller had evaluated the project and thought it justified – more over, Rockefeller was someone with whom he had been consulting for nearly a decade and whom he considered a friend. As a consequence, however, he faced not only attacks on his views but assaults on his personal integrity and suggestions that he was involved for financial gain. His assailant, representing the protesters was Herb Marks, one of Oppenheimer's defence team at the Hearings.*

The work being pursued at Livermore was creating yet more controversy. Ever since the early 1960s and the Starfish Prime shot, the laboratory had continued development of an ABM system based on blasting incoming missiles with nuclear explosions out in space. They were working on the new two-tier system of ballistic missiles, which was initially known as the Sentinel, later to be renamed Safeguard by the Nixon administration. Spartan was the larger of the two missiles, a long-range interceptor capable of achieving velocities ten times the speed of sound and destroying ICBMs at distances up to a hundred miles. The smaller but faster Sprint missile was the back-up, the last-ditch attempt to catch any missile breaking through the Spartan line.

Los Alamos was developing the warhead for the Sprint missile, a low-kiloton design relying on an enhanced neutron flux to 'kill' its target. Livermore had meanwhile begun work on a much larger, more advanced design for a Spartan warhead. This warhead, the W 71, was to yield approximately five megatons, hundreds of times bigger than the Sprint warhead. Instead of relying on a neutron flux to 'kill' its target, the thermonuclear secondary stage of the W 71 was being modified to produce a much-enhanced quantity of high-energy X-rays. On earth, these X-rays would be quickly absorbed by the atmosphere, but out in the vacuum of space they would travel vast distances, where they would act by instantly heating the surface of an incoming missile, causing it to explode and break up.

On 22 November 1967, Livermore received a visit from the new governor of California, Ronald Reagan. Teller had met Reagan through the Hoover Institution on War, Revolution, and Peace on the Stanford campus, a conservative think tank where Teller became a Fellow, and invited him to come and see the laboratory. The governor's 'relaxed manner' impressed and, as Teller recalled, 'the only thing that he obviously worried about was being late for his wife. He addressed all other problems with straight-forward ease.' Reagan was given a tour of the laboratory and then listened to a two-hour presentation covering the progress and problems with the new ballistic missile defence systems. Before becoming governor, Reagan had spent time in public relations for General Electric, and so proved a good listener, asking a number of salient questions.

'We had had a pleasant lunch following the presentation.' Teller recalls, 'By the end of the luncheon the governor had had more friends than he had at the beginning. But no mention was made of the defense system that was so much on my mind. Fifteen years later, I discovered that he had been very interested in those ideas.'

However, the X-ray-enhanced warhead was proving to be a difficult design challenge – indeed, Livermore itself rated it as one of the most difficult in the laboratory's history. One of the main sticking points was how to contain the energy of the burn for as long as possible to enhance the X-ray output of the secondary. One approach being tried was covering the secondary with a layer of gold to retain as much of the radiation for as long as possible. However, such a novel device needed to be tested. The question with such a large device was: where?

Kennedy's Limited Test Ban Treaty allowed for testing underground, but with an estimated five megaton yield, the W 71 was too large to be tested in Nevada. Eyes turned, again, to Alaska, and the Aleutians, a necklace of islands stretching out south-westwards across the channel between Alaska and Russian Siberia. Of these, the island of Amchitka seemed ideal. Lying at the far western end of the chain, it had been uninhabited for much of the time since the Second World War, when it had been a forward fighter-bomber base.

Plans were soon under way to test the island for its ability to absorb a five-megaton blast. The scientists working on the project had a number of concerns, one of the biggest of which was earthquakes. Amchitka is on the Pacific 'rim of fire', at the end of a chain of fault lines connecting, in one direction, with those running right down the western seaboard of North America and in the other with those running down through Japan. The science of plate tectonics was in its infancy in the late sixties, but there were fears that a major test could trigger a whole series of earthquakes in either direction. As well as this, there were still the concerns about the amount of radiation such explosions would vent into the atmosphere – or, perhaps more importantly in this case, underground. The first 'calibration' test, codenamed Milrow, with a projected yield of around one megaton, was to be detonated at a depth of two-thirds of a mile. The main five-megaton shot, codenamed Cannikin, was to be detonated at a depth of one mile. Both these were well below sea level, and given the complex of fissures in the deep rock strata found around earthquake centres, no one knew whether channels might open up, leaking radiation out into the marine environment. Once more it appeared that Livermore was preparing to play dice with global safety.

In her review of *Dr Strangelove*, the American critic Susan Sontag commented, 'Intellectuals and adolescents both love it. But the sixteen-year-olds who are lining up to see it understand the film and its real victims, better than the intellectuals, who vastly over-praise it.'

Those sixteen-year-olds were soon to become the students of the late 1960s who dominated the political protest and counter-culture lifestyle, the group who extended the protests against the 'nuclear age' beyond the technical and the medical concerns of the 1950s to its social and cultural impact; and this was to happen no more profoundly than at Edward Teller's own campus, Berkeley.

The so-called 'Six Years' War' that gave University of California Berkeley a special standing in the annals of 1960s' student revolt began in 1964. Students, hardened from their experiences on Martin Luther King's Civil Rights marches, took exception to the fact that there was no public place on campus where they could speak out on any issue they chose. The so-called Free Speech Movement rapidly began staging sit-ins, occupying the main buildings on the campus. As the atmosphere became increasingly embittered and ugly, the police were called in, arrests made, and teaching became increasingly difficulty.

'The real free speech issue was the one that wasn't talked about,' recalled student administrator Alex C. Sheriffs. 'It was that Edward Teller, for example, couldn't speak in his class because he was one who wanted better behaviour on the campus. And, of course, because of atomic research. Other people made a point of threatening his life and of making a shambles of his class – and did so to quite a number of those on that campus.'

As the years passed, the original Free Speech Movement was augmented by other more extreme and anarchic groups such as the Red Family, Students for a Democratic Society, and Scientists and Engineers for Social and Political Action. Some of their protests were aimed at ridiculing the victim. Teller was to join an illustrious group of high-profile victims such as right-wing commentator William F. Buckley, and members of Nixon's Commission on Pornography, when at one public meeting he received a custard pie full in the face, delivered by leading student activist Jerry Rubin.

However, over the years, Teller had typically learned to contend with the threats to his life and the disruption to his classes and, on occasion, to retaliate gently. The Livermore physicist Stirling Colgate recalled,

> Bill Wattenberg, a local radio talk show host, got Edward to visit the State University at Chico, and the students were rebelling that Edward would come and speak – it was the height of the hippie years. So the whole thing was arranged that the curtains on the stage were closed. And suddenly there was this extraordinary, competent piano playing, and the audience grew quiet, and after a little the curtain slowly opened and there was Edward playing. I mean it was a trick on the whole thing that was beyond the pale – it was all arranged between Wattenberg and Edward – and it was monumentally successful. It started a dialogue with the students and was a great, great success in that the rest of the programme was carried out with intellectual respect. It was typical of his abilities.*

On his home campus, however, the situation was to remain volatile.

'He was practically stoned on occasion,' said Colgate. 'The protesters were anti-nuclear. He was after all "father of the H-bomb", but it was really the same group that was protesting Vietnam, the same drive. He was a victim of it and he lived right on the north side of the campus – just a block or two up the hill – very accessible.'

Protesters had already taken advantage of this accessibility, daubing graffiti on the sidewalk, and on one occasion painting slogans on his secretary Genevieve Phillips's car when it was parked outside; but the Tellers had made no attempt to improve security, partly because Teller himself saw the students as 'aimless, disappointed, confused . . . ripe for any radical solution' rather than in any way dangerous. There had been a massive expansion in university education since 1945 and, in Teller's view, too many students were now simply 'attending college because it is "the thing to do"'. He saw their inspiration as the 'Third World Communism' of Che Guevara and of Mao Tse Tung, but in doing so he overlooked the growing bitterness felt by many over Vietnam and the indignation over Nixon's invasion of Cambodia. By the beginning of the 1970s the mood among student groups was becoming increasingly aggressive.

Some three years earlier Bertrand Russell, the philosopher and nuclear-disarmament campaigner, conducted a mock tribunal for 'war criminals' in Sweden. It was an exercise modelled on the Nuremberg War Crime Tribunals after the Second World War, and Berkeley's own Red Family decided to emulate him by mounting a similar 'tribunal' on campus. The prime objective was to bring the University of California to book over the amount of military research going on in its laboratories. Their victims were the University President, Charles Hitch; the Directors of Livermore and Los Alamos, Mike May and Harold Agnew respectively; Glenn Seaborg of the AEC; John Foster, now Director of Development and Research at the Pentagon – and Edward Teller, who, along with the others, received an invitation to attend a meeting to answer the charges against him. At the same time, flyers and leaflets appeared around the campus listing Teller's 'crimes': his wartime work at Los Alamos, his fathering of the H-bomb and establishing of Livermore, and his advocacy of the arms race and of 'nuclear blackmail'. They also printed a quotation attributed to Teller that read: 'We can win the hearts and minds of men. If the people of the world really want freedom, and if our nuclear forces can stop massed communist man-power, I am convinced that our victory would be assured in any limited war.'

Almost certainly this statement was made about Europe not, as was implied, about Vietnam. In Europe, it certainly was NATO strategy to use nuclear weapons to match the superior ground forces of the communist bloc; but in the guerrilla campaign of Vietnam the use of nuclear weapons made no sense – particularly as North Vietnam was backed by the nuclear resources of China. However, it was yet one more piece of misinformation to be added to the bewildering mix of truth and rumour that had been demonising Edward Teller.

The 'tribunal' was due to take place at 7.30 p.m. on 23 November in the university's Pauley Ballroom. That afternoon a Berkeley physicist, Hardin Jones, became alarmed at the strength of feeling around campus against Teller, and tried to warn the Chancellor's Office that there was a serious threat to Edward and Mici Teller's safety. However, there was a reluctance to confront the students and the Tellers, for their part, refused to move. They turned down an offer to spend the night at Livermore, but at least the security office there had alerted the local police.

The 'tribunal' went ahead as planned, attended by hundreds of students, and by Lowell Wood and another young Livermore scientist, George Chapline. The two went to monitor what was going on and to gauge the level of threat. The meeting was tape-recorded and its mood was set after only a few minutes when one member of the Red Family, Jack Nicholl, reported on the upcoming visit of the Vice-Premier of South Vietnam to San Francisco: 'We don't need to hold a hearing about Ky . . .' he said, 'he deserves to die.'

This was followed by a long list of speakers, who detailed the outrages of the Vietnam campaign, and then a member of the Physics Faculty, Charles Schwartz, focused on Teller as the main culprit before the 'tribunal'. Teller, he said, exemplified 'the Cold War philosophy that holds the Russians . . . the great evil on earth'. As the three-hour session progressed, Teller

was described as 'a leading sparkplug . . . for an even greater military nuclear arsenal' and a 'paranoid anti-communist'. The Livermore was referred
to as 'a scientific whorehouse'.

The criticisms were all familiar ones, but in the intense, angry atmosphere of the meeting they took on a new potency, something of the diabolic.
'Are you going to go home [or] fuck those labs and fuck these people?'
Jack Nicholl asked. 'What are you going to do. Huh? I'm gonna try and
fuck these labs. You want to get some guns, we will do it.'

Passions escalated and within a short while people were shouting, 'Let's
get Teller,' and, 'We want Teller.' Then one voice called out, 'Let's get Teller
tonight!'

A female voice joined in: 'We got a responsibility to expose these people
to the community and not let them live in the same way any longer. Let's go!'

'What's the address?' another voice shouted over the sounds of people
beginning to move out of the hall. There was a last attempt to try to calm
things down, but the people were already leaving, some shouting as they
went, 'Break Teller's windows . . . burn his house . . . kill him!'

By this stage Wood and Chapline had already run from the meeting to
warn the Tellers, but when they arrived at Hawthorne Terrace, the police
were already there, some in riot gear. As both were indistinguishable in
appearance from the protesters, they had some difficulty getting through
the cordon. Not far behind them came the chanting, shouting crowd from
the hall, but they too were brought up short by the police, a block or two
away from the Teller house. When it became apparent that they were not
going any further without a major battle, the jeering crowd set light to an
effigy of their target before returning downhill to wreak havoc around the
campus, breaking down doors and smashing windows.

Wood and Chapline were shocked by the fury that had been unleashed
and urged the Tellers to move out, but to no avail. A short time later the
couple acquired a guard dog named Shah, and not long after that erected
a high, chain-link fence around the property. Within months, however, the
Six Years' War that had made Berkeley synonymous with student dissent
throughout the world had virtually blown itself out. There were no similar
occurrences, and Mici and Edward were to live on at Hawthorne Terrace
until his official retirement from the University five years later.

For Teller, the essential logic behind the development of the ABM system
had remained incontrovertible throughout these troubled years. 'I cannot
tell you how much more I would rather shoot at enemy missiles than to
suffer attack and then have to shoot at people in return,' he told a reporter
in May 1969. 'I want to repeat – with all possible emphasis – that defense
is better than retaliation.' Although he was careful to add a caveat,

qualifying the likely success of the two-tier Safeguard system, Teller still saw its prime role as its potential to 'induce very grave doubts in the mind of any would-be attacker'.

Not everyone, however, believed that an ABM system would serve as a deterrent. Some raised questions about the expense of such a system with its accompanying detection and command-and-control systems, but there was another argument rooted in the psychology of the Cold War. Many of those working closely on defence matters simply did not agree with Teller that the stand-off created by the threat of mutually assured destruction was necessarily a bad thing.

Firstly, the argument of this group went, provided the nuclear arsenals of both countries were currently invulnerable, military balance was insensitive to both the number and the kind of weapons there were. Secondly, there was little chance, under existing circumstances, of either side failing to comprehend the devastating retaliation that would answer a pre-emptive attack. The introduction of ABMs, however, would muddy the water, introducing important but incalculable factors into the existing stark choice. The protection they offered following a pre-emptive attack might, for example, foster the idea that such an attack was worthwhile. This, the thinking went, would lead, in turn, to uncertainties of judgement and therefore greater instability and risk, particularly at a time of crisis. Better, then, to limit defensive measures or do away with them altogether.

When these ideas were first voiced at a meeting with the Soviets, the head of the Russian delegation queried whether there was something wrong with the translation. Was the US really preparing to limit *defensive* weapons? At the time it seemed strange, even perverse, but it represented an opportunity to limit one aspect of the arms race that might grow like Topsy, and the Soviets agreed to think about it. The debate on whether to deploy ABMs continued during the summer on Capitol Hill, leading to a vote in the Senate. It resulted in a 50–50 tie, which was only resolved by Vice-President Spiro Agnew's tie-breaking 'yes' vote. Work went ahead with preparing the warhead tests on Amchitka, but such a near miss served to invigorate the protest groups who opposed the ABM.

By the beginning of October, the AEC were ready with their 'calibration' test shot. The Milrow device, with a yield of 1.15 megatons, was detonated 4000 feet below the surface of the island on 2 October 1969. The following day the *New York Times* carried a front-page headline: 'Aleutian H-bomb Is Fired Without Setting Off Quake'. This was not true. Gene Phillips, Chief of the Seismological Observatory at Barrow, wrote to the Alaskan Senator Mike Gravel, reporting that they had 'detected an alarming influx of earthquakes directly following the test', and warning that 'further tests could trigger many more'.

The Milrow test, coupled with the close vote in the Senate, triggered a wider debate, which took wings as the five-megaton Cannikin shot scheduled for the autumn of 1971 grew closer. The Aleut League, representing the inhabitants of the nearby islands, filed a lawsuit to halt the test. Emperor Hirohito met with Nixon to express Japanese concern about the dangers of earthquakes, and Prime Minister Trudeau objected strenuously on behalf of Canada. Editorials in the main national newspapers urged cancellation, citing danger to the environment, but also picking up on doubts over the probable usefulness of ABMs against anything like a full-scale Soviet attack. Anything less than total effectiveness against a thousand missiles would mean disaster and even with the enhanced X-ray Spartan warhead they were testing, this, Teller himself realised, would not be possible.

It was at this stage that Teller received a phone call from Nixon's National Security Adviser, Henry Kissinger, reiterating the administration's support but telling him, 'You must go on the road and defend it.' So Teller did just that, at every opportunity, exposing himself to yet more public criticism. He was to find opposition to the Cannikin test was not only international but deep-rooted at home as well.

No less than five federal agencies, including the Environmental Protection Agency, recommended that Livermore's Cannikin test should be cancelled or postponed. Scientific witnesses from the Council on Environmental Quality testified that 'sizeable fractures and fissures' would allow 'contaminated groundwater' to escape in concentrations up to '100,000 times the permissible concentration.' However, rather than accept their recommendations, President Nixon invoked an Executive Order to suppress their comments. The AEC then classified them as 'restricted'. In response to this, thirty-three members of Congress filed a suit seeking to release the agencies' reports, the 'Cannikin Papers', as they became known.

In the months leading up to the test there were demonstrations throughout the US and Canada and a fishing boat, the *Phyllis Cormack*, set sail with twelve protesters on board to 'bear witness to the test'. This was the first action to be taken by the newly founded Greenpeace organisation – one of many environmental action groups inspired to protest by the AEC's activities in Alaska. Then, at the beginning of November, the US Supreme Court sat to consider the future of the test.

Meanwhile, at the Amchitka site, some 700 scientists and engineers from Livermore were working in freezing Arctic conditions, finishing the final preparations. They had dug a shaft a mile deep, and lowered down the 380 ton module that contained the bomb and its accompanying monitoring equipment. The new chairman of the AEC, James Schlesinger, had arrived with his wife and two daughters to demonstrate his belief that Cannikin was safe. 'It's fun for the kids and my wife is delighted to get away from the house for a while,' he said.

The weather was atrocious. On the eve of the test, the Supreme Court was still deliberating and a delay, at the very least, seemed likely. However, at 6.30 a.m. the following morning the White House hotline on Amchitka rang with news that, by four votes to three, the Court had ruled that the test could proceed.

Just four and a half hours later, at 11.00 a.m. on 6 November 1971, the mighty five-megaton blast from the Livermore device flared beneath the tundra. It was, and still is, the most powerful underground explosion in American history, registering 7.0 on the Richter scale. Although the worst-case scenario of earthquakes and tsunamis elsewhere along the Pacific rim was not realised, the blast did cause twenty-two minor earthquakes in the region. Then, thirty-eight hours after the explosion, the roof of the cavity created by the explosion collapsed, creating a depression a mile wide and sixty feet deep on the surface of the island, which rapidly filled with water. Six months later, AEC investigators found significant quantities of radioactive gases were leaking into the atmosphere. This was evidence that Cannikin had breached its containment, but it was not made public. The spirit of the 'firecracker boys' of the 1950s lived on.

The public outcry did not end once the Cannikin shot had been accomplished. The legal action in pursuit of the Cannikin Papers continued and the AEC was driven to perform various medical and 'human surveillance' studies on the Aleut populations from the neighbouring islands. Elevated concentrations of radionuclides were found in their blood, but no follow-up studies were ever conducted. The Cannikin test was to become part of anti-nuclear weapon demonology, but the most immediate effect resulted from the link forged in the public's mind between Cannikin and the Safeguard anti-missile system. At the time it was planned to deploy it around all the major cities and the notion of Cannikin-sized nuclear missiles – of fearsome 'nukes' in the backyard – met with strong resistance.

This background of protest was a large factor in causing the administration, which had been responsible for forcing through the Cannikin test, to reassess the ABM programme. In particular, they began seriously examining the mounting costs involved in trying to make the programme effective enough to be viable. At the same time the Soviets, like the Americans, came to realise the financial benefits of limiting the development of defensive weapons, and in 1972 the two countries signed the Anti-Ballistic Missile Treaty. This limited the number of defensive missiles on each side to a hundred, and they had to be arranged around just two sites. In 1974 this was reduced to a single site for each country, the Russians choosing to place theirs around Moscow, while the US placed theirs around a

Minuteman emplacement near Grand Forks, North Dakota.

Teller was to write that such 'unreasonable behaviour always produces sorrow' and he criticised as 'moral bankruptcy' the continued reliance on mutually assured destruction. In his view the US 'virtually offered its unprotected citizens as hostages to an accidental or intentional attack by missiles armed with nuclear warheads'.

'The policy claimed that by concentrating on retaliation and ignoring defence,' he wrote, 'the United States could provide security for its citizens. In fact, it had the opposite effect of making a first strike by the Soviet Union seem more attractive. The Soviet Union had a larger territory than the United States, a lesser concentration of its population in urban centers and well programmed civil defense. The Soviet Union would have suffered far less than "assured destruction" following a first strike . . .'

Although ABM technology had moved on since its inception, it still fell far short of providing the meaningful protection Teller was advocating. He himself had expressed reservations, and it was this technological shortfall that had led to the reassessment, both by the US and the Soviets. 'Such a false hope is extremely dangerous,' Herb York testified, 'If it diverts any of us from searching for a solution in the only place where it may be found: in a political search for peace combined with arms control and disarmament measures.'

By the early 1970s the Tellers had spent the best part of twenty years alienated from the liberal influences of erstwhile scientific friends and colleagues.* There were still a few, like Leo Szilard, with whom he maintained something of his old relationships, but time had not treated them kindly. Szilard had made clear his disapproval of what Teller had done but had not rejected him, and for this Teller was truly grateful. All the Hungarians had, in fact, in their different ways, remained loyal, in part because they all shared a common belief that the Russians were totally untrustworthy.

In 1959 Szilard had been diagnosed with terminal cancer and Teller had witnessed his taking control of his own treatment and demanding that the course of radiotherapy treatment he and his doctor wife had worked out should be used.

The treatment had been a success and Szilard, the man who had tried above all to be a reasonable person, was to turn his hand to writing. The result was a fantasy entitled *The Voice of the Dolphin*, in which he contrasted human folly with the reasonable behaviour of the animal. He had also become interested in biology, and had been asked by Jonas Salk to form an institute in La Jolla. In 1964, he died of a heart attack in his sleep. Yet another friend and counsellor was gone.

An even more painful loss was to follow. The correspondence between Edward and Maria Mayer had petered out in the late 1950s but they had still stayed in touch. In 1963 she had been awarded the Nobel Prize for recognising and explaining magic numbers, and in her Nobel speech she had warmly acknowledged her gratitude to Teller for his contribution to her work. However, at about that time she had developed a neurological condition that seriously affected her speech and Edward had begun seeing her much less frequently. She had invited him down occasionally to San Diego, where she was teaching, and he had given lectures there.

'I remember that when I gave my first lecture in San Diego after her illness,' Teller recalled, 'she made a nice three or four-minute introduction of me, clearly and with no errors. But that evening, I couldn't understand her at all. After I had asked her to repeat herself twice, I didn't want to ask her again, so we didn't talk much that evening.'

In 1972 he heard that Maria, the woman who had been his close friend and confidante, was dying. As soon as he could he went down to visit her. 'She lay in the hospital bed and didn't make a sound. I am not even sure whether she recognised me. I sat with her for almost an hour, occasionally offering comments that I hoped she could hear. But she never responded. She died a few days later.'

Some months afterwards, Edward was talking to a mutual friend about Maria, and mentioned how much he had missed talking to her during her last few years. It emerged that the friend had experienced no difficulty. He had spoken to her in her original tongue, German, while Teller had used English.

'I can never think of Maria in the period following her illness without great sadness at my own stupidity ... Our last meeting was sorrowful, but not because we couldn't talk: Even when words were no longer working well, I always felt that in some way we managed to understand each other. Our last meeting was painful because it clearly was the last.'

The loss of these friends severed yet further links with that other era, before he had become so totally enmeshed in arms-race politics. The refuge Edward had since found with the conservatives who had supported him in his battles for the H-bomb and against Oppenheimer had led to his increasing involvement in what became known as the 'Hidden Right'. This was the network of organisations, powerful individuals and industrialists who achieved their objectives through work done behind the scenes. One such group, of which both he and Ronald Reagan were members, was the American Security Council. Originally established in the mid 1950s to counter communist infiltration of American industry, it had amassed FBI-like files on some 6 million individuals, which it had made available to some 3500 companies and organisations nationwide.

Teller had joined as a member of Team B of the council's 'Coalition for Peace Through Strength'. Eugene Wigner, Teller's old Hungarian friend, was also a member. By the 1970s, the council was deeply rooted in Washington, counting some 190 members of Congress as supporting its central strategy of ensuring superiority over the Soviets. In the eyes of the ASC, this entailed not only huge military build-up, but also increasing intelligence capabilities both at home and abroad, and a strong opposition to all arms-control agreements. The *Washington Post* described it as one of the most successful, and most dangerous, of lobbying groups. Through these associations Teller had gained considerable power and influence, and he had relished both. But this compact was part of Edward Teller's tragedy.

'People do betray themselves . . .' said George Cowan, the radiochemist who had known Teller since Los Alamos. 'Potentially Edward was a great man in the highest sense, but he was betrayed by his obsession for power. Early on he was ambitious, which led to frustration, and then with success came the hubris and the power. And then he was lost. He made a mistake. He knows.'

Marvin Goldberger, one of Teller's students from Chicago, had a parallel view:

Edward felt that Oppenheimer got too much adulation both from his students before the war and from the public after it. What really galled him, in my opinion, was that he felt he was a much better physicist than Robert. I think that is true: Teller was a great physicist with an uncanny sense of how things worked. His lack of recognition resulted from a chronic inability to sit down and work something out in detail, milking an idea for all it was worth and articulating it clearly in print or in a seminar. He had little patience and flitted from challenge to challenge. He didn't have the taste that Oppenheimer had for what was most exciting or his quickness or his ability to inspire students. But I think Edward was much deeper in his understanding of physics.

Ever since the Oppenheimer affair, Teller's right-wing political stance and his willingness to speak out in public had undoubtedly caused much of the demonisation he was now experiencing. In all his struggles, he was having to fight not just the issue in hand but also the prejudices that had built up about him. 'He . . . was becoming so wildly hawkish,' Goldberger said, 'that no one wanted him around except the extremists in the Pentagon.' He had taken himself out on a limb and now had no way of retracing his steps. His behaviour, his sensitivity to the views and feelings of others, seemed to have coarsened, as Goldberger found when organising a conference in 1972.

In the late 1960s the University of Miami had established a Center for Theoretical Physics. Among the centre's sponsors was Alfred Sklar, a former student of Teller's, and a highly successful businessman, and Teller himself was a member of the centre's advisory board. Each year they awarded a prize, established as a memorial to Oppenheimer, for the young scientist whose work had stimulated new directions in research. That year, Goldberger was chairman of the selection committee, which also included Robert Serber, Oppenheimer's former colleague and Goldberger's old teacher. This committee had little difficulty deciding that the recipient should be the young physicist Stephen Weinberg, for his new theory uniting the interactive forces in the nucleus. (Weinberg was to share a Nobel Prize for the work in 1979.) Goldberger had written to the centre's director, explaining the committee's decision: 'And then I made what turned out to be a critical mistake. I added a sentence which said that in addition to all his purely scientific work, Weinberg had been very active in international security and arms control issues. Doesn't sound very provocative but just you wait.'

The award was to be presented at the centre's annual conference and on his arrival Goldberger was greeted at the hotel entrance by Edward Teller, who insisted that they must talk. They met at breakfast the following morning:

> He said, 'We are not going to give the Oppenheimer Prize to Weinberg but instead are giving it to Serber.' I was absolutely flabbergasted, if for no other reason, the fact that Serber had been on the Committee that had overwhelmingly selected Weinberg. I asked what was the basis for rejecting our unanimous recommendation. Teller replied, 'I don't think that would enhance our friendship if I told you.' I left the restaurant spluttering in rage . . .

Goldberger was able to discover that the reason that his committee's decision had been overruled was indeed his reference to Weinberg's arms-control activities. What he may not have known was that Teller may have suspected Weinberg had been involved in the war crimes tribunal that had 'tried' him a short time before. This provides a specific rationale for Teller's anger at the prize committee's decision, but seems rather poor justification for changing it. The following day, Goldberger used his chairmanship of a session to attack the centre for allowing issues unconnected with science to overturn the decision. He ended by saying that, until the centre had got its act together, he would boycott any further events.

Outside the hall Teller was waiting for him and, according to Goldberger, the two went at it rather hot and heavy and he launched into a peroration about the deadly threat from the Soviet Union which I would appreciate if I knew what he knew. (At the time I held clearances that he didn't even know existed.) I interrupted to say that I thought Richard Nixon posed more of a threat to the US than the Russians. At this point he said, 'I can only hope you are just being a fool,' implying I guess, that the only alternative was that I was a traitor. And I said, 'OK Edward, that's it! I'm not going to talk to you anymore.' I walked away and left him and literally did not speak to him for seven years.

This, however, was not the end of the affair. A year later, Goldberger was proposed as Director of the Lawrence Berkeley Laboratory (the old Radiation Laboratory) by Luis Alvarez. Very soon friends, including Alvarez, began contacting him, asking what he had done to enrage Teller. Teller, it appeared was making it clear to everyone that Goldberger's appointment would be a disaster. Goldberger did not get the job. 'Perhaps I owe Edward thanks,' Goldberger wrote. 'If I had become director of the Lawrence Berkeley Lab, I might never have become president of Cal tech!'

In spite of disagreeing with him on almost every issue,' Goldberger continued, 'in spite of his disgraceful and vindictive actions towards me personally, I still have a soft spot for the warm, charming, kind and thoughtful Edward Teller I first met in 1945.'

At the end of 1976 Gerald Ford lost out to Jimmy Carter in the presidential elections. Carter was neither sympathetic to Teller's aims nor, with his nuclear engineering experience, did he believe he had any need of scientific advice. Teller was in the wilderness, but he never stopped lobbying, arguing against the Strategic Arms Limitation Treaties (SALT), warning continuously about the Soviet threat. As he reached his seventieth birthday, Teller still harboured his vision of a defensive shield, still burned to achieve something more. He may have been deeply frustrated, but at an age when most people are retired, he was by no means exhausted. In 1979 an event occurred that provided him with a cause, something on which to focus his extraordinary energies – and in the service of nuclear power. It was also to involve him in yet another public controversy.

On 16 March 1979 an unusual thriller opened in New York. In *The China Syndrome* Jane Fonda and Michael Douglas played journalists who were making a routine film story about a nuclear reactor, when the reactor went berserk and was only just saved from meltdown. In this meltdown, the 'China syndrome' of the title, the white-hot reactor would have burnt through the floor of its containment and into the ground below – heading towards China on the other side of the globe. The film was an immediate box-office success, topical, well made, convincing, and tapping into a basic phobia about nuclear power. Twelve days later that phobia was to become a reality.

There were six men in the control room of the Three Mile Island nuclear power plant on the Susquehanna river in rural Pennsylvania on the night of 28 March. Around 4 a.m., one of them inadvertently cut off one of the water supplies that both fed the steam turbine and cooled the 800 megawatt pressurised-water reactor. The plant's safety system automatically shut down the turbines and, as the temperatures and pressures in the reactor rose, a relief valve opened – as it should have done – at the top of the reactor. It should also have shut after only thirteen seconds, but it stuck in the open position, allowing vital cooling water to drain away. Unfortunately, as the indicator light had gone out once the valve motor had stopped working, the engineers assumed that the valve had shut.

The reactor core quickly began to overheat. Struggling with contradictory and baffling information, the engineers did precisely the wrong thing, shutting off the emergency water system that would have helped cool the core. Within minutes the mammoth control console was 'lit up like a Christmas tree', as one operator described it. Hundreds of flashing lights were accompanied by piercing horns and sirens.

Two hours and twenty-two minutes after the start of the emergency, the engineers realised what had occurred, but by that time the damage was irreversible. The top of the core had been exposed and had reached temperatures of 4300 degrees Fahrenheit – dangerously close to meltdown. The zirconium alloy coating the fuel rods had also interacted with the steam, producing a hydrogen bubble inside the reactor, which could have exploded at any time and remained a threat for much of the five-day emergency.

At this point – early morning on 28 March – the operators of the plant, Metropolitan Edison, made their first announcement: that there had been a problem, which was now under control. Lieutenant Governor Bill Scranton relied on this statement for his own reassurance to the public, but as he came from the podium after speaking to the press, he learnt that there had actually been a release of radioactive material. Metropolitan Edison had reacted with the industry's time-honoured cover-up and, as Scranton put it, 'from that point we could not rely on Met Ed for the kind of information we needed to make decisions'.

With big-business interests as portrayed in *The China Syndrome* very much in their minds, the press gave Metropolitan Edison's assurances short shrift. Headlines like the *New York Post*'s 'Race With Nuclear Disaster' were the standard fare. The catchphrase 'We all live in Pennsylvania now' united feelings not only in the US but in Europe as well.

On the third morning, 30 March, the helicopter monitoring the plant from above picked up a sudden surge in radioactivity. A radioactive cloud, one of low intensity, was moving out over the town of Middleton. Pennsylvania Governor Richard Thornburgh, who up till then had resisted evacuation for fear of the panic, announced that all pre-school children and their mothers within a five-mile radius should be moved out. The convoy of yellow school buses and cars packed with bedding and toys, which carried them out of Middleton, was the advance guard of some 140,000 who would eventually evacuate the area before the crisis was over.

In the midst of this, Edward Teller determined to defend the nuclear industry's reputation as vigorously as he could. In the late 1940s, he had been the first chairman of the Reactor Safeguards Committee, a committee that had laid down the original safety procedures for an industry that could allow for no margin of error. It was widely agreed that he had done an excellent job. Teller himself was enormously proud of this early work and he took very personally any challenge to the industry's safety record. He had recently written *Energy from Heaven and Earth*, a book that, among other things, sought to rebut any argument ever raised by the critics of nuclear power. Now, in quixotic mood, he took to the road to promote the industry's record. No one had been killed, he argued. Although there might have been equipment failures and human error, vital safety systems had worked. There had been no meltdown and the amount of radioactive material released had been small. He accused the opponents of nuclear power of fanning public fears. 'It just so happens that the anti-nuclear movement, lacking a real accident, has latched on to this,' he told *Playboy* while the emergency was still running.

Teller was so upset at the damage being done to the industry by the public outcry that he even considered entering mainstream politics to continue the fight. He had his eye on the seat of the Californian Senator Alan Cranston, a powerful opponent of nuclear power.

On 3 April the reactor stabilised and the core temperature began slowly to decrease. The immediate crisis was passed, but this did not halt Teller's campaigning. It culminated in his appearance before the House Science Committee six weeks later, on 6 May. 'Nuclear reactors are not safe,' he argued, 'but they are incomparably safer than anything else we might have to produce electric energy.' He went on, 'Zero does not exist. I don't expect zero probability from nuclear plants or anything else.' He then referred to the existing situation as he saw it: 'Zero is the number of proven cases of damage to health due to a nuclear plant in the free world.'

The day after his appearance before Congress, Teller was hospitalised with a heart attack. Fortunately, it proved not to be a serious one, but it was to curtail his activities for some months, and nipped in the bud his plans to enter politics. His campaigning continued, however, and on 31 July 1979 there was a headline spread across two pages of the *Wall Street Journal*:

I WAS THE ONLY VICTIM OF THREE MILE ISLAND.

A large photograph of Teller stared out from an advertisement in which he blamed Jane Fonda and the consumer activist Ralph Nader for causing his heart attack. In it he described how he had gone to Washington 'to refute some of the propaganda' of those striving 'to frighten people away from nuclear power'. To prepare for his congressional testimony, he said, he had worked twenty hours a day, overtaxing his heart: 'You might say that I was the only one whose health was affected by that reactor near Harrisburg,' Teller wrote. 'No, that would be wrong. It was not the reactor. It was Jane Fonda. Reactors are not dangerous.'

So soon after the worldwide shock over what had happened at Three Mile Island, his argument fell on deaf ears. An editorial in the *New York Times* accused Teller of propaganda. It also accused him of using outdated risk analyses when he had claimed that the chances of a person being injured within fifty miles of a nuclear power plant were about the same as of being hit by a meteor. It then pointed out something Teller had not mentioned: that the sponsor of the advertisement, Dresser Industries, had manufactured the valve that had stuck open and started the emergency.

In spite of his efforts, during the coming year there were no new orders for reactors, and the cancellation of eleven earlier orders. There was a de facto moratorium on further development of nuclear power, the stigma of which was to be enhanced by the Chernobyl emergency seven years later. It looked as though Teller's apparently unquenchable energy would remain frustrated this time – except that as 1980 dawned the presidential elections promised a close fight between Jimmy Carter and the Republican candidate, Ronald Reagan; and, at Livermore, some recent research had breathed new life into one of Teller's dreams.

25

Bringing up the Props

'There's a quotation Lowell would often use, that, according to Edward Teller, you could never lie about the future. Can we translate that – it's impossible to lie about the future? One version is, you can make whatever promises you like.'

Peter Hagelstein was one of the young scientists who played a crucial role at Livermore at the time it was contributing to the Strategic Defense Initiative, popularly know as 'Star Wars'. When I met him in April 2000, I asked him to assess Teller's role in a programme which was to become yet another of the scandals to emerge from the Reagan era.

'Is Teller microscopically responsible?' he replied. 'Maybe not, but Teller is certainly deeply responsible for all this, the circles within circles.'

In 1978, Peter Hagelstein was a twenty-three-year-old graduate from MIT, working in Livermore's 'O' Group under Lowell Wood. Three years previously, Wood had recruited him as a Hertz Foundation Fellow, impressed by his being 'not only exceedingly competent in a technical sense, but quite creative'.

Hagelstein had a strong antipathy to weapons work – and so his choice of the Hertz Fellowships* and Livermore seems at first sight perverse. However, Hertz offered very attractive grants, and the laboratory did carry out non-military research. '[Wood] said they were working on lasers and laser fusion, which I had never heard of before, and he said there were computer codes out there that were like playing a Wurlitzer organ. It all sounded kind of dreamy.'*

Wood had also expanded a little on another seductive area of research going on at Livermore: the quest for an X-ray laser. Ever since the creation of the first optical lasers at the end of the 1950s, the aim had been to construct them using ever shorter wavelengths. Just as lasers at wavelengths of visible light allowed for 3-D holograms of ordinary everyday objects, so shorter wavelengths would enable microscopic, even submicroscopic, structures to be seen in the same three-dimensional detail. Visible light covers wavelengths between 4000 and 7000 angstroms. X-rays have very much shorter wavelengths – under 100 angstroms – and carry much more energy. If they could be focused in a laser they could penetrate biological structures, nerve cells, brain tissue, cancers, providing researchers with three-dimensional holographic images in a detail never seen before.

At the time Peter Hagelstein was recruited, the laboratory still had funding for such work and it was just the kind of project, with its broad implications, that he wanted to pursue. Shortly after his arrival, however, the money dried up. It was the victim of a priority issue – too little success over too long a period. Nevertheless, Hagelstein went on working, using Livermore's supercomputers to construct models of the experiments he could not perform for real. Elsewhere at the laboratory, however, another X-ray laser project *had* managed to find funding. This one had a very different objective, inspired by another property of X-rays. They are the most energetic form of radiation and, in theory, a beam from an X-ray laser could carry a punch powerful enough to be used against enemy missiles. The problem was to find an energy source powerful enough to energise – to pump – it.

One of those looking into the problem was a senior physicist at Livermore, George Chapline. It was he who, along with Lowell Wood, had kept watch for Teller on the 'war-crimes' tribunal at Berkeley a few years earlier. 'My original idea came from the Russians,' Chapline recalled. 'They did some very interesting work in the early seventies over building the X-ray laser, though not in the context of using a nuclear device – that was the idea I contributed.' What Chapline had in mind was certainly not a candidate for exploring the structure of life in the laboratory. The nuclear explosion creating the energy for the laser would vaporise the lasing equipment a tiny instant after the X-rays, travelling at the speed of light, had pulsed along the lasing rods. However, as a weapon out in space, whose beam would be capable of obliterating enemy missiles in an instant, it did have promise. In fact, it had the potential to qualify as one of the new 'third-generation' weapons Teller had recently conceived. The first generation had been the fission weapons, the second the thermonuclear devices; now the third generation would be weapons systems that could direct, control and enhance the force of nuclear blasts.

In 1977 Chapline had come up with an experimental design for this so-called nuclear-pumped X-ray laser, which, if successful, could be small enough to be taken out into space relatively inexpensively. There would no doubt be endless problems in turning this concept into a weapon, but it was sufficiently novel to capture Teller's enthusiasm. Within a short while, Chapline was given the opportunity of sharing a multi-million-dollar Nevada test with a military project investigating the effects of nuclear blast on the new ten-warhead MX missiles.

The test, codenamed Diablo Hawk and scheduled for 13 September 1978, was set up in a honeycomb of tunnels under the Rainier Mesa. The chamber containing the bomb was connected by pipes that would carry the radiation from the explosion to one chamber containing Chapline's experimental rig and to another containing the missiles under test. During the final preparations the chambers and the connecting pipes had to be pumped free of air to allow the radiation to flow unimpeded, and the whole system was then sealed with concrete. Only at that point was a leak discovered in the vacuum line leading to Chapline's experiment. It was too late to do anything about it and the blast went ahead. Chapline's lasing experiment, as expected, was a failure. If it had not been for some unexpected congressional bounty – a vote of $20 million to be spent on challenging new defence projects – that might have been the end of the X-ray laser; but with Teller's support, Chapline was given a dedicated test of his own, scheduled to take place in some two years' time.

On Thanksgiving Day 1978, some ten weeks after Diablo Hawk, a group of senior Livermore scientists and managers were summoned from family, relatives and roast turkeys to a meeting in Edward Teller's office to discuss Chapline's new X-ray laser experiment. As well as Teller and Chapline, the group included Lowell Wood, Roy Woodruff, the head of 'A' Group, who was responsible for bomb design at the laboratory – and a recalcitrant Peter Hagelstein.

Hagelstein had not wanted to become involved. After three years at the lab, his theoretical understanding of X-ray lasers, based on computer modelling of experiments, was second to none, but he still resisted direct involvement in weapons work. Nevertheless, when Teller was excited about something, he expected his excitement to be shared and, as is illustrated by the timing of this meeting, he was still able to exert his will. As Hagelstein recalled:

> Teller ordered me to review the design calculations other people had done for Chapline and I said I wouldn't do it – that the lab had made an agreement with me and I wouldn't have to do that kind of thing. But – they put a lot of pressure on me. 'Look,' they said, 'all you're going to do is run your code.' Apparently it could do physics the other folk couldn't do. They just wanted me to run my code on it, check it out, that was all, they said. So I did.

Hagelstein knew that he was on a hiding to nothing. Failure to contribute willingly would probably result in his codes – the computer programs he had prepared to model the various combinations of design features for the laser – being requisitioned anyway. In contributing, however, he was helping someone he saw as a direct rival. He hoped against hope that, as he ran the Chapline design through his computer simulation, it would not prove too overwhelmingly successful:

> Lowell came by that night about 2 a.m. and he sensed something. 'H'm, you're in a good mood: something terrible has happened. You've calculated Chapline's design, you've found it doesn't work.' I turned bright red. Lowell had me pegged.
>
> 'So what's wrong?' he asked. 'You promise not to tell a soul?' I asked. 'I shall say nothing to anybody'...
>
> And so I showed him what Chapline and the folks had screwed up, and apparently the next morning. Lowell gets up and says, 'Well, Peter has found that Chapline's screwed everything up and this is what was done.' And this just fanned the flames of antagonism and animosity and so forth.

There was no doubt that the shy postgraduate student, working long hours at his computer, saw Chapline, the senior physicist, a well-regarded figure backed by Teller, as a threat. Chapline in turn found Hagelstein 'difficult. He had a persecution complex and believed I was persecuting him in some way. In fact, I had no problem in him sharing the work.' However, the animosity and antagonism that developed had a great deal to do with the close and competitive atmosphere that existed in 'O' Group.

Ever since he had helped set up the Hertz Fellowships, and the Department of Applied Science at the University of California, Davis (known as 'Teller Tech'), Teller had looked for a way to exploit the originality, the enthusiasm of the talented young scientists recruited into Livermore. 'O' Group had been part of the answer and Lowell Wood had been set to run it. It was a task which well suited his particular talents.

Wood had been a student of Willard Libby's at the University of California, Los Angeles, and Libby had introduced him to Teller, who in turn had brought him to Livermore. There he had rapidly earned a reputation as an iconoclast, at one period in his life shaving his head just once or twice a year to save on haircuts – as Teller himself had done some thirty years earlier. He, too, was a technologist, who also shared Teller's particular blend of political conservatism. 'Lowell is like another son to Edward,' Carl Haussmann, the Los Alamos veteran, commented in 1984. 'Edward has always been an activist. Ten or twenty years ago he was an activist in detail. Now he relies on other people like Lowell.'

Although Wood's own record in research – on laser fusion, for example – had not been exceptional, his ability to persuade and to inspire, as he had done with Peter Hagelstein, unquestionably was. He had an imagination with which to match the enthusiasms of the young members of 'O' Group, and they were an extraordinary group – 'very bright, very hard-working folks', says Peter Hagelstein, 'very motivated, at the very top of their schools from where they came – Cal Tech, MIT, and Carnegie and Stanford. Very top folks. Very interesting.'

All in their twenties, dressed in blue jeans, snacking off peanut butter sandwiches, and drinking Coke, they binged on all-night work sessions. Wood often joined them. At the time he was unmarried, and his lifestyle and his commitment matched theirs. He acted as confidant and taskmaster and had a mood for every occasion.

The journalist William J. Broad spent a week with 'O' Group back in 1984: 'Lowell was the kind of person who inspired either love or hate in those who knew him. I heard both. He had been called rude, crude, sarcastic, arrogant, cocky, crazy, irresponsible, and abrasive. He also had been called clever, smart, loyal, witty, patriotic, creative, sensitive, and deft. From my limited experience, he clearly had a lot of energy. He talked incessantly.'

The projects undertaken by 'O' Group ranged from computers that designed other computers to a solar-power project, from a deep-water sensing device for submarines to X-ray computers. This was Teller's brainchild and Lowell Wood's 'family'. There can have been few groups anywhere working on such a range of projects with so few resources, but Wood was going to ensure that his family beat off any competition – and that they took an idea as promising as Peter Hagelstein's to the limits.

Late in the summer of 1979 Chapline held his final design meeting and, as the design had now been radically altered, Hagelstein was asked to attend. 'I'd been working at night and this was daytime, and I'd been up too long – 20 hours,' he remembered. In the meeting it was as if his subconscious took over, a subconscious fed for four years on every detail of X-rays and laser design. In this dissociated state, viewing himself objectively as if from without, he watched as he talked about a new design concept.

Having provided such a crucial insight, Hagelstein was pushed to perform a detailed computer analysis of how different atoms might undergo laser action at X-ray frequencies. If only he would perform this task, then he could return to his real work, the work on the laboratory X-ray laser, and this time with proper resources. It was the start of a Faustian bargain. An unwilling Hagelstein set to work, hoping to prove that the process would not work. Instead he produced result after result that indicated his approach *would* work – that he had come up with a very good idea.

Before long he had established which metals would produce the best X-ray laser. The heavier the element, he discovered, the shorter would be the wavelengths and the more powerful the laser. Scientifically he was achieving special things, but he himself was despairing at what he was creating. His mood was not helped by a girlfriend at the time who argued that he was being used, that he should resign from the laboratory immediately. He took to playing sombre music in the lab, the requiems of Brahms, Mozart, Beethoven, spiced with the high emotions of Tchaikovsky's 'Pathétique' Symphony; but as he became increasingly depressed, so his work was producing increasingly attractive results. It was indicating a laser of sufficient power to do just what Teller hoped: obliterate missiles across vast stretches of space. Teller himself became increasingly interested, and he put more pressure on Hagelstein. 'The force has a powerful effect on the weak mind,' was how he characterised the effect of Teller's presence.

Within a short time it was decided to develop an experimental test of Hagelstein's invention, which could share the dedicated test the following year with Chapline's device. The test, codenamed Dauphin, was scheduled for November 1980, shortly after the presidential election.

In the early stages of his presidential campaign, Ronald Reagan had been taken to Cheyenne Mountain, the home of the North American Aerospace Defence Command (NORAD). In the 1950s and 1960s, before ICBMs had taken over from bombers as the core of the strategic nuclear force, NORAD had been the main Air Command Centre with responsibility for offensive as well as defensive nuclear strategy. By the time of Reagan's visit, however, things had changed. At this most impressive of facilities, buried in the heart of a mountain, he was told how, in the event of an attack on America, they would be able to accurately track the missiles as they started on their journey. He was also told how they would then alert Washington; but after this, he learnt, they could do nothing – apart from monitor the missiles as they homed in on their targets. It was a salutary experience for Reagan.

Martin Anderson, a colleague of Teller's at the Hoover Institution, was with him on the visit: 'Reagan reflected on the dilemma a President would face in the case of a missile attack. His only options would be either to direct our ballistic missiles to be fired in a counter attack or to do nothing except absorb the attack. President Reagan saw both alternatives as bad.'

Throughout his period as Governor of California, Reagan had always maintained contact with right-wing opinion through his membership of the American Security Council (ASC), and just before his NORAD visit he had joined the executive board of the Committee on Present Danger (CPD). This committee had been established some three years previously by the Washington legend Paul Nitze, and included a number of defence hardliners from both Democrat and Republican ranks among its members. Its formation had been triggered by an ASC report, in which Teller had participated, that had castigated George Bush's CIA for seriously under-estimating both Soviet strategic capabilities and the malevolence of their intent.

Reagan had been fully exposed to the CPD line that the Soviets had undertaken an 'unparalleled military build up' that was 'in part reminiscent of Nazi Germany's rearmament in the 1930s' and that 'the Soviet nuclear offensive and defensive forces [were] designed to enable the USSR to fight, survive and win a nuclear war'. In contrast, his visit to NORAD gave him a vivid impression of just how inadequate were the US's own defences.

He was to use the CPD rhetoric throughout his campaign, a campaign that brought him a landslide victory of such proportions that Jimmy Carter was making his concession speech before some of the polls in the western states had even closed. The landslide extended to Congress as well, where the Republicans took control of the Senate and substantially increased their representation in the House of Representatives. The new administration had enough political freedom to embark on a $1.5 trillion defence spending spree.

On 14 November 1980, a few days after Reagan's election victory, Livermore staged the Dauphin test. The preliminary readings immediately after the test indicated a successful outcome for both Chapline and Hagelstein's experiments – the creation of the world's first X-ray lasers had happened simultaneously. However, it was clear that Hagelstein's laser was much more powerful than Chapline's. Lowell Wood, obviously thrilled by his group's success, rushed back from the test to explain the details to Teller, and then took Hagelstein and other members of the group out to down-town Livermore for Baskin Robins ice cream. The group decided to name the new device after the legendary sword of Arthurian legend: Excalibur.

One of those as excited as anyone by the success of the test was Roy Woodruff. The X-ray laser was a success from his area, and a significant feather in the cap of a man with ambitions to be a future director of the laboratory. Woodruff was just forty at the time of the Dauphin test and the associate director responsible for nuclear design. Against the PhDs from the nation's top colleges who were among his charges, his bache-lor's degree from San Jose State College was ordinary. However he had shown an instinctive ability as a weapon designer as well as good organisational skills and had risen rapidly through the ranks. At one point he had designed the device for one of the last Plowshare tests. 'Roy was pretty much a straight shooter,' said Ray Kidder, one of Livermore's senior physicists. 'What they like to call a key player. They don't want a loose cannon. Too bad but understandable. But Roy did have a tendency, even with large projects at the lab, of telling Teller how it was, of not covering up the warts on a program. He was an enthusiast for the X-ray laser from the start but I don't think he was sufficiently deferential to Teller and he was not a good politician.'

Shortly after the Dauphin test, Woodruff established a new group that would have responsibility for work on the X-ray laser. Chapline and Hagelstein were to be the chief scientists under the leadership of Tom Weaver, another of the 'O' Group wunderkinds. It was a measure of the resources available to Teller that, within a short time there were a hundred scientists working within Weaver's new 'R' Group.

Now that they had something positive to sell, Teller and Wood began lobbying for some of the new funding promised for defence work. In February 1981 they set off for Washington to tout their new invention to Congress, accompanied by Woodruff. Both Teller and Wood were well-known for the way they would talk up a project, and he feared they might go too far. In fact, they behaved impeccably, offering properly qualified briefings to congressmen, exaggerating neither Excalibur's potential nor the speed with which the new X-ray laser could be developed. Woodruff was much relieved; but then, at the end of the month, an article appeared in the trade journal *Aviation Week* (widely known as 'Aviation Leak'), which described the top-secret Excalibur, even specifying the wavelength at which the new laser would work. The report then went on to maintain: 'X-ray lasers based on the successful Dauphin test ... are so small that a single payload on the space shuttle could carry to orbit a number sufficient to stop a Soviet nuclear weapons attack.' The possible agendas behind such a leak were legion, but such talk of a fully developed system after just one pioneering test was just the hype Woodruff had sought to avoid.*

Woodruff, however, was not able to chaperone every visit to Washington. As Teller shuttled back and forth to the capital city over the ensuing months, he steadily widened the groups to whom he talked. Those who ran the defence establishment should have made allowances for Teller's well-known enthusiasms, but those in the new administration were impressed by this scientific star of the Right. They were also in general sympathy with Teller's views, as they knew their President was. In fact, bearing Reagan's background in mind, Livermore and Teller had been relatively slow off the mark in trying to excite interest in a defensive shield.

In August 1979, while he was still building up his campaign, Reagan had received the draft of an article for the journal *Strategic Review*. It had been written by Senator Malcolm Wallop, a conservative from Wyoming. In his article he maintained that technology was rendering the balance of terror 'obsolete' and was promising 'a considerable measure of safety from the threat of ballistic missiles'. The particular technology he had in mind was the chemical laser, which would obtain its energy from the explosive combustion of a fuel mixture similar to those used to power rockets. In his article he boldly quantified just how many would be needed and when they would be ready. Twenty-four such lasers, he said, could 'conceivably destroy a whole fleet of ballistic missiles' – and the first of these could be placed in space as early as the mid-1980s. That summer, while on the campaign trail, Reagan had met Wallop at a barbecue near Lake Tahoe, and the Senator had urged him to make space-based defences a major theme of his election campaign.

Shortly after this another advocate had appeared on the scene. Daniel O. Graham was a retired lieutenant general who had been head of the Defense Intelligence Agency. Short, feisty and tenacious, his initials combined with his temperament lay behind his West Point nickname of 'Little Dog'. Well known for taking up extreme positions and defending them relentlessly, he had already been on Reagan's campaign team in 1976, and in 1979 he was invited to advise him again. At the time he had no specific plan to offer but worked to discover one.

His interest alighted on BAMBI, the acronym for the Ballistic Missile Boost Intercepts, which had first been mooted in the 1950s, during Eisenhower's administration. It was envisaged as a system of hundreds of space stations, each housing several small rocket interceptors. These would use infra-red sensors to track the exhaust of enemy rockets in their boost phase and then smash into them. The scheme had never been tested, but Graham took the basic idea and refined it. The great virtue of his revised concept was, he believed, that it could be assembled relatively cheaply 'off the shelf', modifying and using existing pieces of technology. During the campaign, Graham had the opportunity to discuss the idea with Reagan; then, following Reagan's success at the polls, he set about assembling a group to work on the project and to arrange funding.

One of his key recruits was Karl Bendetsen. The retired chief executive of a forest-products company, he was well known on the West Coast as a successful businessman and as a backer of conservative causes. Graham had first been introduced to him after giving a speech at a defence industry gathering. The two had then met again over dinner, forming an alliance to develop a higher profile for missile defence and to lobby for its inclusion in defence policy.

Bendetsen had just the sort of connections that would ensure that their arguments received proper attention from the White House. He contacted three of Reagan's oldest and closest friends: Joseph Coors, the brewing magnate from Colorado; William Wilson, the oil man who was both Reagan's neighbour in Santa Barbara and trustee of his finances, and Jacquelin Hume, who ran a nationwide groceries chain. Bendetsen also invited Edward Teller to join them. The two men had known each other for many years and, of course, Teller was known to Reagan. In May 1981 they established themselves as a non-profit group, and they gave it the name 'High Frontier', later moving it under the umbrella of the conservative think tank, the Heritage Foundation.

This close networking around Reagan did not stop there. Graham met with Edwin Meese, Reagan's chief counsellor, who agreed to approach more of the President's old friends and contributors for funding. Before doing so, he asked Bendetsen for a short explanation of the group's objectives that he could use. Bendetsen replied mentioning that they would be recommending 'space borne missile defense and other defense systems', which might 'present an historic opportunity for the President to announce a bold new initiative'. He also said that the group would be operating secretively, avoiding publicity, and that they would have their draft 'Reagan initiative' with Meese by November.

The normal way for such a proposal to proceed would have been through the Defense Department, asking them to study the issue; but this was not even considered. The department had never favoured antimissile projects, and so this private group of individuals, with Teller in their midst, opted to bypass it, instead taking their message straight to the White House.

This was a major coup for Teller – and not the only one. He had also served on the science and technology task force that guided the President on how to organise his scientific advice. Three months after Reagan's inauguration, the administration had completed all its appointments except that of scientific adviser. A dozen eminent scientists had been approached, but all of them had declined the position, offering excuses of age or personal problems. The real reason, however, was the growing sense of alienation between the new administration and the scientific community. To begin with, the White House task force had used its influence to close down the President's Science Advisory Committee, which, over the years, had provided the one channel of liberal scientific opinion into the White House. There were also rumours that funding and resources were going to be cut, and that the new adviser would find himself sandwiched in irresolvable disputes between the two camps. It looked to be an extremely unattractive assignment. In March, the two scientists leading the task force, William Baker and Simon Ramo, had gone to see Vice-President

George Bush, and, according to Baker, Bush 'went back to Ronald Reagan and the people around Reagan had suggested that Teller knew about this'.

This was an extraordinarily influential position to be in, and it had been presented to Teller on a plate. His candidate was George Keyworth, the forty-one-year-old head of the physics division at Los Alamos. Keyworth had helped with Teller's re-integration at Los Alamos and had become both an admirer and a protégé. Years later he admitted to the central role Teller had played in his appointment: 'Bluntly, the reason I was in that office is because Edward first proposed me, and the President very much admires Edward.'

The announcement of his appointment 'drew a mixture of surprise and unease from the scientific establishment', reported *Science*, which also noted that Keyworth was 'virtually unknown outside his field'. His supporters counted this a virtue: 'All he doesn't have is 20 years' membership in the club,' was Harold Agnew's comment. Owing Teller so much, he even described him to those he met as a father figure. It was therefore no great surprise that, when Keyworth announced the members of his fifteen-man White House Science Council, which effectively replaced PSAC, Teller was among its members.

The first formal meeting of the High Frontier group was held in July and by 14 September 1981 they felt sufficiently prepared for Bendetsen, Graham and Teller to meet with Edwin Meese at the White House. Meese was accompanied at the meeting by George Keyworth and by Martin Anderson, who had become Meese's Head of Policy Development. In his account of the meeting, Anderson recalled: 'I felt a rising sense of excitement. It became clear that not only did everyone feel we should pursue the idea of missile defenses, but they also deeply believed it could be done.' There was general agreement on objectives, most importantly that it was time to move from total reliance on offence to a balance between offence and defence, but also that a major part of any new weapons system should be based out in space. Initially the group concentrated their discussion on how to provide protection for MX missiles on the ground, but then Teller made a dramatic intervention. The ultimate goal, he said, was not to protect missiles but to protect populations. The goal was 'assured survival' rather than 'assured destruction'.

Once again, it seemed, Teller had 'brought up the props, banged the drum'. There were clear echoes of the H-bomb campaign. He had raised the game and was treading the narrow path between vision and foolhardiness in an attempt to excite the level of interest and support that was essential for such a major technological challenge. The phrase 'assured survival' conjured up the vision of an impenetrable shield, a dream of freedom at last from the nuclear threat. It was to dominate the whole Star Wars debate for years to come.

Some five weeks later, Meese received an extraordinary letter from Karl Bendetsen. Dated 20 October, it was a request for immediate financial support. In it Bendetsen wrote that, because of the 'urgency' of the High Frontier panel's findings, 'we are communicating them at this time, in advance of a full report'.

He then went on to describe Excalibur in some detail, describing how it would direct a fraction of the explosive yield of a small H-bomb 'into one or more tightly focused and independently aimable beams. Each of these beams,' he wrote, 'has about a million times the brightness of the bomb's undirected energy, so that the lethal range of a sub-megaton bomb can readily be extended to distances of thousands of miles.'

By implication all this detail came from the results of the Dauphin test – but, in reality, this was not the case. The test had provided only the most basic information and the letter was largely conjecture, which Bendetsen had nevertheless used to predict a central strategic role for the X-ray laser. 'This initial technology,' he wrote, 'would be useful against Soviet re-entry vehicles as they re-entered the atmosphere, against all Soviet satellites, and against ballistic missiles themselves as they are launched from Soviet submarines off the US Coast.'

In other words this 'initial technology' would not be effective over the long distances required to strike at the ICBMs in their initial boost phase, but Bendetsen was able to report on likely future developments: 'Second generation X-ray laser technology, which is expected to become available three years subsequently, will permit attack of Soviet boosters as they enter space from central Asian silos or from Soviet submarines in distant oceans.'

In the final part of the letter, Bendetsen made his request for funding, urging that $50 million over two years would result in a '15 month reduction in the time before this possibly pivotal technology can be deployed to defend our country from attack from space'. He then went on to conclude 'that X-ray lasers may represent the largest advance in strategic warfare since the hydrogen bomb itself'. The letter may have come from Bendetsen, but it clearly had its genesis with Teller, Wood and Livermore.

Bendetsen's precise time frame promised 'a fully weaponised' laser by 1986 and a second-generation laser by 1989. If there was then an injection of what, by defence standards, was a modest amount of cash, this schedule could be reduced by over a year. In other words, X-ray lasers would be ready for deployment by 1985 – in four years' time. Even by the standards of conventional nuclear weapons development, this schedule was, to say the least, ambitious; and the letter contained no mention of possible pitfalls or any discussion of the radar and other systems required to back up the core technology. This was Teller's guiding notion – that it is impossible to lie about the future – in play. If this proposal had gone through conventional government channels, some of these potentially serious issues might, at least, have been raised, but Bendetsen and High Frontier had circumvented this and gone straight to the President's office.

In doing so, they had also gained an advantage over those working up Graham's Global Ballistic Missile System. This group had, in fact, made good progress, outlining a system much more pragmatic than Excalibur. Two of its main elements were to be civil defence shelters and ground-based ABMs similar to the old Safeguard system. In space they had brought BAMBI up to date and had begun to pin down the detail. They envisaged 432 satellites, each garaging some forty or so miniature homing vehicles, which would intercept the enemy's missiles in their booster phase. Graham always conceded that the system would not be totally effective against a major onslaught, but it was to be assembled from existing technology and so had cost and speed of development to commend it. Bendetsen's letter, however, made no mention of this alternative scheme. This was no coincidence. It was a symptom of a power struggle going on within High Frontier itself.

Once Graham's plan had emerged in detail, a war of words developed at panel meetings between Teller and Graham. Teller sought to stir up doubts by attacking Graham's system as antiquated, expensive, impossibly heavy for delivery into space and, as Graham admitted, by no means totally effective. Graham, in turn, was the only member of the panel to criticise Excalibur. It was not easy to do so, certainly on technical grounds, as Teller had kept the results of the Dauphin test close to his chest, but Graham found reasons other than technical ones.

He countered that, as Excalibur involved an atomic explosion in space, it would simply not be acceptable to the public. He also argued that it had a 'fundamental flaw': its vulnerability to enemy attack. His system could at least defend itself out in space. An X-ray laser, however, carried no defensive weapon and its only defence was self-destruction. Graham linked this technical blind spot to a deep obsession on Teller's part: 'The man is carrying a load and has taken a lot of abuse as the "father" of the H-bomb,' he told journalist William Broad, some four years after the controversy. 'He wants to see nuclear technology turn out to be the answer in the opposite direction, to save the Western world.'

Apparently Teller had not thought of this seemingly self-evident problem, but by the next meeting, he and Wood were back with an answer: 'pop-ups'. The X-ray lasers would be earth-based, perhaps on a submarine, and at the sign of a nuclear attack they would be 'popped up' into space on a rocket. The title of this manoeuvre seemed almost perverse in the way it trivialised the immense engineering task of hoisting an X-ray laser into space and preparing it for action – all within a very short time. However, the panel was content to put their faith in Teller. He had certainly won the round, if not the fight, with Graham. Furthermore, Graham's criticism, with its implicit acceptance of the X-ray laser as a working technology, had reinforced the impression of Excalibur as a viable possibility.

In November the report of the High Frontier panel was, as promised, sent to the President. It too made no mention of Graham's system and, after all the in-fighting, indicated a personal triumph for Teller. He had usurped the competition and won over a close group of the President's old friends. Graham recognised his defeat; in December he split from the other members of the group he had established with Bendetsen and, while still retaining the High Frontier title, Graham and his advisers moved on to progress their scheme by exploiting his contacts in the military.

There was an unnerving silence from the White House after the High Frontier report had been sent. Then, on 7 January 1982, the group were told that there would be a fifteen-minute meeting with the President the following afternoon at 2 p.m. It was extremely short notice, and not every-one in the group could attend. There is some uncertainty over Teller's presence, but whether or not he was there in the flesh probably mattered little.* His ideas and rhetoric permeated both the report and the one-and-a-half-page digest prepared for Reagan.

The threat of Soviet strategic weapons was growing, the report said, echoing one of Teller's continuing refrains. Furthermore, the Soviets were apparently about to deploy 'powerful directed energy weapons' in space. These would allow the Russians to 'militarily dominate both space and the earth, conclusively altering the world balance of power'. The report then said that, in response, America must abandon its current strategy of 'mutual assured destruction' for one of 'assured survival' that relied on defence. The report then called upon the nation's 'directed energy efforts' to be greatly intensified with 'urgent action', noting that the average time for the Pentagon to select and acquire weapons was thirteen years. 'We cannot survive such delays,' it stated ominously. There was, it said, a need for a crash programme of anti-missile research similar in scale to the Manhattan Project.

The meeting developed into a lively one. The President had plenty of questions. The group was preaching to the converted over the need for defence, but Reagan's concern was viability. Would such a system work, and would the cost be affordable? According to Martin Anderson, who was at the meeting, by the end the President seemed to have the answer he was looking for.

Following the meeting, Bendetsen received an encouraging letter from the President, but then there was more silence. A month later he learned that George Keyworth had been asked to set up a panel from members of the White House Science Council. This was chaired by the physicist Edward Frieman, and included Harold Agnew, Charles Townes, the laser pioneer, and a number of others with links to laser projects, so it was hardly an unfriendly panel. Nevertheless, there was a strong chance that the panel would discuss the scheme to death and the project would lose its momentum. Teller, who had quite pointedly not been invited on to the panel, was concerned that they would not support his findings.

With the panel's report not due until the end of the year, 1982 was set to be a quiet one for anti-missile defence; but as the months passed, Edward Teller grew increasingly impatient with the situation. Every time the opportunity arose at meetings of the White House Science Council to talk to the panel chairman Edward Frieman, he pressed his case for Excalibur, reiterating his claim that it could be available in four to five years and would be the focus of an anti-missile revolution. Teller also repeatedly asked George Keyworth if he could arrange a private audience for him with Reagan, but without success.

Keyworth had spent much of his time as presidential adviser trying to reconcile the conflicts and dilemmas of being a place man. He may have been an admiring protégé of Teller's, but as he sat in on the High Frontier meetings, saying little but listening to discussions increasingly dominated by Teller, he had come to question what was being said. He questioned not only the technical viability of each of the strategic defence initiatives but also how they would be seen by the Russians and what their effect would be on the stability and balance of power. He also viewed Teller as too persuasive and Reagan as too vulnerable. A personal bond had always meant a great deal to the President, and if Teller was successful in forging one, anything would be possible. So whenever Teller mentioned a meeting, Keyworth avoided the issue.

Eventually Teller went public to complain. On 15 June he was interviewed for *Firing Line*, the television programme hosted by the conservative commentator William F. Buckley. During the show Teller warned that, because of lack of funds for anti-missile weapons, the Soviet Union was forging ahead. Buckley later picked up on this and suggested that Teller formulate an urgent message to Reagan, drawing a parallel with the warning Einstein had delivered to Roosevelt in 1939. The discussion then went as follows:

TELLER: May I tell you a little secret which is not classified? From the time that President Reagan has been nominated I had not a single occasion to talk to him.

BUCKLEY: Have you sought such an occasion?

TELLER: I have talked to people to whom I am close and who are in turn close to the President. I have tried what seemed to be reasonable to get action on these things. I may have been clumsy in one way or the other, but I am deeply grateful for any opportunity to speak about these things. I have lived through two world wars. I don't want to live through a third one.

Reagan had watched the programme and decided to grant Teller's wish. At 2.30 p.m. on 14 September, the two men met in the Oval Office. 'Mr President,' Teller said as he took Reagan's hand, 'third generation, third generation.' According to Ray Pollock, a National Security staffer present at the meeting, who reported this opening line, it somewhat threw the President. Teller, certainly, was less than confident about how things progressed from that point on:

That thirty minute meeting was far less successful than I had hoped [he wrote]. Quite a sizeable group was assembled in the President's office: President Reagan, Vice President Bush, Judge William Clark [Reagan's new National Security adviser], Attorney General Edward [Edwin] Meese, Jay Keyworth, and Sydell Gold, a National Security staff member responsible for nuclear matters. I did my best to present the information about the developments in defense in a non-technical manner and to emphasise the timeliness of making an effort to pursue research on strategic defense. But Ms Gold injected so many questions and caveats that I felt discouraged about the conference.

Ray Pollock recalled that the President seemed to accept what Teller had told him – but he also recalled William Clark expressing doubts and questioning him closely about his third-generation weapons. Later Keyworth, who must have been extremely nervous about the meeting, called it a 'disaster'. Teller asked Reagan outright for a dramatic increase in funding for the X-ray laser and Clark and Meese felt it necessary to cut the thirty-minute meeting short.

Afterwards Teller met up with Lowell Wood and Greg Canavan, who were both in Washington at the time. 'Their recollection is that my mood was bleak,' Teller recalled. 'I had had my chance and I had failed to convey to the President the gist of the information and the potential that it held.' Nevertheless, Teller followed up the meeting with a letter to the President expanding on details of X-ray laser research and ending: 'I dare to look forward to a decision both timely and favorable regarding American exploration of "Third Generation" nuclear weapons technology ...' He then circulated the advice he had given to the President to other senior figures in the administration. Although it might not have appeared so at the time, it was advice that events across the nation and in Washington were making increasingly apposite.

After some eighteen months in power, Ronald Reagan's administration was fast gaining a reputation for being hawkish and profligate. As it spent its trillions on defence, the country had entered a deep economic depression, one of the deepest since the 1930s. The gross national product was down, while unemployment, at nearly 9 per cent, was the highest since the start of the Second World War. Reagan's personal approval rating was, at 41 per cent, the lowest for any post-war President in his second year of office.

On 12 June 1982 between half a million and a million demonstrators rallied in New York's Central Park, protesting against the nuclear arms race and calling for a halt to the testing and production of nuclear weapons. Then, less than two months later, a resolution expressing support for such a nuclear 'freeze' came within two votes of passing in the House of Representatives.

By the end of the year, a synod of Roman Catholic bishops was drafting what was to become a pastoral letter on war and peace. 'We fear that our world and nation are headed in the wrong direction ...' the bishops said. 'In the words of our Holy Father, we need a "moral about-face". The whole world must summon the moral courage and technical means to say no to nuclear conflict: no to weapons of mass destruction ... and no to the moral danger of a nuclear age which places before humankind indefensible choices of constant terror or surrender.' It was a message that caught the essence of the nation's growing anti-nuclear sentiment.

Then, in December 1982, the House rejected a request from the President for $988 million to produce the first MX missiles for deployment. It was this vote that finally convinced Reagan of just how difficult it was becoming to fund offensive land-based missiles. Matters were reaching an impasse, and his advisers were looking for anything that might soften his hawkish image without alienating the conservatives who had brought him to power. Thus a technological solution that offered the humane notion of 'assured survival' yet came from a well-known scientific hawk and was backed by Reagan's conservative friends was just what they were looking for. At a meeting with the Joint Chiefs of Staff in December the President asked them to give serious consideration to a strategy with a greater reliance on defence.

◆

On 23 June 1982, as part of their year-long investigation, the Frieman panel had gathered in the offices of their chairman in La Jolla, to hear about Excalibur from the Livermore scientists working on it. Tom Weaver, the head of 'R' Group, had led their presentation – one that created a very different impression from the one Teller had been broadcasting.

The High Frontier letter to Meese in October 1981 had suggested that $50 million might be enough to shorten the delivery period for the X-ray laser by over a year. Weaver, on the other hand, talked of a 'technology limited pace' of development, which would be made possible – but not hastened – by continued injections of between $150 million and $200 million a year. This would allow the ten underground nuclear tests considered necessary to establish the device's scientific feasibility to be carried out within six years. At the end of the series, Weaver said, the laser should indeed be a million times brighter than the radiation from an undirected bomb, but this was only the second of six or seven steps on the road to a fully weaponised device. It would be followed by major engineering – much of it pushing technological limits – to ensure firstly that the delicate laser mechanism would survive the stress of the supersonic 'pop-up' into space, and secondly that the cluster of beams could be individually aimed at their distant speeding targets within a matter of seconds. Weaver had estimated that this vital phase could well take up to a decade to complete, making the 1990s the earliest possible delivery of a weapons system. If the financial support proved variable, it might be much later.

Weaver's prognosis was a far cry from the letter to Meese, which talked of deployment by 1985; and so it was that, in the late autumn, the Frieman panel delivered a draft report, describing the X-ray laser as 'blue sky' technology. In fact it was heavily critical of all the strategic-defence projects it considered – the Graham and Wallop projects as well as the X-ray laser – and expressed the view that none of them was mature enough to have an immediate effect on defence strategy.

When Teller heard the panel's findings, he became furious – 'unglued', as one member described his reaction – and threatened to resign from the Science Council if the panel did not review the X-ray laser again. In order to avoid a political storm, particularly among those important conservative figures who were Teller's supporters, Frieman agreed to go back for a second Livermore briefing. In the meantime, Teller focused his anger at Livermore not on Weaver but on Roy Woodruff, under whose auspices 'R' Group operated.

'Roy, I may never speak to you again,' Teller said when Woodruff answered the summons to his office. 'If I do, it certainly won't be this year.' And that was it. Woodruff was dismissed with a wave of his hand. Later that afternoon his invitation to Teller's seventy-fifth birthday celebration a couple of weeks thence was withdrawn: it had suddenly become an intimate affair – just a few friends and top managers. Woodruff had offended and such spats between the two men were to become increasingly frequent, but was Teller's anger justified? Peter Hagelstein, who was witnessing events from within 'R' Group, found his reaction at least understandable:

> It was essential to keep the ball rolling. From Teller and Wood's point of view it's how they did everything else. You tell everybody what they want to hear to keep things going, knowing that what you're after is not precluded by the laws of physics, even though you haven't gotten very far. What's more, you feel there's no reason why you're not going to progress much much further, if only you can put in a little more effort, or a lot more effort, or decades more effort. Who knows how much more effort?

Teller believed that enough of his past success had depended on this type of determined strategy, and that, at the same time, there would be a race with the Russians, who would also be testing the limits of the same laws of physics. That was, after all, what had happened with the H-bomb. But, in spite of Frieman's draft report, he must still have felt that Excalibur was very much afloat. He had just been contacted by the Chief of Naval Operations, who had asked if Teller would come and explain third-generation weapons to him. So a few days later Roy Woodruff's invitation was reinstated. He and his wife attended the party and recalled Teller at his amusing and charming best, the confrontation seemingly forgotten. Someone had given him a toy duck on the end of a stick and Teller moved through the guests waving it around, flapping its wings and laughing at his own silliness.

Admiral James D. Watkins, Chief of Naval Operations, was a devout Catholic who attended mass daily and who had been much affected by the bishops' letter. He had raised the issue at one of the twice-monthly prayer breakfasts the Joint Chiefs shared together, where he had found others were also concerned that they seldom reviewed their responsibilities from a moral standpoint. This had reinforced his growing concern. 'Mutually assured destruction has never been a concept that I could understand,' he confessed. 'I don't think it's morally sound.' He had also learnt that enlisted men and officers had been feeling the same doubts – indeed, they were leaving the Navy because of the bishops' stance.

This was a time when the proposals Teller had widely broadcast among those with top clearances were very much the topic of the moment, and one of Watkins's long-range planners had suggested that the Admiral himself hear what Teller had to say.

The two men met in Watkins's office on 20 January 1983. Teller's presentation of his basic ideas for anti-missile defence lasted almost an hour, with only an occasional question from the Admiral to break his flow. Watkins, an engineer by training, recalls him shaking with excitement as he spoke, like a reed vibrating at its resonant frequency. At one moment Teller dipped into his past, making a pointed comparison between the frustration he was feeling then with the frustration he had experienced when trying to persuade the Truman administration to go for a crash programme on the Super. After a while the two men adjourned to the flag mess across the hall and continued talking over a lunch of shrimp-stuffed sole provençale. Here they discussed what they both knew from their different intelligence sources of Russian activities in the anti-missile field. The consensus view was that, in certain areas, the Russians might be as much as ten years ahead of the US. For Watkins the heady combination of hearing at first hand about a brand new technology, of hearing its inventor pointedly comparing the present situation with the epoch-making moment in the development of the Super, and a shared deep concern about Russian progress, was exhilarating. According to Teller, his final comment was, 'I will tell the President.'

A few days later Watkins sat on the porch of his house in the Washington Navy Yard enjoying the winter sun with Robert McFarlane and John Poindexter, two of William Clark's assistants at National Security; but although McFarlane was encouraging Watkins to persuade the Chiefs to pursue anti-missile defence seriously, he was not motivated by the same moral qualms. Rather, he shared with Teller a much more political and pragmatic view of Excalibur's usefulness. He saw a full-blooded pursuit of such a grand design as a way of convincing Moscow of the sheer financial power and technical superiority of the US. Excalibur would be wonderful technology when – if – completed, but in the intervening five, ten, twenty years it could be the iron fist in a velvet glove, sharpening and breathing new life into dialogue with the Soviets.

On Saturday, 5 February, the Joint Chiefs of Staff listened to Watkins as he laid out his arguments for a shift towards strategic defence. He himself was by no means sure that his arguments would break the impasse over the siting of the MX missile that had blighted the Chiefs' previous forty-two meetings. However, when he finished, much to his surprise, he received a unanimous vote in favour. In fact, the other Chiefs were at their wits' end, desperate to avoid having to tell the President that they were still deadlocked over future plans. They happily agreed to adopt Watkins's position as their own and present it to Reagan.

The meeting on 11 February, therefore, between Chiefs of Staff who saw strategic defence as a tactical means of breaking a deadlock, and a President who was also sold on the concept of a defence initiative – partly for public-relations reasons, partly out of personal conviction – had something of a scripted quality about it.

The briefing paper General Vessey read out followed fairly closely the lines of the Watkins paper, saying that the aim of defence was 'to protect the American people, not just avenge them'. (The President particularly appreciated the rhetorical qualities of this line.) Watkins himself then spoke in support, saying: 'It seems to me that it's possibly within reach that we could develop systems that would defeat a missile attack.'

McFarlane, who was present at the meeting, then came in on cue, adding helpfully that he believed 'Jim [Watkins] is suggesting that new technologies may offer the possibility of enabling us to deal with a Soviet missile attack by defensive means.'

'I understand,' the President replied; 'that's what I've been hoping.'

Each one of these groups had been briefed by Teller – the Chiefs through Watkins, the President and his aides directly. They may have differed in their interpretation of just how mature the particular technologies were, but Teller had allowed them to believe in something they each needed, even something that they wanted.

Although the meeting ended in agreement, however, it was agreement of an ill-defined kind. The Chiefs expected their proposal to be worked over, given flesh by staffers. This, too, was what Macfarlane expected – that others would then frame a research project which, however immature, would provide him with his bargaining counter with the Russians. The President, however, had other ideas. He saw their agreement as a mandate for an immediate statement, and in mid-March he told his astonished aides that he was now ready to announce a Strategic Defense Initiative publicly.

In February Hans Bethe had made the journey out to Livermore. He had been invited by his old adversary to come and offer his assessment of Excalibur. Only four months before, Bethe had been provoked into publishing a recently declassified assessment he had made thirty years ago of who had been responsible for what in the development of the Super, and he and Teller had once again locked horns.* In spite of this it was Bethe's opinion, above all, that Teller wanted. He stayed at Livermore for two days and his investigation was thorough and demanding. Teller sat in on some sessions, but Lowell Wood was with him throughout.

'Bethe had hassled a lot of people in the programme and wasn't getting any satisfaction,' Peter Hagelstein recalled. 'It took me a little while to figure out what in the world he was asking . . . but I thought, well, I know the numbers will work out. So I turned to the blackboard and said, "Here's the answer and let's try to work out everything that leads up to it." ' Forty minutes later Peter had finished. 'By the time I was done the equations had come out right. Hans Bethe was very pleased and impressed, and his criticisms had fallen apart.'

Certainly Bethe's curiosity had been satisfied scientifically, but he still maintained strong doubts about how complete a defence Excalibur would ever provide. Following his visit, he told *Time* magazine, 'I don't think it can be done. What is worse it will produce a Star War if successful.' With this last comment, Bethe was heralding a debate that was just beginning.

At the same time as Bethe's visit, the Frieman panel reported on its second Livermore assessment. If anything, their second report was even more negative than the first, but by now the momentum for strategic change, stemming from both the White House and from the Pentagon, was unstoppable.

On 23 March, at 6 p.m., a White House limousine dropped Edward Teller outside the South West Gate and he made his way through the chill dusk to the Blue Room. There he joined a puzzled group of some forty eminent scientists and politicians, none of whom had any idea why they had been invited to dine with the President at such short notice. Teller had an inkling because, in the brief phone call inviting him, George Keyworth had said, 'Edward, I can tell you this. It's what you always wanted.' When, a short while after he had arrived, Keyworth announced to the group assembled that an historic change in national defence policy would be announced by the President in a televised address at 8 p.m., Teller broke into a grin. He now knew what was going to happen.

'I have become more and more deeply convinced,' the President said in his address, 'that the human spirit must be capable of rising above dealing with other nations and human beings by threatening their existence.'

He then used the phrase first employed by Watkins:

Wouldn't it be better to save lives than to avenge them? Are we not capable of demonstrating our peaceful intention by applying all our abilities and our ingenuity to achieving a truly lasting stability? . . .

What if free people could live secure in the knowledge that their security did not rest upon the threat of instant US retaliation to deter a Soviet attack, that we could intercept and destroy ballistic missiles before they reached our own soil or that of our allies?

I know this is a formidable task, one that may not be accomplished before the end of the century. Yet, current technology has attained a level of sophistication where it is reasonable for us to begin this effort . . .

I call upon the scientific community of our country, those who gave us nuclear weapons, to turn their great talents now to the cause of mankind and world peace, to give us the means of rendering these nuclear weapons impotent and obsolete.

This last passage, a clear invitation to laboratories like Livermore, was the President's own. At the end Teller was both moved and excited. He had fed a political need as well as presidential dreams and this had swept a piece of 'blue sky' technology from laboratory obscurity to a point where it was the foundation for a new national-defence initiative; but in a matter of only minutes this initial excitement was dampened by the scepticism he heard expressed all around him.

'Immediately after the President's speech, Secretary of State George Shultz, with whom I was acquainted, asked me: "Can this system be 100 per cent effective?" My response: "Against the largest attack that the Soviets can mount? Probably not." '

26
Excalibur

'Will there be any security?' This was Edward Teller's first question before
he would venture from his car on arrival to speak at the University of
California's Irvine campus on 3 April 1985. A few weeks earlier, during a
debate at New York University, he had again had food dumped on his suit
by a protester. But at the sunny, placid Irvine campus, there were no such
problems, so he went on to warn a standing-room-only audience of some
500 that the Soviets were well ahead in the arms race, a race America had
to win if they were not to be destroyed by or ruled from Moscow. It was
unreconstructed material of the kind he might have delivered a quarter of
a century earlier, unaltered by the fact that the Soviet Union now had a
new leader, Mikhail Gorbachev. The audience greeted it just as warmly as
it had Hans Bethe's far more conciliatory address a few weeks earlier.

Afterwards, around lunch at the University Club, the seventy-seven-
year-old battled with questions from fourteen academics from the university's Institution on Global Conflict and Co-operation.

'Can you have confidence in SDI,' one asked, 'a system that's never been
tested?'

'The answer is yes.'

'You're recommending something that appears to be suggesting we participate in a continuing arms race.'

'Of course, of course,' Teller replied, looking up from his meal. 'Technology continues. If you want to use the words "arms race" in a *negative*
sense, of course it continues.' Further, he considered it a scandal that the
most ingenious among his colleagues were prepared to sit back and do
nothing: 'Bethe and Garwin and others prevent people from joining our
lab.' His thick Hungarian accent, emphasised by his resonant tone, dominated the conversation. At one point there was the relief of laughter when
his use of the word 'threat' was misinterpreted by one of the group as 'sex'.

Another academic, Lawrence Howard, who taught a course on nuclear weapons, tried an ironic little scenario in which he asked the President: 'Why not begin a programme to achieve immortality? I can't share all our secrets. But all we need is a few billion and we can guarantee immortality by the end of the century.'

Teller looked at Howard warmly. 'You and I seem to be on the same wavelength,' he said smiling. 'You're throwing back the same argument I am using.'

'Would you be convinced?' Howard asked.

'I'm convinced by what you're saying, but I would be even more convinced if you could tell the secrets,' Teller responded.

'But,' Howard asked, 'doesn't it come down to the credibility of the person who is promising you immortality?'

So many unspoken issues hung in the air during this exchange captured by the journalist Susan Cohen. Howard's scenario had been carefully calculated to be provocative. His parallel between the dream of immortality and the defensive shield Teller was offering reflected the deepening scepticism about SDI, two years on from Reagan's Star Wars speech. His last comment on credibility was a direct dig at Teller's own. So why didn't Teller rise to the bait, as he so often did? Did he really fail to see the irony?

There was perhaps a hint of the reason in an aside he had made in his speech that morning, which had passed unnoticed by most of his audience. He had said that the X-ray laser would be highly accurate in hitting and destroying targets. This was something he must have said many times, but on this occasion he had gone on to say more: that such lasers 'exist[ed] not on paper' but were feasible. 'Three weeks ago,' he had added, 'I couldn't have said that.' Something had happened, about which, in his excitement, he had probably already said too much – something that had lifted the pall that had hung around Excalibur ever since Reagan's speech.

There was little doubt that the X-ray laser was the jewel in the crown of the Strategic Defense Initiative, and that without it the score of other projects that now made up the Initiative's portfolio would not have got off the ground. When *Newsweek* reported Reagan's speech, an artist's impression of an X-ray laser in action, its rods firing furiously at enemy missiles, adorned its front cover. Excalibur had no equal in generating the feeling new technological frontiers were being breached.

But as well as inspiring, the project could also disappoint. On 26 March 1983 Livermore detonated its second X-ray laser test and the first anyone had carried out since the Dauphin test two and a half years previously. Codenamed Cabra, the test had been trailed in advance as the largest underground test that year and took place only three days after Reagan's speech – but it was a failure. New sensors failed to produce any meaningful result that would help convince the sceptics and the project was in danger of stalling. It was Roy Woodruff who saved the day by extracting enough funding from the Energy Department for the experiment to be repeated.

The failure of Cabra did not prevent Teller from appearing, a fortnight later, before the research subcommittee of the House Armed Services Committee. 'A great change in the national defence situation is impending,' he stated sweepingly at the outset, giving credit to young colleagues such as Lowell Wood and Tom Weaver, both of whom were with him at the hearing.

Knowing that one of the major factors in the public's mind counting against Excalibur was that it seemed like a sophisticated nuclear bomb, he was at great pains to explain that third-generation directed weapons involved only small explosions but achieved their effect from 'the enormous concentrations of energy and temperature' they produced. 'Thus the third-generation weapons can be used not for mass destruction but to destroy very specific targets such as offensive weapons in action.'

He also rationalised his concerns about the advances made by the Soviets in the laser field by pointing out what had happened to the three scientists who had shared the Nobel Prize for discovering the laser, one of whom was an American: 'Charlie Townes publishes beautiful papers about what is happening in the centre of our galaxy. Basov and Prochorov are working with the Soviet military on lasers. Your guess who is currently ahead.'

He ended by openly criticising both his competitors and his opponents, attacking Wallop's chemical laser and Graham's rocket interceptors for their use of space stations permanently in orbit around the earth. 'To develop great battle stations for use in space is an outlandish idea,' he said, warning that they would be open to attack. This, on the other hand, would not be the case with Excalibur, which would be light enough to be 'popped up' into space just before it was needed. In so saying, Teller neatly turned the tables on his critics. He had re-used the 'pop-up' notion, which had been devised to fend off Graham's initial criticism of Excalibur, to point up the vulnerabilities in Graham's and the others' schemes.

He then went on to target the critics among his fellow scientists, in particular Hans Bethe. Teller described how, after Bethe's two-day visit to assess the X-ray laser, he had withdrawn his technical objections: 'He said in front of me, "You have a splendid idea." But did he change his public position? No . . . For every Bethe there are a hundred others who speak up and who don't even know the basics of what they're talking about.'

It was a strong, aggressive performance, but it was immediately and publicly eclipsed by a powerful whiff of scandal. As Teller left the Senate hearing room, he was quizzed by a *New York Times* journalist not about his comments on SDI but on his stockholding in a high-tech company,

Helionetics, which specialised in laser technology. A damaging article followed, pointing to the handsome profit that Teller had made on the market. It described how Helionetics' owner had given stock worth millions to 'leading scientific and military experts', how Teller had helped to persuade President Reagan to adopt SDI, and how the stock had risen 30 per cent immediately prior to the President's Star Wars speech.

Initially Teller issued a straightforward refutation. In it he said that he had bought the stock before Reagan came to power, that he had not known the contents of Reagan's speech in advance, and that the stock had increased in value because of the technical advances the company had made. At first the *Times* ignored his explanations, but some three weeks later it included a short item reporting that Teller had done nothing improper in his dealings with Helionetics.

Frustrated and angry at the way the *Times* article had blunted the impact of his Senate appearance, Teller believed he had been the victim of a spoiling tactic. So he sought the backing of Accuracy in Media, an offshoot of the right-wing American Security Council, which sought to counteract what its supporters saw as a liberal bias in the media. With their help, he published another full-page advertisement in the 31 May edition of the *Wall Street Journal*.

In a play on the title of his previous advertisement concerning Three Mile Island, the inch-high headlines this time proclaimed: 'I was NOT the only Victim of the New York Times'. The text was a refined version of his earlier refutation, but he had added the fact that the day after the *Times*'s first story, Radio Moscow had broadcast it worldwide with each misstatement and innuendo embellished. 'The *New York Times*, of course, is not directly responsible for this,' the text acknowledged, 'but when the *New York Times* – through misinformation – totally deflects attention from information of utmost national importance, discredits testimony given to Congress, and provides grist for the Soviet propaganda mill, the list of victims includes more than those directly attacked ... Indeed I was not the only victim of the New York Times.'

Throughout this period there is an embattled feeling to Teller's actions. Perhaps the situation carried echoes of the time he had been pushing for the Super and surrounded by opposition from much the same groups as he was now. For a whole year, in 1950, Teller had struggled under the burden of Truman's directive for a crash programme, without having a real solution for satisfying the hopes he had raised. With SDI, Reagan himself had spoken of the end of the century as a possible time frame, but Teller must have known just how tenuous Excalibur was, supported by just one test result staged three years previously.

Only a week before his advertisement had appeared in the *Journal*, he and the key members of High Frontier, Bendetsen, Coors and Wilson, had met with Reagan to urge rapid action on the goals set out in his speech. They had spent much of the time discussing how they might bypass the federal bureaucracy. Then, the following day, the President had awarded Teller the National Medal of Science, the country's highest scientific honour, for his 'leadership in science and technology'. Nevertheless, however well Reagan thought of Teller personally, SDI remained materially poorly funded. As one official said to the *Washington Post*, 'He's low-keyed the program'; and that was to remain the situation for the next eighteen months: Reagan hardly mentioned anti-missile defence publicly at all.

What did happen, however, was that Caspar Weinberger, Reagan's Defense Secretary, initiated three study groups on SDI to assess both the existing schemes and also the new proposals that had inevitably begun flowing in. One panel was chaired by Fred S. Hoffmann and the brief for his panel was to lay the strategic foundation for SDI. Another was chaired by a former director of NASA, James Fletcher. His co-director was Harold Agnew, who recalled the bemusement and confusion of the time:

> They had leased this building and had gotten 200 people involved, and as a chairman I would go in. They had all these committees on detection and on warheads and a whole going organisation. I was amazed. I put up with it for a year, then I resigned. The thing that triggered me was they were looking at a warhead they were confident was going to be an anti-proton warhead. It was just scientific nonsense, nonsense, and a waste of money. The whole thing was, as far as I was concerned. But it was what the President wanted, so everybody saluted.

The members of both the Hoffmann and Fletcher panels came from either universities or industries involved in defence work, so they might have been expected to be friendly to new technologies. However, the Hoffmann panel came close to rejecting the entire project, their unclassified executive summary reporting: '... nearly leak-proof defences may take a very long time or may prove to be unattainable in a practical sense against a Soviet effort to counter the defense.' They relegated the President's goal to some indefinite future, calling instead for a ground-based interceptor system with the limited objective of defending missile silos.

Taken at face value, the Fletcher panel's seven-volume report was far more supportive, concluding, 'The scientific community may indeed give the United States the means of rendering the ballistic missile threat impotent and obsolete.' Following up on this summary, both Weinberger and Reagan's scientific adviser, George Keyworth, publicly stated that the President's 'dream' was a realistic goal. However, a journalistic investigation revealed that the summary did not in fact reflect the content of Fletcher's report, which was actually deeply pessimistic about the chances of over-coming even basic problems. When asked about this discrepancy years later, Fletcher himself stated that the summary had been written by no one in the committee, but by 'someone in the White House'.

One of the panel's major concerns was that X-ray laser beams would not penetrate the Earth's atmosphere. They certainly had the potential to work over great distances in the vacuum of space but, given the curvature of the earth, this meant that unless the enemy missiles rose high above the atmosphere during their booster phase (see figure 6), the positioning of the laser had to be either high in the atmosphere or close to its target; but the enemy missiles could simply avoid the laser beams entirely by flying on a lower trajectory, using the Earth's atmosphere as a protective blanket. As to the 'pop-up' principle, Gerold Yonas, the physicist chairing the sub-committee investigating Excalibur, was quite definite: 'Fast burn boosters would outwit it. The X-ray lasers could never get there [out into space] in time – period.'

Yet in spite of its damning conclusions the Fletcher panel paradoxically recommended an expenditure of no less than $26 billion on the whole SDI venture over the seven remaining years of the 1980s, with $1 billion of this to be spent on the X-ray laser. So what lay behind the paradox? The reason, essentially, was Cold War paranoia.

However fundamental the objections they had found, the panel did not feel that they had necessarily been totally exhaustive. They worried that an enemy might still find an effective, even crucial, role for the laser, which they had missed. For instance, X-ray lasers could well be used to devastating effect against other anti-missile weapons stationed out in space. An X-ray laser stationed in the upper atmosphere would be protected from attack from space by the atmosphere, and yet its freshly generated beam could 'bleach' out into space and there damage the enemy. There were just so many possibilities that the panel felt that they had to cover themselves.

And so it was that the X-ray laser not only survived the scrutiny of the Fletcher panel but was the one project among the host of ideas, hopeful of consideration for funding, that they recommended for priority treatment.

On 27 March 1983, just four days after the Star Wars speech, the Soviet leadership lashed out in direct response. Uri Andropov accused the United States of preparing a first-strike attack and President Reagan of 'deliberately lying' about Soviet military power to justify SDI. Space based defence, he said, 'would open the floodgates of a runaway race of all types of strategic arms, both offensive and defensive'.

His attack was unprecedented. He broke a long-standing taboo by describing both US and Soviet weapons in considerable detail and, for the first time since 1953, he told his nation that the world was on the verge of a nuclear holocaust.

Ever since Reagan had come to power, the Soviets had been subjected to provocation after provocation, from Reagan's own 'evil empire' rhetoric to the US determination to deploy intermediate-range missiles across Europe, but had remained relatively calm. Alexander Haig, Reagan's first Secretary of State, declared himself 'mind-boggled by their patience'; but the Star Wars speech had been a provocation too many, particularly as it had highlighted the problem of Russia's lagging high technology. In an interview with an American journalist, Chief of the General Staff Marshal Ogarkov admitted that 'We will never be able to catch up with you in modern arms until we have an economic revolution. And the question is whether we can have an economic revolution without a political revolution.'

That September, the Soviets' shooting down of a Korean Airlines plane with sixty American citizens on board – even though the CIA showed it was an accident – intensified US belligerence. Secretary of State George Shultz denounced it as "deliberate murder and the President called it 'an act of barbarism.' The situation was then further aggravated two months later, when US and NATO forces began their annual exercise, codenamed Able Archer 1983. The Soviets were familiar with such exercises but this one had two unusual ingredients. It was initially to involve high level officials including the Secretary of State and the President and Vice President and it was to entail a full-scale simulated release of nuclear weapons. According to the double agent, Oleg Gordievsky, the KGB concluded 'that the American forces had been placed on alert – and might even have begun a countdown to war ...' US intelligence picked up on the volume and urgency of communications traffic and on the fact that Soviet fighter aircraft with nuclear weapons in East Germany and Poland were placed on alert.

Through his contacts in UK intelligence, Gordievsky warned the British Prime Minister, Margaret Thatcher, of the genuine fear of war in the Eastern bloc and she, in turn, spoke to the President. For him it was an epiphany. 'Do you suppose they really believe that?' he asked Robert McFarlane, his National Security Adviser.

The Soviet fear of impending nuclear attack was in fact so real that, according to Hungary's last Communist Foreign Minister Gyula Horn, a group of Soviet Marshals had openly advocated an attack on the West 'before the imperialists gained superiority in every sphere.'* It was certainly the closest East and West had come to nuclear conflict since the Cuban Missile Crisis and Mikhail Gorbachev wrote, 'Never, perhaps, in the post-war decades was the situation in the world as explosive, and hence more difficult and more unfavourable, as in the first half of the 1980s.'

This was the situation at the end of 1983. SDI had played a crucial role in creating a crisis of confidence between East and West. While other contributing factors would fade as time passed, it would remain a major item – for good or evil – in the power play between the two. It was at this time that George Shultz began tentative approaches to the Russians and discovered just how serious their concerns were about the 'development of large scale ABM systems' – a reference to SDI. Over the next six months, during the build up of Reagan's 1984 re-election campaign, Presidential rhetoric softened, as he asked for a better understanding with the Soviets, while Shultz worked patiently at bringing the Russians back to the negotiating table.

This period was an extraordinary test of Edward Teller's optimism. He may have dismissed the findings of the Fletcher panel as weak, but he knew the weight they carried. As a possible antidote he formed a group of long-time Livermore colleagues to assess the X-ray laser; but they, too, were critical. Ray Kidder, a veteran Livermore scientist specialising in assessing new projects was a member of the group:

> When you start looking at it, you find there is a sequence of logical steps which are unavoidable. So how long is a rocket in its boost phase, how long have you got? The boost phase of an ICBM, if I remember, is a couple of minutes. How far away is it? Then you have the curvature of the earth. You're going to have to be 50 miles up, maybe more, and the target has got to be 50 miles up too because you can't shoot at it in the atmosphere. So you already find that the time line is pretty close, but then the first countermeasure they can employ is go to a faster booster. Their boost phase can be speeded up. So the next question is, does he have the rockets to do this? The next thing he can do is to go up until he gets to a low enough atmospheric pressure and then he can start angling over on to a lower trajectory so he gets more protection from the atmosphere. So he's got plenty of evasive manoeuvres. So I reported this to Teller, how unfavourable it all was, but at this point he was very reluctant to accept what I was saying. I was very much the skunk at the party.

Figure 6: The Stages of a Missile's Flight
During the boost phase (1–2 minutes), the rocket drives its warhead payload into space. There, the protective nose cone falls away and the 'bus' carrying the multiple warheads continues on its mid-course trajectory (20–25 minutes). As it approaches its target, the warheads, which are shrouded in radar-reflecting balloons and accompanied by similar decoy balloons and 'chaff', separate from the 'bus'. During the final re-entry stage (1–2 minutes), the warheads home in on their target, travelling at some ten times the speed of a bullet.

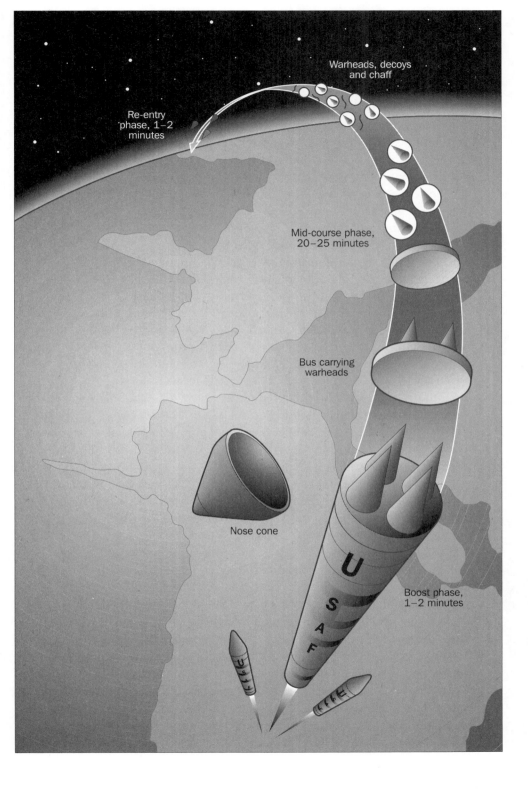

Warheads, decoys
and chaff

Re-entry
phase, 1–2
minutes

Mid-course phase,
20–25 minutes

Bus carrying
warheads

Nose cone

USAF

Boost phase,
1–2 minutes

What Teller needed desperately was positive experimental evidence, and his hopes were pinned on the Romano test, scheduled for 16 December. Romano was the fourth test in the Livermore series and had been specifically designed to prove X-ray lasing. The amplification of the X-ray beam along a lasing rod depended on the length of the rod – the longer the rod the greater the chance for amplification and the more powerful the resultant beam. The Romano device had been constructed with a variety of rod lengths, and if the strength of the beams varied accordingly, it would be proof that lasing was taking place.

The test was carried out at the Nevada test site at 10.30 a.m. By late afternoon an excited Lowell Wood was back at Livermore with the first results, poring over them with Teller, Weaver and Woodruff. Even though they would take weeks to analyse fully, the overall picture was clear: 'We knew we had a tremendously successful experiment,' Woodruff recalled.

On 22 December 1983, obviously in elated mood, Teller wrote to George Keyworth. The letter was on Livermore paper and it was marked 'SECRET': 'Dear Jay, Merry Christmas. This may be the first classified Christmas greeting you have received,' Teller wrote, adding that his Christmas present to Keyworth was news of the first proof of the existence of the X-ray laser. Then came a phrase which was to echo down the years. The three key parameters measured in the test, Teller wrote, were in 'essentially quantitative agreement' with predictions, showing that X-ray scientists understood the principles underlying the laser's action. As to the length-versus-gain signals, there was 'no other theory except that of the laser which could explain these results'. Teller then added a carefully judged comment on their success: 'What our results may mean is not that we are geniuses at Livermore, but that too many people may have overestimated the difficulty of the job.' In that one sentence he both justified Livermore's persisting with the work and suggested the ease with which the Soviets might achieve the same level of success.

He then appealed for funding as, so far, the grant recommended by the Fletcher panel had not materialised. 'I agree that science cannot be sped up by throwing money at it,' he wrote, 'but we are now entering the engineering phase of X-ray lasers where the situation is different.' As Tom Weaver, the leader of 'R' Group, had told the Frieman group a year earlier that the phase of scientific prior to this exploration was going to take at least five years, Teller's statement had considerable significance. He went on to say, 'We have also developed the diagnostics by which to judge every step of the engineering progress,' indicating that they even had new techniques to hand that would help keep the engineering phase on course. It was not a risk-free scenario he was offering, but it was an optimistic one. All that was needed, he said, was the funding already recommended by the

Fletcher panel to make it happen – $50 million more for the current year and $100 million for the following year.

His letter made no direct mention of the caveats expressed by the various panels or the group at Livermore, but his summary did carry an unspoken understanding of criticisms that would be known to Keyworth: 'I do not believe that the X-ray laser is clearly the only means, the best means, or even the most urgent means for defense,' he wrote. 'It is clear, however, that it is in this field that the first clear-cut scientific breakthrough has occurred. It is necessary to draw all the possible consequences from this fact.' Finally, he offered a reminder: 'We are now in the stage where money talks.'

Teller sent the letter, but even though he made substantial – even out-landish – claims for a Livermore project and had asked for money, he did not copy the letter to anyone at the laboratory or elsewhere. His aide, Robert Budwine, was very concerned. 'Roy, the old man's done it this time,' he confided to Woodruff.

Woodruff was furious, and went straight down to confront Teller in his office. There he argued that the engineering phase was still years away, and that even after the Romano test there were still far too many scientific uncertainties for his claims to be backed. The tests, so far, had not even shown that the laser would ever be powerful enough to be used as a weapon – and Teller was wrong to assert that there was precise quantitative agreement between lab predictions and the field results obtained – quali-tative, yes, perhaps, but not quantitative.

Teller was caught off guard and did actually agree that parts of the letter contained exaggerations, but he refused to send any kind of corrective letter. It would, he argued, be damaging to his main mission on behalf of the laboratory: to obtain the necessary funding for Excalibur. According to Woodruff, Teller also said, 'My reputation would be ruined.'*

When Woodruff returned to his office, he drafted a corrective letter, which he intended to send to Keyworth; but Livermore's director, Roger Batzel, refused to let him send it. Keyworth was well aware that Teller was a 'technical optimist', Batzel said, and he had also always been one of the project's sceptics, however much he felt he had to support the President's dream. However reluctantly, Woodruff accepted Batzel's decision. 'At this point it was not a fall-on-your-sword kind of issue,' Woodruff said.

A month later, in February 1984, Woodruff sent Keyworth the full results of the Romano test. He reported that two out of the three parameters measured in the test – the laser beam's colour and shape – had been found to be in 'excellent quantitative agreement with predictions'. True, the results for the third parameter, the beam's brightness, were not so positive and were described by Woodruff as only 'in solid qualitative agreement with

predictions'. Nevertheless, the original differences that had fuelled his confrontation with Teller seemed to have narrowed considerably. The criticism
from others, however, continued to grow.

In the spring of 1984, the Livermore physicist George Maenchen, a specialist
in sensing devices, began examining the interaction between the high-
powered X-ray beam and the sensors used on the Romano test. He started
to suspect that the interactions were such that it might be producing a false
reading for brightness. His fears were borne out by some erratic results
produced in another test, codenamed Correo, which had been run by the
Los Alamos laboratory. Here too the scientists had achieved positive results,
but further analysis had shown that their sensors were giving a false reading.
The question was whether these results were the exception or the rule. If the
latter, the whole programme might be moving forward on a false premise.

Outside Livermore, a scientist working in another defence laboratory,
the Argonne, near Chicago, had used basic unclassified information about
the X-ray beam to estimate just how much the laser might diffuse and
spread over 2000 kilometres – the kind of distance over which it would
need to operate. Edward Walbridge had used information about the wavelength of the laser and the length of the lasing rod to calculate that, over
that distance, the beam would diverge by 60 metres, weakening its intensity
considerably. This meant that the bomb used to pump a device with fifty
lasing rods, producing fifty beams powerful enough to be lethal against
missiles, could not be the tiny device talked about. It would have to be big
enough, in fact, to produce a blast of 3.7 megatons. Further, if he was right,
at least twenty such devices would be needed to take out a 1000 missile
attack – so Excalibur would need no less than 74 megatons to stand a chance
of achieving its objective. Walbridge had conjured up a space Apocalypse of
terrifying proportions.

When Lowell Wood called Walbridge up after he had read his findings
in the 19 July 1984 issue of the science journal *Nature*, 'He was aggressive
and hostile,' Walbridge recalled; but while he criticised the findings, he
offered no hard evidence in support of his comments. It was, of course, all
classified.

So acrimonious did matters become that the dispute surrounding SDI
became known as the 'Science Wars'. One group, the Union of Concerned
Scientists, of which Richard Garwin and Hans Bethe were members, made
its own highly critical study of the whole range of strategic defences on
offer. They concluded that overall there was little prospect of either a
perfect or even near-perfect defence system. In one section the Union's
report revisited earlier criticisms of the chemical laser and again performed
calculations that ridiculed the project. According to them, an effective

system would need, not the ninety battle stations its advocates were proposing, but 2400, each weighing up to 100 tons and costing $1 billion apiece – approximately 2.4 trillion dollars.

This was an accusation that had to be countered. When the scientists at Livermore and Los Alamos looked at these figures, they found an error and the Union had to reduce its estimate, first to 800 battle stations and then to 300. Robert Jastrow, a physicist close to Livermore and to Teller, accused them of 'shoddy work' aimed 'in one direction only – toward making the President's plan seem impractical, costly and ineffective'. Garwin, however, responded by offering a recalculation of his own in which he did not make the improbable assumption, made in the first calculation, that the system would work perfectly. This time he assumed a level of failure and his estimate returned to 2263 battle stations. He also jibed at Jastrow for having made 'a career of hyena-like behaviour'. The Science Wars were bringing alive all kinds of old enmities.

There were still sharp divisions over SDI in Washington too, but these had nothing to do with the practical considerations exercising the scientists. By now, the political implications of SDI had taken on a life of their own. Ever since the Star Wars speech, the notion of a defensive shield had provided the substance for a particular kind of political rhetoric used by Reagan. Everyone now knew that the President was for defence, for making nuclear weapons obsolete. Reagan himself believed he could now argue with the Russians from a position of strength, both moral and technical, and encouraged Secretary of State Shultz in his efforts to mend bridges with the Soviet Union.

In the months following his re-election in November 1984, members of Reagan's administration, including those who had been most disturbed by his 1983 speech, fell in line with him. They referred to 'the President's vision' and to his 'ultimate goal'. Even Defense Secretary Weinberger described SDI as 'the only thing that offers any real hope in the world'. But behind this seeming act of blind faith, there were still factions at work, each with a very different reason for maintaining SDI, whatever its technical validity.

Weinberger probably had the most straightforward reason for supporting the Initiative. While he professed a belief that it might eventually provide a reliable defence, he could also see a value in the partial defences it might offer in its intermediate stages. No one was going to spend the vast sums envisaged on something that might only provide a defensive circle for existing missiles, but this was what, at some stage, SDI might be able to do. Weinberger was happy to exploit anything from its development that might prove useful.

George Shultz had a different agenda. As he had worked, with considerable success, to re-establish a dialogue with the Russians, he had found that they continued to be spellbound by the SDI programme. The Soviet leader, Constantin Chernenko, had written to Reagan protesting about 'the development of large scale ABM weapons', while the spectre of 'space weapons' and 'the militarisation of space' were constant themes at meetings with Soviet officials. Shultz had begun to realise that this fear of a US breakthrough in space-based weaponry might encourage the Soviets to trade their offensive missiles for a reduction in this new threat.

Richard Perle, a well-known hardliner in Weinberger's office, was one of those who had opposed the 1972 ABM treaty. He believed the country's best interests would have been served by a continuing arms race involving both defensive and offensive weapons. Now that the Soviets were taking SDI seriously, its very existence could block any plans Shultz might have for an arms agreement if it could be turned into a long-term development programme. Its existence could also be used to wreck the ABM treaty. Perle was to become one of those most clamorous supporters of SDI, always reporting great progress and turning meetings with other agencies into sales pitches.

Throughout his life, Edward Teller's overstatements, his exuberance and optimism had always been a crucial element in his achieving his goals. Now he was in a special position, playing a central role both in guiding the X-ray laser research at Livermore and also in having access and credibility in Washington. He knew that a project had to maintain its technical momentum if it was not to lose its way politically. Thus, ever since the Fletcher panel's report, Livermore had been trying to increase the laser's brightness so that it would be able to cut through the protective blanket of the Earth's atmosphere. They had already been working on increasing the size of the nuclear device pumping the laser, but they had also begun studying another possibility: reducing the spread of the beam by focusing it.

Lowell Wood had set 'O' group working on the problem and they had first produced the concept known as Excalibur Plus, which would create beams not a million but a billion times brighter than the undirected radiation from the bomb. This proposed thousand-fold increase in brightness was followed shortly by a further development, known as Super Excalibur. The paper that outlined this plan was written by four members of 'O' Group, including Peter Hagelstein and Tom Weaver, and it described a laser which would produce a beam that was a further thousandfold brighter, a breathtaking trillion times brighter than the undirected bomb radiation.*

Super Excalibur would have such extraordinary firepower that it could be placed in geostationary orbit 22,300 miles above the equator, permanently positioned over its target area. Even from this distance, its creators believed, it could cut through the atmosphere to destroy missiles in the vulnerable boost phase shortly after launch. Being so far out in space would also confer on Super Excalibur much greater security than other space systems that were in orbit close to earth.

On 21 September 1984 Teller was sent a top-secret memorandum describing the new plan. It was just the kind of development he needed, an exciting advance to sidestep the problems plaguing the original concept. When, in October, he spoke to the ferociously anti-Russian Hungarian community from around Silicon Valley as part of his support for Reagan's re-election campaign, he made it clear that he believed the US now had the means to defend itself against a full-blown missile attack. 'I wish I could tell you how ingenious they are,' he told them, 'how promising.' Then, on 15 October, Lowell Wood travelled to Washington to meet with the head of the Pentagon's SDI Organisation, Lieutenant General James Abrahamson, to present Super Excalibur to him. It was the first of a continuing round of visits Wood was to make over the next year, enthusing and informing administration officials, keeping up the momentum.

On Reagan's re-election, Teller himself made contact with senior defence figures in Washington about the new developments at Livermore. On Friday, 28 December, he phoned Paul Nitze, still the senior arms-control adviser both to the State Department and to the President, and told him to expect important news in a letter that was being brought to him by Lowell Wood. No chance, then, of it languishing in a pending tray. In fact, Wood not only delivered the letter personally to Nitze, but also arrived equipped with visual aids to give Nitze a full presentation on Super Excalibur.

The letter itself, again bearing 'SECRET' stamps both top and bottom, began with a more than averagely optimistic account of Excalibur, and a typical warning that the Russians were ahead in laser technology; but then Teller came to the meat of the letter, in which he stated that he would not have written 'in so urgent a manner' unless there had been 'a final consideration which [was] very little known in Washington': Wood's team, he explained, had achieved a breakthrough that promised 'a real prospect of increasing the brightness' by as much as a trillion times, making 'X-ray lasers a really telling strategic defence technology. For instance, a single X-ray laser module the size of an executive desk which applied this technology could potentially shoot down the entire Soviet land-based missile force, if it were to be launched into the module's field of view.'

The dichotomy between this heady vision and the reality in the laboratory could not have been more marked. While Livermore scientists were arguing among themselves about whether the laser beam would ever be bright enough to be useful as a weapon, Teller was expanding his vision into the fantastic. It might be 'devastatingly effective', he wrote, not just against missiles in the boost phase but also against literally tens of thousands of warheads and hundreds of thousands of decoys that would emerge from the missiles once out of the boost phase and on their journey through space. 'It might be possible to generate as many as 100,000 independently aimable beams from a single X-ray laser module,' he wrote, 'each of which could be lethal even to a distant hardened object in flight.'

In concluding his letter to Nitze, Teller wrote, 'I felt that you should be aware of the possibilities of such striking advances, both the ones already in hand and the even more impressive ones in reasonable near-term prospect, before you go to Geneva.' However much Teller may have known about the Soviet concerns about SDI, he certainly wanted to provide sustenance for its use in the Geneva talks, which were to be the prelude to a summit later in the year; and in sending the letter to Nitze he had targeted the most experienced arms-control negotiator of them all. In addition, Paul Nitze had developed the basic 'criteria' used to assess, on a cost-effectiveness basis, whether new defensive technologies were worth developing: if they could be made more cheaply than a set of counter-measures against them, then they were worth considering. It was an approach to overall strategy that was soon to be adopted as national policy, and Teller's vision was, in theory, an answer to Nitze's prayer: a single module that would shoot down the whole Soviet land-based missile force. But would Nitze find it credible? – or would he be yet another who would disregard the technological fantasy and still use SDI in discussions as Teller intended?

On the same day as he had written to Nitze, Teller had also written to another carefully chosen target, Robert McFarlane, who had recently become Reagan's National Security Adviser. McFarlane had always regarded SDI as a useful bargaining counter, an emblem of America's technological invincibility. In his letter to him, Teller again offered news 'of urgent importance' and sketched in the technical advances, from Excalibur to Super Excalibur, that had been made since he had briefed Reagan in 1982. 'Assuming even moderate support, together with considerable luck,' he wrote, 'this [Super Excalibur concept] might be accomplished in principle [within a few years].' Again it was a prediction that seemed much at odds with the situation at Livermore.

This time Teller copied both his letters, again couriered by Lowell Wood, to Roger Batzel. For the second time Woodruff found out what happened from Teller's aide, Robert Budwine, and he drafted another corrective letter, complaining in essence that Teller and Wood were 'overly optimistic' about both the time frame and the budget. As to Teller's vision of a single module capable of shooting down the whole land-based missile force, he wrote: 'Will we ever develop a weapon close to the characteristics described? Not impossible but very unlikely.' As before, Batzel refused to let Woodruff send his letter, but he did allow him to visit Nitze. In February, they met for two hours in Nitze's State Department office and Woodruff stressed his view that Super Excalibur was a fantasy. 'He had real doubts about whether the project was worthwhile,' Nitze recalled, saying that he took Woodruff's reservations seriously. 'I thought he was a bright guy. It's always good to get a bright sceptical mind on a problem.'

In March 1985, after less than two years in power, Constantin Chernenko died and was replaced by Mikhail Gorbachev. At fifty-four he was a full generation younger than his predecessors and some twenty years younger than Reagan. Articulate, energetic and well dressed, he often appeared in public with his attractive wife, Raisa, and he had an affinity for the media. After having met him some three months before he took office, Margaret Thatcher had famously commented, 'We can do business together.'

When Gorbachev came to grips with SDI, he found a conflict of opinion among his own scientists and military that very much echoed the dispute in the US. There were those, for example, who pointed to the vulnerability of space-based weapons, and argued that they could be dealt with cheaply and effectively using counter-measures. On the other hand there were those who saw the whole strategy as an attempt by the US not to defend their own but to obtain, over time, a dominating first-strike capability. For them, SDI was seen as part of a new policy of aggression, a view supported by the Reagan administration's heavy financial investment in it. They wouldn't be spending billions of dollars for nothing at all – would they?

Early on, Gorbachev had been exercised by these conflicting views and, according to his US ambassador, Dobrynin, had expressed this view privately: 'Maybe it is time to stop being afraid of SDI? The United States is counting on our readiness to build the same kind of costly system, hoping meanwhile that they will win this race using their technological superiority. But our scientists tell me that if we want to destroy or neutralise the American SDI system we only would have to spend 10 per cent of what the Americans plan to spend.'

Dobrynin reported that, perhaps predictably, it was the Soviet military–industrial complex that persuaded him against this plan and instead to concentrate his efforts on removing SDI. It was this objective that came to dominate US–Soviet arms-control discussions for the next two crucial years.

In January 1985 Lowell Wood sent a memo to an engineering colleague at Livermore, Fritz Ritmann. Using the time-honoured means of introducing a memo's subject, he had titled it 'Re: CHARGE!'

He then went on in effervescent prose to pass on the good news that Batzel had 'provisionally' decided to feature Super Excalibur in the upcoming round of congressional budget hearings. The project, he wrote, 'needs to be accelerated to the maximum extent practicable'. The memo was not just sent to Ritmann but used to advertise Wood's buoyant mood to twenty-eight other senior Livermore officials. In closing, Wood referred skittishly to the group working on the project as the 'High Brightness Conspiracy'. Was this title meant to be taken as a college-style joke, or was it the symptom of something more obsessive? Opinions varied at Livermore from outright disbelief that Super Excalibur would amount to anything, to expressions of hopeful interest. Outside Livermore, opinions were even more negative. At Los Alamos, the failure of their Correo test, particularly given their initial optimistic findings, had provoked deep misgivings about the whole X-ray laser programme. This was something that emerged during the aftermath of a meeting between senior officials from both Livermore and Los Alamos at the Nevada test site at the end of January.

When business discussions were over, the group went for a steak dinner together and then returned, already well lubricated, to the Los Alamos dormitory for further social drinking. The two laboratory directors, Roger Batzel and Donald Kerr of Los Alamos, found themselves sitting opposite each other and the conversation veered rapidly and dangerously towards Excalibur. Kerr, a tough, straight-talking manager who was trying to revitalise his own laboratory, began by being heavily critical of Teller for overstating the virtues of Excalibur. 'We fired him when he was a group leader at Los Alamos,' Kerr said, 'and you, sir, rehired him.'

Batzel was stunned by the directness of the attack, while Woodruff responded by asking Kerr, sarcastically, how old he had been during the Manhattan Project. But a fired-up Kerr continued, saying that it was unconscionable that Batzel allowed Teller and Wood to continually and widely misrepresent Livermore's work at governmental level.

'I would fire Lowell Wood for what he has said,' Kerr continued, adding that Teller should go as well. Woodruff again responded, saying that Kerr must really mean he would put restraints on the couple.

'No,' Kerr had replied. 'They should be fired.'

Woodruff, who shared Kerr's criticism of Teller's and Wood's actions, if not his proposed solution, found himself having to defend them. Their creativity was an asset to the nation's security programmes, he said; between them they had brought new blood to the laboratory, which was doing remarkable work. In the end, however, he had to admit to his own concerns about the overselling of Excalibur. There should be more discussion, he said.

At that point the group, embarrassed by this public display of animosity and rivalry, broke up, the Los Alamos group heading off to a bowling alley while Roger Batzel went in search of a beer. He found the bar closed. The true feelings of those who knew enough about Excalibur had broken surface. Kerr's anger reflected not only a personal antagonism towards Teller and Wood but also a real fear of long-term damage that was being inflicted on the credibility of both the national laboratories.

The discord within Livermore continued as Wood, backed by Teller, maintained his claims for Super Excalibur while bidding at review meetings for the extra funding needed to turn the vision into reality. Much to Woodruff's annoyance, he proposed that part of the funding should be diverted from funds devoted to basic nuclear-weapons research – work that was under Woodruff's control.

'These are very real possibilities pointed out by Lowell,' Teller said at one review meeting held on 30 January 1985. 'The X-ray laser has produced entirely unexpected results ... This work should be supported at much higher levels ... and we can go to Washington and talk about what's been done ... There can be no argument among us.' But later that afternoon, Woodruff had weighed into Wood's claims, saying that 'he was heading off on a divisive path', and concluding that his plan would wreak havoc. 'There's a dark side to him that needs to be kept under control.' But there were others at the meeting who felt much more in sympathy with Teller. Carl Haussmann, a veteran from Livermore's early days, felt that his vision was being overshadowed by short-sightedness: 'Edward was right when he observed that the real crux is that X-ray lasers deserve grossly more funding than they're getting. This is an example of the malaise that's present today in this country.'

This gave Teller the opportunity to underline his vision: 'The whole point is whether defence can be more economical than the offsetting measures.' And while Reagan was President, he had the best option to see that vision through.

In the financial tug of war that followed, Woodruff stood his ground and warded off Wood's claim on his budget. At every opportunity he also tried to bring a sense of proportion to the consideration of the various Excalibur concepts. At the Senate's SDI Working Group in mid-February, he listed the best possible 'concept validation dates' – the end of the pre-engineering phase – assuming full funding. Excalibur was listed as 1992, Excalibur Plus as 1995, but against Super Excalibur there was a question mark. With SDI's central role in the political agenda of US–Soviet relations, however, such technical niceties were not foremost in Washington's mind; and Woodruff's conclusions were about to be challenged by the latest Excalibur test.

◆

The test, codenamed Cottage, was one of the most elaborate underground shots ever. It was an agglomeration of experiments whose diversity reflected the growing confusion and discord surrounding Excalibur. There were two experiments using different methods to improve the focusing and therefore the brightness of the beams. Alongside these were others aimed at testing new sensors and improving the ones already used. There was also one aimed at answering George Maenchen's concerns that the existing sensors had so far delivered false brightness readings. The date fixed for the firing was Saturday, 23 March, the second anniversary of Reagan's Star Wars speech. Unusually, Teller accompanied Lowell Wood to the Nevada test site.

From their initial analysis of the results, the two scientists were euphoric. They believed the focusing experiments had been a success, and would provide the hard evidence needed to demonstrate that Super Excalibur was a realistic possibility. Wood and Teller flew into Livermore airport the next day, to be met with a champagne reception; then, on Monday, Wood was en route for Washington to brief SDI officials. Later in the week Teller also travelled to Washington, keen to use the results from Cottage to lobby the new Energy Secretary, John Herrington, for extra funding.

It was a few days after this, on 3 April, that Teller made his appearance before the audience at the University of California's Irvine campus, where he boasted that the X-ray laser existed 'not on paper' but in reality, and ignored the carefully devised taunts about his own credibility. As had happened with the hydrogen bomb and Polaris, he must have felt that his extraordinary persistence and optimism had paid off once more.

A short time later, a number of key journalists reporting on defence matters received an invitation to a special party. 'O Group's High Brightness Conspiracy' invited them to a Gala Celebration of 'High Brightness Events coinciding with the second anniversary of President Ronald Reagan's Strategic Defense Speech of 23 March 1983'. The event was to take place on 20 April at 'The Sty in the Sky', Wood's pet name for his home in the hills above Livermore. Given that Excalibur was a top-secret project, both Teller's and Wood's lack of discretion was indicative of just how bullish they both felt.

In June, Teller had another opportunity to enthuse the President about Super Excalibur as well as to personally lobby for more funds. Reagan was sufficiently impressed to promise him $100 million extra on the spot, an exceptional gesture of good faith. While in the White House, Teller had called upon Robert McFarlane, whom he had not yet met, again to make his case for extra funding. The National Security Adviser was not favourably impressed. 'He was very excited about the experiment and made very extravagant claims,' McFarlane recalled. 'He said with additional infusions of money, we could demonstrate the feasibility of [Super Excalibur] in

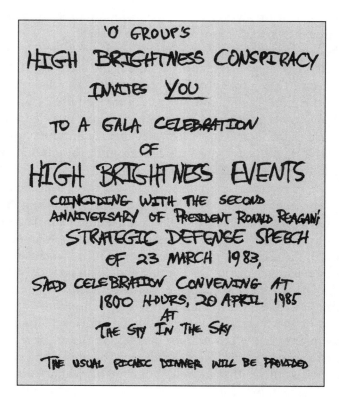

'O GROUP'S
HIGH BRIGHTNESS CONSPIRACY
INVITES YOU

TO A GALA CELEBRATION
OF
HIGH BRIGHTNESS EVENTS
COINCIDING WITH THE SECOND
ANNIVERSARY OF PRESIDENT RONALD REAGAN'S
STRATEGIC DEFENSE SPEECH
OF 23 MARCH 1983,
SAID CELEBRATION CONVENING AT
1800 HOURS, 20 APRIL 1985
AT
THE SPY IN THE SKY

THE USUAL PICNIC DINNER WILL BE PROVIDED

high Earth orbit. My instinctive reaction was that he was not an impartial analyst. He was an advocate.'

But Teller's salesmanship, now backed by experimental data from the Cottage test, ensured a strong funding base both for the project and for Livermore in the years ahead. In the fiscal year 1986, the Department of Energy's funding for directed weapons, the bulk of it for Excalibur, rose from 1985's figure of $215 million to $270 million – plus an extra $100 million from the President, and in 1987 the figure was to be $560 million. In the six years to 1991 the project was scheduled to have received a total injection of $3.7 billion.

The collapse began a month after Teller had met with the President. It was instigated by Los Alamos, who had pursued problems with the sensors shown up by the Correo test with a vigour inspired by the years of jealous rivalry between the laboratories, a rivalry sharpened by the fact that this project was Teller's own brainchild. Their conclusions, published in a report dated 10 June 1985, centred on the reflectors, made of the hard metal beryllium. These were positioned towards the end of the evacuated pipe that carried the beam away from the explosion, in order to reflect a small part towards the various sensors lying off to the side.

Beryllium has molecules of oxygen occluded on its surface, and the Los Alamos scientists suggested that these were being excited by the radiation from the beam. They glowed brightly at exactly the wavelength of the laser itself and therefore provided a strong source of false brightness. This, however, was Los Alamos's best-case analysis. At least this scenario accepted the existence of the laser beam – however dim. In another, however, they argued that it might be the raw radiation from the bomb that excited the oxygen and that no lasing had occurred at all.

At Livermore there was anger and a sense of betrayal over the Los Alamos findings. 'This issue had been addressed and dealt with during the very first shot,' Peter Hagelstein explained. 'This claim, that it made data from a wide range of our experiments null and void was nuts. Los Alamos was never significant in the science, and this was one of several wild claims that had nothing to do with science and everything to do with politics.' Frustratingly, George Maenchen had experimented with new sensors which had used hydrogen gas to deflect the beam into nearby sensors, thus avoiding the distortions of the metallic mirrors. Thus far, however, his results were too crude to answer the Los Alamos charges. So the anger at the mounting criticism turned inward, exacerbating a growing feeling among those working on Excalibur that Lowell Wood in particular was overselling the project. In fact, his attitude in general was causing problems. In May 1985 Woodruff's assistant, George Miller, reported on the situation in 'O' Group: 'An explosion of personal resentment may be imminent with an attendant loss of manpower and consequent failure of the program. The problem is that his style is absolutely monolithic, uncompromising and frequently abusive personally to the other scientists in the programme. He has little tolerance for those less brilliant and less well-spoken than he.'

This view is backed up by Peter Hagelstein:

At the time I felt compromised. It seemed that all the sacrifices that were made, the blood that was spilled was my blood. The lack of ethics, permeating the upper parts of all these programs – I couldn't believe people behaved in that way.

One issue was to do with friendship. I thought Lowell was my friend, but I was ostensibly befriended such that I would do what folks wanted – or whatever. Lowell kept promising things I was interested in, but he knew he couldn't deliver – the laboratory X-ray laser program for instance. And when there was eventually a laboratory program, I was excluded.

I realise now that my expectations were too high ... but at the time I was very bitter.

It was just at this time, as the disillusion and anger among the staff and the serious technical problems began to threaten the future of Excalibur, that the first summit meeting in Geneva between President Reagan and Premier Gorbachev was taking place. Envisaged as the first in a series at which the two men were to take the measure of each other, it lasted for only two days in mid-November 1985. Little hard negotiating took place, the main achievement being the apparent 'bond' that was quickly established between the two men. Such a bond was inevitably something of a diplomatic construct, but Gorbachev was so much younger and more flexible of mind than his predecessors and Reagan was such a past master at affability that they achieved a plentiful dialogue, if not a true meeting of minds.

On the first day they touched on SDI. This had continued to obsess the Soviets, and the accounts of the success of the Cottage test had been read with considerable alarm. In as offhand a manner as possible, Gorbachev immediately went to the core of the matter, telling Reagan that he should have no illusions about bankrupting the Soviet Union, or about achieving military superiority through space weapons. Reagan replied tangentially, talking about the origins of the Cold War and putting the blame squarely with the Soviets.

The following day they returned to discussing SDI, and this time Gorbachev made his case in more detail. Did Reagan realise, he said, how costly it was going to be? He would be putting a trillion dollars into the hands of the military–industrial complex, and to what end? Was he seeking a first-strike capability? If so they, the Soviets, would match the threat with programmes of their own. 'You are trying to catch the firebird with technology,' Gorbachev said. 'How can we go before the world and say we lost the chance for 50 per cent reductions because we wouldn't stop research on space weapons?'

'How can you defend a chance for 50 per cent reductions just because you were stubborn [about a] research [program]?' Reagan responded, before embarking on an impassioned speech about the virtues of defence rather than relying on the ability to 'wipe each other out'. When he finished, there was a long silence.

Robert McFarlane has written that Gorbachev 'had to conclude one of two things: either Reagan was being cynical with all his preaching about eliminating nuclear weapons, and his real intention was to bankrupt the Soviet Union; or he was incredibly ignorant'. Whatever his conclusion, McFarlane says, Gorbachev knew he had to put up with Reagan. 'Mr President,' Gorbachev finally said, 'I don't agree with you, but I can see you really mean what you say.'

They then moved on to talk about Afghanistan. The next time they would talk about SDI would be a year later at Reykjavik.

◆

Three weeks before Reagan and Gorbachev had met in Geneva, Livermore hosted the annual Nuclear Explosive Design Physics Conference held by the National Laboratories at the end of October. At the conference the Los Alamos scientists presented a report reiterating their claims about oxygen fluorescence and questioning the very existence of the X-ray laser.* This was the first time that many of the Livermore scientists had heard the details of their argument, and the effect was explosive. It became the talking point of the whole meeting, as many faced the possibility that they had been working on a dead-end project for five years.

Los Alamos did not spare anyone's blushes. Paul Robinson, their head of weapons research, handed George Miller a letter, saying that they doubted if even the existence of the X-ray laser had been proven and they also said that Livermore managers were losing their credibility for not facing up to Teller and Wood.

For Woodruff it was the last straw. He had battled with Teller and Wood for years and received little support. The following day he handed his letter of resignation to Roger Batzel. In it he described the deep frustration of having no direct responsibility for Teller or Wood and yet being criticised for not standing up to them – and being held ultimately responsible for their false promises. He then criticised Batzel himself, bluntly telling him that he had not kept promises to moderate Wood 'to the detriment of national security'. There had been a vacuum at the top, he said, and ended by stating: 'I can no longer support you as Director.'

The resignation was announced the following afternoon at a meeting of senior managers, where Batzel and Woodruff simply said that they had irreconcilable differences. No details were given and few scientists at Livermore knew why Woodruff had resigned. The reasons, however, were soon to emerge from articles in both the national papers and the scientific press.

The *Los Angeles Times* reported the tensions between Woodruff and Wood. Although not formally in charge of X-ray laser research, Wood, it reported, nevertheless had 'immense indirect influence' over the project 'buttressed by his connection to Teller and the high-level access both men enjoyed within the Reagan administration'. Woodruff, however, did have responsibility for the project and 'said he was tired of Wood's end runs to Washington'.

The *Times* article also described the essence of the false-brightness problem, saying it had been known about for some time, and that scientists had urged that the upcoming $30 million Goldstone test, scheduled for the end of the year, be postponed; but there had been a fear that any delay would have 'unfavourable political repercussions'.

The higher echelons of the Reagan administration sought to play down the problem as much as possible, indeed Teller himself backed away from the concept of total protection, while at the same time offering a justification for his shift of position. 'A 99.9% defense cannot be obtained soon. It may not ever be obtained ... But a defense does not need to be a tight umbrella. It suffices if it makes the success of an attack uncertain.' Teller did not draw attention to the fact that this was exactly what the existing defences were offering. For the administration it was of cardinal importance to maintain the political credibility of SDI throughout the planned summits with the Soviets. Energy Secretary Herrington, therefore, described the arguments between the scientists as 'one group of scientists arguing with another group of scientists on how you measure the intensity of a light beam'; and when thirty congressmen lobbied Defence Secretary Weinberger for a delay in the Goldstone test, he refused their request. The congressmen, however, did not let the matter rest and instead asked the General Accounting Office to look into the project's administration. Excalibur stood on the verge of becoming a public scandal.

The Goldstone test was the disaster its critics had predicted. In the process of its assembly, one of the canisters containing many of the sensors had been bent as it had been moved into position in the ground. In their rush to meet their deadlines, they had also not had time to replace the beryllium sensors with hydrogen capsules. Only three out of the ten experiments produced any result at all and these were not favourable. They indicated that the laser beam was something like a tenth the brightness assumed so far. It was a grim picture, adding to the anger and frustration. 'Pressure to go faster', one scientist said, 'means making mistakes like relying on a calibration system they didn't fully understand, which gave a false large signal, so after five years, we still don't know what they have.' The accusing finger pointed at Teller and Wood, but primarily at Wood.

Scientists began to consider leaving the project and a number did. George Chapline looked back with the regrets of a pioneer. 'The work was very beautiful from a physics point of view – such extraordinary experiments and using a nuclear explosion and my regret is that it's still classified and it needn't be. There are scientific aspects which are very important, like how does space and time form in the early universe. If people had access, the results we got might have a profound effect, of long-term importance for mankind.'

Others were angry and bitter. Peter Hagelstein was one of the first to go, deeply wounded by his experiences. Another, Lowell Morgan, has said, 'To lie to the public because we know that the public doesn't understand all this technical stuff, brings us down to the level of hawkers of snake oil, miracle cleaners and Veg-O-Matics.'

27
Reykjavik

Even if they had relied solely on the foreign press, the Soviets would have been well aware of the storms brewing around Excalibur and Livermore, but knowledge of these difficulties was not sufficient to moderate Gorbachev's determination to block the move into space-based defence. Detailed planning began for a further summit, and in August 1986 Eduard Shevardnadze, Gorbachev's Foreign Secretary, made contact with the US to propose another meeting between the two leaders. The Icelandic capital, Reykjavik, was chosen as the venue and the date was set for that November.

On the one hand, the Reagan administration saw the meeting as, at most, a planning session for a later summit in Washington. On the other, they saw it as a very useful public-relations boost – much as Geneva had been – for Reagan and the Republicans in the upcoming mid-term elections. As major developments were not expected, the US delegation arrived in the rainswept city briefed to hold to their previous policies. They had certainly not discussed any new moves with the Joint Chiefs or with Congress, so what happened on the first morning was a true shock.

The two leaders met at the Hofdi House, on a bleak peninsula overlooking the north-west Atlantic several miles outside Reykjavik. Straightaway Gorbachev offered substantial cuts in offensive weapons, not only in intermediate-range missiles but also in the ballistic missiles that were the Soviets' strongest asset; but these substantial cuts were offered in conjunction with restrictions on SDI: they wanted the Americans to cease testing and confine research to the laboratory.

It was an extraordinary offer. During the lunchtime recess the American delegation met in the US embassy's tiny security bubble, where the veteran negotiator Paul Nitze expressed his delight at what had happened. 'It's the best Soviet proposal we have received in twenty-five years,' he said. But the Soviets' continued attempt to limit SDI did not please Reagan, and that afternoon he read from a seven-page document which, in effect, reiterated an unchanged position. Afterwards, a furious Gorbachev responded by saying that the President was offering 'the same old moth-eaten trash ... from which the Geneva talks [were] already choking'.

In an effort to break the stand-off, the two leaders agreed on their delegations working on into the evening to try to achieve some agreement. During their discussions nearly all the concessions were made by the Soviets. There were concessions on the number of ballistic missiles, and the Soviets made a major one on the counting of bombers, as well as others on a host of smaller issues; but as the two groups worked on through the night and beyond, the impasse on SDI remained. The US position was cast in stone. Reagan's arms-control adviser, Ken Adelman, wrote later, 'We merely echoed the Chinese expression, "No problem, we can't do that." '

Early the following morning, the US delegation was optimistic, but the Soviets were, not surprisingly, disappointed. To Gorbachev it seemed that the Americans had come empty-handed to 'gather fruit in their basket'. Although further concessions emerged from both sides during the morning session, on the number and placing of intermediate missiles, Gorbachev maintained that none of these cuts could be agreed without movement on SDI. Reagan's offer to turn over the problem to the negotiators in Geneva was sneered at by Gorbachev as being 'kasha for ever' – eating porridge for ever. At one point Gorbachev made as if to leave, but he stayed, and the negotiations spilled over into a further and unscheduled afternoon session.

By now the Soviets were desperate for a settlement. 'You're a creative person – can't you think of something?' Shevardnadze asked one of the US delegation. Further shifts and changes were made, but no progress was achieved beyond a proposal that would get rid of the Soviets' best weapons and still leave SDI ripe to continue in development. The session that followed was fractious and confused. After a break the Soviets had come back with an extraordinary offer – that the ABM treaty be observed by both sides for ten years, during which they would both eliminate *all* strategic offensive weapons, and SDI would be confined to the laboratory. The Soviets had raised the game, increasing the cuts in offensive weapons from 50 to 100 per cent.

Still this did not satisfy the US, and various groups disappeared in last-minute attempts to come up with a form of words that satisfied everyone. Two members of the American delegation took over a bathroom, using a board over the bath tub as their desk. By early evening Reagan was restless. He had a family dinner date, which his wife Nancy expected him to attend. At 6.30, after a series of irritable exchanges, he stood up. 'Let's go, George,' he said to Shultz. 'We're leaving.'

As the two leaders put on their coats and moved towards their limousines, Gorbachev made one last attempt at mending fences. 'Can't we do something about this?' he asked.

'It's too late,' an angry Reagan replied.

In the weeks after the summit, there was a clear dichotomy of views on what had happened, but nobody was happy. The defence community, for their part, could not believe that Reagan had actually come as close as he had to bargaining away the weapons and the strategy that had held for more than thirty years. They believed that the impasse over SDI had saved America and its allies from disaster.

Anyone who believed in negotiation as the only way to end the arms race, however, saw the end result as a tragedy. A President with a fantasy of ultimate defence, nourished by reports of technical advances that were increasingly being shown to be at best grossly overinflated and at worst non-existent, had allowed that fantasy to block a real chance of abolishing strategic weapons.

'I told [Gorbachev] that I had pledged to the American people that I would not trade away SDI,' Reagan told the nation the day after he returned to Washington. Nevertheless he remained optimistic, he said, that a way would be found to 'begin eliminating the nuclear threat'. There were many who believed that he had passed on his best opportunity.

Ever since the early days of Excalibur, Edward Teller had met for breakfast on a monthly basis with his two protégés Lowell Wood and Greg Canavan. 'We all had our eye on what the President wanted,' Canavan said:

> From early on we knew that Excalibur was going to be difficult to realise and so, often, our sessions would turn into brainstorming sessions, trying to think of a project which would realise his dream in a reasonable time frame.
>
> It was at one of these meetings in the autumn of '86, about the time of Reykjavik, that I came up with a suggestion which looked as if it had real possibilities.

The project Canavan had mentioned bore a striking resemblance to General Graham's initial development of the 1950s' BAMBI project. It consisted of what Teller called 'pre-deployed arrays of explosive devices' that would scatter projectiles into the path of a rising attack missile. Given that the satellite carrying the projectiles would be in space orbit travelling at five miles per second and the ballistic missile target would be travelling at four miles a second (an ordinary bullet travels at half a mile per second), the extra one mile per second would give the projectile enough kinetic energy to smash into and destroy its target. Teller wrote:

I objected that once again we were looking at pre-deployed stations in space, which are cheaper to destroy than to deploy. Greg countered that the satellites could be made very small and therefore would be hard to find ... The components of such a satellite – microprocessors and rockets to power the device – were small, light, and increasingly inexpensive. A constellation of a few thousand such devices could defend the Northern Hemisphere.

Lowell Wood picked up the idea and spent time trying to prove or disprove its viability. He and members of 'O' Group worked on it for ten months, by which time it had acquired a name – Brilliant Pebbles – as in the pebbles used by David to slay Goliath. Each Brilliant Pebble satellite would be just over 3 feet in length and weigh some 120 pounds. At any one time, only a fifth of the total constellation would be likely to be in range of the target missiles when they were launched, but those in range would, on command, be able to move into position, firing projectiles into their path. 'Brilliant Pebbles, we believed, was just what the President was looking for,' said Canavan.

On 23 December 1986 Andrei Sakharov returned to Moscow from internal exile in the city of Gorky. Gorbachev had personally ordered his release, a gesture that heralded *glasnost* – openness – and the massive changes soon to sweep through Soviet society.

The release of the man who had created the Soviet H-bomb but who, in 1980, had been exiled without trial for speaking out against the invasion of Afghanistan and against violations of civil rights, was a major event in the city. Admirers crowded into his apartment and journalists from the West besieged him for interviews. Sakharov once again had both the personal profile and political platform from which to operate as an opinion former. On 28 December, only five days after his return, he used this position to speak with several reporters. Predictably, he pressed for the release of other political prisoners but, less predictably, he took the opportunity to criticise Gorbachev's policy of trading arms reductions in order to gain restrictions on SDI research.

His view on SDI was the one that was often heard in the West, but also one heard in the Soviet Union as well: that strategic defences would never be able to stop the full onslaught of a powerful opponent. He believed, therefore, that it was wrong to compromise other arms-control issues by linking them to SDI, which was still in the research stage and had such an uncertain future.

In February 1987 he attended a forum on disarmament initiated by Gorbachev and, during a closed session, expanded his argument. SDI

would be like a Maginot Line in space, he said, expensive and vulnerable
to counter-measures. It could never serve as a defence for the population, or even as the shield behind which a first strike could be launched,
because it could so easily be breached. In trying to explain US strategy,
he said that proponents of SDI were counting on an accelerated arms race
to ruin the Soviet economy. If that was the case, they were mistaken,
because the counter-measures would be far less expensive than SDI itself.
(When he said this, Sakharov might well have known about the long-
standing SDI projects but not about Canavan, Wood and Teller's new
brainchild.)

'I believe,' he concluded, 'a significant cut in ICBMs and medium-
range missiles and battlefield missiles and other agreements on disarmament should be negotiated as soon as possible, independently of SDI,
in accordance with the understanding laid out at Reykjavik. I believe that a
compromise on SDI can be reached later. In this way the dangerous
deadlock in negotiations on SDI can be overcome.'

Initially Sakharov believed his speech had made no impact on his colleagues, and indeed he was criticised by some of them. Furthermore a
number of others had already said much the same thing and to little effect.
However, his clarity and uncompromising moral lead had more effect than
he imagined. At the end of February, a matter of days after the forum,
Gorbachev proposed an agreement to eliminate intermediate-range missiles, separate from any agreement on SDI. The spell that the initiative had
cast on the leadership of the Soviet Union had been broken. From this
point on, the Reagan administration would be unable to use SDI to extract
serious concessions from the Soviets. 'We came to realise,' said Roald
Sagdayev, head of the Institute for Space Research, 'that we had not helped
ourselves by screaming so much about SDI. We had encouraged some
Americans to think that anything the Russians hate so much can't be all
bad. And we had overestimated how much damage SDI could do to
strategic stability in the short and even the medium term.'

Sakharov's own view of the decision was predictably cool. 'This
represented considerable progress,' he wrote, 'but I continued to press
for abandonment of the package with respect to ICBMs as well.' By the
following December, he had achieved his aim.

On 8 December 1987, at 1.45 p.m. precisely, Mikhail Gorbachev and
Ronald Reagan signed the Intermediate Nuclear Force Treaty, eliminating
intermediate-range weapons worldwide. The timing of the ceremony
created a rush over lunch, but Nancy Reagan's astrologer had declared that
moment as the most propitious. Before the ceremony, Reagan had been
advised that the Russians did not wear cufflinks; nevertheless, he gave

Gorbachev a set, similar to some of his own, on which figures were beating swords into ploughshares.

The treaty was a compromise. There were political and psychological benefits, particularly among America's European allies, who no longer had to host the likes of Pershing missiles; but it fell far short of what Gorbachev had hoped for a year previously. It eliminated only 4 per cent of the fifty thousand warheads the two sides possessed between them while, back in Reykjavik, Gorbachev had offered the chance to eliminate them all. That evening there was a dinner in his honour at the White House and one of the guests was Edward Teller. How much Gorbachev must have known about Teller's role in creating and sustaining the myth of SDI, which had so thwarted his efforts to break the half-century-old grip of the arms race, could only be guessed at. How would he react on meeting him?

'I was among the last to go through the receiving line,' Teller recalled. 'President Reagan was standing next to Premier Gorbachev and introduced me to him, saying, "This is Dr Teller." I put my hand out to shake hands, but Gorbachev stood unmoving and silent. Reagan then repeated to Gorbachev, "This is the famous Dr Teller." Gorbachev then said with his hands at his sides, "There are many Tellers." '

Teller was the only person among the senior politicians, arms-control negotiators and scientists present at the dinner to receive this treatment, and said afterwards:

> I was momentarily shocked and hurt. I did not look upon shaking Gorbachev's hand as a particular honour . . . My response was, 'There are, indeed, many Tellers.' I then turned and left the receiving line . . .
>
> I have come to regard the incident as a great compliment . . . I am a little proud that my efforts to protect freedom, and to extend it to those behind the Iron Curtain, were noticed.

28

Brilliant Pebbles

In the autumn of 1988 the Ethics and Public Policy Center, a Washington-based conservative think tank, honoured Edward Teller for his work in the development of nuclear energy and for his contributions to national defence. 'His motto, "Better a Shield Than a Sword", read the citation, 'expresses a timeless truth.'

Just before the black-tie presentation dinner at the Washington Hilton on 16 November, it was heard that Andrei Sakharov was coming to the US for medical treatment and might be able to attend. The Russian scientist was on a tight schedule and had to fly out to Boston that same evening, but it seemed too good an opportunity to miss. Sakharov would be a star speaker and there would be the special moment when the two 'fathers' of thermonuclear weapons met for the first time.

As the government officials, politicians, industrial and military leaders and their wives convened over cocktails in the Washington Hilton reception lounge, the two thermonuclear-weapon pioneers met briefly in an upstairs room. They spoke through an interpreter and, according to Teller, Sakharov said much the same during this conversation as he was to say in the speech he gave to the assembled guests.

Because of his flight arrangements, Sakharov spoke before the dinner. He was brief – even including the interpretation, the speech lasted only nine minutes – and he spoke wearily, with little inflection and few gestures. 'In Dr Teller I see a man who has always acted, his whole life, in accordance with his convictions.' Expressing his 'deepest respect' for Teller, he went on to describe their parallel endeavours on the hydrogen bomb as a 'great tragedy. It was a tragedy which reflected the tragic state of the world which made it necessary, in order to maintain peace, to do such terrible things.' He and Teller agreed on a number of things, he said, mentioning the development of safe thermonuclear power systems, but there were 'spheres in which we disagree', in particular space-based anti-missile defence. 'I consider the creation of such a system to be a grave error. I feel it would destabilise the world situation . . . If such space-based systems are deployed, even before they are armed with nuclear weapons, there will be a temptation to destroy them, and this in itself might trigger a nuclear war.' Having delivered his principal message, he ended as he had begun, expressing his respect for Teller and hoping such meetings would continue. He had received a thirty-five-second standing ovation on his arrival and now, as he left the hall to go straight to the airport, he was accorded another of precisely the same length.

'We honour Dr Sakharov and wish him everything in the world,' said William F. Buckley, the evening's master of ceremonies, who had sensed shock and concern among his audience at the Russian's warning. 'But we don't wish to give him, as a souvenir from America to take back to Moscow, a lapsed resolution to proceed with our SDI.' This comment was greeted with loud applause and cheers from the audience and their solidarity was strengthened when the lights dimmed and on a large screen appeared the image of Ronald Reagan, praising Edward Teller as a 'tireless advocate' of SDI who had insisted 'that American citizens are entitled to protection against enemy missiles'. He ended his tribute by describing the award winner as 'one of the giants of American science, and one of the bulwarks of American freedom'.

In his acceptance speech, Teller further augmented this show of support for SDI by restating his argument 'that we must know what can be known'. This exploration, he argued, must include nuclear energy, lasers and – referring to Brilliant Pebbles – devices small and effective enough to stop missiles. Sakharov's clearance had been revoked twenty years ago, he said, and it was not surprising that he was not familiar with Soviet 'defence systems where we have some reason to believe their accomplishments are years ahead of ours'.

One of the major concerns about Sakharov's speech was that, however independent of thought he was, he would not have been able to make it at such a gathering, or to reiterate it to the President, as he did a few days later, without Gorbachev's personal consent. Thus it was seen as a warning from the highest level in Moscow that SDI was not going to be deployed without a fight. The renewal of the Russians' concerns, particularly Sakharov's own, when, only two years previously, he had been so dismissive, reflected an enormous change during that time in the fortunes of SDI.

In 1986 the technical problems with Excalibur had been fully reported in the press and this had been followed by a major financial scandal centred on activities at Livermore. The scandal had followed on from Roy Woodruff's resignation and the backing he had provided for a growing concern in the press and elsewhere that Teller, Lowell Wood and the Livermore had attempted to mislead the administration about the X-ray laser, resulting in the wastage of large sums of money. A number of senators had taken their concerns to the General Accounting Office, the organisation charged with monitoring the spending of government departments. The result had been a continuing three-year investigation which had eventually exonerated Teller, Wood and the Livermore.

Since then the fortunes of SDI had fluctuated. At one point, a group of thirty defecting or emigrant Soviet scientists had alarmed Congress with an account of Russia's own rapidly developing anti-missile programme. That had worked in SDI's favour; but then, a few weeks later, the Assistant Secretary of Defense, Richard Perle, had made a widely publicised statement. He had said that SDI's prime role was to be the defence of the nation's ability to retaliate – that is, of its missile silos rather than of its inhabitants – something that had dampened enthusiasms considerably.

It was at this point that Greg Canavan had come up with the notion of Brilliant Pebbles at the Boston breakfast meeting. Canavan had freely admitted that the concept was not a new one and had originally been conceived back in the 1950s, when the technology had not been available to realise it. Then, in the early 1980s, Richard Garwin, working at IBM, had in one stroke proposed, and then disposed of, the concept. While accepting that electronic miniaturisation made such tiny missiles technologically feasible, Garwin had believed that literally millions of his so-called 'hornets' would be needed to provide adequate cover, and these would have to be accompanied by anything between ten and a hundred times as many decoys. The result, in his opinion, would be a massive traffic jam in space.

This had not prevented Lowell Wood from taking the concept and trying to turn it into a reality. As he began his explorations he had been delighted to discover that so much of what they required could be obtained off the shelf. In October of 1987, almost a year after their breakfast meeting, he and Teller had gone to see General Abrahamson, director of SDI, to introduce him to their new concept. He had been sufficiently impressed to agree to visit Livermore; and a few weeks later, Teller, Canavan and Wood had met once again:

On this occasion, [Canavan recalled] we all came back with a briefcase full of all these pieces which we laid out on the table. Here's 30 grams of electronics that does all the brain work, here's a telescope that does all the sensing, then the baby rocketry. And it was decided that Dr Teller should go to General Abrahamson – as the one person who had the credibility – to say, look we want you to create a black programme, one that doesn't appear on the books, and give it unlimited funds to see what can be done. So in a year they spent a hundred million bucks – something like that.

The meeting with Abrahamson had taken place at Livermore on 25 November 1987, when the general had also been given a tour of the laboratory, examining hardware, watching demonstrations and talking well into the night. He had accepted the invitation to dedicate a prototype imaging system for the project and, in his speech, he had described the laboratory's work as an inspiration. His commitment to the project was sealed when he had agreed that funding for the project should be increased substantially.

Three months later, at a three-day conference at Washington's Shoreham Hotel to mark the fifth anniversary of Reagan's Star Wars speech, the top-secret Brilliant Pebbles programme had gone public. Edward Teller and Wood had made the presentation and the President himself had been present to give the project his stamp of approval. 'Dr Teller,' he had quipped, 'is proof that life begins at 80'; and he had gone on in optimistic vein to describe how anti-missile technology had come on 'more rapidly than many of us ever dreamed possible'. Then had come his endorsement, his intention to support an initial deployment of an anti-missile system. It was a clear indication to the Soviets that the hopes he had expressed in his speech five years previously were still alive and taking on a new form.

That afternoon, Lowell Wood had conjured up the vision of a system that Soviet technology could not hope to match. 'Each pebble,' he had said, 'carries so much prior knowledge and detailed battle strategy and tactics, computes so swiftly, and sees so well that it can perform its purely defensive mission adequately with no external supervision or coaching.' And the cost? Derisory by defence-programme standards. The off-the-shelf technology meant that each pebble would, he said, cost around $100,000, including the transport into space. And even allowing for the three-fold multiplier so often necessary with defence estimates, the cost of 100,000 such pebbles would only be a relatively modest maximum of $50 billion. 'So in the end we provided the President with what he had asked for,' Canavan recalled; 'and I believe it wouldn't have happened if Teller hadn't forced Lowell and me to go through those interactions again and again, adversarial, punishing for me particularly as the one who always got beat up. And then of course he went and got the programme funded and established.'

In July, Teller and Wood had given the President his own presentation of Brilliant Pebbles. So intrigued was he that he had invited them back again a month later. SDI was riding high again, with a President committed to an anti-missile programme. Two months later, in his short speech at Edward Teller's award dinner, Sakharov had made clear just how credible – and how unacceptable – the Soviets found the new SDI concept to be.

But the Reagan administration was drawing to a close. Even though his likely successor, George Bush, had been Vice-President, it was known that Bush was deeply sceptical about the notion of the protective shield. So would he continue to support it? Or would he, perhaps, trade off SDI as a bargaining counter with the Russians to achieve a more complete arms control agreement? No one was sure.

Teller, then eighty years old and suffering from both heart problems and his continuing colitis, had begun lobbying furiously to keep the project on the road, but it was increasingly a strain. In the run-up to the election, he was called to testify before the Senate Foreign Relations Committee on the Intermediate Nuclear Force (INF) Treaty recently signed by Reagan and Gorbachev and now being presented for ratification. When 'the old lion', as *Time* had recently called him, arrived in the hearing room, he came accompanied by Mici. She had recently been ill and, a few days earlier, he had injured his shoulder while swimming in his pool at home. He was wearing a cardigan over his cast and the two entered the room unnoticed, looking for all the world 'like a couple who might have strayed from a senior-citizen tourist group'. They sat themselves at the back of the spectators' area, and only after the proceedings had begun were they spotted by an aide.

When called to the witness stand, however, Teller seemed to undergo a transformation. 'Let me start out with a simple unqualified statement,' he began in his unmistakable, rich, heavily accented baritone: 'I hope that the INF Treaty will be ratified.' In fact, the view he expressed in private at that time was that the treaty might prove to be 'another Munich'. His scepticism about their Soviet partners was evident when he referred to their 'sophisticated opponent' and stated that there must be a bipartisan policy that 'involved agreements on arms control and arms reduction and at the same time included well-conceived, rational defense ... the Strategic Defense Initiative provides that kind of defense'.

His central point was made. The country must be prepared for the worst. There were no questions and he rose stiffly from the microphone, 'a tired old man finishing another task'. Taking Mici by the arm, he led her slowly to the door – to another meeting for which he was already late. Not once did he waver from his deeply held belief that advanced technology was the essential prerequisite in any dealings with the Soviets.

'He was truly Messianic,' Herb York recalled. 'Shortly after George Bush was elected, in 1989, I felt it was time to mend fences with Teller. He happened to be at Livermore when I was there so I went in and said, "Edward, Peace?" "OK," he replied, "but here's what you have to do." For instance, one of the things I had to do was to get going to Washington and get George Bush to continue with Star Wars. So I had to join him. No suggestion of anything mutual. I had to do it his way.'

On a Saturday morning in April 1989, some three months after his inauguration, President Bush met with a group of his key advisers to decide on the future policy for Star Wars. On taking office, he had rapidly realised just how firmly the defensive potential of Star Wars had gripped the imagination of the right wing of his own party. In addition, at that meeting at Camp David, Bush had to hand two appraisals of the new addition to SDI, Brilliant Pebbles. One was from his Vice-President, Dan Quayle who, as a senator, had been a powerful supporter of Teller and Wood's endeavours. Quayle was very positive, arguing that the new programme should be deployed as soon as possible.

The other was from General Abrahamson. He had decided to retire from his post as director of the SDI Organisation and had written an end-of-tour report. In it he recommended a radical change that would exploit the improved performance and dramatic cost reductions offered by the new programme. Brilliant Pebbles, he said, could be proven in two years, deployed three years after that, and would cost no more than $25 billion.

Along with the pebbles themselves, Livermore was now offering Brilliant Eyes, a space-based surveillance satellite for spotting, tracking and targeting incoming missiles and, according to Abrahamson, this raised the possibility that these two could become the centrepiece of a system integrating all the other developing SDI programmes.

Given this new, cheaper alternative and wishing to avoid potentially damaging confrontation with elements of his own party, Bush moved quickly and formalised a multi-billion-dollar plan with Brilliant Pebbles as the centrepiece. It was to consist of three segments, two of them ground-based and one based out in space. Of the two ground-based segments, one was directed at protecting US forces on the battlefield, while the other would provide protection for the nation from accidental or limited attack from, it was estimated, up to 200 incoming warheads. The third segment would consist of Brilliant Pebbles and Brilliant Eyes, which would be stationed in increasingly distant layers out in space. Between them, they would help in detecting, and attacking incoming missiles at all stages of their flight, and would coordinate the efforts of the other segments. Gone was the dream of a comprehensive shield against all comers, but, with the

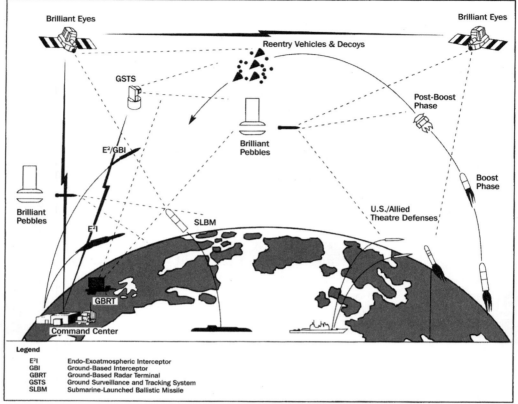

Legend

E²I	Endo-Exoatmospheric Interceptor
GBI	Ground-Based Interceptor
GBRT	Ground-Based Radar Terminal
GSTS	Ground Surveillance and Tracking System
SLBM	Submarine-Launched Ballistic Missile

Source: SDIO Program Office

Figure 7: The Global Protection Against Limited Strikes

GPALS had three segments and was intended to provide protection against a strike
of up to 200 warheads. One ground-based system was to protect US and allied armed
forces, while a second ground-based system would protect the US itself. The space-
based component, made up of Brilliant Pebbles and the surveillance satellite, Brilliant
Eyes, would detect and intercept incoming missiles fired from anywhere in the world.
The GAO were to criticise GPALS for inadequate integration through its battle
management system.

reduction in the threat from the Soviet Union – now the CIS – attention
was focusing much more on smaller-scale attacks from rogue states and
terrorists. This ambitious system, the Global Protection Against Limited
Strikes (GPALS), was to be tested over the next three years (See figure 7).

In February 1990 the President confirmed his personal commitment
when he visited Livermore to see the work in progress on the project. 'If
the technology I've seen today proves feasible – and I'm told it looks very
promising,' he said, addressing the Livermore staff, 'no war planner would
be confident of the consequences of a ballistic-missile attack.' It had been

little more than three years since Greg Canavan, Lowell Wood and Edward Teller had brainstormed their new concept into existence.

Inevitably, as the project increased in importance so the chorus of criticism grew. Yet again there was concern about the vulnerability to attack: how well could the pebbles be hardened to withstand, say, a thermonuclear blast out in space? How easy would it be for the enemy to improve the performance of their missiles such that they could outpace their attackers?

These criticisms were voiced not only by action groups committed to a critical view of arms development but by those associated with or even within the defence industry. Among these were the Jasons, a group of senior defence scientists who advised the government, and scientists within Livermore itself. Furthermore, over the two years since the Shoreham Hotel launch, the size and cost of each pebble had risen steeply. They were no longer conceived of as tiny devices a few inches long and costing $100,000 apiece; they had steadily grown to over a metre in length and the cost had risen to $1.5 million apiece, excluding the launch costs. The intended size of the fleet had also dropped dramatically, from 100,000 to some 4600.

In defence of their project, Greg Canavan and Teller wrote a paper, which was published in the international journal *Nature* and described how the missiles could indeed be hardened and protected for their duties out in space. They argued that each pebble could be protected by up to 400 decoys – radar reflecting balloons – and still be cost-effective.

They also proposed a pilot study, involving the deployment of just 100 pebbles, which, they argued, 'could provide a good (though never complete) defence against missiles fired accidentally or by a terrorist state'. In a concluding paragraph, which was very much a mark of changing times, they offered a reminder that ever since its inception the SDI had 'been intended to protect the whole of mankind not just a single nation. A cooperative demonstration supported in part by contributions from other nations would be a highly constructive step towards assured safety for all.'

This was a world away from the 'Evil Empire' rhetoric of seven years previously and even carried echoes of Teller's early advocacy of world government. It was to receive a big boost, too, from the 1991 Gulf War, during which the stark consequences of missile attacks on the population of cities like Tel Aviv and Riyadh were seen regularly on television. So, too, were the powerful images of Patriot missiles blasting the attacking Scuds out of the skies. The *Wall Street Journal* described these images as 'a great advertisement for SDI', an indication that the time when a real defence could be mounted against attacking missiles was nigh. Moscow, too, realised that it was, if anything, a more likely target than the US for rogue attacks and this led to serious talks about collaboration. 'I think the time has come,' the new leader of the CIS, Boris Yeltsin, said in a speech to the

United Nations, 'to consider creating a global system for the protection of the world community ... Russia considers the United States and the West not as mere partners but rather as allies.'

Furthermore, the technology to make such a collaboration possible was, apparently, already a reality. In August 1990 the Livermore had tested a prototype Brilliant Pebbles against a moving target out in space. The first test had failed because of a launch malfunction, but the following April they had repeated the experiment with '90 per cent success' and 'all of the main objectives of the test' accomplished. Teller and the teams at Livermore, it seemed, were this time delivering, not the totality of the assured survival dream, but a system that, for the first time, offered meaningful protection from nuclear attack.

On 1 December 1990, after fifty-four years, Edward Teller returned to Budapest. He had been invited fifteen years earlier by his close school friend at the Minta Nándi Keszthelyi, but he had had to wait to the end of communist rule before going back. 'Like every other exile, I had dreamed of seeing my home again,' he wrote:

> The effect of being in Budapest again was almost overwhelming. I was conscious at all times of the familiarity and comfort of all my surroundings. I was once again at home. On the day before my departure, the Hungarian Academy of Sciences made me an honorary member. I could hardly believe the warmth of the ovation that I received. For almost the first time in my life, I found giving an acceptance speech difficult: I could scarcely talk for the multitude of emotions I was feeling.

After such an experience, he decided he must return with Mici, and a short time later, in January 1991, they arrived for two weeks' stay. For those two weeks they travelled back in time more than seventy years. They visited their old homes, their schools and Freedom Park, 'where the same clock looked down on sweethearts as it had on Mici and me 60 years before'. They also returned to Mátrafüred, where as a young couple they had shared their 'separate' vacations and where Edward had become convinced that Mici was to be his life partner. 'It is such a great pleasure to see the familiar surroundings of one's youth after an absence of more than 50 years,' he wrote. 'It is an even greater pleasure to see those places with the sweetheart who shared them originally. Mici and I were aware of our immense good fortune. Our native land had survived a terrible nightmare, and we were able to see that it was recovering.' That trip was the last one that Mici's health permitted her to make.

◆

Less than a year later, the smell of scandal once more surrounded Livermore and the GPALS programme. In February 1992 the General Accounting Office issued the first of two reports, entitled 'Changing Design and Technological Uncertainties Create Significant Risk'. It was, in effect, a warning that the SDI Organisation was overreaching itself in its rush to produce the highly complex command-and-control architecture needed to integrate the various segments of GPALS. Further, the degree of coordination necessary between the various segments was taxing 'immature technologies'. If GPALS failed, the report warned darkly, 'millions of people could be killed'.

The GAO's second report claimed that the results from four out of the seven flight tests so far carried out on GPALS technology had been overstated. One of these four was the Brilliant Pebbles test where 'a 90 per cent success' had been reported. According to the report, the interceptor had failed to 'detect, acquire, and track an accelerating target's rocket plume,' one of the main objects of the exercise. To add to this, the GAO had also re-evaluated the reports of the Patriot successes during the Gulf War, which had helped to boost the progress of the Brilliant Pebbles programme. In the midst of the conflict, President Bush had claimed that the Patriots had destroyed forty-one out of the forty-two Scuds they had engaged. By the end of their investigation, the GAO reported that only 9 per cent of the Scuds had definitely been destroyed. It was a depressing picture on the eve of the President's bid for a second term.

Only six years earlier Edward Teller had fought the GAO every step of the way over Excalibur, but he was now eighty-four years old and his health was causing serious problems. More and more, Lowell Wood made the running on Brilliant Pebbles. It was he who now made the rounds of Capitol Hill, wheeling behind him a trolley carrying a scale model of a pebble. While Edward never lost contact with the work at Livermore, the next few years were to be as near a period of retirement as he was to achieve.

He visited Russia. On his first visit, in 1992, he met, and was impressed by, Yuli Khariton, the founding director of the first Soviet nuclear weapons laboratory at Sarov. He empathised with the way that Khariton had had to give up a promising research career to work on nuclear explosives in order to answer what he saw as the Nazi threat; and he admired his leadership of both the Russian fission programme and also their thermonuclear programme, where he had been Sakharov's director. A few years later, in a gesture tinged with the quixotic, Edward nominated Khariton for the Fermi Award. The award had already been given to a number of scientists who were not American citizens, and Edward not only believed that Khariton deserved it but saw it as a gesture that might help in reintegrating the

Soviets into the rest of the scientific community. However, few of his American colleagues supported the nomination.

He revisited Hungary on an almost regular basis – five visits in five years – and inevitably became involved in the various aspects of scientific life there.* He broadcast talks on the radio, wrote articles and became involved in assessing the safety of the country's nuclear reactors. He even found support for one of his long-standing causes: the damaging effects of scientific ignorance. In Hungary, he thought science teaching to be excellent and he believed he could see its impact socially in the response to the 1986 Chernobyl reactor accident. The radioactive cloud had passed over Hungary, as it had over other neighbouring European countries, but the increase in radioactivity outside the Soviet Union, according to Teller, had never been significant. Yet in the period after the accident, abortions in Western Europe rose by 50,000. Teller believed that this was due to an hysterical reaction based on ignorance and, for him, this was borne out by the absence of any rise in Hungary. The usual number of healthy babies were born, 'thanks to Hungary's excellent science education', Teller wrote. 'The unnecessary loss of 50,000 potential lives in Europe was the result of ignorance, not the Chernobyl accident.' As Sakharov had commented, his was a life lived in accordance with his convictions.

Given all the technical uncertainties associated with Brilliant Pebbles and GPALS and the memories of the scandals surrounding Excalibur, it was perhaps not surprising that the new Clinton administration should look askance at all the SDI projects. Within four months of taking office in January 1993, they announced that they would be transmuting the SDI office and its personnel into a new organisation with different objectives. According to the new Defense Secretary, Les Aspin, he was intending to take 'the stars out of Star Wars' and switch the emphasis from space-based systems to those defending US ground forces against short-range battlefield missiles. Whereas 20 per cent of the SDI's budget had gone into battlefield missile defence and 80 per cent into national defence, Aspin reversed these figures, with 80 per cent now going to battlefield defence.

It meant the end of Brilliant Pebbles, whose supporters considered it the best project to emerge from the $30 billion that had been spent on SDI since its inception; and it meant the end of the vision of 'mutually assured survival' and a return to 'mutually assured destruction'. The concept of a global protection system that had been the subject of detailed discussions with the Russians was taken no further, and instead there were moves to control the situation by strengthening the ABM treaty.

The teams at Livermore did their best to maintain the expertise they had developed on Brilliant Pebbles. A group under Lowell Wood used that experience to develop Clementine, a highly sophisticated package of sensors that in late 1996 produced the most detailed map so far of the Moon's surface and discovered water at its south pole. There were other schemes proposed for Clementine, including one that Teller had initiated.

On a second visit to Russia, Teller attended the conference on meteorite and comet impacts with the Earth at Chelyabinsk, the site of the second Russian weapons-research laboratory. While there, he signed innumerable photos of himself standing beside a model of the 58 megaton bomb, the largest ever, that the Russians had exploded in 1961; and he came away with an idea for using Clementine: it could be used to intercept sizeable meteors threatening the Earth and, if necessary, implant explosives below their surface, either to blast them off course or to break them up. He was sure to make clear that this could probably be achieved using conventional explosives; but it was to remain, for the time being at least, just an interesting possibility.

Over the next six years, national missile-defence systems only received spasmodic support and that coincided very much with when the Republicans had sufficient political leverage on Capitol Hill. They had it in 1994, when they had a big success in the congressional elections, and they had it in 1999, when President Clinton was trying to garner support at the time of his impeachment over the Monica Lewinsky affair. However, it would be fair to say that, during those six years, a regular reader of the newspapers could have been forgiven for thinking that there was no longer any missile-defence programme at all.

On most of his journeys abroad at this time, Edward was accompanied by a colleague – often Lowell Wood, as Mici was now seriously ill. 'In the early eighties,' Paul Teller recalled, 'my mother had developed the condition in her brain where a number of the blood vessels are enlarged. With this condition there's a constant chance of having an aneurysm so she had a cubic centimetre of her brain removed. I didn't notice any change in her behaviour at the time, but roughly seven or eight years later she started to become paranoid and demented.'

Rosie and Stirling Colgate recalled how she worried obsessively about everything: 'Edward was very good with her. We discussed whether it was paranoid schizophrenia or paranoid dementia and he was concerned about what he himself could do for her. I told him there was little that he could do and that, however reluctant he was, it was a problem he had to hand over to the medical practitioners.' A smoker right up to the time when dementia took hold, Mici's life was further complicated by emphysema and she was dependent on her carers.

Then, in 1995, Edward himself suffered a severe stroke. He made a good recovery mentally and although, like Mici, he became increasingly dependent, it did not dim his attempts to continue working, writing, proselytising. He still went in to Livermore several times a week and he still did interviews – on one occasion a ten-hour session over four days, encapsulating his life in an interview with a past director of Livermore John Nuckolls; and he worked on his memoirs with his editor, Judy Shoolery.

On 4 June 2000 Mici died; she was ninety-one. 'I cannot overestimate how much her steadfast love and support sustained me for seventy-six years (sixty-six of them as my wife),' Edward wrote. She did not like formalities so, in accordance with her wishes, no funeral services were held. 'Instead, I invited friends and family members to the house, where we exchanged our memories of Mici, beginning with my sister Emmi, who talked about Mici in her early teens. We all recalled her indomitable and generous spirit. She gracefully weathered through determination the many difficulties life placed in her path, and she relished the joys.'

The litany of ailments that now afflicted Edward – arthritis, deafness, macular degeneration of the retina, and abdominal problems – would have conquered a lesser spirit, but he struggled on. He maintained his visits to the laboratory and while at home he spent hours on the phone. At other times he lay covered in a tartan blanket on a couch in the corner of the living-room window. The blinds were nearly always drawn and he would lie listening to his favourite music. In the window stood his piano, and around him were some of his many awards and the furniture Mici had bought at the Congress Hotel sale in Chicago nearly seventy years before. In the year following Mici's death he suffered from abdominal problems from which it was thought he would not recover. It stopped his plans for another trip to Hungary, but by the end of the summer he was going in to the laboratory again.

On 26 August 2003, at a ceremony at the Livermore, Edward received the President's Medal of Freedom, the country's highest civilian award. Less than a fortnight later, on Sunday, 7 September, he suffered another stroke. He recovered consciousness for a time on the Monday, but on Tuesday, 9 September, aged ninety-five, he passed peacefully away.

Epilogue

One of the scientists I met at Los Alamos while I was researching for this book commented that, around his laboratory nowadays, Teller was seen increasingly as a truly significant figure and one whose importance would be seen in time to eclipse that of Oppenheimer. This is a view I have come to share, not so much in terms of his importance compared to Oppenheimer – as the two men have very different historical roles – but rather in terms of the significance of what Edward Teller achieved in his life.

I became familiar with him during the research and making of a drama series on Oppenheimer, which I produced for the BBC in 1980 and in which he was a character. It gave me a fragmented view of him as the *enfant terrible* of wartime Los Alamos, who obsessively 'fathered' the H-bomb and became Oppenheimer's nemesis when the wartime director of Los Alamos tried to block his path. The knowledge I gained saved me from the demonic view of Teller, although when I met him on my researches for the drama he certainly played up to that image. In the home of one of his sponsors just off Hollywood's Rodeo Drive, the screenwriter and I sat in a room lined with original Modiglianis while Teller delivered an impassioned monologue for almost an hour without allowing us to take a single note. I was fascinated by that passion, the monomania that so obviously obsessed someone of the highest intellect. He was one of the most remarkable people I had ever met.

Returning to him after more than twenty years, I find him if anything more remarkable. I've been impressed by the unblinking honesty with which he described his childhood: the over-protectiveness of his mother, his unhappiness over long periods at school, and the sense of insecurity and impending failure that accompanied him as a student as he moved from one scientific centre of excellence to another. I have also been intrigued by the way that he failed to address properly in his *Memoirs* so many of the controversies of his later life, and the fact that he initially refused to write about the Oppenheimer affair.

I have been touched, too, by the way he described his sense of rootlessness, his dissatisfactions with himself, the vulnerability to the opinions of others that emerge from his letters to Maria Mayer. 'If only you will stop being mad at me', he wrote to her on one of the few occasions she was angry with him, 'I am really convinced that whoever is really nice to me must be crazy and you are much too reasonable.' Given what was to follow, it was also touching to see just how much the recognition and respect accorded him during his time with Gamow at George Washington University had so obviously settled his anxieties and self-doubt. He placed great value on the friendships and collaborations he established during what he called this 'wonderful quiet period in my life'. But those same sensitivities, when disturbed once more – as they were when he went to Los Alamos – can be seen to have contributed a great deal to the conflicts and personality clashes that accumulated over the following decade. His relationship with Oppenheimer and Bethe were both damaged early on when he was passed over for the post of head of the new laboratory's Theoretical Division, and his relationship with Norris Bradbury staggered through a decade of mistrust as, time after time, Teller failed to win the support he believed was due to the Super.

The roots of his discontent are understandable and, not infrequently, justifiable. However, time after time one wishes for more wisdom. If only he had found the strength, the equanimity, to accept the logic behind Bethe's appointment to the Theoretical Division post, or the fact that a business manager like Marshall Holloway was essential for a complex project like Mike. If only he could have had more appreciation of the compromises Oppenheimer was forced to make in running the three-ring circus that was Los Alamos. But it was not to be. As knockback followed knockback the mistrust grew until, by the time of the shock announcement of Joe 1, it had become a major problem for both sides. Teller no longer trusted anyone close to Oppenheimer, while the laboratory's senior management were angry about Teller's continual 'end runs' to Washington.

His growing paranoia can also be seen as having contributed to what he described as his 'mental inertia' over the design of the Super. It caused him to block out the suggestions of others about the significance of compression in enhancing the prospects for a fusion reaction. It meant also that he saw Ulam's – and Fermi's – repeated negative results from their Super calculations, not as a problem that needed a solution but as sabotage. They were further errors of judgement that contributed to what he admitted was his simple 'simple, great and stupid' mistake in not recognising a solution to the H-bomb problem earlier. At the time he blamed Oppenheimer for the delays but he himself was at least as responsible. The bitterness and anger created by the delays during that troubled year following Truman's call for a crash programme were never to be assuaged.

However, it is essential to pause at this stage and remember that Teller had spent ten years with the Super concept and that, from the beginning, it had been a mission for him. His critics question his motives, the role of his strong personal ambition, but few question his sincerity. Furthermore, against all the odds, he was eventually successful. Critics may argue that the discovery was inevitable but there were good people, some of them these very critics, who could have found the answer but failed. Bethe it was who, not long after the discovery, compared it to that of fission. And in March 1951, when Teller and Ulam published their joint paper, it's worth remembering that the Russians had by then been working on their own thermonuclear device for some three years.

It was at this time, just after he resigned from Los Alamos, that Teller was at his most bitter, and it was then that he vigorously expressed his frustration, with Oppenheimer and the 'Oppie Machine' in particular, during two interviews with the FBI. He had then found a sympathetic ear for his view of his struggle against the odds in journalists like James Shepley and Robert Coughlan. In personal terms, however, irreparable damage had been done and the stage had been set for the tragedy of the Oppenheimer Hearing, a nadir for both men. All the conflicts and personal animosities of the recent past were to be played out there, to the point that when Borden's claim that Oppenheimer was a spy was read out in court, it seemed a total irrelevance.

As for Teller himself, he had set a trap for himself with his comments to the FBI. He was caught between the angry accusations he had made earlier about Oppenheimer – which were in the Board's possession – and a growing realisation as to how much of a betrayal they would now seem to his colleagues. Teller was nevertheless shocked by the power of their reaction – shocked and deeply wounded. As a consequence he was driven further into the company of the hawks, 'quitting the appeasers and joining the Fascists' as he himself put it; a glib remark that disguised a massive emotional upset.

Teller, however, had already become an accomplished political operator, one who had tasted power and enjoyed it. According to George Cowan, he had learnt what decided policy, 'which, by the way, is like magic. He knew how to conjure very well'. He learned how to use worst case analyses, exaggerations to impress his audience – 'bringing up the props' Cowan called it, borrowing from *The Wizard of Oz*. So skillful did he become that his mastery of the interface between the scientific and the political was second to none. He moved in elevated circles, describing many of the people he met there as 'friends', and could walk into almost any office on Capitol Hill or in the Pentagon. However, in Cowan's opinion, he nevertheless had a great deal of integrity. 'He pursued power because he believed it wasn't going to be used. He was prepared to create that notion of overwhelming power without using it.'

Teller's moral position was clear and distinct. He believed firstly that defence was essential – it always had been and it always would be. He used events in Hungary at the time of Genghis Khan as an example. The Mongol forces under two of his generals descended on Hungary and killed 90 per cent of the population within a few weeks. They had no nuclear weapons but put the country to the sword. 'The limitation is not in the ability but in the intention,' he wrote. 'So there was reason to fear at any time.'

'Aggression is wrong,' he wrote, 'whether carried out by bow and arrow or by the hydrogen bomb. Defence is right, whether it uses a stream of particles or the concentrated energy locked in the atomic nucleus. Without such an agreement [on the morality of self-defence] we cannot count on the survival of the society that holds moral values so strongly that it calls them human rights.'

He also believed that if a 'development is possible, it is out of our powers to prevent it.' The possible, therefore, had to be explored – but not used except in defence. He believed in peace through strength, the second of his moral imperatives. There could be no place for qualms when facing an amoral opponent. He discounted as a 'simplified romantic approach' the renouncing of 'particular weapons as immoral precisely because they are most powerful, [while] at the same time an amoral society deploys and is ready to use those weapons'.

This was a position to which he stuck with absolute consistency. But with statements such as 'if we should fail to explore the limits of human power we shall surely be lost', he ran up against heavy criticism for irresponsibility – that he was ignoring the battle of priorities that controls even defence spending. Furthermore his determination to pursue the ultimate deterrent ran counter to the reactions of other colleagues – including Oppenheimer, Bethe, Fermi and Rabi – who expressed regrets, reservations and fears at the direction events they had catalysed were taking. The position of these men had a supporting logic: a belief that negotiation, compromise, common interests, would triumph even with the Soviets; and it spoke to the world of a group of men who had carried out a necessary task, but now suffered from all too easily understandable remorse and doubts.

Teller, on the other hand, was invoking the development of new technologies with a certainty that seemed to deny he had any scruples. In fact, the more successful he became in creating his illusion of invincibility, the more his actions could be – and were – interpreted as the unnatural obsession with weapons of mass destruction that Kubrick had targeted in his film, *Dr. Strangelove*. Not only members of the public and colleagues saw his actions that way but Presidents as well. At the end of his administration,

Eisenhower warned both of the power of the 'military-industrial complex' and of the influence of the 'scientific-technological elite', a phrase he constructed with Teller in mind. After three years of further battles over a test ban, Kennedy's aides were to draw a parallel between him and a combination of John L. Lewis and Billy Sunday.

A critical element in Teller's thinking was the existence of an 'amoral enemy' – the Soviet Union. Teller never trusted the Russians to negotiate. Nor, it must be said, did his Hungarian compatriots who had shared the same experiences of them; and in the early 1950s, with Stalin in power and Communism fomenting in every corner of the globe, nor did millions of other people. Furthermore, the West's nuclear supremacy was seen as the only politically acceptable answer to the Soviet Union's superior conventional forces. The American public might well abhor nuclear weapons, but the other main choice – higher taxes to support much larger conventional forces and possibly the draft – never won the day. Nor did the other choice, a basic deterrent such as the French and the British have had.

So, particularly at the beginning, Teller was very much in the political mainstream. But as the 1950s progressed and Stalin was replaced by more liberal leaders, there was increasing reason to believe that, for economic reasons if for no other, the Russians wanted to ease the arms race. So the test ban negotiations began, envisaged as the first step along the path to full disarmament. Such moves worried Teller greatly, who believed that this would leave the West very vulnerable, and he was in a unique position to do something about it. For half a century, Livermore provided him with a power base from which to float his technical dreams and solutions – and to fight battles like this one.

Firstly it was the 'clean' bomb. The 'moral' attraction of this idea, eliminating fall-out and reducing nuclear weapons from massive radiological infernos to mere megaton explosives, was enough to deflect Eisenhower from declaring a moratorium on testing. 'Clean' devices then became the *sine qua non* of the Plowshare Project, their reduced fall-out making acceptable the grandiose geographical engineering of Chariot or the new Panama Canal. These schemes, which recalled the visions of the great nineteenth-century engineers, were seriously intended to offer alternative uses for nuclear power, but were also a cover under which testing for military purposes could continue without interruption.

Livermore under Teller was, however, at its most creative in blocking progress towards the test ban agreed at the Conference of Experts in 1958. First they devised hypothetical schemes by which the Russians might cheat a ban, from giant screens in space to the more practical Latter 'holes' underground. Their real success, however was in demonstrating the uncertainties in distinguishing seismologically between earthquakes and tests.

In showing up the confusion, Teller and his laboratories destroyed the hard-won agreement achieved by his fellow scientists. It was these actions which caused one of his sternest critics, Isidor Rabi, to brand him 'an enemy of humanity'.

By the 1960s times were changing. Chariot saw the early activities of Barry Commoner, one of the future leaders of the environmental movement. The attempts to produce a warhead for Teller's earliest anti-ballistic missile, the five megaton device tested on Amchitka Island in 1972, provoked an international outcry and subsequent legal action. The small boat that sailed out to witness the explosion was the first venture by the newly born Greenpeace.

Teller fought what he saw as the hysterical, ill-educated and irrational fear of nuclear activities with great vigour. He battled against the fears of fall-out in the 1950s and his argument – that there was an over-reaction to the dangers of fall-out – still has support from researchers today. He publicly debated and supported other ventures throughout the 1960s and 1970s, culminating in his defence of nuclear power at the time of Three Mile Island. Often he was defending an issue rather than a specific project of his own and his supporters believed that he was being exploited by a number of his right-wing and military associates, many of whom kept a much lower profile than he did. Teller, though, was never frightened to stand up to be counted and he backed friends as well as causes. When Lewis Strauss was under attack during his ratification as Eisenhower's Secretary of Commerce, Teller risked his own reputation to stand up and defend him. As Sakharov said when they met, Teller was a man who 'acted his whole life in accordance with his convictions'. There were many who shared Teller's opinions, but few who held them with such conviction. So it was he who was demonised, whose effigy was burnt in public, whose lectures were disrupted and who had a custard pie thrust in his face.

But there remains the question of whether Teller was ultimately justified in his determination to challenge and deter the Soviet Union at every turn by what happened with the Strategic Defense Initiative. Teller had the foresight to recognise the potential of an X-ray laser beam pumped by a small nuclear explosion to work as an anti-missile weapon out in space; and his ability, on the basis of just one promising test, the Dauphin test, to talk around a whole administration to support not just the X-ray laser but the whole group of such projects that became SDI, was again most impressive. He had then, once more, 'brought up the props and banged the drum'. But this time his exaggerations were so extreme as to become deceptions, as epitomised by his letter in late 1984 to the strategist Paul Nitze, in which he declared that a module 'the size of an executive desk . . .

could potentially shoot down the entire Soviet land-based missile force'. The experimental results never came close to supporting this claim and a wrangle at Livermore led to a major financial investigation.

President Reagan, however, was captivated by the dream of assured survival through defence rather than the mutually assured destruction of deterrence. Teller's marketing of the project helped ensure that it took on a political life of its own, one almost entirely divorced from its reality as a series of speculative and troubled technological ventures. In Washington, various members of the administration interpreted its value in different ways, ranging from George Shultz's wish for additional teeth in disarmament negotiations with the Soviets through to the virtually opposite hopes of Richard Perle that it would rekindle the arms race. But however each saw it, they were all for the President's 'vision'.

As to the Russians, they had initially been profoundly shocked by the President's Star Wars speech. Andropov accused the Americans of 'inventing new plans to unleash a nuclear war' and, exceptionally, went on to warn his nation of an impending nuclear holocaust. It also brought home just how far they lagged behind the West technologically. But three years later, during the preparations for the Reykjavik summit, the problems at Livermore and the 'science wars' were widely reported and Gorbachev's own advisers were able to explain to him how vulnerable various of the SDI projects would be to counter-measures.

Nevertheless, SDI was still at the centre of the negotiations. In common with his American counterparts, Gorbachev had his own use for SDI – the reduction of, and possibly an end to, the arms race which he believed both expensive and dangerous. It was as if SDI itself was a sufficiently abstract notion for everyone to use it as currency in the negotiations. The illusion of a new order of technological warfare had infiltrated decision-making at the highest level and the conjuror creating that illusion had been Edward Teller. Everyone reacted to it accordingly and to the amazement of those taking part, it became the vehicle that brought the elimination of the whole nuclear armoury as close to happening as it has ever been. Paradoxically, it was Reagan's belief in the reality and indispensability of Star Wars defence that stalled matters at the last moment.

It was an outcome that Teller may neither have expected or dreamt of but it is difficult to conceive of it happening without his contribution and he declared himself 'a little proud of his efforts to protect freedom' over the signing of the compromise Intermediate Forces Treaty. He had then continued to involve himself, as much as he was able at eighty-years-old, in yet another technological vision, this time a global system of defence that would protect against accidents or limited strikes from 'rogue' nations.

In 1993, this GPALS programme was closed down by the Clinton administration but, before he died, Edward Teller added his support to the new Bush administration's initiative to dispense with the ABM Treaty and to, once again, set their sights on defence in space. Following the catastrophic events of 11 September 2001, however, the rules of the game changed. A question-mark hangs over any attempt to answer the new threats from global terrorism by the technological means, illusory or real, which were Edward Teller's stock in trade for the past half-century.

Some seventy years ago, as young students in Copenhagen, Teller and his friend Carl von Weizsäcker, wrote down their own sets of Ten Commandments. One of Edward's was 'Live in such a way that some time you shall be able to think of your life as finished'. It is an open question whether he actually thought he had achieved his aim. He had regrets which must have stirred him to think of what might have been. He might actually have preferred a career that paralleled that of the friend he most esteemed, Enrico Fermi, and to have ended as the distinguished Nobel Laureate he might have become. He might have felt, as some of his friends, as well as his enemies believed, that he had known power and to some degree been corrupted by it. This must have been in his mind when he wrote in an open letter to Hans Bethe that, rather than Oppenheimer's view that physicists 'have known sin', he believed that they 'have known power'.

But as to the way he lived his life, there can be no question – he lived it to the full, propelled by a determination to achieve, guided by vision and foresight. Edward Teller was a force of nature.

Appendix 1
The New Physics: the Path that Led to Quantum Mechanics

The special and general theories of relativity

By 1928, when Teller moved to Munich, Einstein's special and general theories of relativity were no longer generating the controversy they had when newly published at the turn of the century. The special theory has many ramifications but one is especially crucial: that matter and energy are inter-convertible, almost as if they were two forms of the same thing, and that inter-convertibility is expressed mathematically in the famous equation $E = mc^2$, where E is energy, m is mass and c is the speed of light.

The constant controlling this relationship between mass and energy is the numerically enormous square of the velocity of light, and clearly showed the huge amount of energy that might be generated from a tiny amount of matter – should such a generation ever be possible.

Quantum theory

Alongside this work of Einstein's, quantum mechanics was very new in 1928, but its origins in Max Planck's quantum theory also went back to the turn of the century. What Planck postulated back in 1900 was that radiation could not be emitted or absorbed in infinitely small amounts but only in discrete packages, which he called *quanta*.

It was the first theory that moved beyond everyday perception, defied our own senses and offered a new perception of the world. By comparison, it is as if water from a tap were delivered only in cupfuls and never in fractions of a cupful. It also extends classical Newtonian physics into the infinitesimally small arena of the individual atoms themselves.

Planck's original work was then reinforced by Einstein, who in 1905 postulated that light was not a continuous wave but a series of energy packages, paralleling Planck's *quanta*. Einstein called the energy of a quanta of light a photon.

The structure of the atom

Then came the next step, a major contribution by the New Zealand physicist Ernest Rutherford, who showed that the atom consisted largely of empty space. This discovery was the launch pad for the Dane Niels Bohr to come up with his classic 'planetary' model of the atom with the nucleus orbited by a complement of electrons. But even at the time, this classic model was recognised as little more than a visual metaphor. As Teller himself expressed it: 'It is more difficult to convey the idea of an atom by a picture than it is to make a drawing of last night's dream.'

Nevertheless, the metaphor served its purpose. Using it, Bohr postulated that by assuming the energy of the electron was quantised, he could explain the spectrum of light emitted by the atom. When it absorbs energy, for example by heating, it then jumps to a further orbit, and if that energy is then lost, in the form of light, it will return to its previous orbit. Electons within their orbit, however, do not emit energy. This notion of quantum orbits and the jumps between them marks the beginning of quantum mechanics – detailing and quantifying the intricacies of this subatomic activity and structure that control the behaviour of matter.

It was to provide insight extending well beyond just the atom itself to the interaction of the atoms within molecules, and to understanding of such processes as superconductivity and those underpinning electronics.

The nature of matter: particle, wave, or both

But back to the orbiting electron. Why did electrons have particular quantised orbits? Why not any orbit? In 1925, only two years before Edward Teller commenced his course with Herman Mark, the French scientist Louis de Broglie offered an explanation. Every particle, including the electron, possesses wave qualities; and only those orbits can exist having a circumference that allows precisely for a whole number of complete waves. It was yet another explanation that ran counter to any notions of common sense: that a particle could also be a wave. But in 1926 another physicist, Erwin Schrödinger, was able to support de Broglie's wave picture with full mathematical detail.

Meanwhile, an absolutely different theory of the atom was proposed by Werner Heisenberg. He bypassed the analogies of orbits and their accompanying waves, and concentrated on the mathematics of what is observable when an electron either absorbs or emits energy – jumping, say, from orbit 1 to orbit 2. He found that the energy levels ranged in blocks, or matrices. This Heisenberg approach came to be called matrix mechanics, and was based on particle rather than wave theory. Yet – and this was the crucial observation – matrix mathematics yielded exactly the same result as Schrödinger's wave mechanics. This mathematical equivalence was demonstrated by the British mathematician Paul Dirac. As physicist George Gamow later wrote: 'It was as if America was discovered by Columbus, sailing westward across the Atlantic Ocean, and by some equally daring Japanese, sailing eastwards across the Pacific Ocean.'

It was a revolution in the understanding of matter, a paradox that needed a resolution. This came from Niels Bohr. He explained that the wave and particle concepts were not contradictory but complementary. Experimental evidence showed that matter would behave either as a wave, or as a particle, depending on conditions.

Heisenberg's uncertainty principle

In addition, Heisenberg showed that the way a particle behaved – as a wave or as a particle – was unpredictable, random, uncertain. Only when dealing with particles en masse, statistically, was it possible to predict likely behaviour. For individual particles, however, behaviour was random. There was no link between cause and effect, no chain of causality. Attempting to measure a system breaks the relationship of causality and with this there comes the implication that, while the past can be defined, the future will always remain totally unpredictable. We cannot have complete knowledge of a system and this is the big break between quantum and classical mechanics. As one scientific wag put it: Heisenberg explained what God had been doing since the last day of creation. He had been contending with uncertainty.

Others took these philosophical issues more seriously. Einstein, for one, could not accept this uncertainty principle, as it became known. It offended his sense of mathematical order, which, he believed, pervaded the universe. 'God does not throw dice,' he insisted. Teller himself did see a new universal structure implicit in quantum mechanics:

In doing so, I will imitate Einstein, by talking about God. Physicists in the last century had to believe that, if God existed, he was unemployed, because 10 billion years ago he created the universe, together with cause-and-effect relationships, and whatever he established at that time would determine the whole future. And God could not change that, except by violating the principles of physics, which we like to call a miracle. So old-fashioned physics, with the future determined, did not allow for any concept like a God who can do something. Now in quantum mechanics the future is not determined.

Appendix 2
Basic Information on the History of Fission

1931

Cockroft and Walton split the atom.

1932

James Chadwick discovers the neutron.

1934

The Joliot-Curies discover artificial radioactivity – by bombarding a target with alpha particles. As a result new elements are produced.

1934

Fermi extends the Joliot-Curies' work and bombards elements, not with alpha particles but with neutrons. When he does this with uranium as the target, he reports finding several artificially radioactive elements as products.

1938

Hahn and Strassmann prove that what Fermi had observed four years earlier was the bursting of the nucleus. The first paper from them describing what was to become known as 'atomic fission' is published internationally on 22 December 1939. Frisch and Meitner describe the mechanism for this fission and indicate that large amounts of energy are given off by the process. The big question is whether neutrons are liberated which can then cause a further fission, a further release of energy, and so on, to produce a chain reaction.

1939

Leo Szilard confirms that the neutrons are produced and an explosive chain reaction is likely.

1939

During March and April various papers are published confirming Szilard's finding and a paper by Joliot-Curie shows that 3.5 neutrons are produced by every fission. The chain reaction is confirmed.

Appendix 3

The Sketch for the 'Super' that Evolved During the Berkeley Conference, Summer 1942

In his initial calculations, Teller had proposed two possible reactions between fusing deuterium (D) nuclei. The first produced helium 3, ejecting a neutron and releasing 3.2 MeV of energy:

$$D + D = He^3 + n + 3.2\,MeV$$

The second reaction produced tritium – the heaviest form of hydrogen, which does not occur in nature, and the proton and 4 MeV of energy:

$$D + D = T + p + 4\,MeV$$

The deuterium had to be hot enough, the individual nuclei excited enough, to overcome the nuclear energy barrier repelling them. The minimum necessary energy thought necessary for this was 35,000 eV, equivalent to 400 million degrees K. At that temperature, both the above reactions were equally probable.

Measured by mass, fusion produced slightly less energy per gram than fission, but it had none of the limiting factors, such as critical mass, which a fission reaction had. The output was dependent simply on the amount of fuel used and, in theory, was infinite.

Deuterium was a much easier and cheaper fuel to acquire than either U-235 or plutonium. The group calculated that a kilogram of deuterium would produce the equivalent of 85,000 tons of TNT. 12 kilograms of deuterium would produce the equivalent of 10 million tons of TNT. On the best estimates at the time, this was equivalent to 500 atomic bombs.

The Ignition Problem

It seemed likely that deuterium fusion would proceed too slowly and would be blown apart by the fission trigger before it gained enough momentum. Konopinski came up with an answer. Using tritium (T), the fusion between deuterium and tritium would require only 3500 eV, equivalent to 40 million degrees K, one-tenth the energy required for fusion between two deuterium nuclei:

$$D + T = He + n + 17.6\,MeV$$

This reaction would produce more than four times the energy of a straight deuterium fusion. However, tritium would have to be manufactured – from a rare isotope of lithium, the lightest metal known, and known as Lithium (Li^6). The Li^6 would split into two tritium (H^3) nuclei. The large flux of neutrons needed for this could only come from a pile like the one Fermi was building and this was still untried technology.

One further possibility was to create the tritium within the bomb itself by packing it with a dry form of lithium, lithium deuteride. This would have to be enriched with Li^6, but this would be easier than enriching a Uranium weapon with U-235.

Notes and References

The following sources are referred to by abbreviations in the notes:

AEC: Records of the US Atomic Energy Commission, RG 326, National Archives.
FRUS: US Department of State, Foreign Relations of the United States.
CIC: Coordination and Information Center, US Dept of Energy, Las Vegas, Nevada.
Foote Collection: Don Charles Foote Collection, Alaska and Polar Region Dept, University of Alaska, Fairbanks.
JRO/AEC: Records of the Personnel Security Board, AEC Division of Security, RG 326, National Archives, College Park, Maryland.
JCAE: Joint Committee on Atomic Energy, RG 128, National Archive.
JRO: J. Robert Oppenheimer papers, Library of Congress, Washington, DC.
JRO/FBI: J. Robert Oppenheimer file no. 100–17828, FBI reading room.
LAHS: Los Alamos Historical Society Archive, Los Alamos , New Mexico
LANL: Los Alamos National Laboratory archives, Los Alamos, New Mexico.
LLNL: Lawrence Livermore National Laboratory, Livermore, California.
LLS/HHPL: Lewis Strauss papers, Herbert Hoover Presidential Library, West Branch, Iowa.
MED: Records of the Manhattan Engineering District, RG 77, National Archives.
NARA: National Archives, Washington, DC, and College Park, MD.
OSRD: Records of the Office of Scientific Research and Development, RG 227, National Archive.
PSAC: Records of the President's Scientific Advisory Committee, RG 359, National Archive.
Science Archive. 34-42 Cleveland Street, London W1P 6LB

The Maria Mayer letters are held in the University of San Diego Library, San Diego, California. The Teller papers are held at the Hoover Institution Library and Archives, Stanford University.

Introduction

PAGE

xix 'FLASH: . . . COLUMBIA THEA-TRE (LONDON)': *Variety,* 5 February 1965, p. 1.

xix '*Variety*'s list of All-Time Top Grossers': *Variety,* 6 January 1965, p. 39.

xx 'Following this approach . . . grisly laugh': Nelson (1982), p. 81.

xx 'You can't fight in here . . . living or dead': *Dr Strangelove,* screenplay by Terry Southern, 1964.

xxii 'scientifically organised nightmare . . . the human race': Lewis Mumford, letter, *New York Times,* 1 March 1964, p. 25.

xxii 'As recently as 1999 . . . the office': 'Infamy and Humour at the Atomic Café', *Scientific American,* October 1999.

xxiv 'a danger to all that is important . . .

without Teller': I. I. Rabi, quoted in Blumberg and Owens (1976), p. 1.

xxiv 'a great man . . . of science': Eugene Wigner (1988), pp. 1, 12.

Chapter 1: War, Revolution, Peace and Maths

1 'was too worried by the dangers . . . capacity for worry': Teller (2001), p. 21.

1 'I hear myself . . . of my parents': ibid., p. 4.

2 'Now, perhaps I have . . . while she is away': Blumberg and Owens (1976), p. 4.

3 'Perhaps as a consequence . . . her love for him': Teller (2001), p. 26.

3 'My mother and her sister . . . I was not surprised by it': ibid., p. 27.

4 'Both my parents were kind . . . worried in imitation': ibid., p. 6.

5 'My love of music . . . of feeling secure': ibid., p. 7.

6 'The Communists overturned . . . in our home': ibid., p. 13.

6 'On one of them . . . wherever I went': ibid., pp. 13–14.

7 'But there was . . . dislike cabbage': ibid., p. 15.

7 'Too many . . . excesses': ibid., p. 15.

7 The story of the 'red train' is still to be found on various websites interested in establishing Jews as the originators of communism, e.g. www.holywar.org

8 'Within my first . . . more complete': Teller (2001), p. 16.

8 'With thoughtless honesty . . . chess together again': ibid., p. 20.

8 'I was never able to talk to him . . . decisions about my life': ibid., p. 20.

9 'more of a friend . . . her optimism': ibid., pp. 17–18.

9 'Helping him to dress . . . Emmi more than him': Blumberg and Owens (1976), p. 12.

9 'a little younger . . . semi-revolt': Teller (2001), pp. 21–22.

9 'I reached adolescence . . . was intolerable': Teller (2001), p. 32.

10 'little for one's . . . social outcast': ibid., p. 22.

10 'Klug could do . . . its own sake': ibid., p. 23.

10 'even when . . . teased me about': ibid., p. 33.

11 Teller's nickname of Coco: Teller refused to reveal its significance to Blumberg and Owens during the research for their 1976 book. When the authors pressed him, arguing that such a mystery could 'make or break' the book, Teller 'facetiously answered, "Break the book."' See Blumberg and Owens (1976), p. 25.

11 'mentally retarded . . . happening to me': ibid., p. 34.

12 'A cosine . . . but I seldom laughed': ibid., pp. 35–6.

13 'the price of my good fortune felt high': Edward was to go hiking with his friends the following summer but as science took over his life he saw less and less of them. They attended his marriage in 1934 and he saw them in 1936 on his last visit to Budapest for half a century. Suki, his brother-in-law, and Tibor were to die in Nazi concentration camps. Only Nándi survived to welcome the Russians. As Edward wrote: 'No nightmare I imagined could have been as awful as the reality these dear friends of my youth lived out.' One of the two students with whom he shared the Eötvös maths prize, Lázló Tisza, was to become a friend and work with Edward on one of his first research projects at Leipzig. Teller (2001), p. 41. See chapter 2, p. 22.

Chapter 2: In the Company of Gods

14 Teller's arrival in Karlsruhe: Blumberg and Owens (1976), p. 33.

14 'My family was . . . with happiness': Teller (2001), p. 45.

14 Teller's early laboratory experiences: ibid., p. 42.

15 'Teller really wasn't . . . tell us was this . . .': Blumberg and Owens (1976), pp. 34–5.

16 Arnold Sommerfeld: Teller's new

professor at Munich; he had modified Bohr's original model to better
fit experimental results by proposing that the electronic orbits were
elliptical rather than circular.

16 'By this time . . . but barely':
Nuckolls interview with Edward
Teller, p. 19. Science Archive.

16 'She did not . . . being sustained':
Blumberg and Panos (1990), p. 28.

16 During his stay in hospital, Edward
became very friendly with his
young surgeon, von Lossow. Lossow
it was who told him that people
used to believe that phantom pain
from an amputated limb was the
result of the missing part having arrived prematurely in hell, and that
if he 'continued to misbehave and
go to hell', he said, 'that's what you
will feel like all over'. Teller (2001),
p. 49.

18 'Heisenberg was only . . . role of
Geheimrat': Teller (2001), p. 52.

18 'Would my . . . looking in?': ibid.,
p. 56.

19 The story about Landau is to be
found on www.anecdotage.com

19 'Wo is der Witz? . . . the joke?': Teller
(2001), p. 57.

19 'They were excellent . . . Can you
solve it?' Blumberg and Owens
(1976), p. 39.

20 'The assignment . . . to the work':
Teller (2001), p. 60.

20 'I particularly . . . more polite':
ibid., p. 61.

20 'I could see . . . make up for it':
Blumberg and Owens (1976), pp.
40-41.

20 'Teller would . . . all back':
Newhouse (1989), p. 38.

Chapter 3: Twilight of a Golden Age

24 'spared me . . . barber shop': Teller
(2001), p. 77.

24 'amazed me . . . in Leipzig': ibid.,
p. 75.

25 'But far worse . . . as love or war
. . .': ibid., p. 78.

25 'He was a man . . . at least':
Blumberg and Owens (1976), p. 47.

25 'He was also . . . a collaborator':
Teller (2001), p. 78.

25 Teller's cousin George: George was
the embittered elder of the two
Dobo brothers, the younger of
whom, Stephen, had committed
suicide at the age of fourteen—see
Teller (2001); chapter 1, pp. 3–4.

26 'At the game's end . . . left-handed':
Teller (2001), p. 80.

26 Harold Nicolson's description, in
his 1932 book *Public Faces,* of an
'atomic bomb', written seven years
before the discovery of fission, is
truly prescient. In it he described an
element 'so unstable that beside it
radium would be as dull as lead'
and which would 'transmute itself,
as radium transmutes itself into
lead, but with infinitely more vio-
lence; in fact, with an explosion
that would destroy all matter
within a considerable range . . .'
Quoted in Teller (2001), p. 87.

27 'Throughout the city . . . was to
come': Teller (2001), p. 86.

27 'The most important . . . arbitrary
additions': Blumberg and Owens
(1976), p. 49.

28 'a period of beauty . . . can change
the fact': Teller (2001), pp. 90–1.

28 'unique . . . golden years': ibid.,
p. 93.

29 'As I climbed . . . answer was yes':
ibid., pp. 94–5.

29 'Carl von Weizsäcker . . . against in
my youth . . .': ibid., p. 97.

30 'I missed Mici . . . change of plans':
ibid., p. 100.

30 'The official . . . ceremony was
touching': ibid., p. 102.

31 'Bohr invented . . . never met as human beings': ibid., p. 108.

33 'London in 1934 . . . in the world':
ibid., pp. 120–1.

Chapter 4: America the Beautiful

35 'There's this one man . . . just everything': Blumberg and Owens
(1976), p. 65.

35 'Gamow chose . . . selecting questions': Teller (2001), p. 123. In the

late 1940s, Gamow, along with a colleague, Ralph Alpher, wrote an important paper on the 'big bang' theory of the origin of matter. Its importance, however, did not prevent him from inviting Hans Bethe to become one of the paper's authors, so that he could entitle it 'The Alpher–Bethe–Gamow Theory'.

36 'Edward as a teacher . . . what we had learned': interview with Harold Argo, July 2001.

36 'All relished . . . Teller's forte . . .': Blumberg and Owens (1976), p. 74.

36 The Jahn–Teller effect: Teller had been looking at the carbon dioxide molecule, which was seen as an oxygen atom, a carbon atom and a second oxygen atom arranged along a straight line ($O=C=O$). He and Herman Jahn had shown that only in molecules like CO_2, where the atoms lie alonga straight line, could two electrons exist on the same energy levels, and that this could only result from the rotation of the electrons in eithera clockwise or an anti-clockwise direction about this straight-line molecular axis. The discovery of this clockwise and anti-clockwise electron rotation, the Jahn-Teller Effect, is now the focus of an annual international symposium, and has been fundamental in understanding such phenomena as colossal magneto-resistance and 'bucky-balls', clusters of carbon atoms grouped in a geodesic structure. These have applications ranging from computer information storage and superconductors to new fuels and drugs.

37 'Gamow was important . . . nucleons can interact': interview with Hans and Rose Bethe, April 2002.

37 'wonderful quiet period': Teller (2001), p. 135.

38 'It was a wide open . . . or things like that': interview with Hans and Rose Bethe, April 2002.

38 'Hans and I . . . quite selfishly': ibid.

38 'His bad foot caused . . . brave about it': ibid.

39 'From my experience . . . and to Mici': ibid.

39 'He wanted . . . everyone to be special': interview with Frank Oppenheimer, September 1978.

40 'I found talking with him . . . handle comfortably': Teller (2001), p. 133.

40 'overpowering': Blumberg and Owens (1976), p. 75.

41 'speed and attention . . . Zllooooh!': Teller (2001), p. 135.

41 'If you want . . . best choice': ibid., p. 137.

42 'Edward is responsible for my fame . . . problem for me': interview with Hans and Rose Bethe, April 2002.

42 'Bohr has gone crazy . . . was talking about': Teller (2001), p. 139.

43 Teller's memory of the subdued discussion is described in Teller (2001), p. 140.

44 'If fission had been . . . influence on history': Teller (2001), p. 141.

44 'When I invited him . . . right conclusions': interview with Luis Alvarez, quoted in Rhodes (1986), p. 274.

44 'Within perhaps . . . of a bomb': quoted in Weiner (1972), p. 90.

44 'We drove to our home . . . depend on it': Teller and Brown (1962), pp. 9ff.

45 'I was thrown out of Rutherford's office': Blumberg and Owens (1976), p. 86. As a gesture of defiance, Szilard took out a patent on the chain reaction—for which he was paid a derisory $20,000 by the US government after the war.

45 'Nuclear energy . . . fission process': Teller (2001), p. 142.

45 'New York. Szilard . . . nothing': Nuckolls interview with Edward Teller. p. 76. Science Archive.

46 'strongly appealed . . . of these discoveries': L. Szilard to Arthur Compton, 12 November 1942, p. 3. MED/NARA 201, quoted in Rhodes (1986), p. 292

46 Szilard's struggle to interest the

Navy is described in Rhodes (1986), pp. 292–294; Blumberg and Owens (1976), pp. 90–91; Herken (1992), pp. 7–8.

46 'an explosive that . . . against this': quoted in Fermi (1954), p. 162.

46 John Wheeler: Teller had met Wheeler when both were newly married Rockefeller fellows in Copenhagen.

47 'We argued . . . he declared': Teller (2001), p. 143.

47 'Joliot will not . . . of the neutron': ibid., pp. 143–4.

48 'I drove down . . . agreed not to': ibid., p. 144.

Chapter 5: The Hungarian Conspiracy

49 'There's a WOP outside': interview with Hans Bethe, 1982, quoted in Rhodes (1986), p. 295.

50 'like a painter's assistant': Emilio Segré interview, 1983, quoted in Rhodes (1986), p. 299.

50 'I lectured . . . communication chain': Teller (2001), pp. 144–5.

51 The conspiracy's origins are described in Rhodes (1986), pp. 303–307 and Blumberg and Owens (1976), pp. 92–96.

51 'shared the opinion . . . United States be advised': Szilard (1972), p. 214.

51 'so we began . . . to Germany': Weart and Szilard (1978), p. 82.

51 'I never thought of that!': Clark (1971), p. 669.

51 'very quick . . . a false alarm': Weart and Szilard (1978), p. 83.

52 'he took the position . . . Roosevelt in person': ibid., p. 91.

52 'not only . . . particularly nice': ibid., p. 91.

52 'I entered history as Szilard's chauffeur': NOVA (1980).

52 'Einstein . . . we left': Teller (2001), p. 147.

53 'Alex . . . requires action': quoted in Rhodes (1986), p. 314.

53 It was Alistair Cook in his *Letter from America*, August 2002, who likened the Einstein letter

to the Declaration of Independence.

53 'We realised . . . atomic bombings': Wigner (1945).

53 'then the . . . is abolished': Weizsäcker (1978), pp. 199ff.

54 'His experience . . . chauffeur to messenger boy': Teller (2001), p. 148.

54 Cross sections are a convenient way in which physicists indicate whether a particular nuclear reaction will or will not happen. What they are is neatly explained in an analogy used by the physicist Rudolf Peierls: 'For example, if I throw a ball at a glass window one square foot in area, there may be one chance in ten that the window will break, and nine chances in ten that the ball will just bounce. In the physicists' language, this particular window, for a ball thrown in this particular way, has a "disintegration cross section" of $1/10$ square foot and an "elastic cross section" of $9/10$ square foot.' Cross sections on a nuclear scale are measured in tiny fractions (10^{-24}) of a square centimetre, and there are different sorts. The main one is the capture cross section, describing the likelihood of a neutron being captured by the nucleus—the equivalent, in Peierls's analogy, to the chance of the window being open and the ball going through, Rhodes (1986), p. 282.

54 'At Aberdeen . . . perfectly healthy': Teller (2001), p. 148. The meeting is also described in Rhodes (1988), pp. 315-317 and Blumberg and Owens (1976) pp. 98–99.

55 'The important thing . . . to be ahead': quoted in Rhodes (1986), p. 316.

55 'I said that . . . this is expensive': Blumberg and Owens (1976), p. 98.

55 'The researchers . . . my request' Teller (2001), pp. 148–149.

55 'Wigner, the most polite . . . your money': Hewlett and Anderson (1962), p. 20.

56 'Roosevelt . . . in mind': Science Archive interview with Edward Teller, May–June 1996, p. 85.

56 'I was one . . . protect freedom': Teller (2001), p. 150.

57 'After I received . . . Do I regret my decision? No': ibid., p. 152.

58 The MAUD Committee: the Committee received its eccentric name through a series of complicated misunderstandings. It was the Christian name of the English nanny of Niels Bohr's son.

58 'amazed and distressed . . . of his committee': Oliphant (1982), p. 17.

58 'Conant began to be convinced . . . I'll do it': Compton (1956), pp. 7ff. Oliphant's lobbying is described in Rhodes (1986), pp. 372–378.

59 No one can argue with the courage and determination of the Hungarian 'conspirators' and their accomplices, but many do argue with just how effective their initial efforts were in catalysing the race to produce the atomic bomb. 'At the fortieth anniversary of Los Alamos,' Herb York recalled, 'Serber was the main speaker on history. In answer to a query about the role of the Einstein letter, he said it delayed the project by about six months. It resulted in the wrong people (Briggs and Breit and some military) focusing on the wrong issues.' (Herb York, personal communication with the author, November 2003).

Certainly, it was only after Oliphant's visit, his explanation of the MAUD Committee's work and their involvement that matters began to move apace. Work at Berkeley on plutonium continued independently of the Briggs committee. However, Fermi's experimental work at Chicago was a result of the Einstein initiative and of considerable importance, both in terms of what it achieved and how it prepared the ground.

59 'Fermi asked me . . . to produce fusion': Teller (2001), p. 157.

59 'then, during . . . my explanation': ibid., p. 157. Fermi and Teller were not the first to consider the possibility of a fission reaction producing heat sufficient to create a fusion reaction between two atoms of a light element. In May of that year, Tokutaro Hagiwara, a physicist at the University of Kyoto, had given a lecture entitled 'Super-explosive U-235' (see Rhodes 1986, p. 375). In it he said, 'If by any chance U-235 could be manufactured in a large quantity and of proper concentration, U-235 has a great possibility of becoming useful as the initiating matter for a quantity of hydrogen. We have great expectations for this.'

60 'about the...family matters': AEC(1954), p. 575

60 'I decided . . . in the world': ibid., p. 9.

60 'When he came out . . . solutions': Teller (2001), pp. 157–8.

Chapter 6: Skirmishes

61 'The real problem . . . the enemy lines': Teller (2001), p. 158.

61 'Breit was always . . . became uninformative': Goodchild (1980), p. 48.

61 The Berkeley Conference is described in Rhodes (1986), pp. 415-420

62 'beautiful place . . . a "palace" ': Teller (2001), p. 159.

62 'our best friends in the country': interview with Hans and Rose Bethe, April 2002.

62 'He had set up . . . of graphite': Bernstein (1980), p. 71.

63 'On that visit . . . would work': Interview with Hans and Rose Bethe, April 2002.

63 'We had a compartment . . . the hydrogen bomb': Bernstein (1980), pp. 72ff.

64 'The theory . . . to do much': interview with Hans Bethe quoted in Rhodes (1986), p. 417.

64 'and on a walk . . . I decided to do it': Bernstein (1980), p. 73.

65 In his *Memoirs*, Teller writes: 'The question of igniting the atmosphere, if it was mentioned at all,

was not discussed in any detail at the summer conference. It was not an issue.' He recalled talking with Oppenheimer when he returned, who had told him that he had provided the novel surprise of potential atmospheric ignition, as 'further reasons for establishing a laboratory to work on weapons design'. Compton's account, Teller believed, was simply the result of his confusion over quantum mechanics and over what Oppenheimer had told him. In denying that the concern over atmospheric ignition occurred at this stage, Teller is flying in the face of evidence, not just from Compton but from Bethe, who did comprehend quantum mechanics. See Teller (2001), p. 160.

65 'Oppie took it . . . we knew more': interview with Hans Bethe quoted in Rhodes (1986), p. 418.

65 'I'll never forget . . . must never be made': Compton (1956), p. 127.

65 'I very soon . . . by my arguments': Rhodes (1986), p. 419.

65 'Calculation . . . common sense': Hawkins (1947), pp. 15–16.

66 'My theories . . . positive conclusion': Teller and Brown (1962), p. 39.

66 'However . . . uninteresting': Teller (2001), p. 161.

67 '150 times . . . will be ready': handwritten notes, 26 August 1942, 'Structure of the Bomb', Bush-Conant File f14a, Office of Scientific Research and Development, S-1 (RG 227), National Archive, Rhodes (1986), pp. 420- 421.

67 'We have become . . . terminated': Executive Committee report, 'Status of the Atomic Fission Project', Bush-Conant File f12. Ibid., p. 421.

67 'The physicists . . . [last] report': quoted in Rhodes (1986), p. 421.

68 'He's the . . . understanding': interview with General K. D. Nichols, September 1978.

69 'There is one last thing . . . like that': Goodchild (1980), p. 60.

70 'Mici had been convinced . . . buying furniture': Teller (2001), p. 162.

71 'As we travelled . . . never returned': ibid., pp. 163–4.

71 'Perhaps I over-reacted . . . Oppenheimer's personality': ibid., p. 164.

72 'very good . . . quite close': interview with Hans and Rose Bethe, April 2002.

72 'a mental . . . in conversation': Coughlan (1954), p. 90.

72 The first controlled fission chain reaction is described in Rhodes (1986), pp. 436-442

73 'The Italian . . . friendly': Los Alamos 50th Anniversary website, www.lanl.gov.

74 'I loaded . . . other migrations': Teller (2001), p. 165.

Chapter 7: Maverick on the Mesa

75 'unmarried couples': interview with Priscilla Duffield, September 1978.

75 'doing his usual magic . . . going to be sorted?': interview with Robert Wilson, September 1978.

76 'So I called . . . has not come again': Brode (1997), p. 11.

77 Life at Los Alamos: there were also drama productions on the Mesa. In the laboratory's production of the black comedy *Arsenic and Old Lace,* while Oppenheimer played the first corpse, it was not without significance that Teller played the second.

77 'pleasant friends . . . friendships completely': interview with Alice Smith, 28 May 1992, LAHS.

77 'In those days . . . some of these': interview with Cyril Smith, 28 May 1992, LAHS.

77 'That never . . . with the Tellers.': interview with Alice Smith, 28 May 1992, LAHS.

77 'The Smiths . . . worry about him': Brode (1997), pp. 41ff.

78 'One moonlit . . . piano moved': ibid., p. 45.

78 'There is no question . . . conflicting personalities': ibid., p. 47.

79 'He wasn't much of a speaker . . . all he cared about': Goodchild (1980), p. 80.

79 'a very . . . to come': Rhodes (1986), p. 453.

80 'Rabi told . . . began to change': interview with Hans and Rose Bethe, April 2002.

80 'Look . . . those reactions': ibid.

80 'I suspected . . . brick maker': Teller (2001), pp. 176–177.

80 In a conversation with me (July 2001), Dr Teller referred to 'Herr Geheimrat Bethe', and suggested this title—associated amongst German students with pomposity—should underscore the view I took of Dr Bethe.

81 'If our work . . . again had a major voice': Bethe (1982), p. 44.

85 'blow in a beer can without spattering the beer': Davis (1968), p. 216.

85 'At that moment . . . research around': Teller (2001), p. 175. There are differences between Teller and Bethe over credit for this work. Hans Bethe, along with the official history (Hawkins, 1983, p. 82), attributes the fundamental idea of using shaped charges to a member of the British Mission at Los Alamos, James Tuck, who had worked on shaped charges back in the UK. He does credit von Neumann with the calculations to produce the shaped charges and Teller with calculating the effect of pressure on the critical mass. 'If you compress everything,' Bethe said, 'the uranium critical mass goes down as the square of the density. You are able to reduce the critical mass enormously. Teller devised that calculation' (interview with Hans and Rose Bethe, April 2002).

87 'Both because...originally anticipated': Rhodes (1986),p.544.

88 'Well . . . our friendship': Teller (2001), p. 177.

88 'Only after . . . atomic bomb': Bethe (1982), p. 44.

89 'That arbitrary . . . left the room': Teller (2001), p. 193.

89 'his wish . . . was irksome': Bethe (1982), p. 44.

90 'But there . . . Edward Teller': interview with R. Peierls by Alice Kimball Smith, 1975, quoted in Smith and Weiner (1980), p. 273.

90 'I may be unjust . . . not done': Blumberg and Owens (1976), pp. 136ff.

91 'The scientific community . . . were very great': Teller (2001), p. 183.

Chapter 8: The Little Toe of the Ghost

93 'Oak Ridge . . . how to get there': Jordan (1952), p. 83. Jordan's raid is described in Rhodes (1995), pp. 95–100.

93 'they had only to make known their desires': New York Times, 31 August 1951, quoted in ibid., p. 66n.

93 'mass production': Gouzenko (1948), p. 123.

93 'The only spot . . . October 1942': General Groves, quoted in Jordan (1952), pp. 85ff. Rhodes (1995), p. 100.

93 The incident with Maria Mayer and the security officer is described in Teller (2001), p. 187.

94 'I had . . . long[,] involved operas': Teller (2001), pp. 192 and 192n.

95 'Maria and my father . . . if it's not true': interview with Paul Teller, April 2002.

95 'Edward didn't like . . . Judaic law': interview with Judith Shoolery, October 2003.

95 While by no means a 'womaniser' in the same style as Kennedy, it became widely known that Oppenheimer had maintained contact with Jean Tatlock after his marriage, though this relationship with his former lover had a great deal to do with his concern over her state of mind. Later, however, he earned Ernest Lawrence's disapproval when he had seduced Ruth Tolman, the wife of Caltech physicist, Richard Tolman. Lawrence had passed on the information to Lewis Strauss who wrote that 'according to Dr

Lawrence, it was a notorious affair which lasted for enough time for it to become apparent to Dr Tolman who died of a broken heart' (Strauss to file, 9 December 1957, Box 1, LLS/HHPL). Strauss had passed this information on to Teller (Herken, 2002, p. 290n).

96 'The head of the . . . Landau and myself': Holloway (1994), p. 43.

96 'a tease . . . in others': ibid., p. 43.

97 'For a considerable part . . . for fifteen years': Teller (2001), p. 182.

97 'continuing to . . . were very great': ibid., p. 183.

98 'When the Nazis . . . my admiration': Rhodes (1995), p. 56.

99 The Alsos Mission: 'Alsos' is Greek for 'Groves'—some people felt this somewhat obvious link not only represented an increased risk to security but was also a rather poor joke.

99 'If the Germans . . . then we'll use it': interview with Sam Goudsmit, September 1978.

100 'I personally feel . . . of the war': letter, Szilard to Teller, reproduced in Teller (2001), pp. 204–205.

101 'stay ahead . . . with Russia': Rhodes (1986), p. 646.

102 'no acceptable alternative to military use': ibid., p. 697.

102 'in a polite . . . I did not': Teller and Brown (1962), pp. 13ff.

102 'Oppie's reaction . . . to be made': Teller (2001), p. 206.

102 Teller's letter is quoted in ibid., p. 208.

103 'The more decisive . . . helped it to escape': letter quoted in ibid., pp. 207ff.

103 'For them . . . beginning of that controversy': Palevsky (2000), p. 49.

103 Trinity dedication to Tatlock, Herken (2002), p. 129.

104 'take along a bottle of whiskey': Teller (2001), p. 211.

105 'There were pools . . . this thing off?': Goodchild (1980), p. 160.

105 'We were all . . . not be an experiment': Teller (2001), pp. 211ff.

106 'A few people . . . shatterer of worlds': Oppenheimer quoted in Giovanetti and Freed (1965), p. 197.

106 'Oppie . . . sons of bitches': Wilson (1975), p. 230.

106 'that we had . . . against the Japanese': Truman (1955), p. 416.

Chapter 9: The Legacy of Hiroshima

107 'The mushroom . . . the smoke': Mark (1967), pp. 171ff.

108 'I just could not . . . this was it': Lifton (1967), pp. 22ff.

108 'He entered the . . . the podium': Goodchild (1980), p. 167.

109 'expect a . . . on this earth': White Press Release, 6 August 1945.

109 'If the baby doesn't . . . There will be no refusals': Rhodes (1995), p. 178.

109 'the mountain quiet . . . was over': Teller and Brown (1962), p. 21.

110 'We all felt . . . weapons in peacetime': Bethe (1982), p. 45.

110 'Oppenheimer . . . Oppenheimer's mind': Teller (2001), p. 219.

111 'not merely . . . hearts and spirits': Harrison, 'Memorandum for the Files', 25 September 1945, no. 77. Harrison in Bundy file, MED/NARA.

111 'In that conversation . . . Germany had been': interview with Hans Bethe, quoted in Blumberg and Owens (1976), p. 185.

111 'I assume that . . . atomic bombs . . .': Teller (2001), pp. 218–19.

111 Throughout his life, Edward Teller campaigned vociferously against secrecy. He also summed up his position, which was to change little over the years, in the paper he wrote for Brunauer and the navy: 'Pure scientific data—that is, facts concerning natural phenomena—must not be kept secret. If such secrecy is continued, it will warp the entire research activity of any man who is involved in work on atomic power. He either has to sever relations with the scientific world not involved in the development of atomic power or he has to acquire a split person-

ality, remembering in certain parts of his work only certain parts of the information available to him. Furthermore, scientific facts cannot be kept secret for any length of time. They are readily rediscovered. If we attempt to keep scientific facts secret, it will certainly hinder us but will hardly interfere with the work of a potential competitor' (Teller, 2001, p. 226). In spite of this position, Teller was to rigorously observe security procedures—to the point where both his strict observance, and his apparent lapses, were seen, certainly by his opponents, as taking political advantage of the 'secrets' to which he had access.

112 'I said . . . tests a year': Teller and Brown (1962), p. 22.

112 'Unquestionably . . . throughout his career': Teller (2001), p. 221.

112 'I neither can nor will do so': Teller and Brown (1962), p. 23.

112 'drifted over . . . in any way': Teller (2001), pp. 221–2.

113 'a beautiful pantomime . . . he represented': Teller to Maria Mayer, 25 September 1945.

113 The 'secrets' of the plutonium metallurgy are described in Rhodes (1995), p. 192.

114 'Right from the start . . . influencing the system from within': interview with Frank Oppenheimer, September 1978.

114 'Only three . . . will be eliminated': Teller (2001), p. 224.

115 'Everyone . . . in the morning.' Goodchild (1980), p. 47.

115 'International control . . . atomic bombs': Barnard et al. (1946), p. 8.

115 'There was a . . . international control': interview with Isidor Rabi, September 1978.

115 'Nothing that we . . . world union': Teller (1946), pp. 8, 10, 13.

115 'There is . . . powers to prevent it': Rhodes (1995), p. 208.

116 'The magnitude . . . on a vast scale': Aspray (1990), p. 47.

118 ENIAC, the Electrical Numerical Integrator and Computer: it was

through his work in producing this memory that von Neumann developed the process of digital computing.

118 'hopes for . . . in general': Aspray (1990), p. 47.

119 'What I saw . . . world of 1939': Teller (2001), p. 227.

119 'The Super conference is described in Rhodes (1995), pp. 252–255; Herken (2002), pp. 171–173; Fitzpatrick (1998), pp. 119–124; Hansen 'The April 1946 Super Conference.'

119 'a large scale . . . is justified': Frankel (1946), abstract, and p. 47.

120 'exploring in great . . . usable system': Bradbury press conference, 24 September 1954, LANL archives.

120 'a reaction . . . within one or two years': 'Report of Conference on the Super', LANL, LA-575, pp. 44–5.

121 'I found the report . . . we agreed on': Fitzpatrick (1998), pp. 123–4. Much later, in 1954, Carson Mark was to write a report on the work done on thermonuclear weapons at Los Alamos in post-war years (LA-5647). 'The reason I wrote the report,' he told Judith Shoolery in 1992,'is that Rabi came to Los Alamos for a visit in 1954 and reported that two *Time–Life* reporters [Shepley and Blair] were getting ready to do a hatchet job on the laboratory. They never came, but in getting ready for them, I documented activities prior to 1952 . . . The impression that we were not working on the Super is not justified.'

121 'He maintained that . . . foreseeable future': Teller (2001), pp. 236–7. Teller quotes as evidence from Truslow and Smith (1983), Appendix A, p. 363. On 1 October 1945 Bradbury gave a speech saying that although fundamental experiments to establish the feasibility of a Super would be undertaken, 'this does not mean we will build a Super. It couldn't happen in our time in any event.'

122 Kennan telegram: Kennan (1967), pp. 547ff. Growing international tension is described in Rhodes (1995), pp. 233–238.

122 'Because we . . . what has happened': interview with Isidor Rabi, September 1978.

Chapter 10: Wilderness Years

124 'A place . . . and I do': Teller letter to Maria Mayer, 20 December 1948.

124 'It was a wonderful . . . a bit miserable': interview with Paul Teller, April 2002.

125 'If we . . . surely be lost': *Bulletin of the Atomic Scientists,* December 1947, p. 122.

125 Norris Bradbury: Bradbury had been a professor at Stanford before going to Los Alamos.

125 'I saw that Mici . . . ladyship arrived': Teller letter to Maria Mayer, 31 August 1946.

126 'Pedestrians shall . . . side walk': Teller letter to Maria Mayer, 10 September 1946.

126 'But I can tell you . . . would help': Teller letter to Maria Mayer, September 1946.

127 'since 5.30 pm yesterday . . . from Chicago' (ibid).

127 'Teller had done . . . liking to him': Dyson (1979), p. 88.

127 'That same . . . from my father': interview with Paul Teller, April 2002.

128 'If only he could . . . concentrate on': quoted in Moss (1968), p. 68.

128 'While his ideas . . . a prima-donna': Peierls (1985), pp. 199–200.

128 The scientific problem that Edward had failed satisfactorily to address for Maria concerned the fascinating problem of 'magic numbers'. Maria had spotted that atomic nuclei that had either 2, 6, 14, 28, 50, 82 or 126 protons or neutrons were far more abundant than those with very similar proton or neutron counts. If the total of both proton and neutron counts were magic numbers (like lead 208, the isotope that is the major component of the metal and contains 82 protons and 126 neutrons), then that isotope was particularly abundant. This insight eventually led to an understanding of how atomic structure affected the difference in stability and reactivity among the elements.

128 'the proper . . . I have been': Teller letter to Maria Mayer, early August 1946.

128 'about the absence of . . . thermonuclear research': Teller (2001), p. 242.

128 The Alarm Clock is described in Rhodes (1995), pp. 305–306; Herken (2002), p. 173.

129 'We have been . . . very easily': Teller letter to Maria Mayer, undated, autumn 1946.

129 'One may . . . atomic warfare': The Russian Atomic Plan is described in Herken (2002), p. 187.

130 'Despite summer's distractions . . . war psychology': Sakharov (1990), p. 96.

130 'I did not start . . . about it less': Teller letter to Maria Mayer, August 1948.

131 'He is a good . . . an idealist': Dyson (1979), p. 88.

131 'would just as soon be the President of the United States': interview with Harold Agnew, July 2001.

131 'scientifically the most . . . stimulating': Blumberg and Owens (1976), pp. 109–10.

131 'You know . . . a lazy bum': Teller letter to Maria Mayer, 20 August 1947.

131 'I know I am . . . fool him completely': Teller letter to Maria Mayer, February/March 1948.

132 'Your letter . . . virus pneumonia': Teller letter to Maria Mayer, February/March 1948.

132 'and so . . . a theatre. Edward': Teller letter to Maria Mayer, April 1948.

132 'You know . . . After that I slept': Teller letter to Maria Mayer, August 1948.

132 The March 1947 FBI document and its impact is described in Herken (2002), pp. 178–181; Goodchild (1980), pp. 185–188.

132 'relative to . . . Oppenheimer': Hoo-

ver to Lilienthal, 8 March 1974, Box 2, JRO/AEC.

133 'would have . . . the project': Wilson to file, 11 March 1947, Box 2, JRO/AEC, quoted in Herken (2002), p. 178.

133 'US ATOM . . . loyalty or ability' *Washington Times-Herald,* 12 July 1947.

133 'Afterwards . . . Strauss would initiate': Teller (2001), p. 279.

134 'unfriendly hands . . . national security': AEC Minutes, Meeting No 95, 19 August 1947.

134 'No man . . . no part at all': JCAE investigation into the US Atomic Energy Project, Part 5, 8 June 1949, pp. 224–7.

134 'Well Joe . . . too well': Goodchild (1980), p. 195.

135 Sandstone is described in Rhodes (1995), pp. 320–321.

135 'If I am asked . . . (sure I am)': Teller letter to Maria Mayer, August 1948.

135 'Norris was rather . . . (. . . back to Los Alamos)': Ulam (1991), pp. 192–3.

135 'not want . . . last one': Teller letter to Maria Mayer, August 1948.

136 'happy about my . . . a personal affair': Teller letter to Maria Mayer, 10 October 1948.

136 'I suspect . . . make it': Teller (2001), p. 259.

136 'Los Alamos . . . look into cross sections': Teller letter to Maria Mayer, 20 January 1949.

137 'The conference . . . shake myself': Teller letter to Maria Mayer, early summer 1949.

138 'Roughly at that . . . McMahon': Norberg interview with John Manley, 1980, Bancroft Library. p. 82.

138 'Freddie is . . . will be unfulfilled?': Teller letter to Maria Mayer, summer 1949.

138 'Rabi came through . . . Three guesses': Teller letter to Maria Mayer, summer 1949.

139 'Please remember . . . You will have need of it': Teller (2001), p. 277.In 1962 Chadwick remembered the

conversation well. 'I agree that I said what you report, but I hoped I had said something about the future as well as the past. But however that may be, I am very happy to know that some part of our conversation was helpful' (ibid., p. 278n).

139 'Feel it . . . went away': Holloway (1994), p. 203.

139 '. . . and incidentally . . . an atomic bomb': Teller (2001), p. 278.

Chapter 11: The Taking of Washington

141 'We have evidence . . . in the USSR': *Los Angeles Times,* 24 September 1949.

141 'Doctor what . . . our friends': Goodchild (1980), p. 197.

141 'What shall we . . . shirt on': Teller and Brown (1962), p. 33.

142 'We shall stay . . . the same way': Teller letter to Maria Mayer, late summer 1949.

142 Details of William Borden's early career: AEC (1954), pp. 832–3. Rhodes (1995), pp. 357–358.

142 'streaming red sparks and whizzing past us': Borden (1946), p. x.

143 'Let Stalin decide—atomic peace or atomic war': Herken (1985), p. 39.

143 'step towards a thermonuclear bomb': AEC (1954b), p. 20. The immediate reaction to the Soviet bomb is described in Rhodes (1995), pp. 379–381.

143 'quantum jump': quoted in Strauss (1962), pp. 216–17.

144 'He said . . . and fast': Sidney Souers Oral History interview, 16 December 1954.1f. Harry S. Truman Library, quoted in Rhodes (1995), p. 381.

144 'might get there . . . impossible': AEC (1954), p. 774. The lobbying for the Super is described in Rhodes (1995), pp. 383–386; Herken (2002), pp. 201–204.

144 'polite call . . . ears sharpened': Teller (2001), p. 280.

144 'They give . . . tritium available': AEC (1954), p. 775.

145 'In the present . . . must go ahead': Teller (2001), p. 281.

145 'What you . . . travelling possible': ibid., p. 282.

145 'Russia may be ahead of us in this competition': quoted in Rhodes (1995), p. 384.

145 'very quickly started . . . sense of urgency': interview with George Cowan, July 2001.

145 'turned his chair . . . the matter': AEC (1954), pp. 777–8.

145 'Ernest Lawrence . . . have to offer?': Lilienthal (1964), p. 577.

146 'if the Russians. . . situation will be hopeless': JCAE (1953), p. 28.

146 'To my great pleasure . . . hydrogen bomb': Teller (2001), p. 283.

146 'equally undecided . . . had given': AEC (1954), p. 328.

146 'over my dead body': AEC(1954), p. 215.

146 'I am pretty sure . . . I am still coming': AEC (1954), p. 715.

147 'I was quite impressed . . . with my wife': AEC (1954), p. 328.

147 'Yes, we talked . . . as it was in '49': interview with Hans and Rose Bethe, April 2002.

147 'Hans Bethe . . . go back there': quoted in Bernstein (1980), p. 93.

147 'that after . . . we were fighting for': AEC (1954), p. 328.

147 'I am not sure . . . full of dangers': AEC (1954), pp. 242–3.

148 'Fermi did not tell me . . . and for a delay': Teller letter to Maria Mayer, October 1949.

149 'unfortunate if . . . among scientists': AEC (1954), p. 717.

149 'What happened . . . a strain, I'd say': Norberg interview with John Manley, 1978. Bancroft Library, p. 82.

149 At the end of October 1949, McMahon wrote to both Lilienthal and the Pentagon about the status of Soviet thermonuclear research: 'As you know, there is reason to fear that Soviet Russia has assigned top priority to development of a thermonuclear superbomb.' The speculative notion Teller had raised in his 'Russian Atomic Plan' a year earlier had apparently achieved the status of fact (Herken, 2002, p. 204).

149 'In retrospect . . . national security': Teller (2001), p. 286.

150 The GAC meeting is described in Rhodes (1995), pp. 395–406; Herken (2002), pp. 206–210.

150 'there were better things . . . this super programme': AEC (1954), p. 518.

150 'mostly psychological . . . around the house': Lilienthal (1964), p. 581.

151 'We built . . . for the second time': ibid., p. 581.

151 visceral response: Anders (1987), p. 59.

151 'an imaginative . . . its development': GAC 29–30 October 1949 reports reproduced in Seaborg (1990), III, p. 317Aff.

151 'We believe . . . hope of mankind': ibid.

151 'one of . . . physical means': ibid.

152 'rather violent discussion': Manley diaries, 1 November 1949, LANL Archives.

152 'What he says . . . much time': Lilienthal (1964), p. 584.

152 Klaus Fuchs's penetration of the Manhattan Project: Herken (2002), pp. 213–214. The FBI had been suspicious of Fuchs since August 1949, when they had discovered that the Russians were in possession of a top-secret paper on gaseous diffusion, written by Fuchs. Ibid., p. 385, n19.

153 '[The Frank–Haakon] . . . Robert was involved': Groves to Strauss, 4 November 1949, Strauss folder, Groves Papers/RG200, National Archives, quoted in Herken (2002), p. 214.

153 'Very unusual': Manley (1985), p. 5.

153 'Fermi's characteristic . . . ready to harvest': Teller (2001), p. 288.

154 'I wish . . . of sanity': Teller letter to Maria Mayer, 30 November 1949.

154 'helpless to aid [its] friends': Lilienthal (1964), pp. 616–17.

154 'a frenetic campaign to obtain converts': Manley (1987), p. 15. Lobbying by the various groups is described by Rhodes (1995), pp. 404–406; Herken (2002), pp. 216–217.

155 'Edward promised . . . didn't appal them': interview with Carson Mark, 3 June 1994, quoted in Rhodes (1995), p. 400.

155 'You know . . . "by example?"' Arneson (1969), p. 29.

155 'That is . . . believe them': Minutes, Jan. 9, 1950, no. CXXV, JCAE quoted in Herken (2002), p. 215–216.

155 'made a . . . we should do': Dean Acheson file memorandum, 19 January 1950, Harry S. Truman Library, quoted in Rhodes (1995), p. 407.

155 'What the hell are we waiting for? . . . with it': interview with Sidney Souers, 16 December 1954, p. 8, Harry S. Truman Library, quoted in Rhodes (1995), p. 407.

155 'necessary to . . . very beginning': ibid., p. 7 and ibid., p. 407.

155 '[Truman] said there . . . on the H-bomb': Eben Ayers, Truman's assistant press secretary, diary entry quoted in Rhodes (1995), p. 407.

155 'It is a part . . . super-bomb': quoted in Blumberg and Owens (1976), p. 231.

156 'I never . . . could have done': Bernstein (1975).

156 'You don't . . . plague of Thebes': Goodchild (1980), p. 204.

156 'I simply . . . fight if it must be': Teller letter to Maria Mayer, December 1949–January 1950.

Chapter 12: Unholy Alliances

157 'The roof . . . human race': Lilienthal (1964), pp. 634–5. The reaction to Fuchs's arrest is described in Rhodes (1995), pp. 411-413, Herken,(2002), pp. 218-219

157 'Is it all . . . put it there': Davis (1968), p. 324.

157 'Quite a few . . . a witch hunt': Teller letter to Maria Mayer, February 1950.

158 'But who shall . . . what is the right approach?' Teller letter to Maria Mayer, February 1950.

158 'Both Gamow and I shared . . . very much': Ulam (1991), p. 212.

159 'Sonny . . . daddy of them all': Herken (2002), p. 233.

159 That's what . . . hired Edward': interview with George Cowan, July 2001.

159 Carson Mark confirmed how widespread Teller's influence now was. '[AEC Chairman] Dean didn't understand the first thing about the problems of thermonuclear reactions, but he was being advised by LeBaron, Finletter, McMahon and his sidekick Borden, who were without understanding—and whose people were being advised by Edward.' (Mark interview with Judith Shoolery, 1992.)

159 'The idea . . . the AEC': Norberg interview with Darol Froman, 1980, p. 66., Bancroft Library.

159 'I feel that . . . of my life': Teller letter to Maria Mayer, February 1950.

159 'that great troubles would . . . his overwhelming ambition': Ulam (1991), p. 212.

159 'He was convinced . . . atomic weapons': interview with Hans and Rose Bethe, April 2002.

160 'There was . . . seven years': interview with Marshall Rosenbluth, July 2001. Ted Hall was a nineteen-year-old brought to Los Alamos in 1944 to work on the implosion bomb. He had immediately volunteered his services to the Soviets and passed back information many think was as significant as Fuchs's material. He was never arrested and only exposed in 1995 at the time the Venona transcripts were released. He died in 1999.

160 'on a honeymoon . . . is over': Teller (1950), p. 72.

160 'shocked by the . . . our efforts': Fitzpatrick (1998), p. 274. Herken (2002), p. 225.

160 'into the lowest class . . . similar vermin': Teller letter to Maria Mayer, March 1950.

161 'that Russia's . . . may be in actual production': memorandum Loper to Le Baron, 16 February 1950, Basis for Estimating Maximum Soviet Capabilities for Atomic Warfare, 20 February 1950, Box 211, Harry S. Truman Library.

161 'It may be stated . . . amounts of tritium': Teller's assessment of the Super, LA 643 in the LANL Archive, is quoted in Rhodes (1995), pp. 417-418.

161 'We shall see . . . disastrous effects': ibid.

161 'Nobody will blame . . . very incomplete': Bethe (1982), p. 47.

162 'Teller kept on . . . scheme': Ulam (1991), p. 213.

162 Carson Mark witnessed the growing friendship between Ulam and Gamow at this time, a friendship that, according to Mark, survived, 'even when Gamow was very difficult to be around, as an alcoholic at the end in Colorado'. (Carson Mark interview with Judith Shoolery, February 1992)

162 The extract from Ulam's personnel file comes from 'Stan, Teller and the H-Bomb, 1949–52', Françoise Ulam unpublished MS.

162 Ulam's short-sightedness: interview with Jay Wechsler, July 2001. As to Ulam's work-shyness, Teller wrote to Maria Mayer: 'Just before I left Los Alamos, Stan Ulam told me he wants to do some work. He wants to work himself, personally, not by proxy, as much as two hours per diem. I should have liked to know how Stan looks when he is working. No one ever caught him red-handed.'

162 'You'd be sitting . . . from it': interview with Marshall Rosenbluth, July 2001.

163 The Super calculations are described in Rhodes (1995), pp. 422–424; Herken (2002), pp. 222–224; and Hewlett and Duncan (1969), p. 440.

163 'then with us . . . not going to work': 'Stan, Teller and the H-Bomb, 1949–52', Françoise Ulam unpublished MS, p. 18.

163 'that the model considered is a fizzle': Hewlett and Duncan (1969), p. 440.

163 'was not . . . started immediately': Mark (1974), p. 8.

163 'Inside the Tech Area . . . Stan's figures': 'Stan, Teller and the H-Bomb, 1949–52', Françoise Ulam unpublished MS, p. 18.

163 'Teller thought . . . disliked each other': interview with Herb York, July 2001.

164 'He was always . . . should not be politicians': interview with George Cowan, July 2001.

165 'to keep Teller . . . Schintlemeister report': Arnold Kramish quoted in Rhodes (1995), p. 465.

165 'Now my only . . . Goddam Administrator': Teller letter to Maria Mayer, March/April 1950.

165 'the thing . . . away from going': Hewlett and Duncan (1969), p. 440.

165 'Teller was not easily . . . programme project': Ulam (1991), p. 216.

165 'intuitive . . . by Fermi': ibid. p. 219.

166 'Fermi filling in . . . would explode': interview with Richard Garwin, April 2002.

166 'You can't get...was optimistic': Rhodes (1995), p. 456

166 'I can't believe that Korea really is it': Teller letter to Maria Mayer, mid-July 1950.

166 'Look let us . . . ought to stop it': quoted in Cumings (1990), p. 756.

167 'It is wonderful . . . agreeing with me in advance': Teller letter to Maria Mayer, early summer 1950.

167 'it was like . . . to light a match': Jastrow (1983), p. 27.

167 'It turned out . . . about the same thing': interview with Carson Mark quoted in Rhodes (1995), p. 456.

168 The FBI's interest in Teller as a potential espionage agent is described in Rhodes (1995), pp. 429–30. If, however, the investigations had continued then they could have established the link between Teller's refusal to do further work on implosion at Los Alamos and the arrival of Klaus Fuchs in the laboratory. His persistence with the unworkable classic Super, and his refusal to accept the significance of compression, all arguably contributed to the delay in the programme.

168 'This was . . . was very important':
interview with Carson Mark in
Teller (2001), p. 302.

169 'As long as my father . . . proved
fruitless': ibid., p. 307.

169 'I was becoming truly . . . electronic
computer': ibid., p. 302.

169 'I'm an ultra-conservative . . . conservative policies': Terrall interview
with Willard Libby, p. 88, Bancroft
Library.

170 'so negative . . . squash the project':
Teller (2001), p. 303.

170 'In his briefing . . . the triumph':
Hewlett and Duncan (1969), p. 530.

170 'a successful test . . . important
questions': Teller (2001), p. 303.

171 'I went as an extra consultant . . .
wanted to do it': interview with
Hans and Rose Bethe, April 2002.

171 'I overheard . . . began to smoulder': 'Stan, Teller and the H-Bomb,
1949–52', Françoise Ulam unpublished MS, pp. 18–19.

171 'Gamow placed . . . the hydrogen
bomb': McPhee (1974), p. 90.

172 'that the hydrogen . . . at Los
Alamos': AEC (1954), p. 788.

172 'In fact . . . Los Alamos laboratory':
Goodchild (1980), p. 208.

172 Oppenheimer was . . . freshman
course': Goodchild (1980), p. 208.

Chapter 13: A 'Simple, Great and Stupid' Mistake

174 'Things . . . turned out no-how':
Teller (2001), p. 308.

175 'Engraved . . . someone else first':
Françoise Ulam, 'Stan, Teller and
the H-Bomb, 1949–52', Françoise
Ulam unpublished MS, pp. 19–20.

175 'I know that . . . talking about':
ibid., p. 20.

175 'For the . . . enthusiastically': Ulam
to Seaborg, 22 March 1962, LANL
Archives; Rhodes (1995), p. 466.

175 'By that time . . . repercussions':
Teller (2001), p. 311. Teller commented in a note: 'Unfortunately,
Carson passed away in the late
1990s, without my ever telling
him about the role he had played.'
This raises the question of why
this crucial anecdote was never

corroborated while Mark was
alive.

176 'The absorption . . . stupid': ibid.,
p. 313.

177 'Within an hour . . . mental inertia':
ibid., p. 314.

177 'Bradbury could hardly . . . I was
stupid': ibid., p. 314.

177 'It almost . . . had been met': ibid.,
p. 315.

177 'would serve . . . for many years':
ibid., p. 314.

178 'in the hope . . . at Los Alamos':
ibid., p. 315.

178 'an imaginative suggestion': Teller
(1955), p. 273.

178 'He [Ulam] announced . . . were
impractical': Teller (2001), p. 316.

178 'But Stan . . . through radiation':
ibid., p. 316.

178 'The next day . . . thermonuclear
explosion': ibid., p. 316.

179 'Edward said . . . point of collapse':
Arnold Kramish, quoted in Rhodes
(1995), p. 467.

179 'I wrote . . . these principles': Ulam
(1991), p. 220.

179 'The scheme . . . in the mass':
Teller–Ulam report, LAMS-1225;
Rhodes (1995), p. 467.

179 'My impression . . . he withdrew':
Ulam (1991), p. 311.

180 'Edward is . . . they will not work':
Hewlett and Duncan (1969), p. 537.

180 'To me the authorship . . . changed
his mind': Teller (2001), pp. 324,
324n.

180 'It's sort of . . . radiation implosion':
Rhodes (1995), p. 469.

180 'Had you sat down . . . than had
Ulam': Rhodes (1995), p. 470.

181 'the new concept . . . in 1939': Bethe
(1982), p. 49.

181 'I cannot tell . . . I believe so': Teller
interview with George Keyworth, 20
September 1979, pp. 14–15.

181 'The George Shot . . . depression':
interview with Hans Bethe, April
2002.

182 'I see it . . . quite clearly': interview
with Marshall Rosenbluth, July
2001.

182 'Ulam's idea . . . it was entirely him':
interview with Herb York, July 2001.

In his *Memoirs,* Teller (2001), p. 324, writes that 'one of the mathematicians working on the [Teller–Ulam] design claims that, after several months of work, Ulam did not understand the new design'. Norris Keeler disputes this. 'I don't believe this for a minute. I was a new hire from Berkeley when I was shown this scheme and understood it immediately—and I was far from the sharpest knife in the box—at least in that box' (Norris Keeler, personal communication).

182 'Once it was . . . not an option': interview with George Cowan, July 2001.

182 'outlined more . . . theoretical work': Foster Evans, letter to Judith Shoolery, 17 August 1990.

183 'help promote . . . of an organisation': Teller (2001), pp. 317–18.

183 'Following our . . . principle equipment': Teller to Gordon Dean, 20 April 1951. Teller Papers. Teller had copied his report to Norris Bradbury and he estimated that 'Christmas 1951 might see some experimental equipment in operation'—only eight months after the letter.

184 'He is one . . . them very strongly': interview with Paul Teller, April 2002.

184 After twenty-four years 'almost as a family member', Judith Shoolery, who co-authored Teller's memoirs, has a very clear view of his honesty and morality: 'I have seen him react to difficult situations within his work, misrepresentations by the media, public attack by purported friends, to grave ill-health, and to the ill-health and deaths of people he loved . . . He was in all situations a man of great moral character, a man who considered deeply the moral implications and effect of his actions . . . During that time he has not once misrepresented, much less lied about any current situation, nor have I in the course of substantiating and trying to disprove his accounts of events in the past ever

succeeded. It is of course, possible that I did not find the right question to tackle. But I believe that in the fourteen years that I was specifically engaged in searching for evidence to contradict his account, I have not found any.'

184 'A whole atoll . . . no racial problems': Teller letter to Maria Mayer, May 1951.

185 The preparations for the Greenhouse series are described in Rhodes (1995), p. 473.

185 'the vindication or non-vindication of [Teller's] major contribution' interview with Louis Rosen, July 2001.

185 'Why did one . . . fishy in my life': Teller letter to Maria Mayer, May 1951.

185 'but he never . . . from me': interview with Louis Rosen, July 2001.

185 'We were . . . recall those moments': York (1987), pp. 57–8.

186 'We walked . . . had been a success': Teller (1987), p. 80.

186 'Early next morning . . . the plane took off': interview with Louis Rosen, July 2001.

186 'high time . . . on H-weapons': AEC (1954), p. 305.

Chapter 14: Technically So Sweet

187 'Eniwetok . . . next one': Anders (1987), p. 144.

187 'This big . . . do this right now': interview with Herb York, July 2001.

187 'Although I was . . . next step': Teller (2001), p. 325.

188 'becoming aware . . . couldn't contain himself': interview with Hans and Rose Bethe, 26 April 2002.

188 'I was amazed . . . I insisted on being heard': Teller and Brown (1962), pp. 52–3.

188 'He was convinced . . . to be angry': interview with Herb York, July 2001.

188 'I asked . . . ". . . listen to Teller?"': Teller (2001), p. 325.

189 'Edward may have . . . and Mark': interview with Hans and Rose Bethe, 26 April 2002.

189 Oppenheimer's summary of deci-

sions at Princeton meeting: AEC (1954), p. 84.

189 'that everyone . . . was gone': AEC (1954), p. 305.

189 'so contrary . . . surprise to emerge': Teller (2001), p. 326.

190 'I told him . . . a second shot': Teller interview with George Keyworth, 20 September 1979, p. 18.

190 'Teller made an offer . . . program only': Borden's draft, 15 July 1951, JCAE Classified Document No. 2283, NARA, quoted in Rhodes (1995), p. 477.

190 'Bradbury was . . . much more serious': interview with Herb York, July 2001.

190 The politics behind the Mike test is described in Rhodes (1995), pp. 477–479.

190 'Teller would never be completely happy': Anders (1987), p. 161.

190 'available for crying [on]': ibid.

191 'Somewhat negative . . . project or not': Teller (2001), p. 327.

191 'None of the project leaders . . . more collegial': interview with Harold Agnew, July 2001.

191 'Marshall was . . . keep on thinking': interview with Jay Wechsler, July 2001.

192 'wonderful illogic . . . faced my opponents': Teller (2001), pp. 334–5.

193 'listened with . . . of that conversation': Teller (2001), p. 336.

193 'He asked me . . . nuclear weapons laboratory?': interview with Herb York, July 2001.

194 'There was time . . . personal feelings': interview with Hans and Rose Bethe, April 2002.

194 'the University of California . . . of pigs': Teller letter to Maria Mayer, August 1950.

194 'Since the days . . . Nylan . . .': Teller letter to Maria Mayer, September 1950.

195 'So there were . . . did it': interview with Herb York, July 2001.

195 'Acheson said . . . it is all about': Anders (1987), p. 204.

195 'bought Teller hook line and sinker': Walker and Borden to McMahon, 4 April 1952, no.

CDXCIX, JCAE, quoted in Herken (2002), p. 248.

196 'might have . . . Oppenheimer': San Francisco FBI to Hoover, 5 April 1952, JRO/FBI; Herken (2002), p. 249. Rhodes (1995), pp. 536–537.

196 'In his youth . . . cross examination': Albuquerque FBI to Hoover, 27 May 1952, McCabe to Hoover, 12 June 1952. JRO/FBI. In Teller (2001), p. 372, Edward described his conversation with the FBI as follows: 'As far as I was concerned, the advice of the committees Oppie had chaired had come close to shutting down the programme on several occasions, first after the war, then after the first Soviet atomic bomb test, and then a year later when the calculations suggested that the Super design would not work. Most recently, Oppenheimer had opposed a second laboratory that was intended to refine the new weapon as rapidly as possible. "But," I added pointedly [to the FBI agent], "it is improper to think of a man as a spy because of his opinions. There is nothing subversive about giving advice, no matter how much I disagree with the advice."'

197 Teller's meetings with the FBI are described in Rhodes (1995), pp. 537–538; Hewlett and Holl (1989), p. 77; Herken (2002), p. 249.

197 'There is . . . too much': Teller letter to Maria Mayer, June/July 1951.

197 The Teller family's deportation in Hungary described in Teller (2001), p. 326, and Blumberg and Owens (1976), pp. 173–6.

198 'a paranoid': a description of Griggs and Oppenheimer's meeting is in Griggs's testimony at the hearings, AEC (1954), pp. 757–8.

199 'Edward says . . . was just critical': interview with Jay Wechsler, July 2001.

200 'At one meeting . . . radiation to flow': interview with Hans and Rose Bethe, April 2002.

200 'This was . . . went away': interview with Jay Wechsler, July 2001.

200 'vitriolic talk . . . contingent':

Gordon Dean's diary, 19 May 1952, quoted in Herken (2002), p. 249.

200 'He was . . . unsettled by it': Teller (2001), p. 341.

201 'Ernest would like me to be the new director': ibid., p. 340.

201 'The issue . . . one more deception': interview with Herb York, July 2001.

202 'He was obviously paranoid . . . happy with': ibid.

202 'I have . . . joined the fascists': Herken (2002), p. 256. According to Rabi, Teller made this remark to journalist Stephen White. He subsequently denied making the remark but he wrote of the move to Livermore: 'I turned away from my original choice—to work on pure science—with my eyes open.' Ibid., note 32.

Chapter 15: Mike

203 'He called . . . official drawing': interview with Jay Wechsler, July 2001.

205 The Mike test is fully described in chapter 24 of Rhodes (1995), pp. 482-512.

205 'ended the possibility . . . no more tests': AEC (1954), p. 562.

205 'an incoming President . . . thereafter': ibid.

205 'may well . . . the Maelstrom': Arneson to Acheson, quoted in Herken (2002), p. 254.

205 'undoubtedly give . . . propaganda machine': Hans Bethe to Gordon Dean, 9 October 1952, quoted in Rhodes (1995), p. 497.

205 'to see to it . . . official documents': Herken (1992), p. 63.

206 The Zuckert episode is described in Hewlett and Duncan (1969), pp. 590-2, and ibid., p. 63.

206 'I ranked . . . to go ahead': Hess (1971), pp. 60-1.

206 'I think . . . very much moved at the time': AEC (1954), p. 562.

207 'I watched . . . "It's a boy"': Teller (2001), pp. 351-2.

208 'I was surrounded . . . extraordinary': interview with George Cowan, July 2002.

209 'I said . . . asked about it': interview with Jay Wechsler, July 2001. While logistically Mike had been a great success, there was concern about the reason for its excessive yield. It emerged that there had been a failure to appreciate second order effects, where the neutron flux generated in the early stages of thermonuclear reaction went on interacting.

209 'As we sat . . . about that': Teller (2001), p. 353. There is no corroborating evidence for Teller's anecdote. However, during the 1954 hearings, Rabi confirmed that, at the time of the Korean War, Oppenheimer did consider the notion of preventive war. AEC (1954), p. 470.

210 'contest in producing . . . by both sides': Foreign Relations of the United States, vol II, pt 2, pp. 1038, 1056-91. Herken (1992), p 64.

210 'a weapon . . . other weapons': ibid., pp. 1049-56. ibid., pp. 64-65.

211 'a horrendous . . . however foolish': ibid., pp. 1051-5. ibid., pp. 64-65.

211 Truman's speech was generalised, without disturbing detail. It referred to 'a new era of destructive power, capable of creating explosions of a new order of magnitude, dwarfing the mushroom clouds of Hiroshima and Nagasaki'.

Chapter 16: 'Soled' to the Californians

212 'learned . . . thermonuclear "device"': Time magazine, 12 April 1954, pp. 21-4.

212 'You gave to our laboratory . . . best of our abilities': Teller (2001), p. 368.

213 'He'd been hostile . . . you do it': interview with Jay Wechsler, July 2001.

213 'They were busy . . . recruited virtually none': interview with Herb York, July 2001.

213 Description of Livermore's early days: Herken (2002), p. 260.

214 'soled [sic] . . . and delivered': Teller letter to Maria Mayer, April 1953.

214 'a wonderful white elephant . . . beautiful view': Teller letter to Maria Mayer, summer 1953.

214 'No fights . . . sit and work': ibid.

214 'I am no longer . . . immorality and similar': ibid.

215 'We could not . . . not have had otherwise': York (1987), p. 70.

215 'Edward worked at . . . represented it': interview with Stirling Colgate, July 2001.

216 'Edward is . . . who had seen it all': interview with Herb York, July 2001.

217 'sickeningly small': Herken (2002), p. 262.

217 'We were, indeed, embarrassed': Teller (2001), p. 354. In spite of the rivalry between the two laboratories there had been some co-operation. Stirling Colgate was one of those asked to make diagnostic measurements on the Los Alamos Bravo test and the collaboration went very smoothly (Stirling Colgate, private communication).

217 'could have . . . reminded himself': interview with Hans and Rose Bethe, April 2002.

218 The incident is described in Hewlett and Holl (1989), pp. 37–40, Rhodes (1995), p. 532, Herken (2002), pp. 258–260.

218 'Atomic Program . . . been right' Herken (2002), p. 258. 218 'like errant schoolboys': Gordon Deans' diary, 17 February 1953, Dean papers.

218 Eisenhower briefing is described in Hewlett and Holl (1989), pp. 3–5; Herken (2002), p. 262.

219 'Lewis, let us . . . of nuclear war': Strauss (1962), p. 336.

219 'a highly influential . . . mover': Hewlett and Holl (1989), pp. 47–8. Murphy's principal source for the *Fortune* article had been Teddy Walkowicz, Head of the Air Force's Special Study Group and a friend of Teller's. Strauss reviewed the article and he and Murphy were in constant phone contact. Herken (2002), pp. 398n85.

220 'In the meantime . . . out of it.' Teller letter to Maria Mayer, spring 1953.

220 'Joe 4 was . . . pre-emptive war': interview with George Cowan, July 2001.

Chapter 17: Bravo

222 'This is . . . for its furtherance': BBC Reith lectures 1953.

222 'You know . . . disloyalty to a trust': interview with K. D. Nichols, September 1978.

223 'frequent contact . . . diplomatic policy': text of Borden letter in AEC (1954), pp. 837–8. Borden made clear that his suspicions about Oppenheimer had been aroused by information from Klaus Fuchs, 'indicating that the Soviets had acquired an agent in Berkeley who informed them about electromagnetic separation research during 1942 or earlier'.

223 'blank wall': Eisenhower, Memorandum from the Attorney General, 3 December 1953, Eisenhower Library.

223 'and when . . . avoid it': Teller (2001), p. 367.

224 'not much . . . court': JCAE minutes, 25 August 1949, no. 1203. Volpe's comment was made in relation to the reinvestigation of the Berkeley scientists.

224 'Have you seen that . . . horse is gone': interview with Harold Green, September 1978.

225 'conduct and loyalty', 'veracity': K. D. Nichols letter to Oppenheimer, 23 December 1953, AEC (1954), p. 6.

225 'Strauss let . . . was amiable': Nichols meeting with Oppenheimer, 21 December 1953, AEC; Belmont to Ladd, 21 December 1953, JRO/FBI.

225 'the Bureau's technical . . . was contemplating': Hewlett and Holl (1989), p. 81. Goodchild (1980), p. 227.

226 'When he saw . . . I said I would': Blumberg and Owens (1976), p. 360..

226 'they devoted . . . loyal citizen': Teller (2001), pp. 373–4.

226 'He did not . . . as a witness': Stern (1969), p. 516.

227 'There were . . . I ever knew': interview with Roger Robb, September 1978.

227 'could lose . . . were irrelevant': Goodchild (1980), p. 231.

228 The Bravo test is described in Hewlett and Holl (1989), pp. 172–175; Rhodes (1995), pp. 541-542.

228 'I've always . . . something amazing': interview with Harold Agnew, July 2001.

229 'terror stricken . . . the third time': Shukan Asahi, 17 March, 7 April 1954, quoted in Hewlett and Holl (1989), p. 176.

229 'must have . . . danger area': Strauss, quoted in 'The H Bomb and World Opinion', Bulletin of the Atomic Scientists, 31 March 1954, p. 163. Privately, Strauss told Eisenhower that the Lucky Dragon must have been a 'red spyship'. (Hewlett and Holl, 1989, p. 177.)

229 'That accident . . . suffered directly': Teller (2001), p. 357.

230 'It was . . . denying clearance': interview with Harold Green, September 1978.

230 The results of Bravo: the reason Bravo had outstripped all expectations was a failure to appreciate an additional reaction involving the lithium in the lithium hydride, which was being used for the first time. (George Cowan. Private Communication.)

231 'was built . . . to Livermore': interview with Herb York, July 2001.

231 'We had experienced . . . worked perfectly': Teller (2001), p. 357.

231 'After three . . . closed down': ibid., p. 358.

231 'not only . . . general's respect': ibid., p. 368.

Chapter 18: The Hearing

232 'DR OPPENHEIMER . . . alleged': New York Times, 13 April 1954.

233 'indicate to him in any way': AEC (1954a), p. 54.

233 'calm-voiced' . . . 'Metropolitan area, yes': Time magazine, 12 April 1954, p. 21.

233 'I wouldn't . . . that way Lewis': quoted in Herken (2002), p. 285.

234 'Let us cease all . . . our mental balance': Lewis Mumford, New York Times, 28 March 1954.

234 The public reaction to Bravo is described in Divine (1978), pp. 18–22.

235 'ROBB: Did he tell you . . . No': AEC (1954), p. 130.

235 'ROBB: Now let us . . . Yes': ibid., p. 137.

235 'His statements to Pash . . . I felt sick': interview with Roger Robb, September 1978.

236 'ROBB: So that . . . I was an idiot': AEC (1954), p. 137.

236 'ROBB: Isn't it a . . . Right': ibid., p. 149.

236 'I came home . . . as cheerful as ever': interview with Roger Robb, September 1978.

236 'the typical . . . on a friend': AEC (1954), p. 167.

237 'would not clear Oppenheimer today': ibid., p. 171.

237 'ROBB: Would you agree . . . job of promotion': ibid., p. 243.

238 'shove it . . . go on': interview with Joseph Volpe, July 1979.

238 'it would apply . . . General Nichols': AEC (1954), p. 389.

238 'We have an H-bomb . . . a man's life': ibid., pp. 468–70.

239 'I feel that . . . a terrible thing': ibid., p. 565.

239 'My feeling . . . I ever saw': interview with Roger Robb, September 1978.

239 'Of course . . . great torture really': interview with Lloyd Garrison, September 1978.

239 'announced emotionally . . . about to start now': Alvarez (1987), p. 180.

240 'was interested only . . . enthusiasm for the program': Charter Heslep to Lewis Strauss, Conversation with Edward Teller at Livermore on 22 April 1954, 3 May 1954, LLS/HHPL. Blumberg and Owens (1976), pp. 359-360.

241 'As he describes it . . .

Oppenheimer's loyalty': Teller (2001), p. 374.

241 'He was clearly . . . went ahead': interviews with Marshall Rosenbluth, July 2001.

242 'We were still . . . As simple as that': interview with Hans and Rose Bethe, April 2002.

242 'He was looking . . . were clear': Dyson (1979), p. 90.

242 'I considered it . . . actual facts': Teller, quoted in Blumberg and Owens (1976), p. 361.

243 'ROBB: To simplify . . . in other hands': AEC (1954), p. 710.

244 'was to make . . . earlier allegations': Green (1977), p. 59.

244 'GRAY: Do you feel . . . grant clearance': AEC (1954), p. 726.

244 'I'm sorry . . . what you mean': Stern (1969), p. 340.

245 'As a result . . . regard to Chevalier': Teller (2001), p. 383.

245 'I think . . . making decision': interview with Hans and Rose Bethe, April 2002.

245 'Many of our boys . . . over those young men': AEC (1954), p. 660.

245 'The main thing . . . not, sir': AEC (1954), pp. 802–3.

246 'but that it . . . Chevalier was a friend of mine': ibid., p. 888.

246 'rough time' . . . past tense': Teller to Strauss, 13 May 1954.

247 'Commissioner Strauss . . . see the Director': Heinnrich to Belmont, 20 May 1954, sec 40, JRO/FBI.

247 'I thought . . . nincompoop on the Board': interview with Roger Robb, September 1978.

247 'true' . . . '. . . been affected': AEC (1954), pp. 1016–17. During the Nixon administration Teller became a member of the President's Foreign Intelligence Advisory Board. Gordon Gray was also a member and at one time discussed the Oppenheimer case. According to Teller (2001), p. 392, Gray told him that his original inclination was to clear Oppenheimer. 'The Chevalier Affair,' Gray told him, 'changed my mind.' Teller went on to say that even before this comment, he sus-

pected 'that neither Oppenheimer's opinion on the hydrogen bomb nor my ambivalent testimony had much to do with the decision of either the Gray Board or the AEC'.

248 '[Oppenheimer] did not . . . subversive tendencies': AEC (1971), p. 1021.

248 'criminal dishonest conduct': ibid., p. 1043.

248 'the Oppenheimer . . . by the week': Teller letter to Maria Mayer, 30 May 1954.

248 'on the underlying reason . . . Lawrence, or I': Teller (2001), p. 398.

249 'The exile I was . . . my entire life': ibid., p. 399.

249 'Dr Luis Alvarez . . . best that I could': Strauss file memorandum, 23 June 1954, LLS/HHPL.

250 'Coughlan was there . . . later for the picnic': interview with Rose and Stirling Colgate, July 2001.

250 'Coughlan's friendly presence . . . exile had begun': Teller (2001), p. 401.

250 'If a person . . . affected her health': Blumberg and Owens (1976), p. 365.

Chapter 19: Aftermath

251 'Sir . . . anxiety': J. M. Jauch to Teller, 1 July 1954.

251 'In my testimony . . . concern to me': Teller to Strauss, draft note, 2 July 1954, Teller papers.

251 'Unless you think . . . proud of [him]': Robb to Teller, 8 July 1954, Teller papers.

252 'I came back from Los Alamos . . . Edward': Teller letter to Maria Mayer, July 1954.

253 'Dr Teller's Magnificent . . . the globe': Coughlan (1954), pp. 61–9.

253 'wonderful' . . . '. . . about anybody': Teller to Coughlan, 30 September 1954, Teller papers.

253 'insulting . . . possible embarrassment': Blumberg and Owens (1976), p. 369.

253 'Gordon Dean . . . of the book': Alsop and Alsop (1954), p. 12.

254 'bound to be interpreted . . . the book says . . .': Brunauer to Teller, 10 October 1954.

254 'It might be worthwhile . . . owe to

you': Shepley to Teller, 8 December 1954.

254 'It was a great pleasure . . . with my colleagues . . .': Teller to Shepley, 15 December 1954, Teller papers.

255 'Isn't this a dirty trick on me?': Teller (2001), p. 405.

255 'Fermi whether . . . wonderful man': Blumberg and Owens (1976), pp. 374-375.

255 'According to Enrico . . . close to suicide': interview with Jean Argo, July 2001.

256 'the leadership . . . imaginative suggestion': Teller (1955), pp. 267–75.

256 'She had fainted . . . problem was': interview with Paul Teller, April 2002.

256 'a large residual . . . been reversed': Robert C. Cowen (2003).

257 'At the present M-second . . . for myself': Teller letter to Maria Mayer, December 1954.

257 'Life at . . . some improbable way': Teller letter to Maria Mayer, June 1955.

257 'He had the . . . poor substitute': interview with Paul Teller, April 2002.

257 'My own view . . . a tragedy': interview with Marshall Rosenbluth, July 2001. In Shakespeare's play, *Coriolanus,* the Roman general wins a great battle for Rome against the Volscians. His pride in his achievement and his military skills create jealousy and make him enemies among both the people and the Senate, who conspire to banish him. He joins forces with the Volscians to seek revenge against the Romans.

258 'Look let me . . . at any time': Palevsky (2000), p. 53ff.

258 'that with . . . getting involved': interview with Marshall Rosenbluth, July 2001.

259 The Livermore crisis is described in Herken (2002), p. 300.

259 'Going to California . . . new existence': Teller letter to Maria Mayer, late 1954.

259 'The brilliant new ideas have not appeared': Bradbury to Fields, 22

September 1954, no. 125192, CIC/ DOE, quoted in Herken (2002), p. 300.

259 'an advertising stunt': quoted in Herken (2002), p. 300.

259 'And every new bomb . . . defective individuals': Divine (1978), p. 33.

260 'Fifty per cent . . . change you make': interview with Herb York, July 2001.

261 'to be increased . . . free world': Divine (1978), pp. 38–9.

261 'Now this alarmed me . . . and so on': interview with Sir Joseph Rotblat, October 2003.

Chapter 20: 'Almost like Ivory Soap'

263 'Mici is becoming . . . has assurance': Teller letter to Maria Mayer, summer 1955.

263 'I am not . . . he will say': Teller letter to Maria Mayer, April 1956.

263 '1) Five days . . . single job': Teller letter to Maria Mayer, summer 1956.

264 'He would read . . . for a child': interview with Paul Teller, April 2002.

264 'smiled . . . a little less sweetly': Blumberg and Owens (1976), p. 370.

265 The description of Project Nobska: Hansen, Submarine Launched Ballistic Missile Warheads, pp. 5ff, Chuck Hansen private collection. The possibility of mounting much lighter missiles on submarines represented a major threat to the balance of power between East and West. Submarine-based missiles would greatly reduce the effectiveness and the wisdom of any first strike, because there would always be submarines, hidden and undetected, that could and would retaliate.

265 'Then Carson . . . your estimate': Teller (2001), p. 420–1.

265 Similar radical thinking had led to developments in other areas. Not only had Teller laid down the foundations for reactor safety with his work on the Reactor Safeguard Committee, but in the mid-1950s he also played an important role in

initiating work on a small foolproof reactor. Freddie de Hoffmann, his aide at Los Alamos, had been financed by John Jay Hopkins, the founder of the weapons technology company General Dynamics, to establish a company to enter the 'atoms-for-peace' field. General Atomic, as the company was called, was to explore the use of fusion as a power generator, but Teller also persuaded de Hoffmann to develop a reactor sufficiently safe to be used by young graduate students and in medical research. The problem was 'by no means an easy one', Teller explained, 'because fools are extremely ingenious in conducting their folly: this has been demonstrated over the years both in Three Mile Island and at Chernobyl'. The result was the Triga reactor (an acronym of Training, Research and Isotope (production) and the company's name General Atomic), designed by Freeman Dyson and incorporating a fail-safe mechanism. Its ingenious design combined the U-235 in the fuel rods with zirconium hydride which, should the temperature in the reactor rise, would absorb increasing numbers of neutrons and slow down the fission process to the point where the reactor would eventually shut itself down. These were used in their hundreds worldwide for several decades. See Teller (2001), pp. 423–424.

266 The Livermore approach to Polaris is described in Hansen: Submarine-Launched Ballistic Missile Weapons, p. 14.

266 Gavin's testimony to Congressional Committee: *New York Times,* 29 June 1956, Hewlett and Holl (1989), p. 345.

267 'much of importance . . . humanitarian aspect': Divine (1978), p. 82.

267 'Part of the . . . an H-bomb': ibid., p. 83.

267 'a catastrophe . . . every circumstance': ibid., p. 122.

267 'go the second . . . leukaemia': ibid., p. 139.

267 'We cannot trust . . . anything': ibid., p. 144.

268 'magic moment': Norman Cousins, quoted in Divine (1978), p. 157.

268 'with the progress . . . foolish': Transcript of Hearings on Plutonium and Tritium Requirements Before the JCAE, Military Applications Subcommittee, 20 June 1957, AEC. Hewlett and Holl (1989), p. 399.

268 'it would be . . . the people': transcript of Hearings on Technical Aspects of Inspection before the JCAE, June 21, 1957. AEC. Quoted in Hewlett and Holl (1989), pp. 400–401.

269 'could truly be a crime . . . clean weapons': The meeting with Eisenhower is described in Hewlett and Holl (1989), pp. 400–1; Divine (1978), pp. 148–50; Herken (1992), pp. 97–8.

270 The 'clean' status of American bombs: the President had obtained his figure from Lewis Strauss, and was based on a recent test in the Plumbob series. The low figure for fall-out had resulted, in part, from the removal of the bomb's uranium tamper. However, what Strauss had not told the President was that the bomb had been carried aloft by a balloon and detonated there, thus avoiding a dust cloud.

270 '[American weapons] . . . to the Soviets': Herken (1992), p. 98.

270 'How can . . . dirty things?': Divine (1978), p. 151.

270 'At the time . . . along this road': interview with Herb York, July 2001.

271 The confrontation between Strauss and Rabi is described in Herken (1992), pp. 103–104 and Herken (2002), p. 316.

271 Bethe's findings on pre-initiation: by the spring of 1958, it was clear that the Soviets had corrected the fault—if it had even been there in the first place. Rabi's plan for a nuclear shield went no further.

271 'a great mistake . . . Lawrence':
Herken (2002), p. 316.

271 'I learned . . . practically nil': Ferrell
(1981), pp. 348–9; Herken (2002), p.
316.

272 'The Mesa shivered . . . was complete': Teller (1962), p. 83.

272 Consequences of the Rainier test
see Divine (1978), pp. 157–158.

Chapter 21: A Matter of Detection

273 'worked hardest . . . Russians were
coming': *Time* magazine, 18 November 1957, p. 21.

273 The description of Teller's lifestyle
and his 'atomic alphabet': ibid.,
pp. 21–25. Four of the five letters of
Edward's 'atomic alphabet' that he
had found the time to compose so
far, went as follows:

A stands for Atom; it is so small
No one has ever seen it at all.

B stands for bombs; the bombs are
    much bigger.
So, brother, do not be too fast on the
    trigger.

H has become a most ominous
    letter:
It means something bigger, if not
    something better.

S stands for secret; you can keep it
    forever—
Provided there's no one abroad who
    is clever.

274 Kulp's research is described in Divine (1978), p. 184.

274 'a terrible poison . . . few years': Divine (1978), p. 186.

275 'World wide fall-out . . . two
months': ibid., pp. 187ff.

275 'exaggerated statements . . . confusion': ibid., p. 187.

275 'Linus Pauling and . . . All right?':
Terrall interview with William
Libby, pp. 113–14, Bancroft Library.

276 'I would say . . . was a threshold':
interview with Professor Sir Joseph
Rotblat, 10 October 2003.

276 'The standards . . . decades of research': Kolata, *New York Times,* 27
November 2001. In the US the issue
of permitted radiation levels is still
an ever-present one, with concerns
over relicensing nuclear power stations and the growing problem of
disposing of nuclear waste. The Nuclear Regulatory Commission set its
acceptable level of radiation exposure for an individual from any one
source at 25 milli-rem a year. A single chest X-ray gives about 10 milli-
rem to the chest, equivalent to 1–2
milli-rem to the whole body. In
contrast, the average natural level of
background radiation, from sources
such as cosmic radiation, is about
350 milli-rem a year, with some areas exceeding this figure many
times. Two-thirds of the population's exposure to radiation is due
to natural sources, some 15 per cent
from medical sources and less than
2 per cent from machines and the
research industry. The cost of improving radiation standards beyond
present levels rapidly becomes pro-
hibitively expensive. As to long
term effects, a third of the population will suffer from cancer at some
time in their lives. The problem has
been to establish how many of
those cases could be prevented by a
reduction in background radiation
levels. Even among the 80–90,000
survivors of the Hiroshima and Nagasaki bombs, who were exposed to
thousands of milli-rem, it has been
hard to identify excess cancers. In
recent years, of the 12,000 of those
survivors who had died of cancer,
only 700 could be regarded as excess cancers likely to be attributable
to the two weapon attacks.

277 'technical . . . and supervision': The
Bethe Panel is described in Herken
(1992), p. 108, Hewlett and Holl
(1989), p. 471–472, 477. Yet another
comparative study at this time, by
the CIA, showed the US to be technically well in advance. By 1958 the
Russians had tested only eleven H-
bombs, a fraction of the number

tested by the US, and the efficiency of at least four had been poor. A Pentagon report corroborated this, showing the US was ahead in almost every technical area except 'ultra high altitudes'. Herken (1992), pp. 253–254, n24.

277 'On the Bethe . . . own reading': interview with George Cowan, July 2001.

278 'When Ernest died . . . growing up': Teller (2001), p. 414. Another friend Teller was to lose at much the same time was John von Neumann, who was diagnosed with terminal prostate cancer towards the end of 1956. Over the next four months, Edward visited him in hospital in Washington some ten times, during which they continued to talk science and to partner each other in making puns in three languages, English, German and Hungarian. Eventually it became painfully obvious that von Neumann's mind was affected, discouraging him and causing Edward 'great anguish'. Von Neumann died in February 1957.'I believe that if a mentally superhuman race ever develops,' Edward noted, 'its members will resemble Johnny von Neumann.' Ibid., p. 410.

278 Strauss's resignation is described in Divine (1978), pp. 217–218, Herken (1992), p. 111.

278 'It was . . . underground tests': interview with Hans Bethe, April 2002.

279 'Livermore was . . . without being found out': Freeman Dyson letter to his parents, autumn 1958, quoted in Teller (2001), p. 436.

280 'some wild schemes': The description is taken from an interview with Albert Latter, quoted in Herken (1992), p. 114.

280 'Now we thought . . . doomed our efforts': interview with Hans and Rose Bethe, April 2002.

280 Decoupling and Latter 'holes' are described in Divine (1978), p. 254–255; Herken (1992), pp. 114–115.

281 'In the competition . . . will win': Teller (1958), p. 141.

281 Russian resistance to on-site inspection: years later, Khrushchev admitted that a major reason behind this was the fear that these would reveal just how backward their nuclear armoury was.

281 'Dr Albert Latter . . . discredit them': Killian (1977), p. 168.

281 'The notion . . . based on them': Kistiakowsky (1976), pp. 17–18.

282 'convinced . . . testify in opposition': ibid.

282 'I've never seen . . . meet his objections': Blumberg and Owens (1976), p. 407.

282 JCAE meeting is described in Divine (1978), pp. 306–309, Herken (1992), p. 121.

283 'flared up' . . . '. . . presidency': Kistiakowsky (1976), p. 375.

283 'the acquisition . . . scientific-technological elite': Eisenhower valedictory address, 17 January 1961. Quoted on the Avalon Project of the Yale Law School website, www.yale.edu/lawweb/avalon.

283 'Eisenhower spent . . . von Braun and Teller': interview with Herb York, July 2001.

Chapter 22: Plowshare

284 'we looked . . . do the job': Fairbanks Daily News-Miner, 15 July 1958. O'Neill (1994), pp. 31–32.

284 'one will probably . . . might cause': Teller, quoted in Zodner (1958), p. 7.

285 Number of devices to blast the new canal: this estimate was later revised upwards to a staggering 262 devices with an aggregate yield of 270.9 megatons. O'Neill (1994), pp. 23–24.

285 'So you want . . . plowshares': Teller and Brown (1962), p. 82. The original quote can be found in Micah, chapter 4, verse 3.

285 'And that ended . . . Alaska': interview in O'Neill (1989), **p. 28.**

285 'blast will not . . . be used': Fairbanks Daily News-Miner, 15 July 1958. O'Neill (1994), p. 33.

286 'firecracker boys': A. Johnson letter to L. Vierek, 21 February 1961, p. 3, Foote collection: Don Charles Foote

Collection, University of Alaska, Fairbanks, Archive.

286 'autonomous . . . to do it': O'Neill (1989), p. 29.

286 'The thing . . . no one cared': interview with Gerald Johnson, University of Alaska archives; O'Neill (1994), p. 74.

287 Details of the Neptune shot: O'Neill (1994), pp. 55–6.

287 'in recognition . . . the menace of tyranny': quoted in O'Neill (1994), p. 90.

287 'If your mountain . . . card': *Anchorage Daily Times,* 26 June 1959, p. 11; O'Neill (1994), p. 92.

287 'That is like . . . up to you': ibid., p. 11.

288 'My testimony . . . main topics': Teller (2001), p. 456.

288 'It was a lovely show . . . wish you were here!': Bernice Brode to Oppenheimer, 6 May 1959, Robert Brode folder, box 23, JRO.

289 'One man . . . Oppenheimer': Blumberg and Owens (1976), p. 418–419.

289 'You were good....in touch with each other': ibid., p. 419

289 'as well read . . . good New Englander': D. Foote to AEC (E. Campbell), 26 November 1959; Foote Collection, quoted in O'Neill (1994), p. 106.

290 'for any reason at any time': Point Hope Village Council to AEC, 30 November 1959, in doc #16878 CIC; O'Neill (1994), p. 106.

290 'no significant questions': Ibid., p. 116.

290 'Activity in the . . . rapidly decay': Soundtrack of LLNL film, *Industrial Applications of Nuclear Explosions,* University of California Radiation Laboratory (1958), Film A-81.

290 '[At Eniwetok] . . . scared she gets': excerpts from a tape recording made by Rev. Keith Lawton of the Episcopal Mission in Point Hope, Alaska, during a public information meeting given by members of USAEC and Environmental Committee for Project Chariot on 14 March 1960, UAF archives, Alaska and Polar regions department, University of Alaska, Fairbanks, Archives Department. The meeting is described in O'Neill (1994), pp. 115-130.

291 The economies of truth deployed by the AEC representatives at the Point Hope meeting were considerable. a) At Bikini, after the 1946 Baker test, a Navy radiological monitor reported fish so radioactive that, when laid on a photographic plate, they darkened it. b) The claim that the *Lucky Dragon* was within the exclusion area at the time of the test was in error on two counts. Firstly the boat's log showed quite clearly that it was outside the area. Secondly, editions of the AEC's own publication, Glasstone's *The Effects of Nuclear Weapons,* stated that a much larger area of 7000 square miles was contaminated to the extent 'that survival might have depended upon evacuation of the area or taking protective measures'. The crew could have been hundreds of miles beyond the exclusion zone and still have been in danger (Glasstone and Dolan, 1977, p. 423). c) As to the blasts having 'no effect on the Indian people any place', evidence to the contrary is legion. The inhabitants of Rongelap, the Marshall Island close to the Bravo test site, initially suffered serious radiation burns, though a press release at the time denied it. Years later, government studies showed that of Marshallese children receiving in excess of 1000 rad doses to their thyroid, eighteen out of nineteen died (Wasserman, 1982, p. 86). d) As to the failure to find evidence of damage at the Nevada test site, a Senate sub-committee on Oversight and Investigations concluded that the government had been negligent and that all evidence 'be it on sheep or on people, was not only disregarded but actually suppressed' (Hilgartner, Bell and O'Connor, 1983, p. 84).

The Bio-environmental

Committee's relationship with the AEC is described in O'Neill (1994), pp. 150–5. One member Robert Rausch is quoted as saying: 'I recall quite distinctly a quite high-up AEC official coming up to us and saying if you go along with this you will have all the money you need for research.' The research is described in O'Neill (1992), pp. 131–149.

292 Bill Pruitt's work on lichen and strontium 90 and the reaction to its publication is described in O'Neill (1994), pp. 210–18.

292 'We all feel . . . acting quickly': L. Viereck to J. Haddock, 2 March 1961, quoted in ibid., p. 186.

292 'It was my introduction . . . environmentalism': interview, quoted in ibid., p. 210.

292 The environmental press's reaction to the *Nuclear Information* article is described in ibid., pp. 215–18.

293 'roles as investigator . . . this experiment': quoted in ibid., p. 245.

293 The AEC's press statement on Sedan was released in Las Vegas, Nevada, at 11 a.m. PDT, 6 July 1962.

293 'deposited nearly . . . been predicted': report of Defense Nuclear Agency (1983), p. 23.

293 The Sedan test and the subsequent placing of Chariot 'in abeyance' are described in O'Neill (1984), pp. 252–5.

294 'It is clear . . . under this programme': minutes of AEC meeting, no. 1511, 22 May 1959, Doc #753, CIC, quoted in ibid., p. 40.

Chapter 23: Confounding Camelot

296 'John L. Lewis . . . in one': Senator Fulbright in a telephone conversation with President Kennedy, 23 August 1963. Quoted in Herken (1992), p. 269, note 87.

297 John McCone: after the Bay of Pigs fiasco, Kennedy was convinced that the CIA had deliberately deceived him and he had dismissed its director, Allen Dulles. McCone had been his replacement.

297 'unproven:' The account of the Panofsky panel, ibid pp. 132-133.

297 'undue weight': interview with Glenn Seaborg, quoted in Herken (1992), p. 131.

297 'I agreed . . . policy adopted': Teller (2001), p. 462.

297 'new aspects . . . that we test': Rand report, Disarmament–Test Ban Negotiations, folder, President's Office files, JFK Library, quoted in Herken (1992), p. 131.

298 'completely overwhelmed . . . Chairman': quoted in May and Zelikow (1997), p. 30.

298 'We have been . . . Children': Kennedy's notes are in Disarmament–Nuclear Test Ban Negotiations, 7/30/62 meeting, Box 100A, President Office files, JFK Library, quoted in Herken (1992), p. 134..

298 Khrushchev's attitude: the extreme belligerence demonstrated by Khrushchev was of great concern in Washington. There had, after all, been moments when a rapprochement had seemed possible, particularly at Geneva in 1958; but now one aggressive act followed another. The Russians were, indeed, not to be trusted, but why? Was it simply that they were testing the resolve of the young President? More complete answers were to emerge later, but at the time one piece of information gave a clue to the underlying situation. In late 1961, the CIA launched their first photo-reconnaissance satellite, 'Corona'. The first pictures it sent back covered more territory than all the previous U2 spy-plane flights put together. They showed that, in spite of all the space flights, and Khrushchev's bragging, Russia had just six ICBMs, and some of these were known to be unreliable. It was information that helped Kennedy to see his foe as more vulnerable and, hopefully, more open, eventually, to new arms-control initiatives. Decades later, it was further revealed that, in 1961, Khrushchev had to acknowledge serious failures in his

agricultural policy. Food prices had risen by between 20 and 30 per cent, but to ensure the continued support of the military, he had reversed heavy cutbacks in their supplies. What he needed most, however, was some kind of success after a string of failures.

The implications of the 'Corona' satellite's photo-reconnaissance and how it was backed up by evidence from the US agent, Colonel Oleg Penkovsky is described in May and Zelikow (1997), pp. 32–3.

299 'taken the . . . gentlemen's agreement': Teller (2001), p. 464.

299 'A retaliatory force . . . United States': Teller and Brown (1962), p. 128.

299 'two scorpions . . . own life': Rhodes (1995), p. 409.

299 'Step by step . . . does not deter': Teller and Brown (1962), p. 233.

301 'I was about . . . as a scientist': Teller (2001), pp. 464–5.

302 'in recognition . . . national security': Blumberg and Owens (1976), p. 421.

302 The description of the dispute over the content of Teller's acceptance speech at the Fermi Awards is in Blumberg and Owens (1976), pp. 420–3.

302 'As President Kennedy . . . political timidity': Teller (2001), p. 466.

302 'sour grin': Blumberg and Owens (1976), p. 423.

303 'pale in comparison . . . towards war': Kennedy speech, July 1962, quoted in Herken (1992), p. 140.

303 'It seemed . . . from that quarter': Teller (2001), p. 469.

304 'privileged . . . no testing': Teller (2001), p. 466.

304 Teller's lobbying against a test ban is described in Herken (1992), pp. 142–145.

305 'the United States . . . peace and war': Teller's testimony to US Congress, Senate Committee on Foreign Relations, Nuclear Test Ban Treaty Hearing 1st session, 20 August 1963, pp. 417–27. Teller (1987), pp. 106-113. In Notes Added in 1985', Teller

stated that this claim for the necessity of atmospheric testing was 'overstated, ingenious underground experimentation allowed for finding out far more about defensive weapons than I originally expected' (Herken, 1992, pp. 269–70, note 87).

305 Teller's and Genevieve Phillips's reaction to Kennedy's announcement of the Partial Test Treaty is described in Blumberg and Owens (1976), p. 414.

306 'propaganda step': Herken (2002), p. 330.

306 'He missed . . . because of her behaviour': interview with Marvin Goldberger, July 2001.

308 'I think it is . . . all our futures': Oppenheimer's original words, when he believed President Kennedy was to give the award included not just 'some charity and some courage' but also 'some humor' (Smith and Weiner, 1980, p. 331, and 350n).

Chapter 24: Struggling Uphill

309 'Dr. Strangelove . . . enemies': McQuade, New York Times, 1 March 1964, p. 25.

309 'indulges . . . do its job': Getler, ibid.

309 'When virtually . . . picture proves': Boxen (1995), p. 4.

309 'it is a . . . warmongering': ibid.

309 'which Kubrick . . . perfectly plausible': ibid.

309 'it contains . . . absolutely true': ibid.

310 'I didn't . . . speeches': Terrell, interview with Willard Libby, 1978, p. 79, Bancroft Library.

310 Teller had met Rockefeller in the 1950s and had been invited to join his Prospects for America study group, set up to prepare him for the presidency. Teller served on two committees, one on national security and the other on energy, but Rockefeller's presidential bid was thwarted by public reaction to his divorce. Teller advised Rockefeller throughout the next two decades— on New York's electricity supply when he was state governor and, in

the mid-seventies, on energy policy when he served as Gerald Ford's vice-president. Rockefeller's ideas on energy conservation helped to frame the Ford administration's energy policy. Rockefeller died in 1979.

311 Marks quizzed him about his financial interests in the project. His answer was that he had none, but Marks's line of questioning again raised the spectre of a commitment to big business and close associations with figures of power in the Republican Party like Rockefeller.

312 'relaxed manner . . . '. . . those ideas': Teller (2001), p. 509.

313 'Intellectuals . . . over-praise it': Susan Sontag, quoted in Hoberman. 'When Dr No Met Dr Strangelove' *Sight and Sound,* December 1993.

313 'The real free speech . . . those on that campus': interview with Alex C. Sheriffs, student administrator, Media Research Center, Moffat Library, University of California at Berkeley.

314 'Bill Wattenberg . . . very accessible': interview with Stirling and Rosie Colgate, July 2001. Wattenberg has been both host to one of the most popular talk shows in the Bay Area and a member of staff at Livermore, a physics PhD working on the design of nuclear weapons.

314 'aimless . . .' '. . . Third World Communism': Teller, quoted in Blumberg and Owens (1976), pp. 441–2.

315 'We can win . . . limited war': Teller (2001), p. 504.

315 The Berkeley Tribunal on 23 November 1970 is described in Blumberg and Owens (1976), pp. 441–6.

315 'We don't need . . . kill him': Blumberg and Owens (1976), p. 444–5.

316 'I cannot tell you . . . would-be attacker': *US News and World Report,* 26 May 1969, p. 87.

317 'Aleutian H-bomb . . . Quake': *New York Times,* 3 October 1969.

317 'detected an alarming . . . many more': Greenpeace (1996), part 1,

p. 3. www.greenpeaceusa.org/media/publications.

318 'sizeable fractures . . . concentration': David Evans, affidavit to Congress, 9 November 1971, p. 40222.

319 'It's fun . . . for a while': Greenpeace (1996), part 1, p. 3.

320 'unreasonable behaviour . . . first strike': Teller (2001), pp. 510–511.

320 'Such a false hope . . . disarmament measures': York (1987), p. 239.

320 When Harold Agnew took over as director of Los Alamos at the beginning of the 1970s, he eased Teller's isolation, inviting him back for lectures and discussions. 'It was like Jesus Christ coming,' Agnew recalled, 'People would go home and say, "Hey, today I talked to Edward Teller." I thought it was good for the lab and for the people. He's an intellectual giant. Whatever had happened didn't bother me at all. Edward was clearly on our side as far as the US was concerned and that's that.' Interview with Harold Agnew, July 2001.

321 'I remember that . . . was the last': Teller (2001), pp. 512–13.

322 'People do betray . . . He knows': interview with George Cowan, July 2001.

322 'Edward felt . . . extremists in the Pentagon': interview with Marvin Goldberger, July 2001.

323 'And thenI made . . . I first met in 1945': Goldberger, 'The Wrath of Teller', unpublished MS, July 1990.

324 Background to and events at Three Mile Island: Daniel Mark, 'Three Mile Island'; Report of the President's Commission, 'The Accident at Three Mile Island'; Pringle and Spigelman (1982), pp. 422–4.

325 'from that point . . . ". . . make decisions" ': WGBH American Experience Website, Meltdown at Three Mile Island. www.pbs.org/wgbh/amex.

325 'Race With Nuclear Disaster': *New York Post* headline, Quoted in Pringle and Spigelman (1982), p. 424.

326 'It just so happens . . . latched onto

this': Teller interview, in *Playboy*, August 1979, p. 59.

326 'Nuclear reactors . . . free world': quoted in *Washington Post*, 8 May 1979.

327 'I WAS THE ONLY VICTIM OF THREE MILE ISLAND . . . are not dangerous': *Wall Street Journal*, 31 July 1979.

327 New York Times editorial criticising Teller: *New York Times*, 17 August 1979, p. 24.

Chapter 25: Bringing Up the Props

328 'There's a quotation . . . circles within circles': interview with Peter Hagelstein, April 2000.

328 'not only . . . but quite creative': Broad (1985), p. 103.

328 The Hertz Foundation was initially set up in the late 1940s by John D. Hertz, of yellow cab and car rental fame, to help challenge the technical advances being made by the Soviets. Its board members have included FBI Director, J. Edgar Hoover, head of Strategic Air Command, General Curtis Le May, the nuclear strategist Herman Kahn, and Hans Mark, a particular friend of Teller's and onetime Secretary of the Air Force. Alongside these served several financiers, including Robert Lehman of Lehman Brothers. Teller and Lowell Wood were on the board and responsible for recruitment. Because the aim was to recruit the most talented individuals, there was no insistence on defence work, though new recruits did have to 'morally commit themselves to make their skills and abilities available for the common defense, in the event of a national emergency': quoted in Broad 'The Scientists of Star Wars' Granta 1985 p. 87. For much of the time, however the twenty-five or so Foundation Fellows at Livermore in the 1980s worked on both defence projects and on unrelated research—a scientific free-for-all attractive enough to keep good people from

succumbing to the attractions of nearby Silicon Valley. As another way of easing the recruitment problem, in the early 1960s Teller had a major role in establishing the Department of Applied Science at UC Davis. Nicknamed 'Teller Tech', the courses were aimed at providing a blend of science with engineering to produce what Teller called 'inventive engineers'.

328 '[Wood] said . . . kind of dreamy': Hagelstein, quoted in Broad (1985), p. 105. Laser fusion uses lasers instead of fission bombs to ignite small parcels of hydrogen isotopes to produce tiny 'controlled' fusion reactions for generating power.

329 Angstroms: an angstrom is 100 millionth of a centimetre.

329 The description of Hagelstein's work. Broad (1985), pp. 96–107, and interview.

329 'My original idea . . . I contributed': interview with George Chapline, April 2000.

329 The description of Chapline's work, Broad (1992), pp. 76–78 and interview.

330 'Teller ordered me . . . and so forth': interview with Peter Hagelstein, April 2000 and Broad (1992) p. 82.

331 'difficult . . . sharing the work': interview with George Chapline, April 2000.

331 Wood is described in Broad (1992), pp. 79-80.

331 'Lowell is like . . . people like Lowell': interview with Carl Haussmann, 9 May 1984, Ibid., p. 80.

332 'very bright . . . Very interesting': interview with Peter Hagelstein, April 2000.

332 'Lowell was . . . talked incessantly': Broad (1985), p. 26.

332 'I'd been working . . . twenty hours': interview with Peter Hagelstein, April 2000.

333 'The force has . . . weak mind': Hagelstein, quoted in Broad (1985), p. 114 and Hagelstein interview.

333 'Reagan reflected . . . alternatives as bad': Anderson (1988), pp. 80–6.

334 'unparalleled military . . . a nuclear war': Fitzgerald (2000), p. 84.

334 Dauphin test: Broad (1992) p. 87–88 and Chapline and Hagelstein interviews.

335 'Roy was . . . a good politician': interview with Ray Kidder, April 2000.

335 'X-ray lasers . . . weapons attack': Clarence A. Robinson, Jr, 'Advance Made on High Energy Laser', *Aviation Week and Space Technology*, 23 February 1981, p. 21, Broad (1992), p. 92. George Chapline was to discover that the *Aviation Week* article had reawakened Soviet interest in the X-ray laser, 'and they too started doing experiments with nuclear devices. They didn't know exactly how we did it and, to my knowledge, they never did exactly what we did. They did different things—interesting things but not as interesting as us. But the fact they had so much interest shows it had quite an impact on their strategic thinking.' Chapline interview, April 2000.

336 'obsolete . . . ballistic missiles': Wallop (1979), pp. 13–22. Fitzgerald (2000) p. 121.

336 Graham's initiation of High Frontier is described in Fitzgerald (2000), pp. 124–127 and Broad (1992), pp. 100–102.

337 'space born . . .' '. . . Reagan initiative': Baucom (1992), pp. 149–50. Broad (1992), pp. 105–107; Fitzgerald (2000), pp. 132–134.

337 'went back to . . . knew about this': interview with William Baker, quoted in Herken (1992), p. 200.

338 'Bluntly, the reason . . . admires Edward': *New York Times*. 18 February 1982, p. A 12, quoted in Broad (1992), p. 104, Herken (1992), p. 201.

338 'drew a mixture . . . in the club': *Science*, 22 May 1981, pp. 903–904, quoted in Herken (1992), p. 201.

338 'I felt . . . could be done': Anderson (1988), pp. 94–5.

338 'assured survival' . . . 'assured destruction': Summary of Remarks by Edward Teller, addendum to a memo by Bendetsen to Secretary of Defense, 14 September 1981, Bendetsen Collection. The memo is quoted and the meeting is described in Broad (1992), p. 107–108.

339 'urgency' . . . 'hydrogen bomb itself': Karl Bendetsen letter to Edwin Meese, October 1981. See Broad (1992), p. 109–111.

340 'fundamental flaw . . . Western World': The dispute between Teller and Graham is described in Broad (1992), p. 108 and Fitzgerald (2000), p. 135.

341 Teller's presence at the January 1982 meeting with the President: Bendetsen recalls that Teller was at the meeting, while others do not remember him as present. He is not quoted in the notes of a meeting that he would have been very likely to dominate.

341 'powerful directed energy . . . such delays': Broad (1992), p. 114; Baucom (1992), p. 153; Fitzgerald (2000), p. 136.

343 'TELLER: May I . . . a third one': transcript of Firing Line, 15 June 1982.

343 'Mr President . . . the conference': Teller (2001), p. 530. *Aviation Week* reported this meeting. It cited no source but reported that Teller had asked for increases in X-ray laser funding of $200 million a year 'over the next several years,' Broad (1992), p. 119.

343 'Their recollection . . . it held': Teller (2001), p. 530.

344 'I dare . . . weapons technology': Teller to President Reagan, 25 September 1982. Quoted in Broad (1992), p. 119.

344 'We fear . . . or surrender': Catholic Bishops' Pastoral Letter, 3 May 1983. The letter is quoted on www.nuclearfiles.org.

345 The Livermore presentation to the Frieman Panel, on 23 June 1982, is described in Broad (1992), pp. 127–8.

345 'unglued': Robert Scheer, *Los Angeles Times*, 10 July 1983, Section 6, p. 6.

345 'Roy, I may . . . won't be this year': interview with Roy Woodruff, 16 December 1989, quoted in Broad (1992), p. 129.

346 'It was essential . . . how much more effort': interview with Peter Hagelstein, April 2001.

346 The description of Teller's birthday party is from an interview with Sandy Woodruff, 20 January 1990, quoted in Broad (1992), p. 129.

346 'Mutually assured . . . morally sound': Watkins, quoted in *New York Times,* 6 May 1983, p. A 32. The Teller/Watkins meeting is described in Broad (1992), pp. 122–124; Blumberg and Panos (1990), pp. 8–9; Fitzgerald (2000), p. 196.

348 'to protect . . . been hoping': Fitzgerald (2000), p. 197; Baucom (1992), pp. 191–2; Cannon (1991), p. 329; Broad (1992), p. 125; Weinberger (1990), p. 304.

348 Bethe's publication of his declassified assessment: Bethe's provocation had, in part, been due to the version of events surrounding the crash programme for the Super that had appeared in my then recently published biography of Robert Oppenheimer *Shatterer of Worlds.* In his assessment, Bethe had held Teller himself responsible for the delay in advancing the Super, rather than others such as Oppenheimer (see my chapters 12 and 13 above).

348 'Bethe had hassled . . . had fallen apart': Broad (1985), pp. 123–124.

349 'I don't think . . . if successful': ibid., p. 124.

349 'I have become . . . impotent and obsolete': Teller (2001), pp. 531–2.

350 'Immediately . . . Probably not': ibid., p. 532.

Chapter 26: Excalibur

351 'Will there be any security? . . . promising you immortality?': Susan Cohen (1985).

352 'exist not on paper . . . said that': *Los Angeles Times,* 4 April 1985.

353 'A great change . . . what they're talking about': Teller testimony before House of Representatives Armed Services Committee, 98th Congress, 1st session, pt. 5 of 8, 28 April 1983, pp. 1353–71. Described in Broad (1992), pp. 143–144, Blumberg and Panos (1990), pp. 230–233.

354 'leading scientific and military forces': *New York Times,* 28 April 1983.

354 'I was NOT . . . the New York Times': *Wall Street Journal,* 31 May 1983.

355 'leadership in science and technology': Broad (1992), p. 144.

355 'He's low-keyed the program': *Washington Post,* 15 June 1984.

355 'They had leased . . . everybody saluted': interview with Harold Agnew, July 2001.

355 'nearly leak-proof . . . counter the defense': Hoffmann Report, 'Ballistic Missile Defenses,' p. 1, 2, 3, quoted in Fitzgerald (2000), p. 253, 257.

355 'The scientific . . . impotent and obsolete': Rosenberg (1986), pp. 26–30. Fitzgerald (2000), p. 244.

356 'someone in the White House': *New York Times,* 11 November 1986, p. A 25.

356 'Fast burn . . . period': interview with Gerald Yonas, 23 February 1985, quoted in Broad (1992), p. 145.

356 'deliberately lying . . . and defensive': *Pravda,* 27 March 1983.

357 'mind-boggled by their patience': Fischer (1997), p. 14.

357 'We will . . . political revolution': Gelb, *New York Times,* 27 August 1992, p. 27.

357 'that the Americans . . . to war': Fischer (1997), p. 18.

357 'before the . . . every sphere': *Der Spiegel,* 2 September 1991, pp. 110–11. Horn says he first heard this from a Soviet Committee official. His own experiences in Moscow, he says, convinced him that the anecdote was true.

357 'Never, perhaps . . . of the 1980s': Fischer (1997), p. 2.

358 'development of large scale ABM systems': Shultz (1993), pp. 473–4.

358 'When you start . . . at the party':
interview with Ray Kidder, April
2000.

360 'We knew we . . . experiment': Description of Romano test in Broad
(1992), p. 151.

360 'Dear Jay . . . where money talks':
extracts of this letter are taken from
a declassified version prepared by
the Department of Energy for Senators Brown and Markey and released on 1 August 1988. Other
declassified versions released at the
same time include letters to Ambassador Nitze and to Robert
McFarlane. The letters were analysed for their accuracy by the General Accounting Office, and the
analysis published in 'Accuracy of
Statements Concerning DOE's X-
ray Laser Research Program' (GAO,
June 1988). Quoted in Broad (1992),
pp. 151–152..

361 'Roy . . . this time': Broad (1992),
p. 153.

361 'My reputation would be ruined':
ibid. In later testimony, Teller was
to deny ever having made this comment.

361 'At this point . . . kind of issue': Interview with Roy Woodruff, quoted
in Broad (1992), p. 154.

361 'excellent quantitative . . . with predictions': GAO (June 1988), p. 5.

362 Wallbridge's paper on beam
strength is in Nature, 19 July 1984,
p. 180.

362 'He was aggressive and hostile':
Broad (1992), p. 161.

363 'shoddy work . . . ineffective':
Jastrow, Washington Post, 3 March
1985. Fitzgerald (2000), p. 247.

363 'a career of hyena-like behavior':
Ibid., p. 247.

363 'The President's vision' . . . '. . . in
the world': 'The War for Star Wars',
New York Review of Books, 11 April
1985.

364 'the development . . . of space': Fitzgerald (2000), pp. 257–258.

364 The Super Excalibur concept was
set out in a letter from Rod Hyde
and Lowell Wood to Edward Teller,

9 September 1984, listed in
Livermore documents for 1988
GAO study. It is described in Broad
(1992), pp. 162–163.

365 'I wish I could . . . how promising':
Cohen (1985), p. 8.

365 'in so urgent . . . to Geneva': Teller
to Paul Nitze, 28 December 1984,
declassified version with deletions
(1 August 1988).

366 'of urgent . . . [few years]': Teller to
Robert McFarlane, 28 December
1984, declassified with deletions (1
August 1988).

366 'overly optimistic . . . very unlikely':
GAO (June 1988), pp. 7–8.In pursuing the probe initiated by Congress
into SDI, the GAO investigators
canvassed the views of other
Livermore scientists with 'specific
knowledge' of the X-ray laser on
the 'accuracy of Dr. Teller's statements'. They reported finding 'no
uniformity of opinion'. As to Roger
Batzel's reasons for not forwarding
Woodruff's clarification, the reasons he gave the investigators were
that 'there was nothing in Dr.
Teller's letters that violated any laws
of physics'. Teller had also identified
the Super- Excalibur concept as 'in
principle'.

367 'He had . . . a problem': Broad
(1992), p. 171.

367 'We can do business together':
Smith (1991), p. 146.

367 'Maybe . . . plan to spend':
Dobrynin (1995), p. 620.

367 'Re: CHARGE . . . Conspiracy':
Lowell Wood memo, to Fritz
Rittman, 24 January 1985. Quoted
in Broad (1992), pp. 174–175.

368 The meeting between the Los
Alamos and Livermore scientists is
described in Broad (1992), pp. 175–
6.

369 'These are very real . . . offsetting
measures': notes, Physics Program
Review, LLNL, 30 January 1985, p. 3.
Quoted in Broad (1992), p. 177–179.

369 'concept validation dates': Woodruff's viewgraph, quoted in Broad
(1992), p. 181.

370 Cottage Test; ibid., p. 182–183.

370 'He was very . . . an advocate': interview with Robert McFarlane, 5 December 1989, quoted in Broad (1992), p. 188.

372 'This issue . . . do with politics': Overall Peter Hagelstein was—and still is—confident of their results. 'Lowell's viewgraphs, which he took to Washington, were full of good results. It was the claims he made that worried everyone. He was overselling things.' (Personal communication with the author, November 2003)

372 'An explosion . . . than he': George Miller memo to Roy Woodruff, Lowell Wood and R Program, quoted in Broad (1992), p. 188.

372 'At the time . . . very bitter': interview with Peter Hagelstein, April 2000.

373 The first summit between Gorbachev and Reagan is described in Fitzgerald (2000), pp. 299–313.

373 'You are trying . . . what you say': McFarlane (1994), p. 318; Shultz (1993), p. 603.

374 The report presented by Los Alamos was entitled 'Modelling of Diagnostics Foil Behaviour with a Non-LTE Atomic Kinetic Code', Jack C. Comly et al., October 1985. The meeting and Woodruff's resignation is described in Broad (1992), pp. 199–200. Los Alamos's targets were clearly Teller and Wood. They feared for the credibility of the National laboratories and in a letter to George Miller they blamed Livermore managers for not facing up to Teller and Wood.

374 'to the detriment . . . as Director': Roy Woodruff letter to Roger Batzel, 29 October 1985; ibid., p. 200.

374 'immense . . . unfavourable repercussions': Los Angeles Times, 12 November 1985.

375 'A 99.9 per cent . . . uncertain': Teller, 'SDI, The Last Best Hope', Insight/Washington Times, 28 October 1985; Fitzgerald (2000), p. 254.

375 'one group . . . of a light beam':

Broad, New York Times, 7 December 1985. The goldstone test is described in Broad (1992), pp. 204–205.

375 'Pressure . . . what they have': Los Angeles Times, 12 November 1985.

375 'The work was . . . importance to mankind': interview with George Chapline, April 2000.

375 'To lie . . . and Veg-O-Matics': interview with Lowell Morgan, 5 June 1989, quoted in Broad (1992), p. 195.

Chapter 27: Reykjavik

376 'It's the best . . . in 25 years': Fitzgerald (2000), p. 358. The summit is described in Fitzgerald (2000), pp. 345–369.

377 'the same old . . . already choking': Ibid. p. 358.

377 'we merely . . . can't do that': ibid. p. 359.

377 'gather fruit in their basket': ibid., p. 360.

377 'kasha for ever': Oberdorfer (1991), pp. 195–6. Ibid. p. 360.

377 'You're a . . . of something': ibid., pp. 197–8. Ibid. p. 360.

377 'Let's go . . . We're leaving': Reagan (1990), p. 679.

378 'Can't we . . . too late': Cannon (1991), p. 769.

378 'I told . . . the nuclear threat': Fitzgerald (2000), p. 349.

378 'We all had . . . real possibilities': interview with Greg Canavan, July 2001.

378 'pre-deployed arrays . . . Northern Hemisphere': Teller (2001), p. 535.

379 'Brilliant Pebbles . . . looking for': interview with Greg Canavan, July 2001.

379 Sakharov's return from exile is described in Fitzgerald (2000), pp. 409–11.

380 'I believe . . . can be overcome': Sakharov (1991), pp. 21–4.

380 'We came . . . medium term': Fitzgerald (2000), p. 411.

380 'This represented . . . ICBMs as well': Sakharov (1991), p. 24.

381 'I was among . . . Iron Curtain were noticed': Teller (2001), p. 534.

Chapter 28: Brilliant Pebbles

382 'In Dr Teller . . . a nuclear war': transcript, Teller and Sakharov, pp. 2–6; Blumenthal, 'When Giants Meet: H-bomb Fathers Sakharov and Teller's SDI Dialogue', *Washington Post*, 17 November 1988. Blumberg and Panos (1990), pp. 279–280. Broad (1992), p. 258.

383 'We honour . . . our SDI': ibid.

383 'tireless advocate . . . American freedom': ibid.

383 'defensive systems . . . ahead of ours': ibid.

384 'hornets': interview with Richard Garwin, April 2002.

384 'On this occasion . . . something like that': interview with Greg Canavan.

385 'Dr Teller . . . ever dreamed possible': *Los Angeles Times*, 15 March 1988, p. 1; transcript of Reagan's keynote address, 'SDI: The First Five Years', Institute for Foreign Policy Analysis. Broad (1992), p. 254.

385 'Each pebble . . . or coaching': Lowell Wood, 'Concerning Advanced Architectures for Strategic Defense', reprinted in *Aviation Week and Space Technology*, 13 June 1988, p. 151. Ibid., pp. 254–255.

385 'So in the end . . . and established': interview with Greg Canavan, July 2001.

386 Teller's attendance at the Senate Foreign Relations Committee is described in Blumberg and Panos (1990), pp. 258–60.

387 'He was truly Messianic . . . do it his way': interview with Herb York, July 2001.

388 'If the technology . . . ballistic missile attack': 'The Bush Speech', *Valley Times*, 8 February 1990, p. 4A.

389 'could provide . . . safety for all': Canavan and Teller (1990), p. 699.

389 'a great advertisement for SDI': *Wall Street Journal*, 18 January 1991.

389 'I think the time . . . rather as allies': High Frontier Strategic Issues policy briefing, 86. www.highfrontier.org.

390 '90 percent . . . of the test': GAO (September 1992).

390 'Like every other exile . . . I was feeling': Teller (2001), pp. 551–2.

390 'where the same clock . . . was recovering': ibid., p. 552.

391 'immature technologies . . . could be killed': GAO (February 1992), p. 36.

392 'thanks to . . . Chernobyl accident': ibid., p. 554.

392 'the stars . . . Star Wars': High Frontier Strategic Issues policy brief, no. 84.

392 Teller travelled regularly, not only to Hungary but to Israel. In 1966 he visited for the first time to discuss a Plowshare project, a canal linking the Mediterranean with the Dead Sea which would generate hydroelectric power. Later, in the 1980s, he persuaded the country to join SDI. Throughout his association with Israel, he advised on improving technological and scientific education and broadening the country's technological base.

393 'In the early eighties . . . paranoid and demented': interview with Paul Teller, April 2002.

393 'Edward was very good . . . medical practitioners': interview with Stirling and Rosie Colgate, July 2001.

394 'I cannot overestimate . . . relished the joys': Teller (2001), pp. 552, 552n.

Epilogue

398 'Aggression is wrong . . . those weapons': Teller (1987), p. 226–8.

402 'physicists have known . . . have known power': 'National Defense and the Scientists: An Open Letter to Hans Bethe from Edward Teller', *Policy Review*, March 1987

Appendix 1. The New Physics: The Path That Led to Quantum Physics

405 'In doing so . . . is not determined': Science Archive interview with Edward Teller, May–June, 1996, pp. 31–2.

Select Bibliography

Alsop, Joseph, and Stewart Alsop. 1954. *We accuse!* Simon and Schuster.

Alvarez, Luis W. 1987. *Alvarez.* Basic Books.

Anders, Roger M. (ed). 1987. *Forging the Atomic Shield.* University of North Carolina Press.

Anderson, Martin. 1988. *Revolution.* Harcourt, Brace, Jovanovich.

Arneson, Gordon. The H-Bomb Decision. *Foreign Service Journal,* May 1969.

Aspray, William. 1990. *John von Neumann and the Origins of Modern Computing.* MIT Press.

Barnard, Chester I., et al., 1946. *A Report on the International Control of Atomic Energy.* [Acheson-Lilienthal Report]. Department of State.

Baucom, Donald R. 1992. *Origins of the Strategic Defence Initiative: Ballistic Missile Defence, 1944–1983.* University Press of Kansas.

Bernstein, Jeremy. 20 October 1975. Physicist. *New Yorker.*

———. 1980. *Hans Bethe: Prophet of Energy.* Basic Books.

Bethe, Hans A. Autumn 1982. Comments on the history of the H-Bomb. *Los Alamos Science.*

Blumberg, Stanley A., and Gwinn Owens. 1976. *Energy and Conflict.* G. P. Putnam & Sons.

Blumberg, Stanley A., and Louis G. Panos. 1990. *Edward Teller. Giant Of the Golden Age of Physics.* Charles Scribner's Sons.

Boffey, Philip M. et al. 1988. *Claiming the Heavens: The New York Times Complete Guide to the Star Wars Debate.* New York Times Books.

Borden,William Liscum. 1946. *There Will Be No Time.* Macmillan.

Boxen, Jeremy. 1995. *Just what the Doctor Ordered: Cold War Purging, Political Dissent, and the Right Hand of Dr. Strangelove.* Film Course, Queen's University, Canada.

Broad, William J. 1985. *Star Warriors.* Simon and Schuster.

———. 1992. *Teller's War.* Simon and Schuster.

Brode, Bernice. 1997. *Tales of Los Alamos.* Los Alamos Historical Society.

Bromberg, Joan Lisa. 1982. *Fusion: Science, Politics, and the Invention of a New Energy Source.* MIT Press.

Canavan, Gregory and Edward Teller. Strategic Defence for the 1990s. *Nature,* 19 April 1990, p. 699.

Cannon, Lou. 1991. *President Reagan: The Role of a Lifetime.* Simon and Schuster.

Carlson, Bengt. How Ulam set the stage. *Bulletin of the Atomic Scientists.* July/August 2003.

Caute, David. 1978. *The Great Fear: the Anti-Communist Purge Under Truman and Eisenhower.* Simon and Schuster.

Clark, Ronald W. 1971. *Einstein.* Avon.

Cohen, Susan. The Man Who Made Reagan a Space Warrior. *West* magazine. San Jose Mercury News. 19 May 1985.

Compton, Arthur Holly. 1956. *Atomic Quest.* Oxford University Press.

Conquest, Robert. 1991. *Stalin.* Penguin.

Coughlan, Robert. Dr Edward Teller's Magnificent Obsession. *Life,* 6 September, 1954.

Cowen, Robert C. A long-ago moment with Edward Teller. *Christian Science Monitor,* 12 September 2003.

Cumings, Bruce. 1990. *The Origins of the Korean War.* vol. 2. Princeton University Press.

Divine, Robert. 1978. *Blowing on the Wind: the Nuclear Test Ban Debate, 1954–1960.* Oxford University Press.

Dobrynin, Anatoly. 1995. *In Confidence: Moscow's Ambassador to America's Six Cold War Presidents (1962–1986).* New York Times Books.

Drell, Sidney D., Philip J. Farley, and David Holloway. 1985. *Reagan's Strategic Defence Initiative: A Technical, Political and Arms Control Assessment.* Balinger Publishing.

Dyson, Freeman. 1979. *Disturbing the Universe.* Harper and Row.

Fermi, Laura. 1954. *Atoms in the Family: My Life with Enrico Fermi.* University of Chicago Press.

Ferrell, Robert. (ed). 1981. *The Eisenhower Diaries.* Norton.

Fischer, Benjamin B. 1997. The Cold War Conundrum. Center for the Study of Intelligence, CIA.

Fitzgerald, Frances. 2000. *Way Out There In the Blue.* Simon and Schuster.

Fitzpatrick, Anne. 1998. Igniting the Light Elements: The Los Alamos Thermonuclear Weapons Project, 1942–1952. University Microfilms.

Frankel, S. 1946. Prima facie proof of the feasibility of the Super. Los Alamos Scientific Laboratory LA-551.

Gallison, Peter, and Barton Bernstein. 1989. In Any Light: Scientists and the Decision to Build the Super bomb, 1942–1954. *Historical Studies in the Physical and Biological Sciences* 19, no. 2.

General Accounting Office. June 1986. 'SDI Programme: Evaluation of DOE's Answers to Questions on X-ray Laser Experiment'. GAO/NSIAD-86-140BR.

———. June 1988. 'Accuracy of Statements Concerning DOE's X-ray Laser Research Programme'. GAO/NSIAD-88–181BR.

———. September 1988. 'Briefing to Congressman George Brown and Pete Stark on GAO Report on the Accuracy of Statements Concerning DOE's X-ray Laser Research Program'. GAO/NSIAD-88–181BR.

———. February 1990. 'Strategic Defense Initiative Programme: Extent of Foreign Participation'. GAO/NSIAD-90–2.

————. February 1992. 'Changing Design and Technological Uncertainties Create Significant Risk'. GAO/IMTEC-92–18.

————. September 1992. 'Strategic Defense Initiative. Some Claims Overstated for Early flight Tests of Interceptors'. GAO/NSIAD-92–282.

————. 2000. Radiation Standards. 'Scientific Basis Inconclusive, and EPA and NRC Disagreement Continues'. GAO/T-RCED-00–52.

Giovanetti, Len, and Fred Freed. 1965. *The Decision to Drop the Bomb*. Coward-McCann.

Glasstone, Samuel, and Philip J. Dolan. 1957. *The Effects of Nuclear Weapons*. USGPO.

Goldberger, Marvin. The Wrath of Teller. Unpublished MS in the author's possession.

Goncharov, German. 1996. Thermonuclear Milestones. *Physics Today*, 49, no. 11.

Goodchild, Peter. 1980. *J. Robert Oppenheimer: Shatterer of Worlds*. BBC Books.

Gorbachev, Mikhail. 1995. *Memoirs*. Doubleday.

Goudsmit, Samuel A. 1947. *Alsos*. Henry Schuman.

Gouzenko, Igor. 1948. *The Iron Curtain*. E. P. Dutton.

Graham, Daniel O. 1982. *High Frontier: A New National Strategy*. Heritage Foundation, Washington.

Green, Harold P. 1977. The Oppenheimer case: a study in the abuse of law. *Bulletin of the Atomic Scientists*, September 1977.

Greenpeace. 1996. *Nuclear Flashback: Report of a Greenpeace Scientific Expedition to Amchitka Island*.

Groueff, Stefane. 1967. *Manhattan Project*. Little Brown.

Groves, Leslie, R. 1962. *Now It Can Be Told*. Harper and Row.

Hansen, Chuck. 1988. *US Nuclear Weapons: The Secret History*. Crown.

Hansen, Chuck. 1994. The Status of the H-Bomb Program, January 1950. Unpublished MS.

Hawkins, David. 1983. *Manhattan District History, Project Y, The Los Alamos Project*, part I. Tomash.

Herken, Gregg. 1985. *Counsels of War*. Knopf.

————. 1992. *Cardinal Choices: Presidential Science Advising from the Atomic Bomb to SDI*. Oxford University Press.

————. 2002. *Brotherhood of the Bomb*. Henry Holt, New York.

Hess, Jerry. 1971. Eugene M. Zuckert. Oral history interview.

Hewlett, Richard G., and Oscar E. Anderson Jr. 1962. *The New World*. Pennsylvania State University Press.

Hewlett, Richard G., and Francis Duncan. 1969. *Atomic Shield, 1947/1952*. Pennsylvania State University Press.

Hewlett, Richard G., and Jack Holl. 1989. *Atoms for Peace and War, 1953–1961*. University of California Press.

Hilgartner, Stephen, Richard C. Bell and Rory O'Connor. 1983. *Nukespeak: The Selling of Nuclear Technology in America*. Penguin.

Holloway, David. 1983. *The Soviet Union and the Arms Race*. Yale University Press.

————. 1994. *Stalin and the Bomb*. Yale University Press.

Irving, David. 1967. *The Virus House*. William Kimble.

Jastrow, Robert. 1983. *How to Make Nuclear Weapons Obsolete*. Little Brown.

Joint Committee on Atomic Energy (JCAE). 1951. Soviet Atomic Espionage. USGPO.

———. 1953. Policy and Progress in the H-Bomb Programme: a chronology of leading events. National Archives.

Jordan, George Racy. 1952. *From Major Jordan's Diaries*. Harcourt, Brace.

Kaplan, Fred. 1983. *The Wizards of Armageddon*. Simon and Schuster.

Kennan, George. 1967. *Memoirs, 1925–1950*. Pantheon.

Killian, James. 1977. *Sputnik, Scientists, and Eisenhower*. MIT Press.

Kistiakowsky, George. 1976. *A Scientist at the White House*. Harvard University Press.

Koestler, Arthur. 1940. *Darkness at Noon*. Jonathan Cape.

Kolata, Gina. For Radiation, 'How Much is Too Much?' *New York Times*. 27 November 2001.

Kramer, Michael, and Sam Roberts. 1976. *I Never Wanted to Be Vice-President of Anything: An Investigative Biography of Nelson Rockefeller*. Basic Books.

Lakoff, Sanford, and Herbert F. York. 1989. *A Shield in Space?* University of California Press.

Lifton, Robert J. 1967. *Death in Life*. Random House.

Lilienthal, David E. 1964. *The Journals of David E. Lilienthal*. Harper and Row.

McFarlane, Robert C., with Zofia Smardz. 1994. *Special Trust*. Cadell and Davies.

McPhee, John. 1974. *The Curve of Binding Energy*. Farrar Straus and Giroux.

Manley, John H. 1985. Recollections and memories. Unpublished MS. LANL Archives.

———. 1987. Star Wars and the H-Bomb. Unpublished MS. LANL Archives.

Mark, Carson J. 1974. A short account of Los Alamos' theoretical work on Thermonuclear Weapons, 1946–1950. LASL-5647-MS.

Mark, Joseph L. 1967. *Seven Hours to Zero*. G. P. Putnam & Sons.

May, Ernest R., and Philip D. Zelikow. 1997. *The Kennedy Tapes*. The BelknapPress of Harvard University Press.

Mitchell, Peter. 1969. *The Swift Years: the Robert Oppenheimer Story*. Dodd, Mead.

Moss, Norman. 1968. *Men Who Play God*. Harper and Row.

Nelson, Thomas A. 1982. *Kubrick: Inside a Film Artist's Maze*. University of Indiana Press.

Newhouse, John. 1989. *War and Peace in the Nuclear Age*. Knopf.

Nichols, J. D. 1987. *The Road to Trinity*. Morrow.

NOVA. 1980. *A is for Atom, B is for Bomb*. WGBH Transcript.

Oberdorfer, Don. 1991. *The Turn From the Cold War to a New Era. The United States and the Soviet Union, 1983–1990*. Poseidon Press.

Oliphant, Marcus. The beginning: Chadwick and the neutron. *Bulletin of the Atomic Scientists*. December 1982.

O'Neill, Dan. 1994. *The Firecracker Boys*. St Martins.

———. 1989. *Project Chariot: A Collection of Oral histories*. 2 vols. Alaska Humanities Forum.

Oppenheimer, J. Robert. 1953a. Atomic weapons and American policy. *Foreign Affairs* 31.

———. 1953b. The Reith Lectures. BBC.

Palevsky, Mary. 2000. *Atomic Fragments: A Daughter's Questions.* University of California Press.

Pauling, Linus. 1983. *No More War.* Dodd, Mead.

Peierls, Rudolf. 1985. *Bird of Passage.* Princeton University Press.

Pfau, Richard. 1984. *No Sacrifice Too Great.* University Press of Virginia.

Pharr Davis, Nuel. 1968. *Lawrence and Oppenheimer.* Simon and Schuster.

Poundstone, William. 1993. *Prisoner's Dilemma: John von Neumann, game theory, and the puzzle of the bomb.* Doubleday.

Pringle, Peter, and James Spigelman. 1982. *The Nuclear Barons.* Michael Joseph.

Reagan, Ronald. 1990. *An American Life.* Simon and Schuster.

Rhodes, Richard. 1986. *The Making of the Atomic Bomb.* Simon and Schuster.

———. 1995. *Dark Sun: the Making of the Hydrogen Bomb.* Simon and Schuster.

Rosenberg, Tina. The Authorised Version. *The Atlantic.* February 1986.

Sakharov, Andrei. 1990. *Memoirs.* Knopf.

———. 1991. *Moscow and Beyond.* Knopf.

Schwarz, Stephen, et al. 1998. *Atomic Audit.* Brookings Institution.

Schweber, S. S. 2000. *In the Shadow of the Bomb: Oppenheimer, Bethe, and the Moral Responsibility of the Scientist.* Princeton University Press.

Seaborg, Glenn. 1990. *Journals,* volumes 1–4, 19 April 1942–19 May 1946. Lawrence Berkeley Laboratory.

Shepley, James, and Clay Blair Jr. 1954. *The Hydrogen Bomb.* McKay.

Sherwin, Martin. 1975. *A World Destroyed: the Atomic Bomb and the Grand Alliance.* Knopf.

Shultz, George. 1993. *Turmoil and Triumph.* Charles Scribner's Sons.

Smith, Alice Kimball, and Charles Weiner. 1980. *J. Robert Oppenheimer: Letters and Recollections.* Harvard University Press.

Smith, Geoffrey. 1991. *Reagan and Thatcher.* W. W. Norton.

Stern, Philip with Harold Green. 1969. *The Oppenheimer Case: Security on Trial.* Harper and Row.

Strauss, Lewis. 1962. *Men and Decisions.* Doubleday.

Szilard, Leo. 1972. *The Collected Works: Scientific Papers.* MIT Press.

Teller, Edward. 1946. The State Department Report – 'A Ray of Hope'. *Bulletin of the Atomic Scientists.*

Teller, Edward. 1947. Atomic Scientists Have Two Responsibilities. *Bulletin of the Atomic Scientists.*

Teller, Edward. March 1950. Back to the Laboratories. *Bulletin of the Atomic Scientists.*

———. 1955. The Work of Many People. *Science.* 25 February 1955.

———. 1977. *Nuclear Energy in the Developing World.* Metrek Division, McLean, VA.

———. 1979. *Energy from Heaven and Earth.* Freeman.

———. 1980. *The Pursuit of Simplicity.* Pepperdine University Press.

———. 1987. *Better a Shield Than a Sword: Perspectives on Defense and Technology.* Free Press.

———. Interview with George Keyworth, 20 September 1979.

———. Correspondence with Maria Goppert Mayer, Box 3, Mayer papers, University of California, San Diego Library.

Teller, Edward, and Allan Brown. 1962. *The Legacy of Hiroshima.* Doubleday.

Teller, Edward, and Albert Latter. 1958. *Our Nuclear Future: Facts, Dangers, and Opportunities.* Criterion Books.

Teller, Edward, with Judith Shoolery. 2001. *Memoirs: A 20th Century Journey in Science and Politics.* Perseus Press.

Teller, Edward, Wendy Teller, and Wilson Talley, 1991. *Conversations on the Dark Secrets of Physics.* Plenum, New York.

Time. The Atom. The Road Beyond Elugelab. 12 April 1957.

———. Defense, Knowledge is Power. 18 November 1957.

Truman, Harry S. 1955. *Year of Decision.* Doubleday.

Truslow, Edith, and Ralph Carlisle Smith. 1983. *Project Y: The Los Alamos Story,* part 2, 'Beyond Trinity'. Tomash.

Ulam, Françoise. *Stan, Teller and the H-Bomb, 1949–52.* Unpublished MS in possession of the author.

———. Postscript to *Adventures of a Mathematician* (see below).

Ulam, Stanley, M. 1991. *Adventures of a Mathematician.* University of California Press.

United States Atomic Energy Commission (AEC). 1971. *In the Matter of J. Robert Oppenheimer.* MIT Press.

———. 1954. Thermonuclear weapons Programme Chronology. LANL Archives. 22 iv.

US Defense Nuclear Agency. Projects Gnome and Sedan: The Plowshare Program. DNA 6029F. March 1983.

University of California Radiation Laboratory. 1958. *Industrial Applications of Nuclear Explosions.* Film A81.

Walbridge, E. Angle Constraint for Nuclear-pumped X-ray Laser Weapons. *Nature,* 19 July 1984.

Walker, Alexander. 1999. *Stanley Kubrick Directs.* W. W. Norton.

Wallop, Malcolm. 1979. Opportunities and Imperatives of Ballistic Missile Defence. *Strategic Review.* Fall 1979.

Wasserman, Harvey. 1982. *Killing Our Own: The Disaster of America's Experience with Atomic Radiation.* Delacarte Press.

Weart, Spencer and Szilard, Gertrude W. 1978. *Leo Szilard: His Version of the Facts.* vol. 2. MIT Press.

Weinberger, Caspar. 1990. *Fighting for Peace: Seven Critical Years in the Pentagon.* Warner.

Weiner, Charles. 1972. *Exploring the History of Nuclear Physics.* A.I.P.

Weizsacker, Carl F. 1978. *The Politics of Peril.* Seabury Press.

Wheeler, John A. and Kenneth Ford. 1998. *Geons, Black Holes and Quantum Foam: a Life in Physics.* Norton

Wigner, Eugene. *Saturday Review.* 17 November 1945.

Wigner, Eugene. 1988. An Appreciation of the 80th Birthday of Edward Teller. In Hans Mark and Lowell Wood (eds). *Energy in Physics, War and Peace: A Festschrift Celebrating Edward Teller's 80th Birthday.* Kluwer Academic Publishers.

Wilson, Jane ed. 1975. *All in our time.* Bulletin of the Atomic Scientists.

Wood, Lowell and John Nuckolls. 1988. The development of nuclear explosives. In Hans Mark and Lowell Wood eds, *Energy in Physics, War and peace: A Festschrift Celebrating Edward Teller's 80th Birthday.* Kluwer Academic Publishers, Boston.

York, Herbert F. 1976. *The Advisers: Oppenheimer, Teller and the Super Bomb.* Freeman.

————. 1987. *Making Weapons, Talking Peace: A Physicist's Odyssey from Hiroshima to Geneva.* Basic Books.

Zodner, Harlan, (ed). 1958. Industrial Uses of Nuclear Explosives. University of California Radiation Laboratory. (UCRL-5233)

Index

llustration references (e.g. [Fig. 1]) relate to captions